Astronomy and History
Selected Essays

O. Neugebauer

Astronomy and History
Selected Essays

With 133 Illustrations

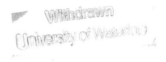

Springer-Verlag
New York Berlin Heidelberg Tokyo

O. Neugebauer
Brown University
History of Mathematics Department
Providence, RI 02912
U.S.A.

AMS Subject Classifications: 01A15, 01A17, 01A20, 01A30, 01A35, 01A75

Library of Congress Cataloging in Publication Data
Neugebauer, O. (Otto), 1899–
    Astronomy and history.
    1. Astronomy—History. I. Title.
QB15.N48   1983   520'.9   83-4729

Printed and bound by Halliday Lithograph, West Hanover, MA.
Printed in the United States of America.

9 8 7 6 5 4 3 2 1

ISBN 0-387-**90844**-7 Springer-Verlag New York Berlin Heidelberg Tokyo
ISBN 3-540-**90844**-7 Springer-Verlag Berlin Heidelberg New York Tokyo

To the memory of

A.J. Sachs
(11 December 1914 – 22 April 1983)

Preface

The collection of papers assembled here on a variety of topics in ancient and medieval astronomy was originally suggested by Noel Swerdlow of the University of Chicago. He was also instrumental in making a selection* which would, in general, be on the same level as my book *The Exact Sciences in Antiquity*. It may also provide a general background for my more technical *History of Ancient Mathematical Astronomy* and for my edition of *Astronomical Cuneiform Texts*.

Several of these republished articles were written because I wanted to put to rest well-entrenched historical myths which could not withstand close scrutiny of the sources. Examples are the supposed astronomical origin of the Egyptian calendar (see [9]), the discovery of precession by the Babylonians [16], and the "simplification" of the Ptolemaic system in Copernicus' *De Revolutionibus* [40].

In all of my work I have striven to present as accurately as I could what the original sources reveal (which is often very different from the received view). Thus, in [32] discussion of the technical terminology illuminates the meaning of an ancient passage which has been frequently misused to support modern theories about ancient heliocentrism; in [33] an almost isolated instance reveals how Greek world-maps really looked; and in [43] the Alexandrian Easter computus, held in awe by many historians, is shown from Ethiopic sources to be based on very simple procedures.

Finally, several of the papers deal with astrological topics. As an excuse for this aberration I may plead not only the usefulness of astrological data for establishing historical chronology (see, for example, [23]), but also the significance of this "wretched subject" for the study of civilization ([1]).

In conclusion, I wish to express my sincerest thanks to Springer-Verlag for making these studies available to a wider public.

O.N.

---

*A complete bibliography (until 1979, compiled by J. Sachs and G.J. Toomer) was published in *Centaurus* **22** (1979), 258–280.

Table of Contents

TABLE OF CONTENTS

Permissions

Springer-Verlag would like to thank the original publishers of O. Neugebauer's scientific papers for granting permission to reprint a selection of his papers in this volume. The following credit lines were specifically requested:

[1] Reprinted from *ISIS* **42,** © 1951 by Univ. of Pennsylvania.
[2] Reprinted from *Studies in the History of Science,* © 1941 by Univ. of Pennsylvania.
[3] Reprinted from *Studies in Civilization,* © 1941 by Univ. of Pennsylvania.
[4] Reprinted from *JNES* **4,** © 1945 by Univ. of Chicago Press.
[5] Reprinted from *Bull. Am. Math. Soc.* **54,** © 1948 by Am. Math Soc.
[6] Reprinted from *Scripta Mathematics* **22,** © 1956 by Yeshiva Univ.
[7] Reprinted from *APS, Proc.* **107,** © 1963 by The American Philosophical Society.
[8] Reprinted from *AFO* **8,** © 1932 by Verlag Ferdinand Berger & Sohme.
[9] Reprinted from *Acta Orientalia* **17,** © 1939 by Pub. House of Hungarian Academy of Sci.
[10] Reprinted from *JNES* **1,** © 1942, by Univ. of Chicago Press.
[11] Reprinted from *Vistas in Astronomy* **1,** © 1955 by Pergamon Press.
[12] Reprinted from *Centarus,* © 1980 by Univ. of Aarhus.
[13] Reprinted from *Zeitschrift der Deutschen Morgenländischen Gesellschaft* **90,** © 1936 by Verlag Franz Steiner GmbH.
[14] Reprinted from *ISIS* **36,** © 1945 by Univ. of Pennsylvania.
[15] Reprinted from *ISIS* **37,** © 1947 by Univ. of Pennsylvania.
[16] Reprinted from *JAOS* **70,** © 1950 by The American Oriental Society.
[17] Reprinted from *Astronomical Journal* **72,** © 1967 by The American Institute of Physics.
[18] Reprinted from *Centaurus* **12,** © 1968 by Univ. of Aarhus.
[19] Reprinted from *AJP* **63,** © 1942 by John Hopkins Univ. Press.
[20] Reprinted from *ISIS* **40,** © 1949 by Univ. of Pennsylvania.
[21] Reprinted from *APS, Proc.* **92,** © 1948 by The American Philosophical Society.
[22] Reprinted from *Rivista Delgi Studi Orientali* **24,** © 1949 by The Universita delgi Studi di Roma.
[23] Reprinted from *AJP* **74,** © 1953 by John Hopkins Univ. Press.
[24] Reprinted from *Scripta Mathematica* **19,** © 1953 by Yeshiva Univ.
[25] Reprinted from *Comm. on Pure & App. Math* **8,** © 1955 by John Wiley & Sons, Inc.
[26] Reprinted from *The Aegean & Near East,* © 1956 by Augustin Pub.
[27] Reprinted from *ISIS* **50,** © 1959 by Univ. of Pennsylvania.
[28] Reprinted from *Scripta Mathematica* **24,** © 1959 by Yeshiva Univ.
[29] Reprinted from *Analecta Biblica* **12,** © 1959 by The Pontificio Instituto Biblico.
[30] Reprinted from *AJP* **84,** © 1963 by John Hopkins Univ. Press.
[31] Reprinted from *APS, Proc.* **116,** © 1972 by The American Philosophical Society.

1. General

The Study of Wretched Subjects

BY O. NEUGEBAUER *

IN the last issue of *Isis* (vol. 41, 125–126, p. 374) there is a short review by Professor Sarton of a recent publication by E. S. Drower of the Mandean "Book of the Zodiac" which is characterized by the reviewer as "a wretched collection of omens, debased astrology and miscellaneous nonsense." Because this factually correct statement [1] does not tell the whole story, I want to amplify it by a few remarks to explain to the reader why a serious scholar might spend years on the study of wretched subjects like ancient astrology.

The great Belgian historian and philologist, Franz Cumont, in August 1898 signed the preface to the first volume of the *Catalogus Codicum Astrologorum Graecorum*, the twelfth volume of which is now in preparation. The often literal agreement between the Greek texts and the Mandean treatises requires the extension of Professor Sarton's characterization to an enterprise which has enjoyed the wholehearted and enthusiastic support of a great number of scholars of the very first rank. They all labored to recover countless wretched collections of astrological treatises from European libraries, and they succeeded in giving us an insight into the daily life, religion and superstition, and astronomical methods and cosmogonic ideas of generations of men who had to live without the higher blessings of our own scientific era.

To the historian of science the transmission of ideas is rightly one of his most important problems. Astrological lore furnishes us one of the most convincing proofs for the transmission of Hellenistic astronomy to India; astrological manuscripts help us to estimate more accurately the combination of Syriac, Arabic, Hindu, and Greek sources in the building up of Islamic science. No Arabic astronomer can be fully understood without a thorough knowledge of astrological concepts. The only hope of obtaining a few glimpses of the astronomical methods of the time of Hipparchus rests in the painstaking investigation of wretched writers like Vettius Valens or Paulus Alexandrinus. Six large volumes of miscellaneous nonsense were published by Professor Thorndike and have become a treasured tool for the study of Mediaeval scientific literature. And the history of the art and philosophy of the Renaissance has gained immensely from the researches carried out by the Warburg Institute on the astrology of preceding periods.

The book by Mrs Drower is only one modest contribution within the larger task of research in the history of the civilization of the Near East. The difficulties of the problems involved are by far greater than in the case of Greek or Latin texts with their so much better known terminology and background. We have to thank Mrs Drower for exposing a new source which may one day furnish the missing link in the transmission of doctrines which have left their imprint in almost all phases of Mediaeval learning, Medicine, Botany, Chemistry, etc.

All these facts are, of course, well known to Professor Sarton. But when the recognized dean of the History of Science disposes of a whole field with the words "the superstitious flotsam of the Near East," he perhaps does not fully realize how much he is contributing to the destruction of the very foundations of our studies: the recovery and study of the texts as they are, regardless of our own tastes and prejudices.

---

* Brown University.
[1] The classification under "Babylonia and Assyria" is misleading because Mandean astrology belongs to the Islamic, and thus ultimately Hellenistic, civilization.

Some Fundamental Concepts in Ancient Astronomy

By

OTTO E. NEUGEBAUER, Ph.D., Ll.D.*

1. THE aim of this lecture is not to give any kind of complete survey of the fundamental ideas or methods of ancient astronomy but, on the contrary, to show how one single fact, the variability of the length of the days, influenced the structure of ancient astronomy. I choose this kind of approach because I am convinced that real progress in the study of the history of science requires the highest specialization. In contrast to the usual lamentation, I believe that only the most intimate knowledge of details reveals some traces of the overwhelming richness of the processes of intellectual life.

The variability of the length of the days connects two fundamental groups of problems: the variability during the year leads to the problem of the determination of the orbit of the sun; the variability with respect to the geographical latitude involves the question of the shape of the earth. Both problems are not only very intimately connected, but both require for adequate treatment the creation of a new mathematical discipline—spherical trigonometry. No one of these three groups of problems—ancient theory of the movement of the sun, determination of the shape of the earth, and history of trigonometry—could be adequately discussed in a single lecture. I will therefore confine myself to a short report about some of the questions involved, which are, I believe, in a certain sense typical for the situation faced by the ancient mathematician, and I will discuss only those methods which are of essentially *linear* character. This means that I shall disregard the mathematical part of the problem, the history of spherical trigonometry,[1]

---

*Professor of Mathematics, Brown University.

[1] The branch of the development where *trigonometric* methods are involved will be discussed in a forthcoming paper by Olaf Schmidt, Brown University. In the following the treatment of these problems by stereographic projection as

and, instead, emphasize an earlier stage of our problem, whose
importance for different problems in ancient astronomy has not
been fully acknowledged.[2]

2. When we talk about the "length of the days" we must
briefly discuss concepts and methods of measuring time. We
all have some feeling of homogeneous time as a kind of equidistant scale, well adapted to measure the events in the observed world. I will not discuss the fact that this a priori concept
of homogeneous time is doubtless due to the fortunate fact
that we are living on a celestial body which moves under almost
the simplest possible conditions (the so-called two-body problem) and that celestial mechanics shows that only with very
little change in the original distribution of masses and velocities our aspect of the sky would be about the same as the
aspect of the lights of a large city from a roller-coaster,[3] where
nobody would create such nice concepts as our day and night
and their smaller parts. But even under the ideal conditions
given on our planet, it took more than two of the four millennia of known history to develop such a simple concept as an
"hour" of constant length.

It is well known that "hour" meant in ancient and medieval times one-twelfth of the actual daylight from sunrise to
sunset, so that "one hour" in January and August, and in Alexandria and Rome, had very different lengths. From our point
of view the first question may be: How is it possible to arrive
at such an obviously inconvenient definition of time? However,

given by Ptolemy in his "Planisphaerium" (opera, vol. II, pp. 225-259) is completely disregarded.

[2] It must be emphasized that Honigmann in his book *Die sieben Klimata
und die πόλεις ἐπίσημοι* (Heidelberg, Winter, 1929) recognized for the first
time the relationship between the problem of the "rising-times" treated here
and ancient geography. Independently Olaf Schmidt discovered the importance
of these questions for the ancient geometry of the sphere, especially in Theodosius. These two sources, together with my own investigations on Hypsicles,
directed my attention to the "linear methods" in Greek and Babylonian astronomy and their relationship.

[3] Cf. e.g., the results of Hill (*Coll. Works*, I, p. 334 f.) and Poincaré (*Méth.
Nouv. Méc. Cél.*, I, p. 109) which show that only a slightly different initial
situation would cause our moon to move in a curve of oval shape in the main
part, but with a loop at each end of the longer axis, such that the moon would
appear half six times during one revolution. In the neighborhood of full- and
new-moon, the moon's velocity would be about the same as now, but around
the loops the movement would be almost zero. I wonder what kind of time
concept would be proved to be a priori by the philosophers of the dwellers on
such a moon.

formulating the question in this way prevents access to the solution. We must not ask who invented the hours of unequal length (the so-called "seasonal hours"), but we must find the causes which finally enforced the creation of such a highly artificial concept as an hour of constant length ("equinoctial hour"). Actually no simple observable phenomenon exists which may give a time-scale with equidistant intervals: vessels of a very special shape only give constant quanta of water-outflow, the shadow changes according to complicated trigonometrical functions, the length of the day changes in rates which are far from being linear, and the stars shift from night to night, and there do not exist ancient clocks exact enough to show the regularity of their movement. However, all those irregularities were just small enough to make *linear* approximations not entirely impossible. The brief discussion of their character and relationship is the topic of this lecture.

3. Apparently the most natural division of day and night is the division into two, three, or four parts. The bisection gives noon and midnight, the thirds are the "watches" in Babylonia,[4] the quarters the "watches" in Egypt.[5] The variation in the lengths of the nights in those southern latitudes is so small that no one needed to worry about the constancy of these watches. How uninterested the Egyptians were in the change of the astronomical seasons is emphasized by the fact that they subdivided their year not in four but in three agricultural seasons: the period of inundation, the reappearance of the fields from the inundation, and the harvest.[6] Obviously primarily agricultural societies do not need any kind of precise definition of homogeneous time; and even in periods where a finer subdivision is required, the older custom of treating day and night separately has been kept in use. Therefore the Egyptian "hours," which can be shown to exist since about 2000 B.C.,[7] are typically seasonal hours of one-twelfth of the day and one-twelfth of the night each. In Greek literature these "hours" do not appear earlier than in Hellenistic times.[8]

[4] Cf. e.g., B. Meissner, *Babylonien und Assyrien*, II, p. 394 (Heidelberg, Winter, 1925).
[5] K. Sethe, *Die Zeitrechnung der alten Aegypter* (Nachr. Ges. d. Wiss. Göttingen Phil-hist. Kl. 1919, p. 287 ff. u. 1920, p. 97 ff., p. 127).
[6] Sethe, *Zeitrechnung*, p. 294.
[7] Sethe, *Zeitrechnung*, p. 111.
[8] The oldest occurrence of "hours" as a well-determined time measure seems

A very different but also very primitive method of counting time has been developed in Mesopotamia. We know that as early as in Sumerian times[9] there existed a distance-unit named *danna*, which may be translated as "mile," corresponding to about seven of our miles. This unit was used for measuring longer distances and became in this way quite naturally also a time-interval: the traveling time for such a distance. If we suppose this slight change in the meaning of the word "mile," it is immediately intelligible how a day or a night could be expressed in "miles." But the origin of these "time-miles" from measuring distance has never been forgotten, and therefore time measurement in "miles" became a homogeneous one, independent of the changing length of the day during the seasons. When later, I may say some time in the first part of the first millennium B.C.,[10] Babylonian astronomy made its first steps to a more systematic recording of celestial phenomena, this length-measure "mile" was transferred to celestial distances too, in the simple way that the number of miles contained in one day was made equivalent to *one* revolution of the sky. Because one day contained twelve of these itinerary miles, the circumference of the sky also became twelve miles. And because the mile (*danna*) has been subdivided in thirty UŠ (the meaning of UŠ is very significant, simply "length"), the length of the main circle of the sky was divided into 12 . 30 = 360 parts. This is the origin of our "degrees" and the custom of modern astronomy of measuring time in degrees.[11]

---

to be in the writings of Pytheas (time of Alexander the Great), quoted by Geminus VI, 9 (ed. Manilius p. 70, 23 ff.); hours are frequently used by Geminus (ca 100 B.C.), Vitruvius and Manilius (time of Augustus). For further literature, see Kubitschek, *Grundriss der antiken Zeitrechnung* (Handb. d. Altertumswiss. I, 7) p. 179. Herodotus (400 B.C.) is often quoted for mentioning the Babylonian "hours" (II, 109) but this sentence has been considered to be an interpolation [recently by J. E. Powell, Classical Review **54**, 1940, p. 69 (without knowing an older attempt in the same direction, mentioned by Kubitschek p. 178, note 1)] but, I think, without sufficient reason.

[9] Oldest example from Tello, period of Agade (about 2400 B.C.) published by Fr. Thureau-Dangin, *Inventaire des Tablettes de Tello conservées au Musée Impér. Ottoman*, Paris, 1910 ff., 11, 1175.

[10] Cf. A. Schott, *Das Werden der babylonischen Positionsastronomie*, Zeitschr. d. Deutschen Morgenländischen Gesellschaft **88** (1934), 302 ff. and his review of Gundel, *Hermes Trismegistos*, in Quellen u. Studien z. Geschichte d. Math., Abt. B., vol. **4** (1937), p. 167 ff.

[11] Cf. O. Neugebauer, *Untersuchungen zur Geschichte der antiken Astronomie III*, Quellen u. Studien z. Geschichte d. Math., Abt. B., vol. **4** (1938), 193 ff.

In modern literature those "miles" are very misleadingly named "double hours" because they correspond actually to two of our time units.[12] But the ancients were well aware of their origin; e.g., Manilius (time of Augustus) speaks correctly about *stadia*, i.e., miles, in his famous astronomical poem.[13] These Babylonian time-distances appear frequently in Greek astronomy and give clear evidence of the important influence which Bablyonian astronomy exercised in the ancient world.[14]

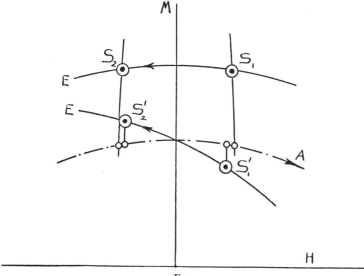

FIG. 1.

4. It may seem that with this (certainly unconscious) creation of a homogeneous time all trouble was over, but the real difficulties begin with the introduction of the concept of homogeneous time: we have to express the natural time intervals, as day and night, by the lengths of some constant time intervals. This problem has two different aspects: first, the practical one

[12] G. Bilfinger, *Die babylonische Doppelstunde*, Stuttgart, Wildt, 1888.

[13] Manilius, *Astronomica III*, 275 ff. (ed. Breiter, p. 74 and p. 88; ed. Housman p. 24 and p. XIII ff.).

[14] They appear e. g., in Herodot II, 109 (ca 450 B.C.), in the "Eudoxospapyrus" (about 3rd cent. B.C.; cf. Pauly-Wissowa, *Real-Enzyklopädie* 6 col. 949), in the Pap. Michigan 151 (*Michigan Papyri*, vol. III, p. 118 ff., a text which I intend to discuss in a forthcoming paper) and implicitly, of course, in the countless places where degrees are used to express time (first instance in Greek is in Hypsicles, about 200 B.C.).

of constructing clocks showing real constant time intervals;
secondly, the theoretical problem of finding the rule by which
the length of the days, expressed in this constant time interval,
changes. In the following we shall be mainly concerned with
this second question.

In order to understand fully the problems involved, it may
be remembered how "one hour" is defined today. The simple
definition, "One hour is the twenty-fourth part of the time
from noon to noon," i.e., from one meridian-passage of the sun
to the next, is not sufficient to obtain hours of constant length
for two reasons. First, the velocity of the sun is not constant.
Secondly, even under the assumption that the sun travels the
same part of its orbit every day, the fact that this orbit (the
"ecliptic" E in fig. 1) has an inclination of more than twenty-
three degrees to the plane of our daily rotation (the "equator"
A) implies that equal parts on the ecliptic do not cross the
meridian (M) in equal lengths of time.[15] In order to avoid both
these irregularities (which combined give a rather complicated
effect) modern astronomy introduced an artificial body called
"mean sun" which moves with the constant *average* velocity of
the "true sun" and which has the *equator* as orbit, and not the
ecliptic. The twenty-fourth part of these *artificial* days lasting
from meridian passage to meridian passage of the *mean* sun is
our familiar "hour."[16]

5. Both sources of this complication in the definition of
"time" were well known to ancient astronomy. The direct observation of the variability in the sun's daily path is of course
far beyond the capacity of any kind of instrument available to
the ancients. However, they realized that the number of days
elapsing between the vernal equinox and the summer solstice
is not the same as the number of days between summer solstice
and autumn equinox, between autumn equinox and winter
solstice, and winter solstice and vernal equinox, and that these
four points divide the year into unequal parts. The method of
taking this observation into consideration is very characteristic
of the ancient astronomical systems.

[15] As an example, in fig. 1, are shown two equal parts S_1S_2 and $S'_1S'_2$ of the
ecliptic which the sun may travel in a day, one at summer solstice (E practically parallel to A) and the other at the vernal point (E maximal inclination
to A).

[16] The difference between true and mean solar time (the so-called "time
equation") reaches a maximum of about ±15 minutes.

The oldest Bablyonian system, which must have been created
earlier than about 200 B.C., introduced an artificial sun, moving in two parts of the year with a different velocity, suddenly
jumping at two well-defined points of its path from one velocity
to the other (fig. 2a). These velocities and jumping-points were

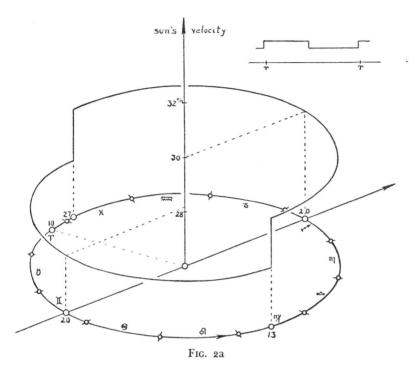

Fig. 2a

chosen in such a way that the time intervals between the four
points in the year, mentioned above, are just the times required
by observations. I think that this purely mathematical construction shows the surprisingly high level of this late-Babylonian astronomy.

Probably very soon[17] after this first attempt to describe
mathematically the movement of the sun, a second theory was
developed in Babylonia, where the required change in the sun's
velocity was represented by an apparently more natural con-

[17] An attempt of P. Schnabel to date the origin of the two Babylonian systems
exactly (*Berossos und die babylonisch-hellenistische Litteratur*, Leipzig, Teubner,
1923, p. 223 ff.) is often quoted in the literature, but it is based on assumptions
which can easily be proved to be wrong.

struction, namely by the assumption of *linearly* changing velocity (fig. 2b). The reason why this model is later than the first
mentioned is a purely mathematical one, because the further
consequences of this second assumption become much more
complicated than the first case.[18]

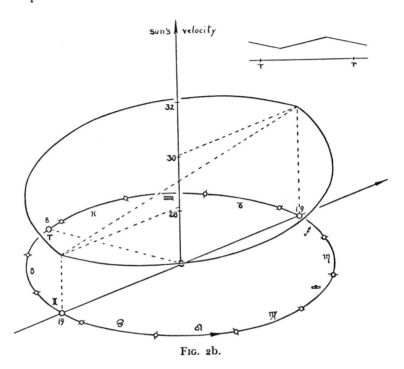

FIG. 2b.

The third solution of the problem, again perhaps only one
hundred or fifty years later than the Babylonian method, was
given by Hipparchus (about 150 B.C.). He interpreted the observed irregularity of the sun's movement as an apparent one
by assuming that the sun moves on a circle with constant
velocity but is observed from an *eccentric* point. This is the
type of astronomical theory which determined the astronomy
of the following 1,500 years, in a certain sense doubtless a re-
gression from a pure mathematical method to assumptions
about the physical nature of our planetary system (fig. 2c).

6. Let us now consider the second part of the questions in-

[18] Expressed in modern terms: The summation processes which are required
in the theory of syzygies become one degree higher in the second theory.

volved in the determination of the length of the days, namely the inclination of the ecliptic. According to ancient custom the "day" began with sunrise or sunset; the second definition was adopted in Babylonia obviously because every new month began with the first visibility of the new moon, which comes just after sunset. The ancient problem is therefore the deter-

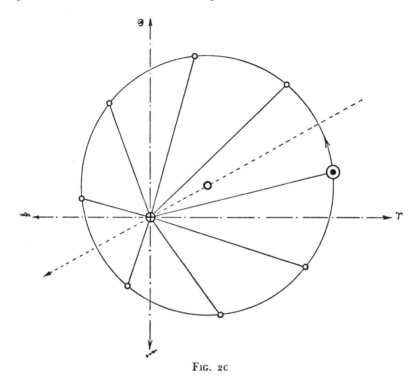

Fig. 2c

mination of the time elapsed between two consecutive crossings of the horizon by the sun. Here the same difficulty occurs as in the above-mentioned case of the crossing of the meridian line by the real sun, moving on the ecliptic and not on the equator; whereas (for a given place) the inclination of the equator to the horizon is constant, the sun's orbit cuts the horizon at continuously shifting angles. The problem is the famous problem of the determination of the "rising-times" (ἀναφοραί) in Greek astronomy: to calculate the equatorial arcs which cross the horizon in the same time as a given arc of the ecliptic.

This problem is obviously a problem of spherical trigonometry. Its complete solution can be found in Ptolemy's *Almagest* (ca 150 A.D.)[19] and has profoundly influenced the earlier treatises on the geometry of the sphere from Autolycos (little before 300 B.C.), Euclid (ca 300 B.C.), and others to Theodosius (ca 100 B.C.) and Menelaos (ca 100 A.D.).[20] According to the limitations of this lecture I shall not discuss the history of this part of the theory. However, I must mention one theorem which shows the direct connection between the problem of the rising-times and the question of the variability of the lengths of the days. This theorem is the following one. Let us consider, for the sake of simplicity, only the twelfths of the ecliptic, the so-called zodiacal signs. Let $\alpha_1, \alpha_2, \ldots, \alpha_{12}$ be the rising times of the first, second, ..., twelfth sign, respectively. Then, if the sun is at the beginning of the i-th sign, the length of the day at this time of the year is equal to the sum of the six consecutive rising times beginning with α_i, i.e., $\alpha_i + \alpha_{i+1} + \ldots + \alpha_{i+5}$. The correctness of this theorem is evident when you remark that after sunrise the i-th sign crosses the east-horizon first, then the following and so on, until the sun comes to the west-horizon, at which moment just one half of the ecliptic has crossed the east-horizon.[20a] During the time from sunrise to sunset just six zodiacal signs cross the east horizon, and this is the proposition of our theorem. This relation explains the high interest of the ancients in the determination of the rising-times: if you know the α's, you know the corresponding length of the day by simple addition.

This relationship is fundamental for the understanding of all ancient discussion of rising-times and day-lengths. It is of course treated in Ptolemy's *Almagest*, where a table of the rising-times for ten different latitudes is given,[21] with intervals of ten degrees. As I mentioned before, this table is calculated by using spherical trigonometry and therefore represents correctly the rather complicated relation between the sun's position in the zodiac and the corresponding rising-times of the ten-degree arc. This relationship is shown in the case of the

[19] Ptolemy, Almagest II, 7 and 8. An excellent treatment of the problem corresponding to our time-equation is given by A. Rome in *Ann. Soc. Sci. de Bruxelles*, sér. 1, 59 (1939), 211 ff.

[20] Cf. notes 1 and 2.

[20a] Here, as everywhere else in this paper, the difference between sidereal time and solar time is neglected for the sake of simplicity.

[21] *Almagest*, II, 8.

latitude of Alexandria in fig. 3. The characteristic property of this curve is the secondary minimum between the two maxima, a kind of asymmetry which gets worse with increasing geographical latitude.

This exact shape of the curve was obtained only by using trigonometry. However, we know very interesting older attempts to describe the rising-times as functions of the sun's

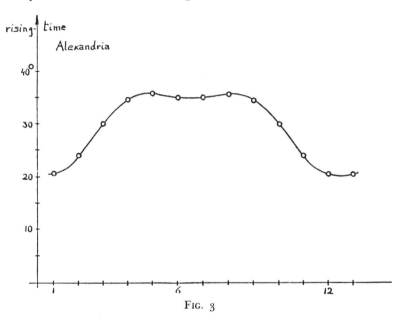

FIG. 3

positions. There exist two different types, one (A) represented by Hypsicles[22] (ca 200 B.C.) for the latitude of Alexandria, the second one (B) by Cleomedes (time of the Roman empire) for the latitude of the Hellespont.[23] Both curves are *linear* approximations of the true curve, with the exception that B inserts in the middle of the slanting lines twice the difference (fig. 4). The corresponding theory about the influence of geographical latitude is given for system A by Vettius Valens (about 150 A.D.),[24] for system B by Pap. Michigan 149,[25] who both

[22] New edition in preparation by M. Krause, V. De Falco, and myself.

[23] Concerning this location see my paper "Cleomedes and the meridian of Lysimachia," accepted for publication in the *Am. J. of Philology*.

[24] I, 7 (ed. Kroll, p. 23).

[25] *Michigan Papyri*, vol. III, p. 63 ff. espec. p. 103 and p. 301 ff. (about 2nd cent. A.D.).

agree on the general method of defining seven "climata" by the assumption that the rising-times increase *linearly* from clima to clima, Vettius Valens starting from Alexandria, the Michigan papyrus with Babylon as main clima.[26]

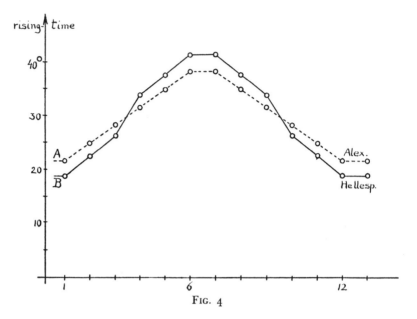

FIG. 4

It is now a very natural question to ask about the corresponding theory in Babylonian astronomy.[27] Here, however, nothing about rising-times was known, but only rules by which the length of the days was calculated during the seasons.[27a] Each of the two systems mentioned above has a scheme of its own. The older one gives (expressed here in degrees) as lengths the following list A, the younger one B:[28]

[26] This follows from a slightly different interpretation of the text, as given by the editors, which requires a much smaller emendation and will be discussed in a forthcoming paper.

[27] Honigmann has already tried to restore the Babylonian rising-times and has discovered that Vettius Valens and Manilius refer to this latitude (*Mich. Pap.*, III, p. 313). He was apparently much disturbed by not quite correct information of Schnabel (p. 314) and a wrong hypothesis of Kugler about a Babylonian scheme of day-lengths (p. 317).

[27a] These rules were discovered by Kugler, *Babylonische Mondrechnung*, Freiburg, 1900, p. 77, p. 99.

[28] Unfortunately Kugler reversed the order of the two systems by calling the older one II, the younger I.

A				B			
10° of	length of the day		10° of	8° of	length of the day		8° of
♈	180	180	♎	♈	180	180	♎
♉	200	160	♏	♉	198	162	♏
♊	212	148	♐	♊	210	150	♐
♋	216	144	♑	♋	216	144	♑
♌	212	148	♒	♌	210	150	♒
♍	200	160	♓	♍	198	162	♓

We must only remember the fundamental relation between the lengths of the day and rising-times in order to find a system of numbers with constant difference from which, by the addition of six of them, the day lengths A and B, respectively, can be derived, namely

A:
$$\alpha_1 = \alpha_{12} = 20$$
$$\alpha_2 = \alpha_{11} = 24$$
$$\alpha_3 = \alpha_{10} = 28$$
$$\alpha_4 = \alpha_9 = 32$$
$$\alpha_5 = \alpha_8 = 36$$
$$\alpha_6 = \alpha_7 = 40$$

B:
$$\alpha_1 = \alpha_{12} = 21$$
$$\alpha_2 = \alpha_{11} = 24$$
$$\alpha_3 = \alpha_{10} = 27$$
$$\alpha_4 = \alpha_9 = 33$$
$$\alpha_5 = \alpha_8 = 36$$
$$\alpha_6 = \alpha_7 = 39$$

Both lists are *linear*, except the double difference in the middle of B, or, in other words, exactly the same, which we knew from the Greek rising-times, mentioned before. The Babylonian list A of rising-times appears explicitly in Vettius Valens[29] and in Manilius.[30]

7. We can summarize our discussion in the statement that the Greek theory of rising-times and variability of day and night is identical with the Babylonian scheme as far as the latitude of Babylon is concerned, and that the Greeks modified these rules in the simplest possible way, namely, linearly, in order to adapt them to geographical latitudes different from Babylon.

It should be mentioned that these *linear* approximations of the complicated actual curve shown in fig. 3 (p. 23) give very satisfactory results for the lengths of the days, at least as far as this can be controlled by the very inaccurate ancient clocks. The proportion 3:2 between the longest and shortest day, adopted in both Babylonian systems, agrees very well with the actual duration of light at Babylon in the summertime,[31] but this custom of characterizing the latitude of Babylon by the

[29] I, 7 and 14 (ed. Kroll, p. 23 and 28).
[30] Cf. note 13.
[31] Kugler, *Sternkunde u. Sterndienst in Babel*, I, p. 174, II, p. 588; Schaumberger in *Ergänzungen*, p. 377.

proportion 3:2 is the reason for a strange deformation of the
ancient world-map, namely, that of placing Babylon at 35° n.l.
(instead of about 33½), a misplacement which affected the
map of the eastern part of the oikumene very much. For this
latitude of 35° is the immediate result of the theory given by
Ptolemy, based on the proportion 3:2, which is trigonometrically correct, but neglects all atmospheric influences in the
duration of the light-day, which are unconsciously included in
the Babylonian values.

8. The theory of the rising-times has one more very important application in ancient astronomy, as far as I know entirely overlooked by modern scholars. This is the question of
determination of the length of invisibility of the moon around
new-moon. This question is of highest importance for the oriental civilizations in which the calendar was regulated by the
actual reappearance of the moon in the evening one or two
days after astronomical new-moon. In order to understand these
relations between rising-times (or here better, setting-times)
and the visibility of the moon, we need only remark that this
visibility not only depends on the distance between sun and
moon in the ecliptic but also on the inclination of the ecliptic
with respect to the horizon. If the ecliptic crosses the horizon
almost vertically, obviously a much smaller distance between
sun and moon is required in order to make the moon's crescent
visible in the dusk than if the ecliptic lies more horizontally and
the sun and moon set almost simultaneously.

We know from investigations by Kugler, Weidner,[32] and
others that Babylonian astronomers were concerned with the
problem of the dependence of the invisibility of the moon on
the seasons at a very early date,[33] when even the variability of
the length of the days was assumed to be linear. Correspondingly, the first attempt to estimate the time between setting of
sun and moon was very unsatisfactory too, namely the assumption of simple proportionality with the duration of the night.

We know very little about the further development of this
question in Babylonia, but I think that a chapter in Vettius
Valens may give some information. At any place where we are
able to check his reports he seems to be very well informed

[32] Kugler, *Sternkunde u. Sterndienst in Babel, Ergänzungen*, p. 88 ff.; E. F.
Weidner, *Alter u. Bedeutung d. babyl. Astron.*, Leipzig 1914, p. 82 ff.
[33] Schott l.c. note 10, p. 310.

about Babylonian sources, which he quotes explicitly.[34] I think, therefore, we may assume his chapter I,14 as essentially Babylonian; here he states that for the latitude of Babylon[35] the elongation of the moon from the sun at the moment when the moon becomes invisible is one-half of the rising-time of the corresponding zodiacal sign.[36] This rule is still a strong simplification of the actual facts, but reveals on the other hand the full understanding of the fact that the problem of the moon's invisibility around new-moon requires the consideration of the change of the ecliptic position with respect to the horizon.[37]

The last step in the development of this theory before the complete solution by spherical trigonometry can also be found in cuneiform texts, but only in the most elaborate system. Here we find an almost perfect solution of the problem, perfect at least as far as observations with very inaccurate instruments are able to control. Here, first of all, the inequalities in the movement of sun and moon are taken into consideration, furthermore the deviation of the moon from the ecliptic (its "latitude"), including an estimate of the influence of the twilight. Finally the rising-times are used in order to transform the ecliptic coördinates of the moon into equatorial coördinates or into "time."[38] We have here a very impressive example of how, by an ingenious combination of linear approximations and their iteration, a very accurate solution of a problem which seems to belong entirely to the realm of spherical trigonometry

[34] IX, 11 (ed. Kroll, p. 354).

[35] This is proved by the fact that the values he gives as examples are exactly the Babylonian values for the rising-times.

[36] I, 14 (ed. Kroll, p. 28). Details will be discussed in a forthcoming paper.

[37] Vettius Valens discusses in chapter I, 13 (ed. Kroll p. 28) the closely related problem of the daily retardation of the moon's rising and setting with respect to sunrise and sunset. The method is purely linear and based on very rough approximations, but mentioned earlier in Pliny H. N. (first cent. A.D.) and later in the *Geoponica* (6th cent. A.D.). These texts have recently been discussed by A. Rome in vol. II, p. 176 of the work of Bidez and Cumont, *Les Mages hellénisés* (Paris 1938), because the Geoponica refers the method to Zoroaster. In the light of the discussion in the work of Bidez and Cumont, a Babylonian origin would be very possible. The method of expressing fractions, however, is the Egyptian one, which speaks strongly for Egyptian origin in spite of the fact that it is more difficult to understand how Egyptian methods could be connected with the doctrines of the mages.

[38] This fact was first discovered by Schaumberger (*Ergänzungen zu Kugler*, cf. note 31, p. 389 ff.) but only by using modern calculations. The relation to the Babylonian rising-times will be discussed in a forthcoming paper.

can be obtained. However, only careful historical investigation of many scattered facts shows that the high building of ancient spherical astronomy and geography is erected on the ground of age-long older attempts and experiences.

9. Our sources are not sufficient, or at least not sufficiently well investigated, to answer the question about the historical origin of the problem of the rising times of ecliptic arcs. It is possible that independent attempts have been made to determine the variability of the lengths of the days directly. One interesting suggestion has been made by Pogo in his investigations on Egyptian water clocks.[39] We have examples of such clocks since the eighteenth dynasty (ca 1500 B.C.) containing inside different scales in order to subdivide the day in twelve parts at the different seasons of the year. Pogo could explain the arrangement of the scales by the following assumptions: let $\triangle$ denote the difference between the longest and shortest day, then the increase of the length of a day in the first month after the winter solstice has been assumed to be $\triangle/12$, in the second $2 \triangle/12$, in the third and fourth $3 \triangle/12$; $2 \triangle/12$ in the fifth, $\triangle/12$ in the sixth and correspondingly in the decrease.[40] It can easily be shown that this rule is equivalent to the newer Babylonian scheme, mentioned above as B, whereas the older one (A) would correspond to the coefficients $\triangle/18$, $3 \triangle/18$, $5 \triangle/18$, respectively.

Pogo's remarks would speak in favor of the assumption that the attempt to characterize directly the rule of the variability of the length of the day was the first step in our group of problems—an assumption which sounds in itself natural enough. But it must not be forgotten, on the other hand, that in the Babylonian astronomy the connection between the length of the days and the visibility of the moon, which involves the rising-times, was established very early, as we have seen above. And finally, one large group of questions has been neglected almost entirely, namely, the methods of determining time and geographical latitude by sun dials.[41] It seems to me

[39] A. Pogo, "Egyptian water clocks," *Isis* **25** (1936), p. 403 ff.
[40] *Isis* **25,** p. 407 ff.
[41] This method is well known from Egyptian, Greek, and Roman sources. Their existence in Babylon has been proved by E. F. Weidner (*Am. J. of Semitic Languages and Lit.*, vol. **40** (1924), p. 198 ff.). I did not realize until recently that texts which I published as "generalized reciprocal tables" are actually "gnomon texts" of a type a little more developed than Weidner's texts

therefore better not to propose any definite solution of the earlier history of our problem but to emphasize the fact that here is a large field open for and deserving our investigation, if we are interested in understanding the creation of our time scale.

(*Mathematische Keilschrift-Texte*, Quellen u. Studien z. Gesch. d. Math. Abt. A vol. **3,** I, p. 30 ff.).

Reprinted from
Studies in the History of Science, 13–29
University of Pennsylvania Press, Philadelphia, 1941

Exact Science in Antiquity*

By

OTTO E. NEUGEBAUER, Ph.D., LL.D.†

IF HISTORY is the study of relations between different cultures and different periods, the history of exact science has a definite advantage over general history. Relations in the field of science can be established in many cases to such a degree of exactitude that we might almost speak of a "proof" in the sense of mathematical rigor. If, for instance, Hindu astronomy uses excenters and epicycles to describe the movement of the celestial bodies, its dependence on Greek astronomy is established beyond any doubt; and the dependence of Greek astronomy on Babylonian methods is obvious from the very fact that all calculations are carried out in sexagesimal notation. However, the fact that the center of interest in the history of science lies in the relationship between *methods* requires a new classification of historical periods. In the history of astronomy, for instance, concepts such as "ancient" or "medieval" make very little sense. The method and even the general mental attitude of the work of Copernicus is much more closely related to that of Ptolemy, a millennium and a half before, than to the methods and concepts of Newton, a century and a half later. It may seem, therefore, a rather arbitrary procedure in the following report on exact science in antiquity to take into consideration only the period before Ptolemy (ca. 150 A.D.). On the other hand, Ptolemaic astronomy climaxed the development of ancient science in its widest sense and we must therefore consider his work at least in a few lines,

---

* In this paper I have, very much against my general principles, refrained from giving any kind of references. The simple bibliographical collection of texts, papers, and books consulted would require about the same space as the text of this paper; and even such a bibliography would be of very restricted use for the reader without often very long discussions in order to justify the special conclusions drawn here. I am still hoping to publish lectures on ancient astronomy which will discuss in detail problems which are touched here.

† Professor of Mathematics, Brown University.

23

in order to be able to understand the influence of the preceding phases on all following development.

Ptolemy (ca. 150 A.D.) was undoubtedly one of the greatest scholars of all time. He left three large works, any one of which alone would place him among the most important authors of the ancient world: the *Almagest*, the *Tetrabiblos* and the *Geography*. The influence which these works exercised on the world-picture of medieval times can hardly be overestimated. Other works, such as his *Music*, *Optics*, investigations on sundials and geographical mapping, in addition to discussions on logic, theory of parallels, etc., show the extremely wide range of his interest.

This is not the place to discuss Ptolemy's works in any detail. It must be remarked, however, that the *Almagest*, for instance, shows in every section supreme mastership and independent judgment, even if he is presenting, as in many cases, results already obtained by earlier scholars. Furthermore, we must emphasize that the modern contempt for the *Tetrabiblos*, the "Bible of the astrologer," is historically very much unjustified. Today we know it to be an error to conclude any influence of the positions of the planets from the obvious influence of the position of the sun on the life on the earth. We must, however, not forget that the instrumental facilities of ancient astronomy were by far insufficient to reveal any idea of the fantastic size of the universe. I, at least, can see no reason why, for example, the theories of earlier Greek philosophers, Plato included, are praised as deep philosophy in spite of the fact that they are hopeless contradictions to facts well known in their own time, while, on the other hand, an attempt to explain the difference between the characters of nations as the result of the difference in the respective inclination of the sun's orbit, the clima, should simply be disregarded as astrological error. The overwhelming historic influence of the *Tetrabiblos* can only be fully understood when we realize that this work is methodically the highest development of the first naturally simple world-picture of mankind, in which earth and universe still have a comparative order of magnitude.

The importance of Ptolemy's *Geography* is generally much more recognized. Hence we do not need to point out the rôle of this work for the knowledge of the inhabited world, but we should, on the contrary, direct our attention to the surprising inaccuracy of the geographical coördinates of almost all places. The method of determining latitude and longitude by astronomical means was known at least as early as Hipparchus (ca. 150 B.C.). The fact that his plan for exact mapping by astronomical methods could never be carried out in practice touches a very essential point in the general situation of ancient science. The determination of geographical longitudes requires the simultaneous observation of a lunar eclipse. All the details of this method are described in a book on optics written by Heron of Alexandria (first century A.D.), but Heron's example shows that not even for Rome were such observations available. Obviously the number of scholars in the ancient world was by far too small to undertake any kind of program based on systematic organized collaboration. One of the reasons for the rapid decline of ancient science lies in the fact that the deeper knowledge of science was then confined to an extremely small number of scholars.

A second element is equally important: the tendency to popularize science in accordance with the taste of the ruling class and to adapt it to the teaching level of the schools. This tendency is clearly evidenced in the extant fragments of ancient scientific literature; I need only to mention the commentaries on the *Almagest* (Pappus ca. 320 A.D., Theon ca. 370 A.D.), the astronomical poem of Manilius (time of Augustus) or the purely descriptive geography of Strabo (same period), which entirely neglected the fundamental problem of exact mapping. Such works were well adapted to create a superficial kind of general education but ill suited for producing an atmosphere of serious research. There was almost nothing left to destroy when the collapse of the Roman Empire fundamentally changed the social and economic structure of the ancient world.

We have not yet mentioned mathematics outside of its applications in astronomy and geography. Actually we have to go back to the Hellenistic period in order to find that kind of mathematics which we have in mind when speaking about "Greek mathematics," and which is most clearly represented by Euclid's *Elements* (ca. 300 B.C.). This type of mathematics covers a very short period indeed, beginning in the time of Plato (Theaetetus and Eudoxus, ca. 400 B.C.), condensed in the *Elements* and appearing for the last time in the works of Archimedes and Apollonius (200 B.C.). The main reason for this early interruption of pure mathematics can be found in the purely geometrical type of expression which was adopted in order to gain the higher degree of generality which the geometrical magnitudes represent, in contrast to the field of rational numbers, which was the exclusive concern of oriental mathematics and astronomy. This geometrical language, however, very soon reached such a degree of complication that development beyond the theory of conic-sections was practically impossible. As a result, the development of theoretical mathematics ended two centuries after its beginning, one century before the cultivated world became Roman.

I think that the influence of this pure mathematics on the general standard of mathematics in antiquity has been very much overestimated. Even Euclid's own works, other than the *Elements*, are on a very different level; this can be simply explained by the remark that the *Elements* are concerned with a very special group of problems, mainly concentrated on the theory of irrational numbers, where the exactitude of definitions and conclusions is the essential point of the discussion. The main part of mathematical literature, however, was less rigorous and represented the direct continuation of Babylonian and even Egyptian methods. The Babylonian influence is, for instance, mainly responsible for the general character of other groups of Greek mathematical literature, as e.g., the work of Diophant (perhaps 300 A.D.). This situation in the field of mathematics corresponds very much to the general character of the Hellenistic culture, with its mixture of very contradictory elements from all parts of the ancient world. One of the most typical elements in this process is the creation of astrology, in the modern sense of this word, and of all kinds of mantic, number-symbolism, alchemy, etc., which became elements of highest importance for both Christian and Arabian thinking.

The different components in the creation of Hellenistic culture are especially visible in the field of astronomy. Mathematical astronomy can be traced back to Apollonius and, in much more primitive form, to Eudoxus. Both men were concerned with the development of kinematical theories for describing the movement of the celestial bodies. The lifetime of both of them is well known as a time of intimate contact between Babylonia and the Greeks. In particular, Apollonius was closely related to the rulers of Pergamon, at whose court one of the Babylonian astronomers, Sudines, well known from Greek sources, lived at the same time.

It is highly probable that this early Hellenistic astronomy was also the source of the Hindu astronomy, from which almost one thousand years later the Arabian astronomy originated. I think that this relationship between the Greek form of Babylonian astronomical computation and the older Hindu decimal number systems explains the creation of a decimal number system with place-value notation, which was transferred by the Arabs to Europe and finally became our number system.

The Babylonian mathematical astronomy which had so much influence on the Hellenistic science is in itself of very recent origin. Although no exact dates can be given, all available source material agrees with a date of about 300 or 250 B.C. for the lifetime of the founder of the oldest form of this theory, Naburianu. The most important feature of this late-Babylonian astronomy is its mathematical character founded on the idea of computing the very complicated observed phenomena by addition of single components, each of which can be treated independently. Here for the first time in history we meet the fundamental method for the investigation of physical problems by using purely mathematical idealizations, a method which determined the course of all future science.

Astronomy of this kind requires highly developed mathematics. Babylonian astronomy contains enormous numerical computations, which could never be carried out with such primitive methods as the Egyptian rules for calculating with fractions or the Roman and medieval abacus methods. Furthermore, every mathematical theory of celestial phenomena must fulfill conditions given by observations or, in other words, requires the solution of equations. The existence of such a mathematical astronomy would therefore be sufficient to justify the conclusion of the existence of corresponding Babylonian mathematics. Hence it is not in itself surprising that we actually have many mathematical texts in cuneiform script which show a development of mathematical methods to the point mentioned above. The surprising fact, however, is that these texts do not belong to the last period of Babylonian culture, as astronomy does, but that they appear as early as in the period of Hammurabi (the so-called First Babylonian Dynasty, about 1800 B.C.).

This leads to one of the most interesting groups of problems in the history of ancient science: Why does the origin of Babylonian mathematics precede the origin of astronomy? Why does such a mathematical astronomy appear at all, if not in direct development from mathematics? And finally: Why is there no parallel development in Egypt, where both mathematics and astronomy never went beyond the most elementary limits? In the following final section I shall try to call attention to some of the conditions which may answer these questions by tracing some main lines of the development of ancient science in chronological order.

The very few old texts of mathematical character which we have from Mesopotamia belong to the latest Sumerian period, the so-called Third Dynasty of Ur (ca. 2000 B.C.). These texts are simple multiplication tables using the already fully developed famous sexagesimal number system. The most important feature of this system is the fact that the powers of sixty, such as 60 itself, or 3600 or 1/60th, and 1 are all denoted simply by "1." This notation makes multiplication or division as simple as in our method of calculation (or even simpler, because the probability of needing infinite fractions is smaller in a system having a base with more divisors). The introduction of this notation is doubtless not a conscious one but is the result of the influence of the monetary system, which was used for the notation of fractions in the same manner as in Roman times. In the beginning the different units were written with number signs of different size, but later this careful notation was omitted and thus the "place value" notation originated. This process is closely related to the economic development of this period, from which we have thousands and thousands of texts which carefully record the delivery of sheep, cattle, grain, etc., for the administrative offices. Hence the first and real decisive simplification in mathematical notations is merely due to the writing practice of generations of business scribes.

The next group of our source material comes from the First Babylonian Dynasty. Those texts are pure mathematical texts, treating elementary geometrical problems in a very algebraic form, which corresponds very much to algebraic methods known from late Greek, Arabian, and Renaissance times. The origin both of a mathematics obviously independent of direct practical needs and of its algebraic form can be explained by the same historical event, namely the complete replacement of the Sumerians by a Semitic population, although in very different senses. The main point is the fundamental difference between the languages of the two types of populations and the fact that the Semites used the Sumerian script to express their own language. The Sumerian script operates with single signs for single concepts (so-called ideograms), derived from a picture script. The Semites used these signs in two different ways: first, in their old sense as representations of single concepts, and secondly, as pure sound symbols (syllables) for composing their own words phonetically. The first possibility of expression corresponds in the field of mathematics exactly to our algebraic notation: instead of writing "length" by six letters, it is sufficient to write l; instead of writing "plus" or "addition," it is sufficient to use one sign $+$. We see here again how an entirely unconscious external influence caused the second fundamental invention of "Babylonian mathematics," the "algebraic" notation. Without such a deep linguistic difference such a powerful instrument as ideographic notation for mathematical operations would never have been introduced, as the parallel with Egypt clearly shows.

The second effect of this contrast between Sumerian and Semitic languages was the creation of systematical philological schools, whose existence is made evident by large collections of texts containing word lists, grammatical rules, etc. Exactly the same thing happened at the very same place in Arabian times, at the school of Baghdad. The new rulers had to study carefully the language and script, religion and law of the preceding culture. This school of language and theology created an atmosphere of general learning, supporting large numbers of well-educated scholars. In these circles Arabian mathematics and astronomy were created, and this corresponds certainly to the *milieu* for the origin of mathematics in the First Babylonian Dynasty.

There is no doubt that some kind of astronomy was cultivated in the same period. The unification of the many different local calendars of Sumerian times was accomplished under Hammurabi's rule, just as he reorganized preceding laws. The first lists of stars and the first rough observations of the disappearance and reappearance of Venus belong to this same period.

Babylonian history knows only short periods of comparative peace. The struggle with and between eastern and northern neighbors kept the country in continuous warfare for many centuries, until finally Assyria succeeded in constructing a powerful kingdom reaching from Persia to Egypt. Corresponding to this shift of power from southern Babylonia to Assyria we find an increasing interest in astronomy in Assyrian texts, where the astrological component, in particular, was developed, if not created.

The Assyrian empire paved the way for the Persian empire and its Hellenistic successors. Babylonia itself lost all political influence, but the cultural tradition was still extant and fully recognized in every part of the ancient world to which Assyrian influence reached. The world of Persian times, however, was very different from the world in the little country around the estuary of the Euphrates and the Tigris, in which Babylonian culture originated. Politically powerless, Babylon became an admired cultural center of a world-wide empire, comparable to the position of Rome in medieval times. The thousand-year-old uninterrupted tradition attracted the admiration of the younger cultures and created the myth of Babylonian wisdom; the main object of admiration was astrology, the "Chaldean" science, which opened inexhaustible new possibilities to religious speculation. Now Persian priests, Jews, and Greeks lived in Babylon, and an international idiom written in simple characters, the Aramaic, made general communication easy. Precisely this actually existing internationalism created competition between national cultures. Zarathustra, Abraham, and Pythagoras were each proclaimed as the inventor of all science and creator of astronomy, astrology, and number-wisdom, and each group asserted itself to be the oldest, and consequently, the teacher of mankind.

In this atmosphere of intellectual competition the Babylonian school of scribes and priests had to defend their authority. Thousands of texts of New-Babylonian, Seleucid, and Parthian times are the evidence of a Babylonian renaissance, returning even in linguistic aspects to old Sumerian traditions. This revival of intellectual centers, this new intellectual activity, where Babylonian priests went to Asia Minor to teach their wisdom to the Greeks, resulted in the last period of Babylonian astronomy. The two mathematical achievements of the old Babylonian period, mentioned above, place-value-notation and algebraic symbolism, became the foundation of a theoretical astronomy of purely mathematical character which deserves more of our highest admiration the more we are able to understand its structure. This astronomy is not based on age-old observations of miraculous exactitude, as usually pretended, but on the contrary reduces the empirical dates to the utmost minimum, mainly period relations, which are easy to observe and almost unaffected by the inexactitude of single instrumental observations. The enormous power of purely mathematical construction was fully recognized here for the first time in the history of mankind.

On the background of the remarks made at the beginning of this lecture we may perhaps resume our discussion with the statement that the development of exact science cannot be adequately described as a systematic step-by-step progress. In any case where we are able to disclose the conditions of essential new development, the contact between highly different cultures appears to give the initial impetus. On the other hand "culture" is in itself equivalent to tradition, which unifies large groups of populations into a common type of opinion and action. However, the same force, tradition, which defines a culture as an individual being, becomes an increasing impediment to further independent development and creates the long periods of "dark ages," which cover by far the largest part of all human history.

Reprinted from
Studies in Civilization, 22–31
University of Pennsylvania Press, Philadelphia, 1941

THE HISTORY OF ANCIENT ASTRONOMY: PROBLEMS AND METHODS*

O. NEUGEBAUER

Institute for Advanced Study, Princeton, New Jersey

OUTLINE

* This article was originally published in the *Journal of Near Eastern Studies,* **4**, 1–38, 1945. The author wishes to thank the editors of these *Publications* for their suggestion to republish this article and the University of Chicago Press for their courtesy in agreeing with this proposal. Republication made it possible to add a new section on Jewish astronomy and to amplify the bibliography. The article will be continued in the April issue.

17

Ce qui est admirable, ce n'est pas que le champ des étoiles soit si vaste, c'est que l'homme l'ait mesuré.—Anatole France, *Le Jardin d'Épicure.*

I. Introduction

1. In the following pages an attempt is made to offer a survey of the present state of the history of ancient astronomy by pointing out relationships with various other problems in the history of ancient civilization and particularly by enumerating problems for further research which merit our interest not only because they constitute gaps in our knowledge of ancient astronomy but because they must be clarified in order to lay a solid foundation for the understanding of later periods.

I wish to emphasize from the very beginning that the attitude taken here is of a very personal character. I do not believe that there is any single approach to the history of science which could not be replaced by very different methods of attack; only trivialities permit but one interpretation. I must confess still more: I cannot even pretend to be complete in the selection of topics essential for our understanding of ancient astronomy,[1] nor do I wish to conceal the fact that many of the steps which I myself have taken were dictated by mere accident. To mention only one example: without having been brought into contact with a recently purchased collection of Demotic papyri in Copenhagen, I would never have undertaken the investigation of certain periods of Hellenistic and Egyptian astronomy which now

---

[1] Also the bibliography, given at the end, is very incomplete and is only intended to inform the reader where he can find further details of the specific viewpoint discussed here and to list the original sources.

seem to me to constitute a very essential link between ancient and medieval astronomy. In other words, though I have always tried to subordinate any particular research problem to a wider program of systematic analysis, the impossibility of elaborate long-range planning has again and again been impressed upon me. The situation is comparable to entering a vast mountainous region on a single trail; one must simply follow the winding path, trying to give account of its general direction, but one can never predict with certainty what new vistas will be exposed at the next turn.

2. The enormous complexity of the study of ancient astronomy becomes evident if we try to make the first, and apparently simplest, step of classification: to distinguish between, say, Mesopotamian, Egyptian, and Greek astronomy, not to mention their direct successors, such as Hindu, Arabic, and medieval astronomy. Neither geographically nor chronologically nor according to language can clear distinctions be made. Entirely different conditions underlie the astronomy in Egypt of the Middle and New Kingdoms than in the periods after the Persian conquest. Greek astronomy of Euclid's time has very little in common with Hipparchus' astronomy only a hundred and fifty years later. It is evident that it is of very little value to speak about a "Babylonian" astronomy regardless of period, origin, and scope. And, worst of all, the concept "astronomy" itself undergoes changes in meaning when we speak about different periods. The fanciful combination of a group of brilliant stars to form the picture of a "bull's leg" and the computation of the irregularities in the moon's movement in order to predict accurately the magnitude of an eclipse are usually covered by the same name! For methodological reasons it is obvious that a drastic restriction in terminology must be made. We shall here call "astronomy" only those parts of human interest in celestial phenomena which are amenable to mathematical treatment. Cosmogony, mythology, and applications to astrology must be distinguished as clearly separated problems—not in order to be disregarded but to make possible the study of the mutual influence of essentially different streams of development. On the other hand, it is necessary to co-ordinate intimately the study of ancient mathematics and

astronomy because the progress of astronomy depends entirely on the mathematical tools available. This is in conformity with the concept of the ancients themselves : one need only refer to the original title of Ptolemy's *Almagest*, namely, "Mathematical Composition."

3. The study of ancient astronomy will always have its center of gravity in the investigation of the Hellenistic-Roman period, represented by the names of Hipparchus and Ptolemy. From this center three main lines of research naturally emerge : the investigation of the previous achievements of the Near East ; the investigation of pre-Arabic Hindu astronomy ; and the study of the astronomy of late antiquity in its relation to Arabic and medieval astronomy. This last-mentioned extension of our program beyond antiquity proper is not only the natural continuation of the original problem but constitutes an integral part of the general approach outlined here. Astronomy is the only branch of the ancient sciences which survived almost intact after the collapse of the Roman Empire. Of course, the level of astronomical studies dropped within the boundaries of the remnants of the Roman Empire, but the tradition of astronomical theory and practice was never completely lost. On the contrary, the rather clumsy methods of Greek trigonometry were improved by Hindu and Arabic astronomers, new observations were constantly compared with Ptolemy's results, etc. This must be paralleled with the total loss of understanding of the higher branches of Greek mathematics before one realizes that astronomy is the most direct link connecting the modern sciences with the ancient. In fact, the work of Copernicus, Brahe, and Kepler can be understood only by constant reference to ancient methods and concepts, whereas, for example, the meaning of the Greek theory of irrational magnitudes or Archimedes' integrations were understood only after being independently rediscovered in modern times.

There are, of course, very good reasons for the fact that ancient astronomy extended with an unbroken tradition deep into modern times. The structure of our planetary system is such that it is simple enough to permit the achievement of relatively far-reaching results with relatively simple mathematical methods, but complicated enough to invite constant improvement of the theory. It was thus possible to continue successfully the "ancient" methods in astronomy at a time when Greek mathematics had long reached a dead end in the enormous complication of geometric representation of essentially algebraic problems. The creation of the modern methods of mathematics, on the other hand, is again most closely related to astronomy, which urgently required the development of more powerful new tools in order to exploit the vast possibilities which were opened by Newton's explanation of the movement of the celestial bodies by means of general principles of physics. The confidence of the great scientists of the modern era in the sufficiency of mathematics for the explanation of nature was largely based on the overwhelming successes of celestial mechanics. Essentially the same held for scholars in classical times. In antiquity, mathematical tools were not available to explain any physical phenomena of higher complexity than the planetary movement. Astronomy thus became the only field of ancient science where indisputable certainty could be reached. This feeling of the superiority of mathematical astronomy is best expressed in the following sentences from the introduction to the *Almagest:* "While the two types of theory could better be called conjecture than certain knowledge—theology because of the total invisibility and remoteness of its object, physics because of the instability and uncertainty of matter—. . . . mathematics alone . . . . will offer reliable and certain knowledge because the proof follows the indisputable ways of arithmetic and geometry."[2]

II. Egypt

4. A few words must be said about Egyptian mathematics before discussing the astronomical material. Our main source for Egyptian mathematics consists of two papyri[3]—certainly not

---

[2] *Almagest* I, 1 (ed. Heiberg I, 6, 11 ff.).

[3] Math. Pap. Rhind [Peet *RMP;* Chace *RMP*] and Moscow mathematical papyrus [Struve *MPM*]. For a discussion of Egyptian arithmetic see Neugebauer [1], for Egyptian geometry Neugebauer [2], and, in general, Neugebauer *Vorl.* The most recent attempt at a synthesis of Egyptian science, by Flinders Petrie (*Wisdom of the Egyptians* [London, 1940]), must unfortunately be considered as dilettantish not only because

too great an amount in view of the length of the period in question! Still, it seems to be a fair assumption that we are well enough informed about Egyptian mathematics. Not only are both papyri of very much the same type but all additional fragments which we possess match the same picture—a picture which is paralleled by economic documents in which occur precisely those problems and methods which we find in the mathematical papyri. The Egyptian mathematical texts, furthermore, find their direct continuation in Greek papyri,[4] which again show the same pattern. It is therefore safe to say that Egyptian mathematics never rose above a very primitive level. So far as astronomy is concerned, numerical methods are of primary importance, and, fortunately enough, this is the very part of Egyptian mathematics about which we are best informed. Egyptian arithmetic can be characterized as being predominantly of an "additive" character, that is, its main tendency is to reduce all operations to repeated additions. And, because the process of division is very poorly adaptable to such procedures, we can say that Egyptian mathematics does not provide the most essential tools for astronomical computation. It is therefore not surprising that none of our Egyptian astronomical documents requires anything more than simple operations with integers. Where the complexity of the phenomena exceeded the capacity of Egyptian mathematics, the strongest simplifications were adopted, consequently leading to little more than qualitative results.

5. The astronomical documents of purely Egyptian origin are the following: Astronomical representations and inscriptions on ceilings of the New Kingdom,[5] supplemented by the so-called "diagonal calendars" on coffin lids of the Middle Kingdom[6] and

---

of its disregard of essential source material but also because of its lack of understanding for the mathematical and astronomical problems as such.

[4] The continuation of this tradition is illustrated by the following texts: Demotic: Revillout [1]; Coptic: Crum *CO*, No. 480, and Sethe *ZZ*, p. 71; Greek: Robbins [1] or Baillet [1]. For Greek computational methods in general, see Vogel [1].

[5] Examples: The Nut-pictures in the cenotaph of Seti I (Frankfort *CSA*) and Ramses IV (Brugsch *Thes.* 1) and analogous representations in the tombs of Ramses VI, VII, and IX.   [6] Cf. Pogo [1] to [4].

by the Demotic-Hieratic papyrus "Carlsberg 1."[7] Secondly, the
Demotic papyrus "Carlsberg 9," which shows the method of
determining new moons.[8] Though written in Roman times (after
A.D. 144), this text undoubtedly refers to much older periods
and is uninfluenced by Hellenistic methods. A third group of
documents, again written in Demotic, concerns the positions of
the planets.[9] In this case, however, it seems to be very doubtful
whether these tables are of Egyptian origin rather than products
of the Hellenistic culture; we therefore postpone a discussion to
the section on Hellenistic astronomy.[10] The last group of texts
is again inscribed on ceilings and has been frequently discussed
because of their representation of the zodiac.[11] There can be no
doubt that these latter texts were deeply influenced by non-
Egyptian concepts characteristic for the Hellenistic period. The
same holds, of course, for the few Coptic astronomical documents we possess.[12] It is, finally, worth mentioning that not a
single report of observations is preserved, in strong contrast to

---

[7] Lange-Neugebauer [1].   [8] Neugebauer-Volten [1].

[9] Neugebauer [3].   [10] Cf. below, Section 16.

[11] I know of the following representations of zodiacs: *No. 1* (Ptolemy
III and V, i.e., 247/181 B.C.): northwest of Esna, North temple of Khnum
(Porter-Moss *TB* VI, p. 118); *Nos. 2 and 3* (Ptolemaic or Roman): El-
Salâmûni, Rock tombs (Porter-Moss *TB* V, p. 18); mentioned by L'Hôte,
LE, pp. 86–87. *No. 4* (Ptolemaic-Roman; Tiberius): Akhmim, Two de-
stroyed temples (Porter-Moss *TB* V, p. 20); mentioned by Pococke *DE*,
I, pp. 77–78. *No. 5* (Tiberius): Dendera, Temple of Hathor, Outer hyp-
ostyle (Porter-Moss *TB* VI, p. 49). *No. 6* (Augustus-Trajan): Dendera,
Temple of Hathor, East Osiris-chapel central room, ceiling, west half
(Porter-Moss *TB* VI, p. 99). *Nos. 7 and 8* (1st cent. A.D.): Athribis,
Tomb (Porter-Moss *TB* V, p. 32). *No. 9* (Titus and Commodus): Esna,
Temple of Khnum (Porter-Moss *TB* VI, p. 116). *No. 10* (Roman): Deal-
er in Cairo, publ. Daressy [1], pp. 126–27, and Boll, *Sphaera*, Pl. VI. Five
other representations of the zodiacal signs are known from coffins, all from
Ptolemaic or Roman times. On the other hand, the original Egyptian constellations are still found on coffins of the Saitic or early Ptolemaic
periods.

[12] The only nonastrological Coptic documents known to me are the
tables of shadow lengths published by U. Bouriant and Ventre-Bey [1].—
P. Bouriant [1] did not recognize that the text published by him was a
standard list of the planetary "houses" with no specific reference to Arabic
astronomy.

the abundance of observational records from Mesopotamia. It is hard to say whether this reflects a significant historical fact or merely that we are at the mercy of the accidents of excavation.

Speaking of negative evidence, three instances must be mentioned which play a more or less prominent role in literature on the subject and have contributed much to a rather distorted picture of Egyptian astronomy. The first point consists in the idea that the earliest Egyptian calendar, based on the heliacal rising of Sothis, reveals the existence of astronomical activity in the fourth millennium B.C. It can be shown, however, that this theory is based on tacit assumptions which are very implausible in themselves and that the whole Egyptian calendar does not presuppose any systematic astronomy whatsoever.[13] The second remark concerns the hypothesis of early Babylonian influence on Egyptian astronomical concepts.[14] This theory is based on a comparative method which assumes direct influence behind every parallelism or vague mythological analogy. Every concrete detail of Babylonian and Egyptian astronomy which I know contradicts this hypothesis. Nothing in the texts of the Middle and New Kingdom equals in level, general type, or detail the contemporaneous Mesopotamian texts. The main source of trouble is, as usual, the retrojection into earlier periods of a situation which undoubtedly prevailed during the latest phase of Egyptian history. This brings us to the third point to be mentioned here: the assumption of an original Egyptian astrology. First of all, there is no proof in general for the widely accepted assertion that astrology preceded astronomy. But especially in Egypt is there no trace of astrological ideas in the enormous mythological literature which we possess for all periods.[15] The earliest horoscope from Egyptian soil, written in Demotic, refers to A.D. 13;[16] the earliest Greek horoscope from Egypt concerns the

───────────

[13] Neugebauer [4], Winlock [1], Neugebauer [5].

[14] Sponsored especially by the "Pan-Babylonian" school.

[15] It is interesting to observe how deeply imbedded is the assumption that astrology must precede astronomy. Brugsch called his edition of cosmogonic and mythological texts "astronomische und astrologische Inschriften" in spite of the fact that these texts do not betray the slightest hint of astrology.   [16] Neugebauer [6].

year 4 B.C.[17] We shall presently see that the assumption of a very late introduction of astrological ideas into Egypt corresponds to various other facts.

6. It is much easier to show that certain familiar ideas about the origin of astronomy are historically untenable than to give an adequate survey of our real knowledge of Egyptian astronomy. A. Pogo is to be credited with the recognition of the astronomical importance of inscriptions on the lids of a group of coffins from the end of the Middle Kingdom,[18] apparently representing the setting and rising of constellations, though in an extremely schematic fashion. The constellations are known as the "decans" because of their correspondence to intervals of ten days. He furthermore saw the relationship between these simple pictures and the elaborate representations on the ceilings of the tombs belonging to kings of the New Kingdom.[19]

It can be safely assumed that the coffin lids are very abbreviated forms of contemporaneous representations on the ceilings of tombs and mortuary temples of the rulers of the Middle Kingdom. The logical place for these representations of the sky on ceilings explains their destruction easily enough. The earliest preserved ceiling, discovered in the unfinished tomb of Senmut, the vezir of Queen Hatshepsut,[20] is about three centuries later than the coffin lids. Then come the well-preserved ceiling in the subterranean cenotaph of Seti I[21] and its close parallels in the tomb of Ramses IV[22] and later rulers.[23] The difficulties we have

---

[17] *Pap. Oxyrh.* 804. From this time until A.D. 500 more than sixty individual horoscopes, fairly equally distributed in time, are known to me.

[18] Cf. n. 6.

[19] Some of Pogo's assumptions must, however, be abandoned, because they are based on the distinction of different types of such coffin inscriptions. A close examination of these texts (and also unpublished material) shows that all preserved samples belong to the same type. A systematic edition of all these texts is urgently needed if we are to obtain a solid basis for the study of Egyptian constellations.

[20] Winlock [2], pp. 34 ff., reprinted in Winlock *EDEB,* pp. 138 ff. and Pogo [5]. The final publication has not yet appeared.

[21] Frankfort *CSA.*

[22] Brugsch *Thes.* I opposite pp. 174–75, but incomplete (cf. Lange-Neugebauer [1], p. 90).   [23] Cf. n. 5.

to face in an attempt to explain these texts can best be illustrated by a brief discussion of the above-mentioned papyrus
"Carlsberg 1." This papyrus was written more than a thousand
years after the Seti text but was clearly intended to be a commentary to these inscriptions. In the papyrus we find the text
from the cenotaph split into short sections, written in Hieratic,
which are followed by a word-for-word translation into Demotic
supplemented by comments in Demotic. The original text is
frequently written in a cryptic form, to which the Demotic version gives the key. We now know, for instance, that various
hieroglyphs were replaced by related forms in order to conceal
the real contents from the uninitiated reader. How successfully
this method worked is shown by the fact that one such sign,
which is essential for the understanding of a long list of dates of
risings and settings of the decans, was used at its face value for
midnight instead of evening.[24] It is needless to emphasize what
the recognition of such substitutions means for the correct understanding of astronomical texts. A complete revision of all previously published material is needed in the light of this new
insight into the Egyptian scheme of describing the rising and
setting of stars the year round. One point, however, must be
kept in mind in every investigation of Egyptian constellations.
One must not ascribe to these documents a degree of precision
which they were never intended to possess. I doubt, for example,
very much whether one has a right to assume that the decans are
constellations covering exactly ten degrees of a great circle on
the celestial sphere. I think it is much more plausible that they
are constellations spread over a more or less vaguely determined
belt around the sky, just as we speak about the Milky Way.
It is therefore methodically wrong to use these star lists and
the accompanying schematic date lists for accurate computations,
as has frequently been attempted.

The second Demotic astronomical document, papyrus Carlsberg 9, is much easier to understand and gives us full access to
the Egyptian method of predicting the lunar phases with sufficient accuracy. The whole text is based on the fact that 25

[24] Sethe, *ZÄA*, p. 293, n. 1, and Lange-Neugebauer [1], p. 63.

Egyptian years cover the same time interval as 309 lunations. The 25 years equal 9125 days, which are periodically arranged into groups of lunar months of 29 and 30 days. The periodic repetition of this simple scheme corresponds, on the average, very well with the facts; more was apparently not required, and, we may add, more was not obtainable with the available simple mathematical means which are described at the beginning of this section. The purpose of the text was to locate the wandering lunar festivals within the schematic civil calendar, as is shown by a list of the "great" and "small" years of the cycle, which contain 13 or 12 lunar festivals, respectively.[25] Accordingly, calendaric problems are seen to be the activating forces here as well as in the decanal lists of the Middle and New Kingdom. The two Carlsberg papyri thus give us a very consistent picture of Egyptian stellar and lunar astronomy and its calendaric relations and are in best agreement with the level known from the mathematical papyri.

Before leaving the description of Egyptian science, brief mention should be made of the much-discussed question of the "scientific" character of Egyptian mathematics and astronomy. First of all, the word "scientific" must be clearly defined. The usual identification of this question with that of the practical or theoretical purpose of our documents is obviously unsatisfactory. One cannot call medicine or physics unscientific even if they serve eminently practical purposes. It is neither possible nor relevant to discover the moral motives of a scientist—they might be altruistic or selfish, directed by the desire for systematization or by interest in competitive success. It is therefore clear that the concept "scientific" must be described as a question of methods, not of motives. In the case of mathematics and astronomy, the situation is especially simple. The criterion for scientific mathematics must be the existence of the concept of proof; in astronomy, the elimination of all arguments which are not exclusively based on observations or on mathematical consequences

[25] The "great" and "small" years (already mentioned in an inscription of the Middle Kingdom) have given rise to much discussion (cf., e.g., Ginzel *Chron.*, I, pp. 176–77) which can now be completely ignored.

of an initial hypothesis as to the fundamental character of the movements involved. Egyptian mathematics nowhere reaches the level of argument which is worthy of the name of proof, and even the much more highly developed Babylonian mathematics hardly ever displays a general technique for proving its procedures.[26] Egyptian astronomy was satisfied with a very, rough qualitative description of the phenomena—here, too, we miss any trace of scientific method. The first scientific attack of mathematical problems was made in the fifth century B.C. in Greece. We shall see that scientific astronomy can be found shortly thereafter in Babylonian texts of the Seleucid period. In other words, the enormous interest of the study of pre-Hellenistic Oriental sciences lies in the fact that we are able to follow the development far back into *pre*-scientific periods which saw the slow preparation of material and problems which deeply influenced the shape of the real scientific methods which emerged to full power for the first time in the Hellenistic culture. It is a serious mistake to try to invest Egyptian mathematical or astronomical documents with the false glory of scientific achievements or to assume a still unknown science, secret or lost, not found in the extant texts.

III. MESOPOTAMIA

7. Turning to Babylonian astronomy, one's first impression is that of an enormous contrast to Egyptian astronomy. This contrast not only holds in regard to the large amount of material available from Mesopotamia but also with respect to the level finally reached. Text from the last two or three centuries B.C. permit the computation of the lunar movement according to methods which certainly rank among the finest achievements of ancient science—comparable only to the works of Hipparchus and Ptolemy.

It is one of the most fascinating problems in the history of ancient astronomy to follow the different phases of this development which profoundly influenced all further events. Before giving a short sketch of this progress as we now restore it according

---

[26] See the discussion in Neugebauer *Vorl.*, pp. 203 ff.

to our present knowledge, we must underline the incompleteness of the present state of research, which is due to the fact that we do not yet have reliable and complete editions of the text material. The observation reports addressed to the Assyrian kings were collected by R. C. Thompson[27] and in the editions of Assyrian letters published and translated by Harper,[28] Waterman,[29] and Pfeiffer;[30] much related material is quoted in the publications of Kugler,[31] Weidner,[32] and others. But Thompson's edition gives the original texts only in printed type, subject to all the misunderstandings of this early period of Assyriology, and very little has been done to repair these original errors. Nothing short of a systematic "corpus" of all the relevant texts can provide us with the requisite security for systematic interpretation. The great collection of astrological texts, undertaken by Virolleaud[33] but never finished, confronts the reader with still greater difficulties, because Virolleaud composed complete versions from various fragments and duplicates without indicating the sources from which the different parts came. And, finally, the tablets dealing with the movement of the moon and the planets were discussed and explained in masterly fashion by Kugler;[34] but here, too, a systematic edition of the whole material is necessary.[35] Years of systematic work will be needed before the foundations for a reliable history of the development of Babylonian astronomy are laid.

8. Kugler uncovered step by step the ingenious methods by which the ephemerides of the moon and the planets which we find inscribed on tablets ranging from 205 B.C. to 30 B.C. were computed.[36] It can justly be said that his discoveries rank among the

---

[27] Thompson *Rep.* (1900).   [28] Harper *Letters.*

[29] Waterman *RC.*   [30] Pfeiffer *SLA.*

[31] Kugler *SSB* and Kugler *MP.*

[32] Weidner *Hdb.,* Weidner [1], [2], and numerous articles in the early volumes of *Babyloniaca.*

[33] Virolleaud *ACh.*   [34] Kugler *BMR* and *SSB.*

[35] Such an edition by the present author is in preparation; it is quoted in the following as *ACT.*

[36] The first tentative (but very successful) steps were made by Epping *AB* (1889). Then follow Kugler's monumental works *BMR* (1900)

most important contributions toward an understanding of ancient civilization. It is very much to be regretted that historians of science often quote Kugler but rarely read him;[37] by doing this, they have disregarded the newly gained insight into the origin of the basic methods in exact science. This is not the place to describe in detail the Babylonian "celestial mechanics," as it might properly be called; that will be one of the tasks of a history of ancient astronomy which remains to be written. A few words, however, must be said in order to render intelligible the relationship between Babylonian and Greek methods. The problem faced by ancient astronomers consisted in predicting the positions of the moon and the planets for an extended period of time and with an accuracy higher than that obtainable by isolated individual observations, which were affected by the gross errors of the instruments used. All these phenomena are of a periodic character, to be sure, but are subject to very complicated fluctuations. All that we know now seems to point to the following reconstruction of the history of late Babylonian astronomy. A systematic observational activity during the Late Assyrian and Persian periods (roughly, from 700 B.C. onward) led to two different results. First, the collected observations provided the astronomers with fairly accurate average values for the main periods of the phenomena in question; once such averages were obtained, improvements could be furnished by scattered observational records from preceding centuries. Secondly, from individual observations, for example, of the moment

---

and *SSB* (published between 1907 and 1924), supplemented by Schaumberger's explanation of the determination of first and last visibility of the moon (1935) and continued by the present author with respect to the theory of latitude and eclipses (Neugebauer [8], [9], [20], Pannekoek [2] and van der Waerden [1]). The theory of planets is treated in Kugler *SSB,* to be supplemented by Pannekoek [1], Schnabel [2], and van der Waerden [2]. All previously published texts and much unpublished material will be contained in Neugebauer *ACT.* The whole material amounts to about a hundred ephemerides for the moon and the planets, covering the above-mentioned two centuries.

[37] Abel Rey, *La Science orientale avant les grecs* (Paris, 1930), and E. Zinner, *Geschichte der Sternkunde* (Berlin, 1931), are brilliant examples showing complete ignorance of Kugler's results.

of full moon[38] or of heliacal settings, etc., short-range predictions could be made by methods which we would call linear extrapolation. Such methods are frequently sufficient to *exclude* certain phenomena (such as eclipses) in the near future and, under favorable conditions, even to predict the date of the next phenomenon in question. After such methods had been developed to a certain height, apparently one ingenious man conceived a new idea which rapidly led to a systematic method of long-range prediction. This idea is familiar to every modern scientist; it consists in considering a complicated periodic phenomenon as the result of a number of periodic effects, each of a character which is simpler than the actual phenomenon.[39] The whole method probably originated in the theory of the moon, where we find it at its highest perfection. The moments of new moons could easily be found if the sun and moon would each move with constant velocity. Let us assume this to be the case and use average values for this ideal movement; this gives us average positions for the new moons. The actual movement deviates from this average but oscillates around it periodically. These deviations were now treated as new periodic phenomena and, for the sake of easier mathematical treatment, were considered as linearly increasing and decreasing. Additional deviations are caused by the inclination of the orbits. But here again a separate treatment, based on the same method, is possible. Thus, starting with average positions, the corrections required by the periodic deviations are applied and lead to a very close description of the actual facts. In other words, we have here, in the nucleus, the idea of "perturbations," which is so fundamental to all phases of the development of celestial mechanics, whence it spread into every branch of exact science.

We do not know when and by whom this idea was first employed. The consistency and uniformity of its application in the older of the two known "systems" of lunar texts point clearly to an invention by a single person. From the dates of the pre-

---

[38] Frequently mentioned in the "reports" to the Assyrian court (e.g., Thompson *Rep.*).

[39] A classic example is the treatment of sounds as the result of the superposition of pure harmonic vibrations.

served texts, one might assume a date in the fourth or third century B.C.[40] This basic idea was applied not only to the theory of the moon (in two slightly modified forms) but also to the theory of the planets. In this latter theory the main point consists in refraining from an attempt to describe directly the very irregular movement, substituting instead the separate treatment of several individual phenomena, such as opposition, heliacal rising, etc.; each of these phenomena is treated with the methods familiar from the lunar theory as if it were the periodic movement of an independent celestial body. After dates and positions of each characteristic phenomenon are determined, the intermediate positions are found by interpolation between these fixed points.[41] It must be said, however, that the planetary theory was not developed to the same degree of refinement as the lunar theory; the reason might very well be that the lunar theory was of great practical importance for the question of the Babylonian calendar: whether a month would have 30 or 29 days. For the planets no similar reason for high accuracy seems to have existed, and it was apparently sufficient merely to compute the approximate dates of phenomena, which, in addition, are frequently very difficult to observe accurately.

We cannot emphasize too strongly that the essential point in the above-described methods lies not in the comparatively high accuracy of the results obtained but in their fundamentally new attitude toward the whole problem. Let us, as a typical example, consider the movement of the sun.[42] Certain simple observations, most likely of the unequal length of the seasons, had led to the discovery that the sun does not move with constant velocity in

---

[40] The attempts to determine a more precise date (Schnabel *Ber.*, pp. 219 ff., and Schnabel [1], pp. 15 ff.) are based on unsatisfactory methods. The generally accepted statement that Naburimannu was the founder of the older system of the lunar theory relies on nothing more than the occurrence of this name in one of the latest tablets in a context which is not perfectly clear.

[41] This is shown by a tablet for Mercury, to be published in Neugebauer *ACT*. The interpolation is not simply linear but of a more complicated type known from analogous cases in the lunar theory.

[42] For details see Neugebauer [10] and [9] § 2.

its orbit. The naïve method of taking this fact into account would be to compute the position of the sun by assuming a regularly varying velocity. It turned out, however, that considerable mathematical difficulties were met in computing the syzygies of the moon according to such an assumption. Consequently, another velocity distribution was substituted, and it was found that the following "model" was satisfactory: the sun moves with two different velocities over two unequal arcs of the ecliptic, where velocities and arcs were determined in such a fashion that the initial empirical facts were correctly explained and at the same time the computation of the conjunctions became sufficiently simple. It is self-evident that the man who devised this method did not think that the sun moved for about half a year with constant velocity and then, having reached a certain point in the ecliptic suddenly started to move with another, much higher velocity for the rest of the year. His problem was clearly this: to make a very complicated problem accessible to mathematical treatment with the only condition that the final consequences of the computations correctly correspond to the actual observations —in our example, the inequality of the seasons. The Greeks[43] called this a method "to preserve the phenomena"; it is the method of introducing mathematically useful steps which in themselves need not be of any physical significance. For the first time in history, mathematics became the leading principle for the structure of physical theories.

9. It will be clear from this discussion that the level reached by Babylonian mathematics was decisive for the development of such methods. The determination of characteristic constants (e.g., period, amplitude, and phase in periodic motions) not only requires highly developed methods of computation but inevitably leads to the problem of solving systems of equations corresponding to the outside conditions imposed upon the problem by the observational data. In other words, without a good stock of mathematical tools, devices of the type which we find everywhere in the Babylonian lunar and planetary theory could

---

[43] E.g., Proclus, *Hypotyposis astron. pos.* v. 10 (ed. Manitius, 140, 21).

not be designed. Egyptian mathematics would have rendered
hopeless any attempt to solve problems of the type needed constantly in Babylonian astronomy. It is therefore essential for
our topic to give a brief sketch of Babylonian mathematics.

I think it can be justly said that we have a fairly good knowledge of the character of mathematical problems and methods
in the Old Babylonian period (*ca.* 1700 b.c.). Almost a hundred
tablets from this period are published;[44] they contain collections
of problems or problems with complete solutions—amounting to
far beyond a thousand problems. We know practically nothing
about the Sumerian mathematics of the previous periods and
very little of the interval between the Old Babylonian period and
Seleucid times. We have but few problem texts from the latter
period, but they give us some idea of the type of mathematics
familiar to the astronomers of this age. This material is sufficient to assure us that all the essential achievements of Old
Babylonian times were still in the possession of the latest representatives of Mesopotamian science. In other words, Babylonian
mathematical astronomy was built on foundations independently
laid more than a millennium before.

If one wishes to characterize Babylonian mathematics by
one term, one could call it "algebra." Even where the foundation is apparently geometric, the essence is strongly algebraic,
as can be seen from the fact that frequently operations occur
which do not admit of a geometric interpretation, as addition of
areas and lengths, or multiplication of areas. The predominant
problem consists in the determination of unknown quantities
subject to given conditions. Thus we find prepared precisely
the tools which were later to become of the greatest importance
for astronomy.

Of course, the term "algebra" does not completely cover
Babylonian mathematics. Not only were a certain number of
geometrical relations well known but, more important for
our problem, the basic properties of elementary sequences (e.g.,

---

[44] These texts were published in Neugebauer *MKT* (1935–38) and in
Neugebauer-Sachs *MCT* (1945). A large part of the *MKT* material was
republished in Thureau-Dangin *TMB* (1939). For a general survey see
Neugebauer *Vorl.*

arithmetic and geometric progressions) were developed.[45] The numerical calculations are carried out everywhere with the greatest facility and skill.

We possess a great number of texts from all periods which contain lists of reciprocals, square and cubic roots, multiplication tables, etc., but these tables rarely go beyond two sexagésimal places (i.e., beyond 3600). A reverse influence of astronomy on mathematics can be seen in the fact that tables needed for especially extensive numerical computations come from the Seleucid period; tables of reciprocals are preserved with seven places (corresponding to eleven decimal places) for the entry and up to seventeen places (corresponding to twenty-nine decimal places) for the result. It is clear that numerical computations of such dimensions are needed only in astronomical problems.

The superiority of Babylonian numerical methods has left traces still visible in modern times. The division of the circle into 360 degrees and the division of the hour into 60 minutes and 3600 seconds reflect the unbroken use of the sexagesimal system in their computations by medieval and ancient astronomers. But though the base 60 is the most conspicuous feature of the Babylonian number system, this was by no means essential for its success. The great number of divisors of 60 is certainly very useful in practice, but the real advantage of its use in the mathematical and astronomical texts lies in the place-value notation,[46] which is consistently employed in all scientific computations. This gave the Babylonian number system the same advantage over all other ancient systems as our modern place-value notation holds over the Roman numerals. The importance

---

[45] Incidentally, we also have an example (Neugebauer-Sachs *MCT*, Problem-Text A) of purely number theoretical type from Old Babylonian times (so-called "Pythagorean numbers"); but it should be added that we do not find the slightest trace of number mysticism anywhere in these texts.

[46] Place-value notation consists in the use of a very limited number of symbols whose magnitude is determined by position. Thus 51 does not mean 5 plus 1 (as it would with Roman or Egyptian numerals), but 5 times 10 plus 1. Analogously in the sexagesimal system, five followed by one (we transcribe 5, 1) means 5 times 60 plus 1 (i.e., 301).

of this invention can well be compared with that of the alphabet. Just as the alphabet eliminates the concept of writing as an art to be acquired only after long years of training, so a place-value notation eliminates mere computation as a complex art in itself. A comparison with Egypt or with the Middle Ages illustrates this very clearly. Operation with fractions, for example, constituted a problem in itself for medieval computers; in place-value notation, no such problem exists,[47] thus eliminating one of the most serious obstacles for the further development of mathematical technique.

The analogy between alphabet and place-value notation can be carried still further. Neither one was the sudden invention made by a single person but the final outcome of various historical processes. We are able to trace Mesopotamian number-writing far back into the earliest stages of civilization, thanks to the enormous amount of economic documents preserved from all periods. It can be shown how a notation analogous to the Egyptian or Roman system was gradually replaced by a notation which developed naturally in the monetary system and which tended toward a place-value notation. The value 60 of the base appears to be the outcome of the arrangement of the monetary units.[48] Outside of mathematical texts, the place-value notation was always overlapped by various other notations, and toward the end of Mesopotamian civilization a modified system became predominant. It seems very possible, however, that the idea of place-value writing was never completely lost and found its way through astronomical tradition into early Hindu astronomy,[49] whence our present number system originated during the first half of the first millennium A.D.

10. We now turn to the periods preceding the final stage of Babylonian astronomy which culminated in the mathematical

---

[47] Example: to add or to multiply 1.5 and 1.2 requires exactly the same operations as the addition or multiplication of 15 and 12.

[48] For details see Neugebauer [11] and Neugebauer *Vorl.*, chap. iii § 4. The theory set forth by Thureau-Dangin *SS* (English version Thureau-Dangin [1]) does not account for the place-value notation, which is the most essential feature of the whole system.

[49] Cf. Datta-Singh *HHM* I and Neugebauer [12], pp. 266 ff.

theory of the moon and the planets described above. It is not possible to give an outline of this earlier development because most of the preliminary work remains to be done. A few special problems, however, which must eventually find their place in a more complete picture, can now be mentioned.

In our discussion of the methods used in the lunar and planetary theories, we had occasion to mention the extensive use of periodically increasing and decreasing sequences of numbers. A simple case of this method appears in earlier times in the problem of describing numerically the changing length of day and night during the year. The crudest form is the assumption of linear variation between two extremal values.[50] Two much more refined schemes are incorporated in the texts of the latest period, but it seems very likely that they are of earlier origin. Closely related are two other problems: the variability of the length of the shadow of the "gnomon"[51] and the measurement of the length of the day by water clocks.[52] The latter problem has caused considerable trouble in the literature on the subject because the texts show the ratio 2:1 for the extremal values during the year. A ratio 2:1 between the longest and the shortest day, instead of the ratio 3:2, which is otherwise used,[53] would correspond to a geographical latitude absolutely impossible for Babylon. The discrepancy disappears, however, if one recalls the fact that the amount of water flowing from a cylindrical vessel is not proportional to the time elapsed but decreases with the sinking level.[54] It is worth mentioning in this connection that the outflow of water from a water clock is already discussed in Old Babylonian mathematical texts.[55]

This whole group of early astronomical texts, however, leads to nothing more than very approximate results. This is seen from the fact that the year is assumed, for the sake of simplicity,

[50] E.g., Weissbach *BM*, pp. 50–51.

[51] Weidner [1], pp. 198 ff.

[52] Weissbach *BM*, pp. 50–51; Weidner [1], pp. 195–96.

[53] Schaumberger *Erg.*, p. 377.

[54] Neugebauer [19].

[55] Thureau-Dangin [2] and Neugebauer *MKT*, I, pp. 173 ff.

to be 360 days long and divided into 12 months of 30 days each.[56] This schematic treatment has its parallel in the schemes which we have met in Egyptian astronomy and which we shall find again in early Greek astronomy; we must once more emphasize that elements from such schemes cannot be used for modern calculations, since this would assume quantitative accuracy where only qualitative results had been intended.

The calendaric interest of these problems is obvious. The same is true of the oldest preserved astronomical documents from Mesopotamia, the so-called "astrolabes."[57] These astrolabes are clay tablets inscribed with a figure of three concentric circles, divided into twelve sections by twelve radii. In each of the thirty-six fields thus obtained we find the name of a constellation and simple numbers whose significance is not yet clear. But it seems evident that the whole text constitutes some kind of schematic celestial map which represents three regions on the sky, each divided into twelve parts, and attributing characteristic numbers to each constellation. These numbers increase and decrease in arithmetic progression and are undoubtedly connected with the corresponding month of the schematic twelve-month calendar. It is clear that we have here some kind of simple astronomical calendar parallel (not in detail, but in purpose) to the "diagonal calendars" in Egypt. In both cases these calendars are of great interest to us as a source for determining the relative positions and the earliest names of various constellations. But here, too, the strongest simplifications are adopted in order to obtain symmetric arrangements, and much remains to be done before we can answer such questions as the origin of the "zodiac."

11. Few statements are more deeply rooted in the public mind or more often repeated than the assertion that the origin

---

[56] This schematic year of 360 days, of course, does not indicate that one assumed 360 days as the correct length of the solar year. A lunar calendar makes correct predictions of a future date very difficult. The schematic calendar is in practice therefore very convenient for giving future dates which must, at any rate, be adjusted later.

[57] This name is rather misleading and is merely due to the circular arrangement. Schott [1], p. 311, introduced the more appropriate name "twelve-times-three." Such texts are published in CT 33, Pls. 11 and 12. Cf. also Weidner *Hdb.*, pp. 62 ff. and Schott [1].

of astronomy is to be found in astrology. Not only is historical evidence lacking for this statement but all well-documented facts are in sharp contradiction to it. All the above-mentioned facts from Egypt and Babylonia (and, as we shall presently see, also from Greece) show that calendaric problems directed the first steps of astronomy. Determination of the season, measurement of time, lunar festivals—these are the problems which shaped astronomical development for many centuries; and we have seen that even the last phase of Mesopotamian astronomy, characterized by the mathematical ephemerides, was mainly devoted to problems of the lunar calendar. It is therefore one of the most difficult problems in the history of ancient astronomy to uncover the real roots of astrology and to establish their relation to astronomy. Very little has been done in this direction, mainly because of the prejudice in favor of accepting without question the priority of astrology.

Before going into this problem in greater detail, we must clarify our terminology. The modern reader usually thinks in terms of that concept of astrology which consists in the prediction of the fate of a person determined by the constellation of the planets, the sun, and the moon at the moment of his birth. It is well known, however, that this form of astrology is comparatively late and was preceded by another form of much more general character (frequently called "judicial" astrology in contrast to the "genethlialogical" or "horoscopic" astrology just described). In judicial astrology, celestial phenomena are used to predict the imminent future of the country or its government, particularly the king. From halos of the moon, the approach or invisibility of planets, eclipses, etc., conclusions are drawn as to the invasion of an enemy from the east or west, the condition of the coming harvest, floods and storms, etc.; but we never find anything like the "horoscope" based on the constellation at the moment of birth of an individual. In other words, Mesopotamian "astrology" can be much better compared with weather prediction from phenomena observed in the skies than with astrology in the modern sense of the word. Historically, astrology in Mesopotamia is merely one form of predicting future events; as such, it belongs to the enormous field of omen literature which is so familiar to every student of Babylonian civilization.[58]

Indeed, it can hardly be doubted that astrology emerged from the general practice of prognosticating through omens, which was based on the concept that irregularities in nature of any type (e.g., in the appearance of newborn animals or in the structure of the liver or other internal parts of a sheep) are indicative of other disturbances to come. Once the idea of fundamental parallelism between various phenomena in nature and human life is accepted, its use and development can be understood as consistent; established relations between observed irregularities and following events, constantly amplified by new experiences, thus lead to some sort of empirical science, which seems strange to us but was by no means illogical and bare of good sense to the minds of people who had no insight into the physical laws which determined the observed facts.

Though the preceding remarks certainly describe the general situation adequately, the historical details are very much in the dark. One of the main difficulties lies in the character of our sources. We have at our disposal large parts of collections of astrological omens arranged in great "series" comprising hundreds of tablets. But the preserved canonical series come mainly from comparatively late collections (of the Assyrian period) and were thus undoubtedly subject to countless modifications. We must, moreover, probably assume that the collection of astrological omina goes back to the Cassite period (before 1200 B.C.)—a period about which our general information is pretty flimsy. From the Old Babylonian period only one isolated text is preserved[59] which contains omina familiar from the later astrology. Predictions derived from observations of Venus made during the reign of Ammisaduqa (ca. 1600 B.C.) are preserved only in copies written almost a thousand years later[60] and clearly subjected to several changes during this long time. We are thus again left in the dark as to the actual date of the composition of

---

[58] A comprehensive study of the development of the astrological omina literature by E. F. Weidner is in course of publication (Weidner [2]).

[59] Šileiko [1].   [60] Langdon *VT*.

these documents except for the fact that it seems fairly safe to say that no astrological ideas appear before the end of the Old Babylonian period. Needless to say, there are no astrological documents of Sumerian origin.

The period of the ever increasing importance of astrology (always, of course, of the above-mentioned type of "judicial" astrology) is that beginning with the Late Assyrian empire. The "reports" mentioned previously, preserved in the archives of the Assyrian kings, are our witnesses. But here, again, a completely unsolved problem must be mentioned: we do not know how the "horoscopic" astrology of the Hellenistic period originated from the totally different omen type of astrology of the preceding millennium. It is, indeed, an entirely unexpected turn to make the constellation of the planets at a single moment responsible for the whole future of an individual, instead of observing the ever shifting phenomena on the sky and thus establishing short-term consequences for the country in general (even if represented in the person of the king). It seems to me by no means self-evident that this radical shift of the character of astrology actually originated in Babylonia. We shall see in the next section that the horoscopic practice flourished especially in Egypt. It might therefore very well be that the new tendency originated in Hellenistic times outside Mesopotamia and was reintroduced there in its modified form. It might be significant that only seven horoscopes are preserved from Mesopotamia, all of which were written in the Seleucid period,[61] a ridiculously small number as compared with the enormous amount of textual material dealing with the older "judicial" astrology. It must be admitted, however, that the oldest horoscopes known are of Babylonian origin. On the other hand, at no specific place can all the elements be found which are characteristic for astrology from Hellenistic times onward. Neither Babylonian astrology

---

[61] Two are published by Kugler *SSB,* II, 554 ff., and refer to the years 258 and 142 B.C., respectively. One (probably 233 B.C.) is published in Thompson *AB* 251. Among four unpublished horoscopes, discovered by Dr. A. Sachs, two are very small fragments, one can be dated 235 B.C., and the last was cast for the year 263 B.C.; the last is the oldest horoscope in the world.

nor Egyptian cosmology furnishes the base for the fundamental assumption of horoscopic astrology, namely, that the position of the planets in the zodiac decides the future. And, finally, it must be emphasized that the problem of determining the date and place of origin of horoscopic astrology is intimately related to the problem of the date and origin of mathematical astronomy. Horoscopes could not be cast before the existence of methods to determine the position of the celestial bodies for a period of at least a few decades. Even complete lists of observations would not be satisfactory because the positions of the planets in the zodiac are required regardless of their visibility at the specific hour. This shows how closely interwoven are the history of astrology and the history of planetary theories.

11a. In A.D. 1178 the learned rabbi Moses ben Maimon, commonly called Maimonides, wrote in Cairo a work entitled *Mishneh Torah,* covering in fourteen books a great variety of rabbinical laws. The third book[61a] is devoted to the laws pertaining to festivals, and especially the "Sanctification of the Months." A discussion of this work would lie far beyond the limits of the present article, had not Sidersky shown[61b] that the application of Maimonides' rules for the computation of the visibility of the new crescent leads to excellent numerical agreement with values found in a Babylonian ephemeris for 133/132 B.C.[61c] We

---

[61a] Because the literature concerning this book is nowhere systematically collected, at least to my knowledge, I give here a critical bibliography. Chaps. xii to xvii in German translation (by A. Kurrein) published by Littrow [1] (1872); no commentary, misleading terminology. Chaps. i and vi to xix, omitting examples, Hildesheimer [1] (1881); free German translation and commentary, ignoring Littrow [1]. Chaps. i to x in German translation Mahler [1] (1889) pp. 131–40. Chaps. i to xix Mahler [2] (1889); inadequate commentary, ignoring Hildesheimer [1]. Chaps. xi to xix Baneth [1] (1898–1903); Hebrew text, German translation and very detailed commentary. Chaps. xii to xvii Sidersky [1] (1913) pp. 662–72; free French translation, based on Littrow [1]; no commentary. Followed by Sidersky [2] ([1917] 1923), pp. 182. f.; wrong commentary to chap. xvii, overlooking Baneth [1]. Followed by Sidersky [3] (1919); correct commentary, based on Baneth [1]. A summary of Maimonides' methods is given in Feldman *RMA* (1931), based on Baneth [1].

[61b] Sidersky [3].

[61c] For the dating of these tables cf. Kugler *BMR,* pp. 47 ff., Sider-

herewith obtain an important tool for the investigation of the
Babylonian principles of predicting the length of their months.
If Sidersky's result is borne out by its application to all the
available texts, Maimonides' chapter on the visibility of the
moon would constitute a striking example of the continuity of
astronomical tradition.

Without answer remains the question through what channels
Babylonian methods may have entered Jewish astronomy.
Sidersky pointed out,[61d] that a passage in al-Bērūnī's *Chronology of Ancient Nations,* written about A.D. 1000, indicates
that the Jews about 100 B.C. began to compute the time of the
new moon in order to avoid misleading reports of the visibility
of the crescent.[61e] This date corresponds exactly to the period
of the Babylonian ephemerides discussed above. The first part
of Maimonides' procedure, devoted to finding the exact position
of sun and moon, follows closely Ptolemy's *Almagest,* which
was of course accessible to Maimonides through Arabic sources.
The second and most essential part, however, the problem of
the visibility in its dependence on the moon's apparent position
relative to the horizon, is not discussed in the *Almagest* because
the Greek astronomers were not interested in the problems of a
lunar calendar. For this part, Maimonides had to rely on other
sources, unknown to us. A direct use of cuneiform texts is, of
course, out of the question. Exactly the same difficulty appears
wherever we suspect Babylonian influence, e.g., in Persian and
Greek astronomy. We must probably assume the existence of
intermediate sources, e.g., treatises written in Aramaic. None
of them seems to have escaped destruction.

---

sky [3], p. 34, and Kugler *SSB,* II, p. 584, n. An additional fragment of
this text will be published in Neugebauer *ACT.*

[61d] Sidersky [1], pp. 631 f.

[61e] For the details see Sachau *Alb,* pp. 67 f. The excerpt given by
Sidersky [1], p. 673, is too short.

THE HISTORY OF ANCIENT ASTRONOMY: PROBLEMS AND METHODS*

O. Neugebauer

Institute for Advanced Study, Princeton, New Jersey

IV. The Hellenistic Period

12. Before beginning the discussion of the Hellenistic period, we must briefly describe the preceding development in Greece. Our direct sources of information about astronomy and mathematics before Alexander are extremely meager. The dominating influence of Euclid's *Elements* succeeded in destroying almost all references to pre-Euclidean writings, and essentially the same effect was produced by Ptolemy's works. Original documents are, of course, not preserved—one must not forget that even our oldest manuscripts of Greek mathematical and astronomical literature were written many centuries after the originals.[62] It is therefore not surprising that our present-day knowledge of early Greek science is much more incomplete and subject to conjecture than the history of Mesopotamian or even Egyptian achievements where original documents are at our disposal. One point, however, can be established beyond any doubt: early Greek astronomy shows very strong parallelism with the early phases of Egyptian and Babylonian astronomy, with respect to scope as well as primitiveness. The astronomical writings of Autolycus[63] and Euclid[64] struggle in a very crude way with the problem of the rising and setting of stars, making very strong simplifications which were forced upon them by the lack of adequate methods in spherical geometry. The final goal is again to establish relations between the celestial phenomena and the seasons of the years; the problem is thus of essentially calendaric interest. In addition to these simple treatises, however, we do find one work of oustanding character: the planetary theory of

---

* Continued from the February issue (*Pub. A.S.P.*, **58**, 17–43, 1946).

[62] The oldest preserved manuscript of Euclid's *Elements* was written about twelve hundred years after Euclid (cf., e.g., Heath *Euclid,* I, p. 47).

[63] Autolycus, ed. Hultsch (Leipzig, 1885).

[64] *Euclidis opera omnia,* Vol. VIII, ed. Menge (Leipzig, 1916).

104

Eudoxus, Plato's famous contemporary. He made an attempt to explain the peculiarities of a planetary movement known as retrogradation by the assumption of the superposition of the rotation of two concentric spheres around inclined axes and in opposite directions.[64a] In this way he reached a satisfactory explanation of the general type of planetary movement and thereby inaugurated a new period in the history of astronomy which was marked by attempts to explain the movements of the planetary system by mechanical models. It contains the nucleus for all planetary theories of the following two thousand years, namely, the assumption that irregularities in the apparent orbits can be explained as the result of superposed circular movements. It is only since Galileo and Newton that we know that the circular orbits do not play an exceptional role and that the great successes of the Greek theory were merely due to the accidental distribution of masses in our planetary system. It is, nevertheless, of great historical interest to see how a plausible initial hypothesis can for many centuries determine the line of attack on a problem, simultaneously barring all other possibilities. Such possibilities were actually contained in the approach developed by the Babylonian astronomers in the idea of superposing linear or quadratic periodic functions. These arithmetical methods were, however, almost completely abandoned by the Greek astronomers (at least so far as we know) and survived only in the treatment of certain smaller problems.

One of these smaller problems is again related to calendaric questions but also to a basic problem of mathematical geography: the determination of the geographical latitude by means of the ratio of the longest to the shortest day. We have already mentioned the Babylonian methods of describing the change in the length of the days by means of simple sequences. These "linear" methods reappear in Greek literature and can be followed far into the early Middle Ages[65] in spite of the invention of much more accurate methods.[66] The term "linear" does not refer so much to the fact that the sequences in question form

---

[64a] Schiaparelli *SSAA* 2, 3–112.   [65] Neugebauer [13] and [18].

[66] *Almagest* II, 7 and 8. Cf. also *Tetrabiblos* I, 20 (ed. Robbins, p. 94), 21 (ed. Boll-Boer, pp. 46, 47 ff.).

arithmetic progressions of the first order but is intended to emphasize the contrast with the "trigonometric" method applied to the same problem and explained in the first book of the *Almagest*. Here the exact solution of the problem by the use of spherical trigonometry is given. In contrast thereto, the linear methods yield only approximate results, but with an accuracy which was certainly sufficient in practice, especially when one takes into account the inaccuracy of the ancient instruments used in measuring time. Historically, however, the main interest lies much less in the perfection of the results than in the method employed and in its influence on the further development. A close investigation of early Greek astronomy and mathematics[67] reveals an interesting fact. The determination of the time for the rising and setting of given arcs of the ecliptic, which lies at the heart of the question of the changing length of day and night, appears to be the most decisive problem in the development of spherical geometry. It is typical for the whole situation that a Greek "mathematical" work, the *Sphaerics* of Theodosius (*ca.* 200 B.C.), does not contain a single astronomical remark. The structure and contents of the main theorems, however, are determined by the astronomical problem in question; the methods applied constitute a very interesting link between the Babylonian linear methods and the final trigonometrical methods.

Trigonometry undoubtedly has a very long history. We find the basic relations between the chord and diameter of a circle already in use in Old Babylonian texts which employ the so-called "Thales" and "Pythagorean" theorems.[68] In sharp contrast to the Greek models for the movement of the celestial bodies, which operate with circles and therefore necessarily require trigonometrical functions, we find no applications of trigonometry in the cuneiform astronomical texts of the Seleucid period which are exclusively based on arithmetical methods described above.

---

[67] This investigation has been carried out by Olaf Schmidt (doctoral thesis, Brown Univ., 1943 [unpublished]).

[68] Cf. Neugebauer-Struve [1], pp. 90–91; Neugebauer *MKT*, I, p. 180; and Neugebauer-Sachs *MCT*, Problem-Text A.

So far as we know, spherical trigonometry appears for the first time in the *Sphaeric* of Menelaos[69] (*ca.* A.D. 100). The astronomical background of this work is much more outspoken than in Theodosius, but here, too, much is left to the reader, who must be familiar with the methods of ancient astronomy to understand all the astronomical implications. The modern scholar faces an additional difficulty, namely, the modification of the Greek text by the Arabic editors. The Greek original is lost, and what we possess is only the Arabic version made almost a thousand years later. In this interval falls the gradual transformation of Greek trigonometry, operating with chords, to the modern treatment, which uses the sine function. It is well known that this change goes back to Hindu astronomy, where the chords subtended by an angle were replaced by the length of the half-chord of the half-angle,[70] i.e., our "sin α." It is, however, a much more involved question to separate these new methods from those used originally by Menelaos; this question must be answered if we wish to understand the development of ancient spherical astronomy. This, in turn, is necessary in order to appreciate the contributions made by the Hindu-Arabic astronomers which eventually led to the modern form of spherical trigonometry.

13. It is of great interest to see that the very same problem —the determination of rising times—leads to still other methods which are now known partly as "nomography," partly as "descriptive geometry." We have a small treatise, written by Ptolemy, called the *Analemma*.[71] He first introduces in a very systematic way three different sets of spherical co-ordinates, each of which determines the position of a point on the celestial sphere. Then these co-ordinates are projected on different planes, and these planes are turned into the plane of construction, just

[69] Krause *Men.*   [70] Cf., e.g., Braunmühl *GT*, chap. 3.

[71] Ptolemy, *Opera* II, pp. 187–223. No complete translation of this badly preserved text has yet been published, but an excellent commentary has been given by Luckey [1]. These methods, using descriptive geometry, are of an older date, as is evident from the fact that they are already mentioned by Vitruvius (beginning of our era). Cf. Neugebauer [14] and Luckey [2].

as we do today in descriptive geometry. Finally, certain scales are used to find graphically the relations between different coordinates, again following principles which we now use in nomography. The Arabs used and developed these methods in connection with the construction of sundials.[72] Another method of projection, today called "stereographic," is given in Ptolemy's *Planisphaerium*. The theory of perspective drawing in the Renaissance is directly connected with this work.[73]

The practical importance of the determination of the rising times or the length of the days is not restricted to the theory of sundials. The length of the longest day increases with the geographical latitude, thus giving us the means to determine the latitude of a place from the ratio of the shortest to the longest day. The ratio 3 : 2 accepted by the Babylonian astronomers for the ratio of the longest to shortest daylight led the Greek geographers to determine erroneously the latitude of Babylon as 35° (instead of 32½°). This error seriously affected the shape of the eastern part of the ancient map of the world.[74] The precise relationship can only be established by using spherical trigonometry, but here, too, the "linear" methods were applied to various values of the basic ratio in order to give the law for the changing length of the days for the corresponding latitude. It must be remarked, however, that at this stage of affairs the concept "latitude" does not yet actually appear, but the ratio of the longest to the shortest day itself was used to characterize the location of a place. Zones of the same ratio were considered as belonging to the same "clima," a concept which plays a great role in ancient and medieval geography. The difference in character and behavior of nations living in different climates furnished one of the main arguments for the influence of astronomical phenomena on human life.[75]

[72] Cf., e.g., Garbers *ES* and Luckey [2].

[73] Ptolemy, *Opera* II, pp. 225–59, translated in Drecker [1]; cf. also Loria in M. Cantor, *Geschichte der Mathematik*, IV, p. 582.

[74] For the determination of the size of the earth by Eratosthenes (about 250 B.C.), Marinus of Tyre (about A.D. 100), and Ptolemy (about A.D. 150), see Mžik *EGM*, pp. 96 ff., and, in general, Heidel *GM*, chap xi. Cf. also Letronne [1], Honigmann *SK* and Neugebauer [13].

[75] E.g., *Tetrabiblos* II, 2.

The second geographical co-ordinate—the longitude—caused more trouble. The difference in longitude between two places on the earth is essentially equivalent to the difference in local time. But there existed no clocks or signals to compare the local time at far-distant places. Only one phenomenon could be used as a time signal, namely, records of simultaneous observations of a lunar eclipse from two different places. If each observer took note of the local time at which he observed the beginning and end of a lunar eclipse, a comparison of these records would then furnish the needed information. Hipparchus proposed the use of this method for an exact construction of the map of the world, but his program was never carried out. Only one pair of simultaneous observations seems to have been made, the eclipse of 331 B.C., September 20, recorded three hours earlier in Carthage than at Arbela.[76] Actually the difference in local time between these two localities is much smaller, and consequently the ancient map of the world suffers from a serious distortion in the direction from east to west. Here we see one of the most essential differences between ancient and modern science at work. Ancient science suffered most severely from the lack of scientific organization which is so familiar in our own times. In antiquity, generations passed before a new scientific idea found a follower able to use and develop methods handed down from a predecessor. The splendid isolation of the great scholars of antiquity can only be paralleled with the first beginnings of the new development in the European Renaissance. It seems to me beyond any doubt that even centers like Alexandria or Pergamon during their height would appear very poorly equipped if compared with a modern university of moderate size. And these centers themselves were few and practically isolated at any particular time; and at all times they were dependent upon the mood of some autocratic ruler. No wonder that the great achievements of antiquity are either the result of priestly castes of sufficiently stable tradition or of a few in-

---

[76] Ptolemy *Geographia* i. 4. 2 (ed. Nobbe, p. 11). Cf. also Mžik-Hopfner *PDE*, p. 21, n. 3. For Hipparchus' program see Strabo *Geography* i. C. 7; also Berger *GFH*, pp. 12 ff.

genious men who expended tremendous energy in restoring
and enlarging the structure of a science known to them from the
written legacy of their predecessors. One must not think that
mathematics and astronomy, like the popular philosophical systems or the art of rhetoric, were taught in the same manner from
generation to generation. Three centuries separate Hipparchus
from Ptolemy, one Eudoxus from Euclid, Euclid from Archimedes and Apollonius. To be sure, the literary tradition was
never interrupted between these outstanding men, but most of
the intermediate literature at best merely preserved and commented. This explains not only why ingenious ideas were frequently lost (e.g., Archimedes' methods of integration) but also
why it was so easy to destroy ancient science almost completely
in a very short time. Astronomy alone had a slight advantage
because of its practical usefulness in navigation, geography, and
time-reckoning, supplemented by the fortunate accident that the
Easter festival followed the lunar calendar of the Near East,
thus sanctioning lunar theory when other secular sciences fell
into total desuetude.

The extreme paucity of scientists at almost any given time
in antiquity gave rise to another phenomenon in Greek literature:
the publication of commentaries and popularizing works. A
work like the *Almagest,* written in purely scientific style, was
certainly unintelligible to the majority of people who needed or
wanted to know a modest amount of astronomy. Hence books
were written which attempted to explain Ptolemy's text sentence
by sentence,[77] or which gave abstracts accompanied by explanations of the main principles as far as this could be done without
mathematics.[78] We can observe the same phenomenon in geography. The first chapter of Ptolemy's *Geography*[79] contains a
very interesting theory of map projection, whereas the remaining twelve chapters constitute an enormous catalogue of locali-

---

[77] The commentaries of Pappus and Theon of Alexandria (and presumably of Hypathia) are of this type. For these texts cf. Rome *CPT.*

[78] Represented, e.g., by Theon of Smyrna (second cent. A.D.) or
Proclus (fifth cent. A.D.).

[79] Edited by Nobbe (1843). The first chapter is excellently discussed
by Mžik and Hopfner *PDE.*

ties from all over the then known world and the corresponding values of longitude and latitude to be plotted into the network which was to be constructed according to the method explained in the first chapter. This, again, was not geography for the entertainment of the general reader. To satisfy popular tastes, there was another literature, represented by works like Strabo's *Geography*.[80] These more pleasant writings furnished serious competition to the strictly scientific literature and determined to a large extent the character of the field in late antiquity and the Middle Ages.

14. For the modern historian of ancient astronomy it is therefore of the greatest value to have an additional source of astronomical literature in which the earlier tradition was kept alive without interruption for a much longer period: the astrological texts. We have already mentioned that astrology in the modern use of the word appeared very late in antiquity. The art of casting horoscopes can be said to be a typical Hellenistic product, the result of the close contact between Greek and oriental cultures.[81] We possess Greek papyri from Egypt from the beginning of our era to the Arabian conquest showing us the application of astronomical methods in a great number of specific horoscopes and in minor astronomical treatises.[82] In addition, an enormous astrological literature is preserved, catalogued during the last fifty years in the twelve volumes of the *Catalogus* by Cumont and his collaborators.[83] Finally, Vettius Valens, who wrote shortly before Ptolemy,[84] and Ptolemy himself as the author of the famous *Tetrabiblos,* must be mentioned.[85]

Modern scholars have not yet made full use of this vast mate-

---

[80] Edited and translated in the "Loeb Classical Library" by H. L. Jones (8 vols.; 1917–32).

[81] Cf., e.g., Capelle [1], who shows that only weak traces of astrological ideas in Greek literature can be followed as far back as 400 B.C.

[82] Concerning horoscopes, see above, n. 17. Examples of astronomical treatises are Pap. Ryl. 27, 464, 522/24, 527/28, or Curtis-Robbins [1].

[83] *CCAG.* Cf. also Boll [2].   [84] Kroll *VV.*

[85] Ptolemy, *Opera* III, 1, and "Loeb Classical Library" (ed. F. E. Robbins).

rial. The reason is only too clear: the amount of work to be done surpasses by far the power of a single individual, and the work itself is certainly not very pleasant. The astronomical part must be extracted from occasional remarks, short computations, and similar instances submerged beneath purely astrological matter of a very unappealing character. But this work must eventually be done and will give valuable results. As an example might be mentioned the question of discovering the principle according to which the equinox was placed in the zodiac. This question must be answered, for on it depend our calculations in the determination of constellations, chronology, etc. Moreover, systematic checking of astrological computations will frequently yield information about the character of the astronomical tables used at the time.

We touch here upon a point of great importance for the modern attitude toward ancient astronomy. The usual treatment of ancient sciences as a homogeneous type of literature is very misleading. It is necessary to realize that very different levels of astronomy or mathematics were coexistent, almost without mutual contact or interference. One misses the essential points in the understanding of ancient astronomy if one naïvely considers various documents in their chronological order. Even works by the same person must sometimes be separated from one another. Ptolemy's *Almagest* is purely mathematical, the *Tetrabiblos* (written after the *Almagest*)[86] is purely astrological, and his *Harmonics*[87] contains a chapter on the harmony of spheres employing concepts of the planetary movements which contains such strong simplification of the actual facts that one would try in vain to find similar assumptions in any of the other works of Ptolemy. In other words, it is necessary to evaluate each text in its proper surrounding and according to its traditional style. One cannot, for example, speak without qualification of the contact between Babylonian and Greek astronomy. Such a contact might even have worked in opposite directions in different fields. For instance, we have

---

[86] This follows from the introduction to the *Tetrabiblos*.

[87] Düring *HP* and *PPM*.

already referred to the possibility that Hellenistic astrology returned to Babylonia in the form acquired in Egypt or Syria, whereas observational material from Mesopotamia undoubtedly influenced Greek mathematical astronomy deeply. In general, it can be said that the growth of ancient sciences shows much more irregularity and stratification than modern scientists, accustomed to the fact of the uniform spread of modern ideas and methods, are prone to assume.

The lack of uniformity in the whole field of ancient astronomy in general necessarily interferes also with the investigation of any special problem. We have already mentioned the fact that astrology in the Assyrian age differed considerably from the horoscopic type which prevailed in late antiquity and the Middle Ages. But there exists a third type, standing between the omina type ("when this and this happens in the skies, then such and such a major event will be the consequence") and the individual birth horoscope, namely, the "general prognostication," explained in full detail in the first two books of the *Tetrabiblos*. This type of astrology is actually primitive cosmic physics built on a vast generalization of the evident influence of the position of the sun in the zodiac on the weather on earth. The influence of the moon is considered as of almost equal importance, and from this point of departure an intricate system of characterization of the parts of the zodiac, the nature of the planets, and their mutual relations is developed.[88] This whole astronomical meteorology is, to be sure, based on utterly naïve analogies and generalizations, but it is certainly no more naïve and plays no more with words than the most admired philosophical systems of antiquity. It would be of great interest for the understanding of ancient physics and science in general to know where and when this system was developed. The question arises whether this is a Greek invention, replacing the Babylonian omen literature, which must at any rate have lost most of its interest with the end of independent Mesopotamian rule, whether it precedes the invention of the horoscopic art for

---

[88] For the whole complex of the ancient justifications of astrology, see Duhem, *SM*, II, 274 ff.

individuals or merely represents an attempt to rationalize the latter on more general principles.[89] Thus we see that even in a single field of ancient astronomical thought the most heterogeneous influences are at work; the analysis of these influences has repercussions on almost every aspect of the study of ancient civilizations.[90]

15. The same branching-off into very different lines of thought must also be recognized in the development of Greek mathematics. The line of development characterized by the names of Eudoxus, Euclid, Archimedes, and Apollonius is to be separated sharply from writings like Heron[91] and Diophantus[92] or the *Arithmetic* of Nicomachus of Gerasa.[93] Here, again, the question of oriental influence cannot be discussed as one common phenomenon. Egyptian calculation technique and mensuration were certainly continued in similar works in Hellenistic Egypt and found their way into Roman and medieval practices. At the same time, Babylonian numerical methods influenced Alexandrian astronomy. How Babylonian algebraic concepts eventually reached Greek writers like Diophantus is still completely unknown, but that it did is supported by the strong parallelism in methods and problems.[94] Equally lacking is detailed information as to the revival of these methods in Moslem literature.[95] On the other hand, the problems which emerged from the discovery of the irrational numbers are undoubtedly of Greek origin. It is, however, not correct to consider writings of the same person as equally representative of "Greek" mathematics. Those parts of Euclid's *Elements* (the majority of the work) which deal more or less directly with the problem of irrational numbers are, as we said before, Greek. Most likely of equally Greek origin is Euclid's astronomical

---

[89] This is the assumption of Kroll [1], p. 216, for the tendency exhibited in Ptolemy's *Tetrabiblos*.

[90] Cf. the excellent survey of this situation in Boll [2].

[91] First century A.D.; cf. for this date Neugebauer [14], pp. 21 ff.

[92] Usually dated about A.D. 300; cf., however, Klein [1], p. 133, n. 23.

[93] Greek text ed. Hoche (Leipzig, 1866); English translation: D'Ooge-Robbins-Karpinski *Nic*.

[94] Vogel [2]; Gandz [3].   [95] Gandz [1], [2], [3].

treatise called *Phenomena*,[96] which is written on so elementary a level that nobody would attribute it to the author of the *Elements* if the authorship were not so firmly established. And, finally, Euclid's *Data*[97] contains the treatment of purely algebraical problems by geometrical means—which can be interpreted as the direct geometrical translation of methods well known to Babylonian mathematics.[98] These methods of "geometrical algebra" in turn determine the whole structure of Apollonius' theory of conic sections.[99]

Greek mathematics is by far the best-investigated field of ancient science (and of the history of science in general);[100] the situation with respect to the source material is very good[101] —except where only Arabic manuscripts are preserved.[102] But one must not forget that also this tradition suffers from severe gaps. This is due not only to the destruction of manuscripts over a period of two thousand years but also to the effect of literary influence. I refer not only to the above-mentioned elimination of older treatises by the overshadowing of the great works of the Hellenistic period. The Greeks themselves contributed to the distortion of the picture of the actual development by inventing seemingly plausible stories where the real records were already lost. The oft-repeated stories about Thales, Pythagoras, and other heroes are the result.[103] We should now realize that we know next to nothing about earlier Greek mathematics and astronomy in general and about the contact with the Near East and its influence in particular. The method

---

[96] *Opera* VIII; cf. above, p. 16.   [97] *Opera* VI.

[98] Neugebauer [15].   [99] Zeuthen *KA* and Neugebauer [16].

[100] Best exposition: Heath *GM* and *MGM* and Euclid. A selection of texts is given in Thomas *GMW*.

[101] Most of the texts are edited in the Teubneriana collection.

[102] Menelaos alone is now edited (Krause *Men.*), but Books v, vi, and vii of Apollonius' *Conic Sections* are still unavailable in a modern edition. Archimedes' construction of the heptagon is published in a free translation of the Arabic version in Schoy *TLAB*, pp. 74–91; cf. also Tropfke [1].

[103] As an example might be mentioned the criticism of the story of the Thales eclipse by Pannekoek [3], p. 955; Dreyer *HPS*, p. 12, n. 2; Neugebauer [9], pp. 295–96. Cf. also Frank, Plato, or Heidel [1].

which involves the use of a few obscure citations[104] from late authors for the restoration of the history of science during the course of centuries seems to me doomed to failure. This amounts to little more than an attempt to understand the history of modern science from a few corrupt quotations from Kant, Goethe, Shakespeare, and Dante.

16. Undoubtedly the most spectacular advances in the history of astronomy until very recent times were scored in the theory of the planets. The catch-words "Ptolemaic" and "Copernican" refer to different assumptions as to the mechanism of the planetary movement. This is not the place to underline the fact that the Copernican theory is by no means so different from or so superior to the Ptolemaic theory as is customarily asserted in anniversary celebrations,[105] but we must briefly analyze Ptolemy's own claims to having been the first one who was able to give a consistent planetary theory.[106] This claim seems to contradict not only the existence of pre-Ptolemaic planetary tables in Roman Egypt as well as in Mesopotamia but also Ptolemy's own reference to such texts. What Ptolemy means, however, becomes clear if one reads the details of the introduction to his own theory. He requires an explanation of the planetary movement by means of a combination of uniform circular movements which refrains from simplifications like the assumption of an invariable amount for the retrograde arc and similar deviations from the actual observations. Indeed, in order to remain in close agreement with the observations, Ptolemy had to overcome difficulties which Hipparchus was not able to master and which led Ptolemy to a model which is very close to Kepler's final solution of the problem, by assuming not only an eccentric position of the earth but also an eccentric point around which the movement of the planetary eccenter appears

---

[104] The fragments collected by Diels *VS* not only give an extremely incomplete picture of the lost writings but were certainly very much distorted by the authors from whose works they are taken. One needs only to look at the picture of oriental writings obtained from Greek tradition as compared with the originals.

[105] The correct estimate can be found in Thorndike *HM*, Vol. V, chap. xviii.   [106] *Almagest* IX, 2.

to be uniform. The resulting orbit is of almost elliptical shape with these two points as foci.[107] This whole theory is closely related in method to the explanation of the "evection" of the moon (a periodic perturbation of the moon's orbit discovered by Ptolemy) by a combination of eccentric and epicyclic movements.[107a] Both theories are real masterpieces of ancient mathematical astronomy which far surpassed all previous results.

It is not surprising that Ptolemy's results overshadowed all previous works. All that we know about his forerunners comes mainly from the *Almagest* itself. We hear that Hipparchus used eccenters and epicycles for the explanation of the anomalies in the movement of the sun and the moon,[108] and we learn about theorems for such movements proved by Apollonius.[109] This brings us to the very period (about 200 B.C.) from which the oldest cuneiform planetary texts are preserved—computed, however, on entirely different principles. These cuneiform texts cover the two centuries down to the time of Caesar. A direct continuation, chronologically speaking, but of still another type, are planetary tables from Egypt, written in Demotic or Greek.[110] These tables give the dates at which the planets enter or leave the signs of the zodiac. Such tables were known to Cicero[111] and are most likely the "eternal tables" quoted with contempt by Ptolemy.[112] We do not know how these tables were computed, and their occurrence in Greek as well as in Demotic leaves us in doubt as to their origin—showing us only the degree of interrelation we can expect in Hellenistic times.

The most interesting question would, of course, be to learn more about Hipparchus' astronomy. He is most famous as the discoverer of the precession of the equinoxes. Though this fact

---

[107] Cf. Schumacher [1] for the Ptolemaic theory of Venus and Mercury. For the Greek planetary theory in general, see Herz *GB* I.

[107a] *Almagest* V, 2.

[108] *Almagest* III, 4.

[109] *Almagest* XII, 1 (=Apollonius, ed. Heiberg, II, 137).

[110] Neugebauer [3]. Cf. above, p. 5.

[111] Cicero *De divinatione* ii, 6, 17; cf. also ii. 71. 146.

[112] *Almagest* IX, 2.

cannot be doubted,[113] underlining its importance lays the wrong emphasis on a phenomenon which gained its importance only from Newton's theory, which showed that precession depends on the shape of the earth and thus opened the way to test the theory of general gravitation by direct measurements on the earth. For ancient astronomy, however, precession played a very small role, requiring nothing more than sufficiently remote and sufficiently reliable records of observations of positions of fixed stars. The change in positions must then eventually become evident; and little difficulty was encountered in incorporating this slow movement into the adopted model of celestial mechanics. What we actually need to appreciate in Hipparchus' contribution must be derived from a careful study of all relevant sections of the *Almagest,* not by the schematic method of obtaining "fragments" from direct quotations but by a comparison of Ptolemy's methods and the older procedures which he frequently mentions. That such an approach can lead to well-defined results has recently been shown in the theory of eclipses.[114]

17. One of the most important problems in connection with Hipparchus is, of course, the problem of the dependence of Hipparchus (and Greek astronomy in general) on Babylonian results and methods. Whatever the conclusions derived from a deeper knowledge of Hipparchus' astronomy may turn out to be, one thing is clear: the century between Alexander's conquest of the Near East and Hipparchus' time is the critical period for the origin of Babylonian mathematical astronomy as well as for its contact with Greek astronomy. Since Kugler's discoveries, which showed the exact coincidence between numerical relations in cuneiform tablets and in Hipparchus' theory,[115] no one has doubted Babylonian priority. It is an undeniable fact that the Babylonian theory is based on mathematical methods known already in Old Babylonian times and does not show any trace

---

[113] Schnabel's attempts (Schnabel [1]) to prove that precession was taken into consideration in the cuneiform texts are, to say the least, inconclusive and in part based on mere scribal errors.

[114] Schmidt [1].   [115] Kugler *BMR*, p. 40.

of methods considered to be characteristically Greek. The problem remains, however, to answer the question: What caused the sudden outburst of scientific astronomy in Mesopotamia after many centuries of a tradition of another sort? On what background can we understand, for example, the report[116] that the "Chaldaean" Seleucus from Seleucia on the Tigris[117] completed the heliocentric theory, previously proposed as a hypothesis by Aristarchus? Greek influence on late Babylonian astronomy must not be denied or asserted on aprioristic grounds, if we really want to understand a phenomenon of great historical significance.

These remarks are not intended to make Greek influence alone responsible for the new developments in Mesopotamia. As a matter of fact, this answer would only raise the equally unsolved question why Greek astronomy suddenly emerged from many centuries of primitiveness to a scientific system. The alternative, Greek or Babylonian, might even exclude the right answer from the very beginning. It also seems possible that the rise of mathematical astronomy in Hellenistic times resulted from the suddenly intensified contact between several types of civilization, in some respects to be paralleled with the origin of modern science in the Renaissance. In other words, neither the Greeks nor the Orientals might have been alone responsible for the new development but rather the enormous widening of the horizon of all members of the culture of the Hellenistic age. One result of this process was probably the new attitude toward the relationship between the individual and the cosmos, expressed in the new form of horoscopic astrology. In this case it is quite evident that Egypt and Greece—and perhaps Syria as well—contributed about equally much to the refinement and spread of this new creed. It is equally possible that the contact between Greek scholars, trained to think in geometrical terms which Greek mathematics had developed in the fifth century, and Baby-

[116] Plutarch *Plat. quaest.* vii. 1. 1006 C (ed. Bernardakis, *Moralia,* VI, 138). Cf. also Heath, *AS,* pp. 305 ff.; Schiaparelli *SSAA* I, 361–458, II, 113–77; and Duhem *SM,* I, 423 ff.
[117] Strabo xvi, 739. Seleucus may have lived about 150 B.C.

Ionian astronomers, equipped with superior numerical methods and observational records, brought into simultaneous existence two closely related types of mathematical astronomy : the treatment by arithmetical means in Babylonia and the model based on circular movements in the Greek centers of learning in the eastern Mediterranean. It may well be that competition, not borrowing, was the chief contributor to the initial impetus.[118] At any rate, it is clear that each detail in the development of Hellenistic astronomy which we will be able to understand better will reveal a new aspect in the fascinating process of the creation of the new world which was destined to become the foundation of the Roman and medieval civilizations.

The unique role of the Hellenistic period in the field of sciences, as in other fields, can be described as the destruction of a cultural tradition which dominated the Near East and the Mediterranean countries for many centuries, but also the founding of a new tradition which held following generations in its spell. The history of astronomy in the Hellenistic age is especially well suited to demonstrate that the great energies liberated by the disintegration of an old cultural tradition are very soon transformed into stabilizing forces of a new tradition, which includes about as many elements of development as of stagnation.

V. SPECIAL PROBLEMS

18. Every research program in a complex field will face the need of constant modification and adjustment to unforeseen complications and new ramifications. Problems can arise and results be obtained without having been anticipated in the original question. The context of a mathematical text, for example, can determine with absolute certainty the meaning of a word otherwise only vaguely defined ; sign-forms in a papyrus which is exactly dated by astronomical means may furnish valuable information for purely paleographical problems. From dates and positions given in Demotic astronomical texts, it follows that the Alexandrian calendar introduced by Augustus was used by

---

[118] Neugebauer [17], pp. 30–31.

Egyptian scribes only a few years after the reform,[119] very much in contrast to the common opinion that the Egyptians were especially conservative in general and in calendaric matters in particular. In short, from few, but solidly established, facts we can learn more than from all general speculations.

One of the problems which at first sight lies very much outside the history of ancient astronomy is the study of social and economic conditions of the ancient civilizations. There are, however, several points of contact between these studies and astronomy. We are indebted to Cumont for a masterly investigation of the information contained in the astrological literature from Hellenistic Egypt.[120] His results are not only of interest for the history of ancient civilization but also illustrate very well the background of the men who used and transmitted the astronomical material known to us from the planetary tables or from Vettius Valens. It turns out that the soil in which these practices were rooted was essentially Egyptian, in spite of the use of the Greek language in the documents. This is in perfect harmony with the close parallelism between Greek and Demotic planetary texts mentioned above and shows the constant interaction of Greek and native influences in Hellenistic Egypt. It also shows how dangerous it is to decide the authorship of Hellenistic doctrines or methods simply on the basis of such superficial grounds as the language used.

The analogous question for Babylonia seems to be easier to answer. The Mesopotamian origin of the astrological omina cannot be doubted. We would, however, like to know more about the background of the astronomers of the latest period. It is well known that the names of three Babylonian astronomers appear in Greek literature[121] and that two of them actually were found on astronomical tablets, though in an unclear context. For one particular place, the famous city of Uruk in South Babylonia, we can go much further. It can be shown that the scribes and owners of our texts belong to one of two "families," or perhaps, "guilds," of scribes who frequently call themselves

---

[119] Neugebauer [6], p. 119.

[120] Cumont *EA*. See also Kroll [1].   [121] Cumont [1].

scribes of the omen-series "Enuma Anu Enlil."[122] We can follow
the work of these scribes very closely for almost a hundred years
until the school of Uruk ceased to exist, probably because of the
Parthian invasion of Babylonia in 141 B.C. In contrast thereto,
the school of Babylon survived the collapse of the Greek regime,
as is proved by a continuous series of astronomical texts down
to 30 B.C. This is an interesting result in comparison with the
assumption that Babylon practically ceased to exist after the
Parthian occupation. The grouping of our texts according to
well-defined schools is also of interest from another point of
view. It can be shown that two different systems of computation existed side by side for a long time. Competing schools of
this sort constitute a phenomenon which is usually considered
characteristic for Greek culture.

19. Countless thousands of business documents are preserved
from all periods of Mesopotamian history. For the urgently
needed investigation of ancient economics, a precise knowledge
of the metrological systems is of the greatest importance. Unfortunately, the scientific study of Babylonian measures has
been sadly neglected. Fantastic ideas about the level and importance of astronomy in the earliest periods of Babylonian history led to theories which brought measures of time and space
in close relationship with alleged astronomical discoveries. We
know today that all these assumptions of the early days of
Assyriology must be abandoned and that Babylonian metrology
must be studied from economic and related texts clearly separated according to period and region. For the determination of
Old Babylonian relations between various measures, the mathematical texts are of great value because they contain numerous
examples which give detailed solutions of problems in which
metrological relations play a major role. The consequences of
such relations, established with absolute certainty, are manifold.
For example, we now know from Old Babylonian mathematical
texts the measurements of several types of bricks[123] as well as the

---

[122] For this series cf. Boll-Bezold-Gundel *SS*, pp. 2 ff., and Weidner
[2].

[123] Neugebauer-Sachs *MCT*, Problem-Text O and Sachs [1].

peculiar notation used in counting bricks. It is evident that such information is of importance for the understanding of contemporary economic texts dealing with the delivery of bricks for buildings, thus leading to purely archeological questions. Metrological relations are also needed if we wish to gain an insight into wages and prices.[124] Returning to our subject, it must be said that metrology is of great importance not only for the history of the economics of Mesopotamia but also for purely astronomical problems. Distances on the celestial sphere are measured in astronomical texts by units borrowed from terrestrial metrology. The comparison between ancient observations and modern computations thus requires a knowledge of the ancient relations between the various units. This problem is by no means simple because our astronomical material belongs to relatively late periods, Assyrian and Neo-Babylonian, and the metrological system of these times is much more involved than the Old Babylonian. Mathematical texts would certainly be of great help here, too, but the few tablets from this period are so badly preserved that they present us with at least as many new questions as they answer. Neo-Babylonian economic texts will therefore furnish the main point of departure for the study of the latest phase of Mesopotamian metrology and its astronomical applications.

It might be mentioned, in this connection, that theories about direct relationship between early Mesopotamian metrology and astronomy also gave rise to the rather unfortunate concept of high accuracy in the determination of weights, measures of length, etc. It is of great importance to realize that the absolute values of all metrological units are subject to great margins of inaccuracy and local and temporal variations. The first step in a historical investigation of Mesopotamian metrology must therefore be to establish from economic and mathematical texts the *ratios* between the units; these ratios have an incomparably better chance of showing uniformity than the absolute values deduced from accidental archeological finds.

---

[124] Waschow [1], p. 277, found, in discussing mathematical texts, that the value of the area-measure "še" must be changed by a factor 60 against older assumptions. It is obvious how such facts influence the interpretation of economic texts.

20. Closely related to metrological problems is the question of the accurate identification of ancient star configurations. Much work remains to be done before it will be possible to give a reliable history of the topography of the celestial sphere in general, or even of the zodiacal constellations.[125] In spite of attempts to make Egypt responsible for many forms,[126] the predominant influence of Babylonian concepts on the grouping of stars into pictures must be maintained. But neither Babylonian nor Egyptian developments are known in detail. The identification of Egyptian constellations is especially difficult, mainly because it must be based on relations between the times of rising and setting and therefore depends on elements which are grossly schematized in the texts at our disposal. The situation in Mesopotamia is slightly better because we have actual observations in addition to the schematic lists, at least for the later periods which are of special importance for the Hellenistic forms of the constellations.

For the period following the publication of the *Almagest,* we must take into account the possibility of still other complications. We know from explicit remarks in the *Almagest* that Ptolemy's star catalogue introduced deviations from older catalogues.[127] Astrological works, however, may very well have maintained pre-Ptolemy standards both with respect to the boundaries of constellation and the counting of angles in the zodiac. We have already mentioned the stubborn adherence of astrological writers to methods of computation which were made obsolete by the development of spherical trigonometry.[128] For the modern historian it is therefore of importance to establish the specific standard according to which a given document was written, especially when chronological problems are involved.

21. While metrology is a much-needed implement for economic history and the understanding of ancient astronomy,

---

[125] The best summary is given by the Boll-Gundel article, "Sternbilder," in Roscher *GRM,* Vol. VI (1937), cols. 867–1072.

[126] Cf. esp. Gundel *DD* and *HT* and the criticism of Schott [2].

[127] *Almagest* VII, 4 (ed. Heiberg, p. 37). Cf. also Peters-Knobel *PCS* and Tallgren [1].

[128] Cf., e.g., *Tetrabiblos* I, 20 (ed. Robbins, pp. 94–95).

astronomy itself serves general history in chronological problems. Chronology is the necessary skeleton of history and owes its most important fixed points to astronomical facts. We need not emphasize the use of reports of eclipses, especially solar eclipses, for the determination of accurate dates to form the framework into which the results of relative chronology must be fitted. It must be underlined, however, that the available material is by no means exhausted. A better understanding and reinvestigation of the reports of the Assyrian astronomers will certainly furnish new information of chronological value. It must be stated, on the other hand, that not too much is to be expected from older material. In order to make ancient observations accessible to modern computation, a certain degree of accuracy must be granted; this accuracy seems to be missing in the earlier phases of the development of astronomy. This, for instance, makes the older Egyptian material so ill suited for chronological purposes. For later periods, however, Egypt has furnished and will furnish much information from astrological documents. It is particularly calendaric questions, such as the use of eras and similar problems, which have been illuminated by the dating of horoscopes.

The great variety of calendaric systems, local eras, and older methods of dating raises many difficulties in ancient chronology. This difficulty was clearly felt also by ancient astronomers and was the cause of the early use of consistent eras in Babylonian and Greek astronomy. The Babylonian texts always use the Seleucid Era, whereas Ptolemy reduces all dates to the Nabonassar Era but uses the Old Egyptian years of constant length. This crossing of Egyptian and Babylonian influences is paralleled by the subdivision of the day into hours. The Egyptians divided the day into twelve parts from sunrise to sunset, thus obtaining hours whose length depended on the season. The Babylonian astronomers used six subdivisions of day and night, but these units were of constant length. Combining the Egyptian division into 24 hours with the Babylonian constancy of length, the Hellenistic astronomers used "equinoctial" hours for their computations and solved the problem of finding the relationship between seasonal and equinoctial hours by spherical trigonometry.[129] One sees here again what a multitude of relations, problems, and methods contributed to shape concepts such as a continuous era or the 24-hour day which are so familiar to us today.

Ancient chronology and the accurate analysis of ancient reports have turned out to be of interest even to a modern astronomical problem. In 1693 Halley discovered the fact[130] that the moon's position appeared to be advanced compared with the expected position as computed from positions recorded by Ptolemy. This "acceleration" can be explained by a slow increase in the length of the solar day or by a decrease in the rotational velocity of the earth. Such a decrease is caused by tidal forces,[131] and it is of great interest to determine the amount as accurately as possible. For this purpose, accurate positions of the moon in remote times are of great value, and such positions can, indeed, be derived from records in cuneiform texts.[132] Modern measurements of high precision can thus be supplemented by observations in antiquity.

22. Not only are Hellenistic astronomy and Hellenistic astrology the determining factors for the astronomy and astrology of the Middle Ages in Europe, but their influence is equally important for the development of astronomical methods and concepts in the Middle and Far East. We must therefore at least mention an enormous field which still awaits systematic research: Hindu science. This does not mean that there is not an extensive literature on this subject; indeed, even a small number of original texts are published.[133] The main trouble lies, however, in the tendency of the majority of publications by Hindu authors to claim priority for Hindu discoveries and to deny foreign influence, as well as in the opposite tendency of

[129] *Almagest* II, 9.

[130] Edm. Halley, "Emendationes ac notae Abatênii observationes astronomicas, cum restitutione tabularum lunisolarum ejusdem authoris," *Philosophical Transactions,* 17 (1693), No. 204, pp. 913–21.

[131] Cf., e.g., Jeffreys [1].   [132] P. V. Neugebauer [1].

[133] For the literature until 1899, see Thibaut *AAM*. The best discussion of Hindu astronomy is still Burgess *SS* (1860).

some European scholars. This tendency has been especially strong so far as Hindu mathematics is concerned,[134] and it is aggravated by the inadequate publication of the original documents, from which usually only scattered fragments are cited in order to prove some specific statement. As a result, there is no means today to obtain an independent judgment from the study of the original texts which are preserved in enormous number, though of relatively late date for the most part.

The situation with respect to Hindu astronomy is not much better. There can be little doubt that the original impetus came from Hellenistic astronomy; the use of the eccentric-epicyclic model alone would be sufficient proof even if we did not also find direct witness in the use of Greek terminology.[135] This fact is interesting in itself, but it may very well be that the period of reception lies between Hipparchus and Ptolemy; systematic study might therefore reveal information about pre-Ptolemaic Greek astronomy no longer preserved in available Greek sources. Hindu astronomy would in this case constitute one of the most important missing links between late Babylonian astronomy and the fully developed stage of Greek astronomy represented by the *Almagest*.

The fundamental difficulty in the study of Hindu astronomy lies in the character of the preserved textual material. The published and commented texts consist exclusively of cryptically formulated verses giving the rules for computing certain phenomena, making it extremely difficult to understand the actual process to be followed. It is evident, on the other hand, that no astronomy of an advanced level can exist without actually computed ephemerides. It must therefore be the first task of the historian of Hindu astronomy to look for texts which contain actual computations. Such texts are, indeed, preserved in great

---

[134] Cf., e.g., Datta-Singh *HHM* (reviewed in Neugebauer [12]).

[135] Thibaut *AAM*, pp. 43 ff. The Babylonian ratio 3:2 for the ratio between the longest and shortest days of the year also occurs in India (Thibaut *AAM*, pp. 26–27; Kugler *BMR*, pp. 82 and 195), though it would be suitable only for the latitude of the northern corner of India. For the planetary theory, see Kugler *BB*, p. 120; Schnabel [2], p. 112; Schnabel [1], p. 60.

number, though actually written in very late periods. Poleman's catalogue[136] of Sanskrit manuscripts in American collections lists about a hundred such manuscripts in the D. E. Smith collection in Columbia University in New York. In their general arrangement, these texts are reminiscent of the cuneiform ephemerides from Seleucid times and must reveal many details of the Hindu theory of the planetary movement if attacked by the same methods which have proved so successful in the case of the Babylonian material. The complete publication of this material is an urgent desideratum in the exploration of oriental astronomy.

As mentioned above, the texts in the D. E. Smith collection are of very recent origin, only a few centuries old. This does not mean that the methods used are not of very much earlier date. This is shown by the investigation of one of these texts,[137] which deals with the problem of the varying length of the days during the year. Though written about 1500, the computations are based on methods going back to a much older period. Analogous results can be expected in the remaining material, and there is no reason to assume that the D. E. Smith collection exhausts all the preserved material.

23. In the preceding sections we have frequently touched on methodological questions. In closing, I wish to underline a few principles in a more general way. As is only natural, the study of the development of ancient science began under the influence of the ancient tradition. Herodotus, Diodorus, the commentators of Plato, etc., were the sources which determined the picture of the early stages of Greek and oriental mathematics and astronomy. But while students of political history, art, economics, and law learned in the early days of systematic archeological research to consider this literary tradition about the ancient Orient as nothing more than a supplementary source to be checked by the original documents, the majority of historians of the exact sciences have remained in a stage of naïve innocence, repeating without criticism the nursery stories of ancient popular writers. This is all the more surprising because

---

[136] Poleman *CIM*, pp. 231 ff. See also Emeneau *PIT*, pp. 318 ff.

[137] Schmidt [2].

many of these stories should have revealed their purely fictitious character from the very beginning. Every invention considered of basic importance is attributed to a definite person or nation : Thales "discovered" that a diameter divides the area of a circle into two equal parts, Anaximander and several others are credited with the discovery of the obliquity of the ecliptic, the Egyptians discovered geometry, the Phoenicians arithmetic— and so on, according to an obvious pattern of naïve restoration of facts the origins of which had been totally forgotten. Modern authors then add stories of their own, such as the idea that the construction of the pyramids required mathematics, the assumption of supposedly marvelous skies of Mesopotamia,[138] and the notion of Egyptian Stone Age astronomers industriously determining the heliacal rising of Sirius or carrying out a geodetic survey of the Nile Valley.

It is clear that the replacement of the traditional stories by statements based exclusively on results obtainable from the original sources will not be very appealing. This is the inevitable result in the development of every science ; for increased knowledge means giving up simple pictures. In the history of science, an additional element must be added to the steady increase of complexity resulting from a better understanding of our sources. Not only do we learn to interpret our material more accurately but we also learn to see everywhere the immense gaps in our preserved sources. We will more and more be forced to admit that many, and essential, steps in the development of science are hopelessly destroyed ; that we, at best, are able to sketch mere outlines of the history of science during certain sharply limited periods ; and that many of the driving forces might actually have been quite different from those which we customarily restore on the analogy of later periods.

One consequence of this situation seems to me to be evident : unless the history of science now enters the stage of specialization, it will lose all value in the framework of historical research. It must be clearly understood that the history of

[138] For the poor conditions of actual observation cf. Koldewey *WB*, p. 192 ; Vogt [1], pp. 38–39 ; cf. also Boll [1], pp. 48 and 157.

science must work with methods and must consider its problems from viewpoints which correspond to the methods and standards of other branches of historical research. The idea must definitely be abandoned that the history of science must adapt its level to the alleged requirements of the teaching of the modern fields of science. The intrinsic value of this research must be seen in its contribution to our understanding of the historical processes which shaped human civilization, and it must be made clear that such an understanding cannot be reached without the closest contact with the other historical fields. The call for specialization is not very popular. I am convinced, however, that a well-founded insight into the details of a single essential step in the development is at present of higher value and more fascinating than any attempt at general synthesis. It is ridiculous to believe that we are anywhere able to reach "final" results in the study of the development of human civilization. But the overwhelming richness of all phases of human history can be appreciated only if we occupy ourselves with the real facts as accurately as possible and do not attempt to hide their manifold aspects under the veil of hazy generalizations or let our judgment be guided by the naïve idea of human "progress." Every synthesis written fifty years ago is now completely antiquated and at best enjoyable for its literary style; the careful study of the original works of the ancients, however, will reveal to everyone and at any time the development of their achievements.[139]

The call for specialization must not be misunderstood as a plea for the disregard of the general outlines of the historical conditions. On the contrary, specialized work can be accomplished successfully only if the points of attack are selected under constant consideration of possible interference from other problems and other fields. It is indeed the most gratifying result of detailed research on a well-defined problem that it necessarily uncovers relationships which are of primary inportance for the understanding of larger historical processes. The actual working program, however, needs restriction and minute detail work.

---

[139] An excellent example is Delambre *HAA,* published in 1817 and still not surpassed or even equaled because of its direct contact with the original sources.

The most essential task is that of making the original sources accessible as easily as possible in their best available form. By the indefatigable work of Heiberg, Hultsch, Tannery, and many others, we possess today a great part of the extant writings of the Greek scientists in excellent editions. We owe to Sir Thomas Little Heath many brilliant commentaries and translations of Greek mathematicians.[140] To make Greek and oriental source material more generally accessible, supplemented, of course, by modern translations and commentaries, will be the foremost problem of the future. The extension of this program to include medieval material, on the one hand, and Middle Eastern documents, on the other, appears as a logical consequence, worthy of the serious efforts of all scholars who wish to contribute to the understanding of the past of our own culture.

BIBLIOGRAPHY

AJP	*American Journal of Philology.*
AJSL	*American Journal of Semitic Languages and Literatures.*
Almagest	*See* Ptolemy.
AN	*Astronomische Nachrichten.*
Baillet [1]	BAILLET, J. "Le Papyrus mathématique d'Akhmim," *Mém. publ. par les membres de la mission arch. franç. au Caire,* Vol. **9**, Fasc. 1 (1892).
Baneth [1]	BANETH, E. "Maimuni's Neumondsberechnung," *Bericht über die Lehranstalt für die Wissenschaft des Judenthums in Berlin,* **16** (1898), **17** (1899), **20** (1902), **21** (1903) [197 pp.].
BASOR	*Bulletin of the American Schools of Oriental Research.*
Berger *GFH*	BERGER, H. *Die geographischen Fragmente des Hipparch.* Leipzig, 1869.
Boll, *Sphaera*	BOLL, F. *Sphaera.* Leipzig, 1903.

---

[140] On the other hand, much remains to be done to repair the harm caused by classical philologists who made their editions inaccessible to modern scientists by translating them into Latin instead of a modern language. Great opportunities have been spoiled by this absurd attitude. It has fortunately never occurred to Orientalists to translate their texts into Hebrew. It should be mentioned, however, that the Arabic version of Euclid's *Elements* was published in Latin(!) translation by Besthorn, Heiberg, and others (Copenhagen, 1897–1932).

Boll [1]   BOLL, F. "Antike Beobachtungen farbiger Sterne,"
                Abh. K. Bayerischen Akad. d. Wiss., Philos.-
                philol. u. histor. Kl. **30**, No. 1 (1918).

Boll [2]   BOLL, F. "Die Erforschung der antiken Astrologie,"
                Neue Jahrbücher für das klassische Altertum, **21**
                (1908), 103–26.

Boll-Bezold-Gun- BOLL, F., BEZOLD, C., AND GUNDEL, W., *Sternglaube*
del SS   *und Sterndeutung.* 4th ed. Leipzig, 1931.

Bouriant, P. [1] BOURIANT, P. "Fragment d'une manuscripte copte de
                  basse époque ayant contenu les principes astro-
                  nomiques des arabes [cf. n. 12]." *Journal asia-*
                  tique, 10. sér., **4** (1904), 117–23.

Bouriant, U.-Ven- BOURIANT, U., AND VENTRE BEY. "Sur trois tables
tre Bey [1]   horaires coptes," *Mémoires présentés à l'institut*
                   égyptien, **3** (1900), 575–604.

Braunmühl GT   BRAUNMÜHL, A. V. *Vorlesungen über Geschichte*
                 der Trigonometrie. Leipzig, 1900.

Brugsch, *Thes.* 1 BRUGSCH, H. *Thesaurus inscriptionum aegyptiaca-*
                    rum, I. *Astronomische und astrologische Inschrif-*
                    ten. Leipzig, 1883.

Burgess SS   BURGESS, E. "Translation of the Sûrya-Siddhânta,"
                 JAOS, **6** (1860), 141–498. Reprinted, Calcutta,
                 1935 [with introduction (45 pp.) by P. C. Sen-
                 gupta].

Capelle [1]   CAPELLE, W. "Älteste Spuren der Astrologie bei den
                 Griechen," *Hermes,* **60** (1925), 373–95.

CCAG   *Catalogus codicum astrologorum Graecorum.* Edited
                 by BOLL, CUMONT, *et al.* 12 vols. Bruxelles, 1898–
                 1936.

Chace RMP   CHACE, A. B., MANNING, H. P., AND ARCHIBALD,
                 R. C. *The Rhind Mathematical Papyrus.* 2 vols.
                 Oberlin, 1927–29.

Cicero De div.   CICERO, M. T. *De divinatione.* (English trans. by
                 W. A. FALCONER, "Loeb Classical Library.")

Crum CO   CRUM, W. E. *Coptic Ostraca.* London, 1902.

CT   Cuneiform texts from Babylonian tablets, etc., in the
                 British Museum.

Cumont EA   CUMONT, F. *L'Égypte des astrologues.* Bruxelles,
                 1937.

Cumont [1]   CUMONT, F. *Comment les grecs connurent les tables*
                 lunaires des chaldéens: Florilegium ... dédiés à M.
                 le marquis Melchior de Vogüé ..., pp. 159–65.
                 Paris, 1910.

Curtis-Robbins [1] CURTIS, H. D., AND ROBBINS, F. E. "An Ephemeris
                    of 467 A.D." *Publ. of the Observatory of the Univ.*
                    of Michigan, **6** (1935), 77–100.

Daressy [1]   DARESSY, G. "Notes et remarques 181," *Rec. trav.,* 23 (1901), 126–27.

Datta-Singh *HHM*   DATTA, B., AND SINGH, A. N. *History of Hindu Mathematics.* 2 vols. Lahore, 1935–38.

Delambre *HAA*   DELAMBRE, J. B. J. *Histoire de l'astronomie ancienne.* 2 vols. Paris, 1817.

D'Ooge-Robbins-Karpinski *Nic.*   D'OOGE, M. L., ROBBINS, F. E., AND KARPINSKI, L. C. *Nicomachus of Gerasa.* "Univ. of Michigan Studies: Humanistic Series," No. 16. Ann Arbor, 1926.

Drecker [1]   DRECKER, J. "Das Planisphaerium des Claudius Ptolemaeus," *Isis,* 9 (1927), 255–78.

Dreyer *HPS*   DREYER, J. L. E. *History of the Planetary Systems from Thales to Kepler.* Cambridge, 1906.

Düring *HP*   DÜRING, I. *Die Harmonielehre des Klaudios Ptolemaios.* "Göteborgs Högskolas Årsskrift," No. 36 (1930).

Düring *PPM*   DÜRING, I. *Ptolemaios und Porphyrios über die Musik.* "Göteborgs Högskolas Årsskrift," No. 40 (1934).

Duhem *SM*   DUHEM, P. *Le Système du monde; histoire des doctrines cosmologiques de Platon à Copernic.* 5 vols. Paris, 1913–17.

Emeneau *PIT*   EMENEAU, M. B. *A Union List of Printed Indic Texts and Translations in American Libraries.* "Amer. Oriental Series," Vol. 7, 1935.

Epping *AB*   EPPING, J. *Astronomisches aus Babylon.* "Stimmen aus Maria-Laach, Ergänzungsheft," No. 44. Freiburg, 1889.

Euclid   *Euclidis opera omnia,* ed. J. L. HEIBERG AND H. MENGE. 8 vols. Leipzig, 1883–1916.

Feldman *RMA*   FELDMAN, W. M. *Rabbinical mathematics and astronomy.* London, 1931.

Frank, *Plato*   FRANK, E. *Plato und die sogenannten Pythagoreer.* Halle, 1923.

Frankfort *CSA*   FRANKFORT, H. *The Cenotaph of Seti I at Abydos.* 2 vols. "Egypt Explor. Soc., Memoir," No. 39 (1933).

Gandz [1]   GANDZ, S. "The Sources of al-Khowārizmī's Algebra," *Osiris,* 1 (1936), 263–77.

Gandz [2]   GANDZ, S. "The Algebra of Inheritance," *Osiris,* 5 (1938), 319–91.

Gandz [3]   GANDZ, S. "The Origin and Development of the Quadratic Equations in Babylonian, Greek, and Early Arabic Algebra," *Osiris,* 3 (1937), 405–557.

Garbers *ES*   GARBERS, K. "Ein Werk Tābit b. Qurra's über ebene
                       Sonnenuhren," *QS*, A, **4** (1936).
Ginzel *Chron.*   GINZEL, F. K. *Handbuch der mathematischen und
                       technischen Chronologie.* 3 vols. Leipzig, 1906–14.
Gundel *DD*   GUNDEL, W. *Dekane und Dekansternbilder.* "Stu-
                       dien d. Bibl. Warburg," No. 19 (1936).
Gundel *HT*   GUNDEL, W. "Neue astrologische Texte des Hermes
                       Trismegistos," *Abh. d. Bayerischen Akad. d. Wiss.
                       Phil.-hist. Abt.* (N.F.), **12** (1936).
Harper *Letters*   HARPER, R. F. *Assyrian and Babylonian Letters Be-
                       longing to the Kouyunjik Collection of the British
                       Museum.* 14 vols. Chicago, 1892–1914.
Haskins *MS*   HASKINS, C. H. *Studies in the History of Mediaeval
                       Science.* 2d ed. Cambridge: Harvard University
                       Press, 1927.
Heath *AS*   HEATH, T. L. *Aristarchus of Samos.* Oxford, 1913.
Heath *Euclid*   HEATH, T. L. *The Thirteen Books of Euclid's Ele-
                       ments.* 3 vols. 2d ed. Cambridge, 1926.
Heath *GM*   HEATH, T. L. *A History of Greek Mathematics.* 2
                       vols. Oxford, 1921.
Heath *MGM*   HEATH, T. L. *A Manual of Greek Mathematics.* Ox-
                       ford, 1931.
Heidel *GM*   HEIDEL, W. A. *The Frame of the Ancient Greek
                       Maps: With a Discussion of the Discovery of
                       the Sphericity of the Earth.* New York, 1937
                       (= Am. Geographical Soc., Research Series No.
                       20).
Heidel [1]   HEIDEL, W. A. "The Pythagoreans and Greek
                       Mathematics," *AJP*, **61** (1940), 1–33.
Herz *GB*   HERZ, N. *Geschichte der Bahnbestimmung von Pla-
                       neten und Kometen,* **1**: *Die Theorien des Alter-
                       tums.* Leipzig, 1887.
Hildesheimer [1]   HILDESHEIMER, J. *Die astronomischen Kapitel in
                       Maimonidis Abhandlung über die Neumondsheili-
                       gung.* Jahres-Bericht des Rabbiner-Seminars zu
                       Berlin pro 5641 (1880–1881), 5–64.
Honigmann *SK*   HONIGMANN, E. *Die sieben Klimata und die Poleis
                       episemoi.* Heidelberg, 1929.
JAOS   *Journal of the American Oriental Society.*
Jeffreys [1]   JEFFREYS, H. "The Chief Cause of the Lunar Secu-
                       lar Acceleration," *MN*, **80** (1920), 309–17.
JNES   *Journal of Near Eastern Studies.*
Klein [1]   KLEIN, J. "Die griechische Logistik und die Entste-
                       hung der Algebra," *QS*, B, **3** (1934–36), 18–105,
                       122–235.

90

Koldewey *WB*	KOLDEWEY, R. *Das wiedererstehende Babylon.* 4th ed. Leipzig, 1925.
Krause *Men.*	KRAUSE, K. "Die Sphärik von Menelaos aus Alexandrien in der Verbesserung von Abū Naṣr Manṣūr b. Alī b. Irāq," *Abh. Ges. d. Wiss. zu Göttingen, Phil.-hist. Kl.,* 3. Folge, No. 17 (1936).
Kroll [1]	KROLL, W. "Kulturhistorisches aus astrologischen Texten," *Klio,* 18 (1923), 213–25.
Kroll *VV*	KROLL, W. *Vettii Valentis anthologiarum libri.* Berlin, 1908.
Kugler *BB*	KUGLER, F. X. *Im Bannkreis Babels.* Münster, 1910.
Kugler *BMR*	KUGLER, F. X. *Die babylonische Mondrechnung.* Freiburg, 1900.
Kugler *MP*	KUGLER, F. X. *Von Moses bis Paulus.* Münster, 1922.
Kugler *SSB*	KUGLER, F. X. *Sternkunde und Sterndienst in Babel.* 2 vols. Münster, 1907–24. *Ergänzungen,* in three parts (Part III by J. SCHAUMBERGER). Münster, 1913–35.
Langdon *VT*	LANGDON, S., FOTHERINGHAM, J. K., AND SCHOCH, C. *The Venus Tablets of Ammizaduga,* Oxford, 1928.
Lange-Neugebauer [1]	LANGE, H. O., AND NEUGEBAUER, O. "Papyrus Carlsberg No. 1," *Kongl. Danske Vidensk. Selskab, Hist.-fil. Skrifter,* Vol. 1, No. 2 (1940).
L'Hôte *LE*	L'HÔTE, NESTOR. *Lettres écrites d'Égypte en 1838 et 1839 ....* Paris, 1840.
Letronne [1]	LETRONNE. "Mémoire sur cette question: Les Anciens ont-ils exécuté une Mesure de la Terre postérieurement à l'établissement de l'École d'Alexandrie?" *MAIBL* 6 (1822), 261–323.
Littrow [1]	LITTROW, K. v. "Zur Kenntniss der kleinsten sichtbaren Mondphasen," *Sitzungsberichte d. Kaiserl. Akad. d. Wiss., Math.-Naturwiss. Cl.,* 66 (1872), 459–80.
Luckey [1]	LUCKEY, P. "Das Analemma von Ptolemäus," *AN,* 230 (1927), 18–46.
Luckey [2]	LUCKEY, P. Tābit b. Qurra's Buch über die ebenen Sonnenuhren," *QS,* B, 4 (1937), 95–148.
Mahler [1]	MAHLER, E. *Chronologische Vergleichungstabellen.* Wien, 1889.
Mahler [2]	MAHLER, E. *Maimonides' Kiddusch Hachodesch, Übersetzt und erläutert.* Wien, 1889.

MAIBL

Mémoires présentés par divers savants à l'Académie des Inscriptions et Belles-Lettres de l'Institut de France.

MN

Monthly Notices of the Royal Astronomical Society.

Mžik EGM

Mžik, H. v. *Erdmessung, Grad, Meile und Stadion nach den altarmenischen Quellen* (= Studien zur armenischen Geschichte No. **6**). Wien, 1933. [Reprinted from *Handes Amsorya*, Z. f. *armenische Philologie*, **47** (1933), cols. 283–305, 432–59, 559–82.]

Mžik-Hopfner
PDE

Mžik, H. v., AND Hopfner, F. *Des Klaudius Ptolemaios Einführung in die darstellende Erdkunde*, 1. Wien, 1938 (= *Klotho* **5**).

Neugebauer *ACT*

Neugebauer, O. *Astronomical Cuneiform Texts.* (In preparation.)

Neugebauer *MKT*

Neugebauer, O. *Mathematische Keilschrift-Texte.* 3 vols. (*QS*, A, 1 [1935–38].)

Neugebauer *Vorl.*

Neugebauer, O. *Vorlesungen über Geschichte der antiken mathematischen Wissenschaften*, 1: *Vorgriechische Mathematik.* Berlin, 1934.

Neugebauer [1]

Neugebauer, O. "Arithmetik und Rechentechnik der Ägypter," *QS*, B, **1** (1930), 301–80.

Neugebauer [2]

Neugebauer, O. "Die Geometrie der ägyptischen mathematischen Texte," *QS*, B, **1** (1931), 413–51.

Neugebauer [3]

Neugebauer, O. "Egyptian Planetary Texts," *Trans. of the Amer. Philos. Soc.*, **32** (new ser., 1942), 209–50.

Neugebauer [4]

Neugebauer, O. "Die Bedeutungslosigkeit der 'Sothisperiode' für die älteste ägyptische Chronologie," *Acta orientalia*, **17** (1938), 169–95.

Neugebauer [5]

Neugebauer, O. "The Origin of the Egyptian Calendar," *JNES*, **1** (1942), 396–403.

Neugebauer [6]

Neugebauer, O. "Demotic Horoscopes," *JAOS*, **63** (1943), 115–26.

Neugebauer [7]

Neugebauer, O. Review of Kugler *SSB Erg.* (Schaumberger), *QS*, B, **3** (1935), 271–86.

Neugebauer [8]

Neugebauer, O. "Untersuchungen zur antiken Astronomie, II: Datierung und Rekonstruktion von Texten des Systems II der Mondtheorie," *QS*, B, **4** (1937), 34–91.

Neugebauer [9]

Neugebauer, O. "Untersuchungen zur antiken Astronomie, III: Die babylonische Theorie der Breitenbewegung des Mondes," *QS*, B, **4** (1938), 193–346.

Neugebauer [10]   NEUGEBAUER, O. "Jahreszeiten und Tageslängen in
                      der babylonischen Astronomie," *Osiris, 2* (1936),
                      517–50.

Neugebauer [11]   NEUGEBAUER, O. "Zur Entstehung des Sexagesi-
                      malsystems," *Abh. d. Ges. d. Wissenschaften zu
                      Göttingen, Math.-phys. Kl.* (N.F.), **13**, No. 1
                      (1927).

Neugebauer [12]   NEUGEBAUER, O. Review of Datta-Singh *HHM* I
                      (QS, B, **3** [1936], 263–71).

Neugebauer [13]   NEUGEBAUER, O. "On Some Astronomical Papyri
                      and Related Problems of Ancient Geography,"
                      Trans. of the Amer. Philos. Soc., **32** (new ser.,
                      1942), 251–63.

Neugebauer [14]   NEUGEBAUER, O. "Über eine Methode zur Distanz-
                      bestimmung Alexandria-Rom bei Heron," *Kongl.
                      Danske Vidensk. Selskab, Hist.-fil. Meddel.,* Vol.
                      26, No. 2 (1938).

Neugebauer [15]   NEUGEBAUER, O. *Zur geometrischen Algebra* ("Stu-
                      dien zur Geschichte der antiken Algebra," III)
                      (QS, B, **3** [1935], 245–59).

Neugebauer [16]   NEUGEBAUER, O. *Apollonius-Studien* ("Studien zur
                      Geschichte der antiken Algebra," II) *(QS,* B, **2**
                      [1932], 215–54).

Neugebauer [17]   NEUGEBAUER, O. "Exact Science in Antiquity," *Uni-
                      versity of Pennsylvania Bicentennial Conference:
                      Studies in Civilization,* pp. 23–31. Philadelphia,
                      1941.

Neugebauer [18]   NEUGEBAUER, O. "Some Fundamental Concepts in
                      Ancient Astronomy," *University of Pennsylvania
                      Bicentennial Conference: Studies in the History
                      of Science,* pp. 13–29. Philadelphia, 1941.

Neugebauer [19]   NEUGEBAUER, O. "The Water-Clock in Babylonian
                      Astronomy" (to be published in *Isis* in 1946).

Neugebauer [20]   NEUGEBAUER, O. "Studies in Ancient Astronomy.
                      VII. Magnitudes of Lunar Eclipses in Babylonian
                      Mathematical Astronomy," *Isis,* **36** (1945), 10–15.

Neugebauer-Sachs   NEUGEBAUER, O., AND SACHS, A. *Mathematical
MCT   Cuneiform Texts* ("Amer. Oriental Series," **29**,
                      New Haven, 1945).

Neugebauer-   NEUGEBAUER, O., AND STRUVE, W. "Über die Geo-
Struve [1]   metrie des Kreises in Babylonien," *QS,* B, **1**
                      (1929), 81–92.

Neugebauer-Volten [1]	NEUGEBAUER, O., AND VOLTEN, A. "Untersuchungen zur antiken Astronomie, IV : Ein demotischer astronomischer Papyrus (Pap. Carlsberg 9)," *QS*, B, **4** (1938), 383–406.
Neugebauer, P. V. [1]	NEUGEBAUER, P. V. "Eine Konjunktion von Mond und Venus aus dem Jahre –418 und die Akzeleration von Sonne und Mond," *AN*, **244** (1932), cols. 305–8.
Pannekoek [1]	PANNEKOEK, A. "Calculation of Dates in the Babylonian Tables of Planets," *Koninkl. Akad. van Wetensch. te Amsterdam, Proceedings,* **19** (1916), 684–703.
Pannekoek [2]	PANNEKOEK, A. "Some Remarks on the Moon's Diameter and the Eclipse Tables in Babylonian Astronomy," *Eudemus,* **1** (1941), 9–22.
Pannekoek [3]	PANNEKOEK, A. "The Origin of the Saros," *Koninkl. Akad. van Wetensch. te Amsterdam, Proceedings,* **20** (1917), 943–55.
Pap. Oxyrh.	GRENFELL, B. P., AND HUNT, A. S. *The Oxyrhynchus Papyri.* London : Egypt Exploration Fund, 1898 ff.
Pap. Ryl.	HUNT, A. S., AND ROBERTS, C. H. *Catalogue of the Greek and Latin Papyri in the John Rylands Library, Manchester.* 3 vols. 1911–38.
Peet *RMP*	PEET, T. E. *The Rhind Mathematical Papyrus.* Liverpool, 1923.
Peters-Knobel *PCS*	PETERS, C. H. F., AND KNOBEL, E. B. *Ptolemy's catalogue of stars; a revision of the Almagest. Carnegie Institution of Washington Pub.,* No. **86** (1915).
Pfeiffer *SLA*	PFEIFFER, R. H. *State Letters of Assyria.* "Amer. Oriental Series," No. 6. New Haven, 1935.
Pococke *DE*	POCOCKE, R. *A Description of the East and Some Other Countries.* 2 vols. London, 1743–45.
Pogo [1]	POGO, A. "Calendars on Coffin Lids from Asyut," *Isis,* **17** (1932), 6–24.
Pogo [2]	POGO, A. "The Astronomical Inscriptions on the Coffins of Heny," *Isis,* **8** (1932), 7–13.
Pogo [3]	POGO, A. "Three Unpublished Calendars from Asyut," *Osiris,* **1** (1935), 500–509.
Pogo [4]	POGO, A. "Der Kalender auf dem Sargdeckel des Idy in Tübingen," in Gundel *DD*, pp. 22–26.
Pogo [5]	POGO, A. "The Astronomical Ceiling-Decoration in the Tomb of Senmut," *Isis,* **14** (1930), 301–25.

Poleman *CIM*   POLEMAN, H. I. *A Census of Indic Manuscripts in the United States and Canada.* "Amer. Oriental Series," Vol. 12. 1938.

Porter-Moss *TB*   PORTER, B., AND MOSS, R. L. B. *Topographical Bibliography of Ancient Egyptian Hieroglyphic Texts, Reliefs and Paintings.* 6 vols. Oxford, 1927–39.

Ptolemy   CL. PTOLEMAEUS, *Opera.*

    I. *Syntaxis mathematica,* ed. J. L. HEIBERG. 2 vols. Leipzig, 1898–1903. German translation by K. MANITIUS. 2 vols. Leipzig, 1912–13.

    II. *Opera astronomica minora,* ed. J. L. HEIBERG. Leipzig, 1907.

    III, 1. *Apotelesmatica,* ed. F. BOLL AND AE. BOER. Leipzig, 1940.

    Tetrabiblos, ed. and English trans. by F. E. ROBBINS ("Loeb Classical library" [1940]), and *Opera,* III, 1.

    Geographia, ed. NOBBE. Leipzig, 1843.

    Harmonics. See Düring.

QS   *Quellen und Studien zur Geschichte der Mathematik, Astronomie und Physik.*

RA   *Revue d'assyriologie.*

Rec. trav.   *Recueil de travaux relatifs à la philologie et à l'archéologie égyptiennes et assyriennes.*

Revillout [1]   REVILLOUT, E. *Mélanges sur la métrologie, l'économie politique et l'histoire de l'ancienne Égypte avec de nombreux textes démotiques, hiéroglyphiques, hiératiques ou Grecs inédits ou antérieurement mal publiés.* Paris, 1895.

Robbins [1]   ROBBINS, F. E. "A Greco-Egyptian Mathematical Papyrus," *Classical Philol.,* **18** (1923), 328–33.

Rome *CPT*   ROME, A. *Commentaires de Pappus et de Théon d'Alexandrie sur l'Almagest.* 3 vols. (= "Studi e testi," Vols. **54, 72,** and **106**). Roma, 1931–43.

Roscher *GRM*   ROSCHER, W. H. *Ausführliches Lexikon d. griechischen u. römischen Mythologie.* Leipzig, 1884–1937.

Sachau *Alb.*   SACHAU, C. E. *The chronology of ancient nations, an English version of the Arabic text of the athār-ul-bākiya of Albīrūnī.* London, 1879.

Sachs [1]   SACHS, A. J. "Some Metrological Problems in Old-Babylonian Mathematical Texts." *BASOR,* **96** (1944), 29–39.

Schaumberger *Erg.*   *See* Kugler *SSB.*

Schiaparelli *SSAA*   SCHIAPARELLI, G. *Scritti sulla storia della astrono-mia antica.* 3 vols. Bologna, 1925–27.

Schmidt [1]   SCHMIDT, O. "Bestemmelsen af Epoken for Maa-nens Middelbevægelse i Bredde hos Hipparch og Ptolemæus," *Matematisk Tidsskrift,* B (1937), pp. 27–32.

Schmidt [2]   SCHMIDT, O. "The Computation of the Length of Daylight in Hindu Astronomy," *Isis,* **36** (1944), 205–11.

Schnabel *Ber.*   SCHNABEL, P. *Berossos und die babylonisch-hellenis-tische Literatur.* Leipzig, 1923.

Schnabel [1]   SCHNABEL, P. "Kidenas, Hipparch und die Ent-deckung der Präzession," *ZA,* **37** (1927), 1–60.

Schnabel [2]   SCHNABEL, P. "Neue babylonische Planetentafeln," *ZA,* **35** (1924), 99–112.

Schott [1]   SCHOTT, A. "Das Werden der babylonisch-assy-rischen Positionsastronomie und einige seiner Bedingungen," *ZDMG,* **88** (1934), 302–37.

Schott [2]   SCHOTT, A. Review of Gundel *HT,* in *QS,* B, **4** (1937), 167–78.

Schoy *TLAB*   SCHOY, C. *Die trigonometrischen Lehren des per-sischen Astronomen Abu'l-Raiḥâm Muḥ. ibn Aḥ-mad al-Bîrûnî.* Hannover, 1927.

Schumacher [1]   SCHUMACHER, C. J. *Untersuchungen über die ptole-mäische Theorie der unteren Planeten.* Münster, 1917.

Sethe *ZAA*   SETHE, K. "Die Zeitrechnung der alten Aegypter," *Nachr. Ges. Wiss. zu Göttingen, Phil.-hist. Kl.,* 1919, pp. 287–330; 1920, pp. 28–55, 97–141.

Sethe *ZZ*   SETHE, K. *Von Zahlen und Zahlworten bei den alten Ägyptern* (= Schriften der Wissenschaftlichen Gesellschaft Strassburg, No. **25**). Strassburg, 1916.

Sidersky [1]   SIDERSKY, D. "Étude sur l'origine astronomique de la chronologie juive," *MAIBL,* **12, 2** (1913), 595–683.

Sidersky [2]   SIDERSKY, D. "Étude sur la chronologie assyro-babylonienne," *MAIBL,* **13, 1** (1923), 105–99. Written 1917 according to Sidersky [3], p. 23, note 1.

Sidersky [3]   SIDERSKY, D. "Le calcul chaldéen des néoménies," *RA,* **16** (1919), 21–36.

Šileiko [1]   ŠILEIKO, V. "Mondlaufprognosen aus der Zeit der ersten babylonischen Dynastie," *Comptes-Rendus de l'Académie des Sciences de l'URSS*, 1927, B, pp. 125–28.

Struve *MPM*   STRUVE, W. W. "Mathematischer Papyrus des staatlichen Museums der schönen Künste in Moskau," *QS*, A, **1** (1930).

Tallgren [1]   TALLGREN, O. J. "Survivance arabo-romane du Catalogue d'étoiles de Ptolémée," *Studia Orientalia*, **2** (1928), 202–83.

Thibaut *AAM*   THIBAUT, G. "Astronomie, Astrologie und Mathematik" (art.) in *Grundriss d. Indo-Arischen Philologie und Altertumskunde*, III, 9 (1899).

Thomas *GMW*   THOMAS, I. *Selections Illustrating the History of Greek Mathematics.* 2 vols. 1939–41. ("Loeb Classical Library.")

Thompson *AB*   THOMPSON, R. C. *A Catalogue of Late Babylonian Tablets in the Bodleian Library, Oxford.* London, 1927.

Thompson *Rep.*   THOMPSON, R. C. *The Reports of the Magicians and Astrologers of Niniveh and Babylon.* 2 vols. London, 1900.

Thorndike *HM*   THORNDIKE, L. *A History of Magic and Experimental Science.* 6 vols. New York, 1923–41.

Thureau-Dangin *SS*   THUREAU-DANGIN, F. *Esquisse d'une histoire du système sexagésimal.* Paris, 1932.

Thureau-Dangin *TMB*   THUREAU-DANGIN, F. *Textes mathématiques babyloniens.* Leiden, 1938.

Thureau-Dangin [1]   THUREAU-DANGIN, F. "Sketch of a History of the Sexagesimal System," *Osiris*, **7** (1939), 95–141.

Thureau-Dangin [2]   THUREAU-DANGIN, F. "La Clepsydre chez les Babyloniens," *RA*, **29** (1932), 133–36.

Tropfke [1]   TROPFKE, J. "Archimedes und die Trigonometrie," *Archiv f. Gesch. d. Math., d. Naturwiss. und d. Technik*, **10** (1928), 432–63.

van der Waerden [1]   VAN DER WAERDEN, B. L. "Die Voraussage von Finsternissen bei den Babyloniern," *Berichte d. math. phys. Kl. d. sächs. Akad. d. Wiss. zu Leipzig*, **92** (1940), 107–14.

van der Waerden [2]   VAN DER WAERDEN, B. L. "Zur babylonischen Planetenrechnung," *Eudemus*, **1** (1941), 23–48.

Vettius Valens   *See* Kroll *VV*.

Virolleaud *ACh*   VIROLLEAUD, CH. *L'Astrologie chaldéenne.* 4 vols. Paris, 1908–12.

Vogel [1]   VOGEL, K. "Beiträge zur griechischen Logistik,"
                     Sitzungsber. d. Bayerischen Akad. d. Wiss.,
                     Math.-nat. Abt., 1936, pp. 357–472.
Vogel [2]   VOGEL, K. "Bemerkungen zu den quadratischen
                     Gleichungen der babylonischen Mathematik,"
                     Osiris, 1 (1936), 703–17.
Vogt [1]   VOGT, H. "Der Kalender des Claudius Ptolemäus
                     (= F. BOLL, Griechische Kalender, V), Sitzungs-
                     ber. d. Heidelberger Akad. d. Wiss. Philos.-hist.
                     Kl., 1920, p. 15.
Waschow [1]   WASCHOW, H. Review of Neugebauer MKT III,
                     Archiv für Orientforschung, 12 (1939), 277.
Waterman RC   WATERMAN, L. Royal Correspondence of the As-
                     syrian Empire. 4 vols. "Univ. of Michigan Stud-
                     ies: Humanistic Series," Vols. 17–20. Ann Arbor,
                     1930–36.
Weidner Hdb.   WEIDNER, E. F. Handbuch der babylonischen As-
                     tronomie, I: Der babylonische Fixsternhimmel
                     (= Assyriologische Bibliothek, 23). Leipzig, 1915.
                     [Only 146 pages are published; pp. 147–80 were
                     printed but not published.]
Weidner [1]   WEIDNER, E. F. "Ein babylonisches Kompendium
                     der Himmelskunde," AJSLL, 40 (1924), 186–208.
Weidner [2]   WEIDNER, E. F. "Die astrologische Serie Enûma
                     Anu Enlil," Archiv für Orientforschung, 14
                     (1942), 172–95 [to be continued].
Weissbach BM   WEISSBACH, F. H. Babylonische Miscellen. Leipzig,
                     1903 (= Wiss. Veröffentl. d. Deutschen Orient-
                     Ges., 4).
Winlock [1]   WINLOCK, H. E. "The Origin of the Egyptian Cal-
                     endar," Proc. of the Amer. Philos. Soc., 83 (1940),
                     447–64.
Winlock [2]   WINLOCK, H. E. The Egyptian Expedition, 1925–
                     1927. Section II of the Bulletin of the Metropoli-
                     tan Museum of Art, 1928, pp. 3–58.
Winlock EDEB   WINLOCK, H. E. Excavations at Deir el Baḥri. New
                     York, 1942.
ZA   Zeitschrift für Assyriologie.
ZDMG   Zeitschrift der Deutschen Morgenländischen Gesell
                     schaft.
Zeuthen KA   ZEUTHEN, H. G. Die Lehre von den Kegelschnitten
                     im Altertum. Copenhagen, 1886.

Reprinted from
Publications of the Astronomical Society of the Pacific
58 (340), 17–142 (1946)

MATHEMATICAL METHODS IN ANCIENT ASTRONOMY

O. NEUGEBAUER

CONTENTS

1. **Chronological remarks.** The main source for our knowledge of ancient astronomy is Ptolemy. His *Mathematical composition*, commonly known as the *Almagest*, quotes observations of his own ranging from 127 to 142 A.D. [35; 36; 37].[1] This work seems to be the earliest of a whole series of fundamental works, such as his *Geography* [41], the *Tetrabiblos* [39; 40], and so on, whose influence on mediaeval thought cannot be overrated.

Questions of historical priority will not be discussed here. Nevertheless it must be emphasized that Ptolemy relied heavily on methods developed by his predecessors, especially Hipparchus. Indeed, almost all our information about the latter's work is based on references in the Almagest. However fragmentary our knowledge of Hipparchus' astronomy may be, it is evident that it represents a milestone in the development of mathematical astronomy. Observations of Hipparchus quoted by Ptolemy extend from 162 to 127 B.C.

Finally, we have original Babylonian ephemerides for the moon and the planets covering, with only minor gaps, the years from 227 to 48 B.C. (Kugler [19; 20], Schnabel [47], Schaumberger [20], Neugebauer [31]). Nothing is known about the exact date or origin of the Babylonian methods though it might be a fair guess to assume a date between 400 and 250 B.C.

To our knowledge, Egypt exercised no positive influence on the development of mathematical astronomy. This is in perfect accord

---

An address delivered before the New York meeting of the Society on April 16, 1948, by invitation of the Committee to Select Hour Speakers for Eastern Sectional Meetings; received by the editors May 28, 1948.

[1] Numbers in brackets refer to the bibliography at the end of the paper.

1013

with the fact that Egyptian mathematics never went beyond an extremely elementary level, totally unfit for the description of astronomical phenomena.

2. **Introduction.** It is not the scope of this paper to give even a sketch of the historical development of ancient astronomy. I shall only try to illustrate the close relationship between mathematics and astronomy, a relationship which goes much farther than one might assume at first sight. I shall mention only three problems which seemingly would belong to purely observational astronomy but actually are essentially dependent upon mathematical theories. These problems are (a) the determination of the apparent diameter of the moon, (b) the determination of the constant of precession, (c) the determination of geographical longitude. The first two problems were not solved by direct measurement but by relating them to the theory of the motion of the moon. It requires the whole mechanism of the lunar theory to compute the coordinates of the moon for given eclipses, especially the distance of the center of the moon from the center of the shadow. It is only after these elements were found according to the mathematical theory that the observed magnitudes of the eclipses are used to find the apparent diameter of the moon (*Almagest* V, 14). Similarly, the longitudes of fixed stars are not measured directly but are referred to the moon, and thus eventually to the sun, by means of occultations or close conjunctions. Again the whole lunar theory is required to find the common longitude of moon and star (*Almagest* VII, 3). Finally, the determination of geographical longitude is based on the simultaneous observation of a lunar eclipse, the circumstances of which must be determined from theory. This last problem involves, however, another theoretical consideration. The ancients measured time not by means of clocks of uniform rate but by sun dials and waterclocks which showed "seasonal" hours. Seasonal hours can be simply described as an extremal form of "daylight saving time" because each hour is always the 12th part of the actual length of daylight. Thus the time reckoning is fully adjusted to the variation of the seasons. For civil life, this undoubtedly has its great advantages. For astronomical computations, however, the reduction of seasonal hours to equinoctial hours requires a theory of the dependence of sun dials upon the geographical coordinates and the longitude of the sun.

These examples will suffice to make it understandable that two groups of mathematical theory play a paramount role in ancient astronomy. On the one hand, a detailed celestial mechanics is needed, especially for the theory of the moon; on the other hand, auxiliary problems must be solved, including the theory of celestial and terrestial coordinates, their transformation from one system into another, and their application to the theory of various types of sun dials. In short, we can say that kinematics and spherical astronomy play a much greater role than empirical observations. The ancient astronomers were fully aware of the fact that the low accuracy of their instruments had to be supplemented by a mathematical theory of the greatest possible refinement. Observations are more qualitative than quantitative: *"when* angles are equal" may be decided fairly well on an instrument but not *"how large* are the angles," says Ptolemy with respect to the lunar and solar diameter (*Almagest* V, 15; Heiberg p. 417). Consequently, period relations over long intervals of time and lunar eclipses are the main foundations so far as empirical material is concerned; all the rest is mathematical theory. We shall see that this holds for Greek as well as for Babylonian astronomy.

The fact that ancient astronomy is to a large extent "mathematics" has far-reaching consequences for the history of civilization. The Middle Ages inherited an astronomical system, and with it a picture of the structure of the universe, of a consistency and inner perfection which hardly seemed open to improvement. The bearers of the Christian civilizations, at the very beginning, had lost contact with Hellenistic science; hence the astronomy of Western and Central Europe relapsed for many centuries into a primitive stage of knowledge where a few simple period relations sufficed as the basis of the computation of Easter and similar problems. Though this process was to some extent delayed by the continued use of astronomical tables for astrological purposes, the destruction of the ancient tradition would have been complete had not Greek astronomy found a new and most interesting development among Hindu astronomers. When the Arab conquest reached India, Greek astronomy soon saw a triumphant revival everywhere in the Moslem world, thus preparing the basis for the new development of astronomy and mathematics in the Renaissance. The "Ptolemaic system" has often been blamed for the preservation for almost 1500 years of a narrow, yet much too complicated, picture of the world.[2] It is only fair to underline the fact that this system pre-

---

[2] It should be remarked that "Ptolemaic system" is often used in a rather unhistoric fashion. Actually the Aristotelian version of the homocentric spheres of Eudoxus determined the cosmological ideas of the philosophers and theologians of the Middle Ages. Thus a system 500 years older than the Almagest should take most of the blame.

served for the same length of time the tradition of mathematical methods which became most powerful tools in the hands of Copernicus, Tycho Brahe and Kepler.

A. CELESTIAL MECHANICS

3. Greek astronomy. The cornerstone of Greek celestial mechanics is an essentially "dynamical" principle, however metaphysical its actual formulation in Greek philosophical literature may sound. It is the idea that the circular movement of celestial bodies is the only movement which can last eternally. This "principle of inertia" is the guiding principle of all Greek astronomical theories. The fortunate accident that the orbits in our planetary system deviate very little from circles made it possible to construct geometric models whose gradual improvement corresponds to the addition of new Fourier terms, each of which has a certain physical significance. Modern scientists have often declared simplicity to be the criterium of truth, and a whole philosophy of "economy of reasoning" has offered its guidance to the researcher. This school of thought could rightly claim the early Greek astronomers as its first followers. Nothing "simpler" and more natural could have been assumed than the preference of celestial bodies for circular movements. And the remarkable successes of this assumption could only strengthen confidence in its correctness. Nobody could foresee that the simplicity of the circular movements is due to an accidental distribution of masses or that this also causes the simplicity of Newton's law, conveniently hiding from us the effects of a general gravitational space.

Here is not the place to describe the development of astronomical hypotheses, based on the combination of circular movements, or, originally, of movements of spheres, following an ingenious idea of Eudoxus. This development is described in masterly fashion by Duhem in his *Système du monde* [18]. We shall here restrict ourselves to a short discussion of some points in the theory of eccenters and epicycles in its application to the lunar movement. For the theory of Mercury and Venus see Boelk [3] and Schumacher [48].

Assuming that all circular movements proceed at constant angular velocity, it is obvious that an observer who is not located at the center gets the impression of a variable velocity. We know that already Apollonius, about 200 B.C., knew that an eccentric movement can be replaced by an epicyclic movement, where the center of the epicycle moves on the deferent with the mean angular motion around the observer whereas the object moves on the circumference of the epicycle with the same angular velocity in the opposite direction.

The radius of the epicycle is identical with the eccentricity of the eccenter. Similar relations hold for more general cases and it is therefore a matter of choice which hypothesis is used in a specific case (*Almagest* XII, 1 and III, 3). We know, for example, that Hipparchus used an epicycle for the description of the solar anomaly (Theon Smyrnaei, *De astronomia*, XXXIV [50]) while Ptolemy preferred the eccenter because of its greater simplicity, using only one motion (*Almagest* III, 4).

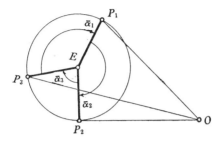

FIG. 1

The determination of the parameters of models of this type requires great ingenuity. As an example I might quote the case of the lunar theory in its simpler (Hipparchian) form which only accounts for the "first anomaly," that is, the eccentricity with uniformly progressing apsidal line. To this end an epicyclic model is assumed and three lunar eclipses are observed, giving the true longitudes $\lambda_1, \lambda_2, \lambda_3$ at given moments t_1, t_2, t_3. The mean motions in mean longitude ($\bar{\lambda}$) and mean anomaly (α) are known from period relations. Thus it is possible to find the positions of the moon on its epicycle, expressed by the differences of their anomalies (cf. Fig. 1). Furthermore the corresponding mean longitudes $\bar{\lambda}_i$ can be found for each t_i. The differences $c_i = \lambda_i - \bar{\lambda}_i$ are the corresponding values of the equation of center and can be interpreted as the angles under which the radii EP_i appear from the observer (*Almagest* IV, 6). The solution of this problem, undoubtedly known to Hipparchus and discussed in detail by Ptolemy, is often needed in surveying; one then speaks of the "Pothenot" or "Snellius" problem (Tropfke [51, V, 97]; the identity of these problems has been seen by Delambre [4, II, 164] and Oudemans [33]). As a result the radius of the epicycle can be found in terms of the radius of the deferent. Thus the eccentricity of the lunar orbit is known.

Before we proceed to Ptolemy's addition to Hipparchus' theory of the moon we might for a moment discuss the often underlined contrast between the simplicity of Kepler's theory and the clumsiness of eccenters and epicycles. Actually one compares here theories of different level. If one accepts the statement that the earth or the moon travels on an ellipse, one disregards perturbations and describes the true longitude in first approximation by

$$\lambda = \bar{\lambda} + 2e \sin \alpha,$$

where α is the mean anomaly. But exactly the same relation holds for an eccenter, or the equivalent epicycle, if its eccentricity is $2e$. If we want to take the movement of the apsidal line into account, again both models are equivalent because a uniform rotation of the apsidal line must be added in both cases. The error of the ancient theory does not at all lie in its unnecessary complication but in its simplicity, which leads only to correct longitudes whereas the distances are very badly represented, especially near the apogee. Because the observation of distances requires the measurement of very small angular differences for parallax or apparent diameters, this part of the theory was very difficult to check. Consequently one distrusted the observation of small angles even in cases where one should have found the discrepancy and preferred a simpler model to added corrections. For the longitudes, however, the ancient theory is exactly as simple and as efficient as Kepler's theory within the same first order approximations.

As was mentioned in the preceding remarks, the fundamental parameters of the lunar theory were obtained from eclipses, thus for syzygies. Already Hipparchus started to test the theory also for intermediate quadratures, and observations of his, quoted by Ptolemy (especially *Almagest* V, 5; Heiberg p. 369), showed that the "mean apogee" of the epicycle, from which α has to be counted, cannot be considered fixed. Ptolemy further investigated this situation and constructed a model which coincides with the simple theory in the syzygies and shows the proper fluctuation of the longitudes in the quadratures, depending on mean anomaly α and double mean elongation 2ϵ. The corresponding term $a \cdot \sin (2\epsilon - \alpha)$ in the modern theory is called "evection." Because the value of the constant a is about $1;16°$ whereas $2\epsilon = 6;17°$, we have for the syzygies and quadratures respectively[3]

$$\lambda = \bar{\lambda} + 6;17° \sin \alpha \mp 1;16° \sin \alpha$$

which shows that the maximum of the equation of center varies be-

---

[3] I use the notation $a,b;c$ for $a \cdot 60 + b + c \cdot 60^{-1}$.

tween about 5;1° and 7;33°. The Ptolemaic values are 5;1° and 7;39° respectively (*Almagest* IV, 10, and V, 8). Also for intermediate elongations and anomalies Ptolemy's representation of the longi-

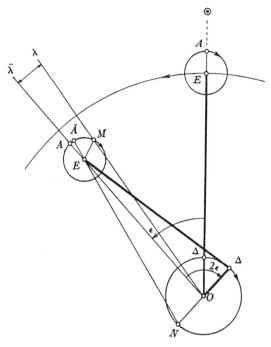

Fig. 2

tudes is very good (Kempf [15]). His model is described by Fig. 2. In order to obtain the necessary increase of the equation of center from conjunction towards quadrature, he moves the lunar epicycle closer to the observer O, by letting Δ travel on a circle around O such that Δ has the elongation 2ε from the mean moon E. Simultaneously the oscillations of the anomaly are obtained by introducing a "mean apogee" $\overline{A}$ from which α is to be counted. The position of $\overline{A}$ is defined by means of the rule that $\overline{A}E$ always points towards the point N which is diametrically opposite to Δ on the circle with center O.

If we want to compare this model with a geometric representation of the modern theory, using the same degree of approximation, we have to represent the evection as follows (Möbius [24]). The center C of the elliptic orbit rotates around the mean center $\overline{C}$ on a circle of radius $\bar{e}/5$ (where $O\bar{C}=\bar{e}$) with an angular velocity which is twice the velocity of the elongation of the mean sun from the mean apsidal line of the lunar orbit. Hence not only the apsidal line but also the eccentricity varies according to position and length of OC. Translating these movements into the language of eccenters and epicycles one obtains a model which closely resembles Ptolemy's construction, though with the essential difference that the radius of the epicycle must be made a function of α. Again one sees that Ptolemy's model is not too complicated but too simple because it represents only the longitudes correctly at the expense of the distances.

4. **Babylonian astronomy.** The final goal of Greek and Babylonian astronomy is, of course, identical. Starting from a few empirical elements one wishes to be able to compute the positions of the celestial bodies for any given moment. One may say with equal right that the progress, or the error, of the Greek method consisted in the invention of an intermediate step, namely the construction of a "dynamical" model, based on circular movements. From this model the Greeks derived their numerical tables exactly as a modern "Nautical Almanac" is derived from computations which are determined by consequences of dynamical rules and certain empirically determined initial values. As far as we can say from the material available to us, it seems that no such theoretical model existed in Babylonian astronomy. One apparently tried to obtain, on purely mathematical grounds, the rules for the computation of the tables from the empirical data. To use modern terminology the Babylonian procedure is very close to harmonic analysis. Given periodic phenomena of a rather complex character; find simple periodic functions, whose combination describes, within given limits of accuracy, the observed phenomena. While in our harmonic analysis the basic periodic functions admit, at least in principle, a direct geometric interpretation by simple harmonic oscillations, no such interpretation is obvious for the functions used in Babylonian astronomy. We do not know what were the Babylonian concepts about the physical structure of the universe. It seems safe to say that whatever concepts might have existed they were not directly reflected in the mathematical methods for the lunar and planetary movements.

The fundamental tools of Babylonian astronomy are periodic difference sequences of first and higher order. The simplest case is represented by tables in which each line represents a certain moment, these moments being equidistant, for example, one mean synodic month apart. The tabulated values can therefore be represented as a function $f(n)$ for integer values of the argument. Linear interpolation then leads to a continuous function $f(n)$ whose graph looks like Fig. 3. We shall call such functions "linear zigzag functions." The period of the function $f(x)$ may be called P and it is obvious that

$$P = 2\Delta/d$$

where Δ is the amplitude and d the slope of the increasing branch.

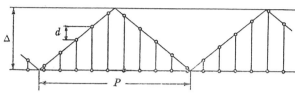

FIG. 3

The tabulated values $f(n)$ have, of course, in general a much larger period than P because P is in practice never an integer. If

$$P = 2\Delta/d = \pi/Z$$

where π and Z are relative prime integers, we see that $\pi = PZ$ is the smallest integer after which the sequence $f(n)$ repeats all its values. We thus call π the "number period," Z the "wave number," that is, the number of waves of length P contained in π. Ordinarily π is a large number. If we, for example, know that one year contains 12;22,8 mean synodic months we have 12;22,8 = 46,23/3,45 which shows that $\pi = 46,23 = 2783$ and $Z = 3,45 = 225$. This means that "2783 mean synodic months = 225 years" is the smallest period relation on which a zigzag function with $P = 12;22,8$ is built.

   Functions of this type are especially used in the lunar theory. Two coexisting "systems" are known. One, called "System B," uses for the variable solar velocity a linear zigzag function; "System A," however, assumes a constant solar velocity for an arc of the ecliptic, with a discontinuous change to another value, constant on the remaining arc. The assumption of system B seems much more natural than the very bold assumption of sudden jumps in the solar velocity. Actually, however, the theory of system B is much more involved in its purely mathematical consequences. The reason can be outlined as follows. Let $f(t)$ be a periodic function of t, for example, the latitude of the moon, represented by a linear zigzag function of period p_0. An "ephemeris" is a table of values of $f(t)$ for equidistant values of t, for example, for all mean conjunctions. If the solar velocity varies from a value w to W then the lunar velocity must also vary from

360+w to 360+W in order to maintain the same mean distance between conjunctions. On the other hand the anomaly of the sun must not influence the mean distance between consecutive nodes. Developing the consequences of this requirement one finds that the tabulated values of $f(t)$ again form a linear zigzag function but with difference d where the solar velocity is w, and with difference D where the solar velocity is W. It then holds that

$$D - d = \frac{\Delta}{3,0 \ p_0} \cdot \frac{Y}{Y + 1} (W - w)$$

where Δ is the amplitude of $f(t)$ and Y the length of the solar year. This formula shows immediately the essential difference between the systems A and B. If the solar velocity is restricted to two values only, fixed in their relation to the ecliptic (system A), then two differences suffice for the computation of $f(t)$. If, however, the solar velocities form a linear zigzag function (system B) then the same holds for the differences of $f(t)$ and $f(t)$ is a difference sequence of second order (Neugebauer [27]). Because we are actually dealing in these tables with arithmetical functions, instead of continuous functions, it is not at all trivial to determine the parameters of the zigzag function $g(n)$ of the differences of $f(n)$. The solution of this problem is found by considering mean periods and mean differences, taken over a whole number period. The mean value μ_g of g is then defined by

$$\mu_g = 2\Delta \ \frac{1 - p_0}{p_0} = d_0$$

where d_0 is the mean slope of $f(n)$. Arithmetical problems of this type were obviously the reason for assuming in system A a discontinuous change in the solar velocity at two points only instead of adopting the model of system B with its much greater mathematical complications.

In dealing with difference sequences of second order, Babylonian astronomers came close to problems whose importance was again seen only in the early days of the development of calculus. As an example may be quoted two sequences in system B, called H and J, whose astronomical significance is here without interest. H is a linear zigzag function with minimum 0 and maximum Δ_H. The values of H are the differences of J. The mean value of J is 0 and its amplitude is determined in such a way that increasing branches change to decreasing branches and vice versa whenever H is zero. In other words H behaves like the derivative of J. Fig. 4 shows in the lower part the function H, in the upper part the function J. I have chosen an example where an error has occurred at the decreasing branch of the second wave of J. The correct value (dotted line) would have been negative. Instead, the computer took the positive sign and continued thereafter according to the rule. Our graph shows how strongly the shape of the curve is affected by this error. On the other hand, it underlines the agreement between the extrema in J and the zeros in

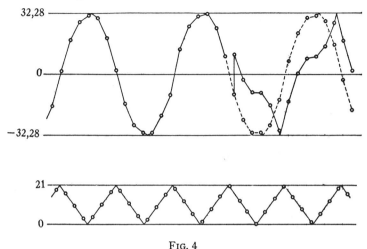

FIG. 4

H in the normal case though we are dealing only with discrete values and not with continuous functions. The question arises how one could determine the amplitude Δ_J of a sequence of discrete values to a given function H such that nevertheless a behavior of this type was achieved. The answer can be given as follows. We simply require that the relation between H and J be correct "in the mean" over the whole number period π of H. In an interval of length π the function J must oscillate, for example, from maximum to maximum $Z/2$ times because two waves of H correspond to one oscillation of J. The total change of values of J, caused by H, is the total of all differences, thus $\sum H = \mu\pi$ where μ is the mean value of H. This total change for a single wave of J is $2\Delta_J$. Thus we have

$$2\Delta_J \cdot Z/2 = \mu\pi \quad \text{or}$$
$$\Delta_J = \mu\pi/Z.$$

This is actually the relation which is satisfied by the parameters of the sequences found in lunar ephemerides.

Many examples could be quoted for the extensive use which Babylonian astronomers made of difference sequences to represent periodic functions. We have examples of modified zigzag functions. The main part increases or decreases linearly. In order to avoid, however, the sharp change of direction at the extrema, sequences of second order which bridge the neighborhood of the extrema are constructed.

The idea of operating with interpolations obviously influenced the whole procedure of describing planetary movements. Apparently no attempt was made to find the longitude λ of a planet directly as function of t. The following very clever indirect method was invented. Instead of dealing with the total motion, single "phenomena" were considered as if they were independent celestial bodies. In the case of Mercury, to mention a specific example, four phenomena are considered: first and last appearance as morning star, and first and

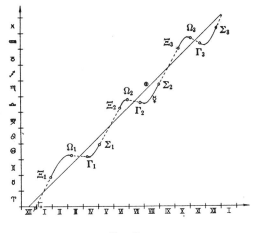

Fig. 5

last appearance as evening star. Now, for example, the first appearance as evening star is taken independently, and rules are given to compute tables for the dates and longitudes of this single phenomenon. Here again, there exist different "systems" for the description of the dependence of these coordinates on the zodiac, that is, implicitly on the longitude of the sun. We have cases of step functions or linear zigzag functions, obviously several systems being used for the same planet. After the coordinates of one phenomenon were found for a period of time, the coordinates for the remaining phenomena were computed in a similar way, though generally with different functions, for the same period. The result is a complete list of all phenomena which now could also be read in their natural order. Finally, interpolation is used for the coordinates of the planet for intermediate dates.

This whole procedure can best be illustrated in an example for

Mercury (cf. Fig. 5). In the first step all points Ξ_i were computed. Three different zones of the zodiac are distinguished such that the distances between consecutive points Ξ depend upon the zone to which they belong. Different zones determine the distribution of the points Γ_i. Eventually, the points Ω_i and Σ_i can be found from the corresponding points Ξ_i and Γ_i respectively. Thus we end up with a sequence of irregularly distributed points near the straight line which represents the movement of the sun. The variability of the distances ·between these points causes also variable differences between consecutive points for different phenomena, for example, Ω and Γ. Because the planet is retrograde between Ω and Γ, this indicates different velocities of retrogradation. In order to find these velocities, schemes of interpolations are used. Fig. 6 shows an example; the

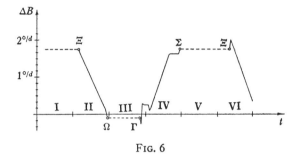

FIG. 6

retrograde motion is represented by the constant velocity $-0;6$ degrees per day. Another section of constant velocity $(+1;45^{\circ/d})$ is assumed from Σ to Ξ. In both sections the planet is near the sun and therefore invisible. For the other parts of the orbit, linearly variable differences are used (except for small irregularities near the end points, probably due to the practical requirement of relatively round numbers for the interpolation). As a result, day-to-day positions (Fig. 7) which represent a very satisfactory representation of the actual movement are obtained. It is clear that these methods require the knowledge of the summation formulae of arithmetic progressions.

These examples suffice to characterize the principle methods of Babylonian astronomy. For details of the planetary theory one may consult Kugler [20], Pannekoek [34], van der Waerden [52; 53]. The historical influence of these arithmetical methods is very great. They opened, for the first time, the way for a consistent numerical treatment of astronomical phenomena. This influence is still felt to-day in the use of sexagesimal units in the measurement of time and angles. In antiquity this influence was much more strongly visible in many cases which have left no trace in the modern development. Ptolemy's table of refraction, for instance, assumes that the angles of the reflected ray form a second degree sequence as function of the angle of the incident ray (Lejeune [21]). Babylonian methods are predominant in astronomical computations for astrological purposes

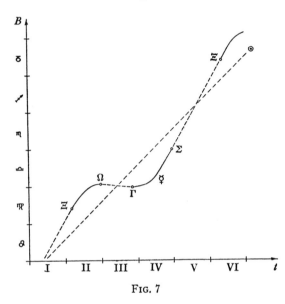

Fig. 7

for the obvious reason that they are easier to handle and do not require full understanding of complicated geometrical models. Greek papyri (Knudtzon-Neugebauer [17], to be supplemented by P. Ryl. 27 [14]) show the use of Babylonian period relations. Ancient and mediaeval geography was deeply influenced by concepts based on Babylonian methods. We shall return to these "linear methods" in the next section.

B. NUMERICAL AND GRAPHICAL METHODS

5. **Introduction; the linear methods.** The fundamental role of the eccenters and epicycles for Greek astronomy and the dramatic history of the discussion of these geometric assumptions in the Renaissance have left us with the impression that Greek astronomy was mainly geometrical in character. Though it is evident that the Greek astronomers must have felt that their models somehow reflected physical facts, at least in theory the geometric models were only the

tools for the computation of tables. Consequently we find 'many tables in the *Almagest* whose construction involved an enormous amount of numerical work. According to the contents they fall naturally into tables for spherical astronomy (chords, rising times, zenith distances, and so on), tables for the sun (mean motion, equation of center), for the moon (mean motion for longitude, anomaly, argument of latitude, anomaly, equation of center, evection, parallax, mean conjunctions, eclipses) and for the planets (mean motion, anomaly, stationary points, elongation, latitudes, heliacal risings and settings).

The Babylonian influence on Greek numerical methods is obvious from the use of the sexagesimal division of all units. Unfortunately this influence did not fully penetrate all computational steps. Thus often we find fractions expressed by means of unit fractions, replacing, for example, 0;42 by 1/2+1/5. Though this notation never occurs in tables, it influenced the accuracy of single calculations because frequently results are rounded off in order to obtain simple unit fractions. Thus, for example, Ptolemy finds in *Almagest* IV, 6 the value 5;13 or 5;14 for the radius of the lunar epicycle (Heiberg [35, pp. 313 and 322]) but he later always uses 5 1/4(=5;15) as a round value. In general it must be said that the ancients were little concerned about the influence of rounding off and accumulated errors. Often the errors are of the same order of magnitude as the effect under consideration. Apparently it was only under the influence of modern analysis that we have learned to consider the evaluation of errors as an essential part of numerical methods. On the other hand it must be said that, for example, excellent approximations of square roots were developed very early. The Babylonians of the second millenium B.C. already used alternating geometric and harmonic means (Neugebauer [26, p. 33 ff.]) and many values in the *Almagest* can be explained by this method. It would not be surprising if this technique reached the Greeks together with the sexagesimal system and it might be significant that the value $2^{1/2}=1;21,50,10$ is not only found in cuneiform records (Neugebauer-Sachs [32, p. 32]) but is used by Ptolemy in the computation of the chord of 90° (Heiberg [35, p. 35, 15]).

The arithmetical methods of the Babylonian astronomers are also discernible in another field of ancient astronomy. We have already mentioned the "seasonal hours" for ancient time measurement which thus requires a knowledge of the law of variation in the length of daylight whenever astronomical computation with equinoctial hours was needed. For the Babylonian astronomers the length of daylight

113

was of great importance also because one of the main goals of their lunar theory was the prediction of the evening of first visibility of the new crescent of the moon after conjunction. This evening defined the beginning of a new month and consequently the whole calendar depended on this problem. For its solution the knowledge of the moment of sunset is, of course, required. Because the length of daylight is the time from sunrise to sunset, one also may ask for the time it takes the semicircle of the ecliptic from λ to $\lambda+180$ to rise if λ is the longitude of the sun at the given moment. A month later the longitude of the sun will be given roughly by $\lambda+s$ where $s=30°$ is the length of one zodiacal sign. Now the length of daylight is the rising time of the semicircle from $\lambda+s$ to $\lambda+s+180$. Its value can be obtained from the previous value by adding the rising time of the zodiacal sign $s+180$ and by subtracting the rising time of s. Hence we see that the length of daylight can be found if the rising times of ecliptic arcs are known. The rising time of a given arc of the ecliptic depends obviously on its variable inclination to the horizon. To obtain an insight into the relationship between rising times, seasons and geographical location can be called the central problems of early Greek spherical astronomy. Its complete solution is found in the tables of the *Almagest* where the rising times are given for every 10 degrees for all latitudes whose longest daylight varies between 12^h and 17^h in steps of $1/2^h$.

Ptolemy already makes full use of spherical trigonometry. The Babylonian astronomers, however, used also here arithmetical schemes to describe the values of the rising times of the zodiacal signs as function of the longitude. Two methods were developed: a crude approximation related to system A of the solar theory, and a more refined scheme in system B. Both are built on arithmetical progressions. The Greeks expanded these methods by varying the parameters linearly, thus introducing geographical zones of given length of daylight, known as "climates." This concept remained fundamental for ancient and mediaeval mathematical astronomy (Honigmann [13], Neugebauer [29]). Ptolemy himself uses the rising times of system A in the Tetrabiblos ([40, I, 20 p. 94–95]; [39, I, 21 p. 46]) and thus contributed to securing the survival of Babylonian methods for many centuries.

6. **Spherical trigonometry.** The lack of a convenient algebraic notation prevented the Greeks from condensing the solution of a general triangle into a single formula instead of solving two right triangles. Of greater consequence was their use of chords in a circle of radius 60 instead of the trigonometric functions. Consequently their plane

trigonometry is expressed in the relations

(1)
$$a = \frac{c}{120} \operatorname{crd}(2\alpha),$$

(2)
$$b = \frac{c}{120} \operatorname{crd}(180 - 2\alpha),$$

(3)
$$\frac{a}{b} = \frac{\operatorname{crd}(2\alpha)}{\operatorname{crd}(180 - 2\alpha)},$$

(4)
$$c^2 = a^2 + b^2.$$

Tables for crd α are given, for example, in *Almagest* I, 11 in steps of $1/2°$ and with an accuracy of two sexagesimal places. Their computation is based on the so-called Ptolemaic theorem for a quadrilateral and its diagonals inscribed in a circle.

As plane trigonometry is described by the four above relations (1) to (4), so spherical trigonometry contains four similar relations which we may describe in our symbols by

(1a)
$$\sin \alpha = f(a, c),$$

(2a)
$$\cos \alpha = g(b, c),$$

(3a)
$$\tan \alpha = h(a, b),$$

(4a)
$$\cos c = \cos a \cos b.$$

Menelaus, about 100 A.D., already knew that a spherical triangle is determined by its angles (Spherics I, 18 Krause [18, p. 138]). Yet relations of the type

(5a)
$$\cos a = \phi(\alpha, \beta),$$

(6a)
$$\cos c = \psi(\alpha, \beta)$$

do not seem to have been discovered before the Arabs.

In general, it is my impression that spherical trigonometry was completed rather late in the development of Greek science. Its problems and methods were exclusively determined by astronomical needs and astronomical concepts. This is obvious in the earliest treatises, about 300 B.C., by Autolycus [2] and Euclid [9], but it also holds for Theodosius ([49], Schmidt [46]), who probably was a younger contemporary of Hipparchus (Ziegler [56]). The astronomical importance of parallel circles certainly contributed to obscuring for a long time the necessity of restricting oneself to great circles, a discovery which very well may have been Menelaus' great contribution. Even Ptolemy did not yet have a clear concept of the possibility of replacing spherical triangles by plane triangles. In the computation of the components of parallax in *Almagest* V, 19 (Heiberg [35, p. 456 f.]), he treats a right spherical triangle with one very small angle, but with two large sides, as if it were a plane triangle in order to find the second angle at the small base, though this is just the case where also the second angle is very close to 90°. It has often been conjectured that already Hipparchus was able to solve problems of spherical trigonometry. This seems rather implausible in view of the above-stated facts, and indeed we shall see (p. 1036) that there existed methods to avoid completely spherical trigonometry. Hence I see no reason for assuming that the central theorem of ancient and Arabic spherical trigonometry, the "Menelaus theorem," was known before Menelaus. That its two forms are not independent was remarked by Theon ([43, p. 569], Rome [44]).

7. **The "Analemma."** At least since the early part of the third century B.C., astronomers were able to predict solar and lunar positions with a high degree of accuracy. In particular, the longitude of the sun could be considered as known for any given date. Ecliptic coordinates, however, are not directly visible in the sky. The daily rotation moves all celestial objects around the pole of the equator, and thus relates the measurement of time to equatorial coordinates. Yet local noon is again determined by the sun and thus, by the relation of ecliptic coordinates to the local coordinates of the observer, horizon and meridian. All these coordinates play a role in the practical measurement of time by means of sun dials, the simplest form of which is the vertical "gnomon" on a horizontal plane. Finally, different observers had to establish their relative positions through the determination of their geographical coordinates. It is therefore no great wonder that we can observe that ancient astronomers concentrated a great deal of attention on the theory of spherical coordinates in their relation to celestial and terrestrial objects and the theory of sun dials.

Ptolemy's role in this branch of astronomy can be well appreciated because we have not only a work of his own, called the *Analemma* ([38, p. 187 ff.], Luckey [22]) but we also know a little about his predecessors. We can see that Ptolemy had all the essential methods, inherited from earlier times, but loaded down with historical relics which made their application unnecessarily clumsy. Ptolemy rationalized the whole procedure, down to the smallest details. First of all, he introduced coordinates whose mutual relations are independent of the geographical position. We consider the octant of the celestial sphere which contains the sun Σ, and whose vertices are the zenith Z, South S, and East E (cf. Fig. 8 left). From each of these points we draw a great circle to the sun, calling the respective arcs "descensivus" ($=$ zenith distance), "horarius" and "hectemoros." Each pair of these arcs may be used to determine the position of the sun.

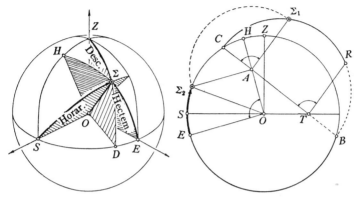

FIG. 8

This perfectly symmetric arrangement can be compared with the older system. There the position of the sun was given by the horarius and the arc of the equator from E to the plane of the horarius. Only for the equinoxes does this arc end at the sun (then being identical with the hectemoros), but ordinarily it meets the horarius in a point which is not only different from the sun but which also depends on the geographical latitude of the observer. Another pair of "old" co-ordinates was descensivus and "antiskion," that is, the azimuth of the direction of the shadow counted from S to D.

Ptolemy now treats in great detail the methods to determine his coordinates for a given solar position. The right part of Fig. 8 describes his solution for the hectemoros. Let the circle around O be the plane of the meridian. The geographical latitude of the observer determines the inclination of the equator, and from the solar tables we can find the longitude of the sun for the given moment. As we shall see presently, this also determines the position of the path of the sun, henceforth called its "daily circle." Its intersection with the plane of the meridian may be CB. Ptolemy now proceeds in the typical fashion of "descriptive geometry." He revolves the plane of the daily path of the sun about its trace CB into the plane of the meridian. C is the culminating point; vertically above A, we have the given position Σ_1 of the sun which rose at R, the arc BR being below the horizon. Because A is the orthogonal projection of the sun onto the plane of the meridian, we know that A belongs also to the plane of the hectemoros. Because also O belongs to both the plane of the meridian and the plane of the hectemoros we have in OAH the trace of the plane of the hectemoros in the plane of the meridian. Perpendicular to this trace is the line OE where E represents the East point of the horizon, revolved about OH into the plane of the meridian. The distance $A\Sigma_1$ is a true distance. Using it as a radius we find on the celestial sphere Σ_2, which gives the position of the sun in the plane of the hectemoros, turned into the plane of the meridian. Thus the arc $E\Sigma_2$ is the true "hectemoros" we wanted to find.

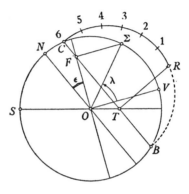

Fig. 9

A similar procedure can be followed for the other angles. The method consists, in general, in constructing first the Cartesian coordinates of the sun with respect to the fundamental planes of the octant ZSE and then revolving the Cartesian coordinate planes about the coordinate lines in order to get the angles in true size.

It follows from the preceding discussion that the whole construction of all Ptolemaic coordinates requires only the main circle with center O and the parallel circle whose diameter CB is determined by the time of the year, that is, by the longitude of the sun. Ptolemy thus proceeds to construct a nomogram whose central part is the great circle with center O. Added are half circles of the type BC corresponding to the solar path from month to month. This system of circles is mounted on a turntable with various scales for the reading of the angles. This makes it possible to bring the equator in the proper position towards a given horizon. Rectangular plates placed on the proper points of the scales then allow the direct determination of the angles required (cf. for details and for a figure Luckey [22]). A mechanized age can only be delighted with the extremal efficiency of this apparatus, of which nothing has come down to us except for an incomplete description in a Greek palimpsest and a poor Latin translation by William of Moerbeke.

Ptolemy is not the inventor of descriptive geometry. More than a century before his time, Vitruvius describes in his *De architectura* the Analemma in connection with the construction of sun dials ([54]). From Vitruvius we also learn how to determine graphically the path of the sun for a given longitude λ. Let the circle around O again be the meridian (Fig. 9); then the trace ON of the equator is given for a given place. If ϵ denotes the obliquity of the ecliptic, we also know the trace OF of the ecliptic. Revolving the plane of the ecliptic into the plane of the meridian we find the vernal point V on the end point of the radius OV perpendicular to FO. Hence Σ is the sun for given λ, and FB the trace of its path. If we divide RC into six equal parts we have the position of the sun for each seasonal hour from sunrise (R) to noon (C).

The theory of sun dials leads to the solution of two problems in the history of mathematics, problems whose fame is inversely proportional to their interest: the date of Heron of Alexandria and the origin of the conic sections.

It is in itself of very little importance to establish accurately the date of a rather mediocre author whose role in the history of science is due only to the fact that so much else is lost. Yet it is somewhat unsatisfactory to know no more about an often quoted writer than that he lived sometime between -200 and $+300$. It is therefore a pleasant side result of the study of an analemma, described by Heron in his *Dioptra* ([12], Rome [42], Neugebauer [28]), to see that the elements which he quotes for a lunar eclipse fit exactly one and only one eclipse between -200 and $+300$, namely the partial eclipse (magnitude 8 digits) of A.D. 62 March 13. It is extremely plausible to assume that the writing of the Dioptra fell close to the occurrence of this eclipse. This is indeed the only excuse for selecting an example whose date coincides almost with equinox, because the main problem in the analemma discussed by Heron consists in reducing local seasonal time of one place (Rome) to the local seasonal time of another locality (Alexandria); this problem loses its importance only twice a year, namely at the equinoxes, when seasonal time and equinoctial time coincide. Only the desire to quote a real eclipse, which had occurred recently, can have led to quoting as an example a date which fell only a few days before equinox. If we thus must place Heron into the end of the first century A.D. we support at the same time an argument of J. Klein [16, p. 135, note] who conjectured that Heron and Diophantus were contemporary, both belonging to the time of Nero.

For us the interest in Heron's analemma consists in the direct relationship it establishes between astronomy and mathematical geography. The only method in antiquity to determine the geographical longitude of a place consisted in using lunar eclipses as time signals. The comparison of two local seasonal times is made by Heron by means of a hemispherical sun dial combined with an analemma of the Vitruvian type. Its main idea is to use the analemma for each place and thus to find the position of the local meridian with respect to the sun. Transferring the results into the hemispherical dial gives directly the angular difference between the meridians. This is one of several examples which show the intimate combination of geometrical construction and the direct use of globes or hemispheres in Greek astronomy. In general it may be said that Greek mathematics is often much less "pure" than is generally assumed.

This latter remark might be kept in mind when we try to relate one of the most interesting subjects of Greek geometry, the conic sections, to an astronomical origin. It is well known that these curves were defined by Apollonius (about 200 B.C.) as the intersection of a circular cone by planes of variable inclination, whereas the "old" geometers considered only right circular cones, intersected by a fixed plane perpendicular to one generating line. The different types of these curves were obtained by varying the angle at the vertex (cf., for example, Heath [1]). This definition suggests immediately that its origin is to be sought in the fixed right angle between the gnomon and the receiving plane of the shadow. Indeed, we have only to point the gnomon towards the culminating point to obtain exactly the configuration required (Neugebauer [30]). The sun travels on its "daily circle," its rays form the right circular cone, whose vertex is the tip of the gnomon. At noon the gnomon falls into one generating line and the receiving plane is perpendicular to it. The shadow describes a hyperbola. The natural question, namely, how these curves depend on the declination, is equivalent to asking how the angle at the vertex of the cone influences the shape of the curve. This is exactly the form in which the conic sections were studied by Menaechmus (about 350 B.C.).

8. **The "Planisphaerium."** The importance of spherical astronomy is reflected in the manifold of mathematical tools developed in order

to solve its main problems. Eventually spherical trigonometry superseded all other methods which were, in all probability, invented before spherical geometry was sufficiently far advanced to reach numerical results. At that early stage the methods of descriptive geometry,
reflected in the *Analemma*, might seem the most natural way to transform arcs on the sphere into arcs of one plane. Another work of

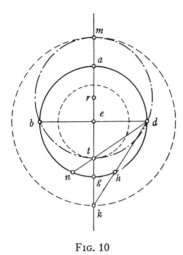

FIG. 10

Ptolemy, the *Planisphaerium* ([38, p. 227–259]; Drecker [5]), shows,
however, that also stereographic projection was known, in particular
its important quality of mapping circles into circles, straight lines
included. No proof is given for this fact by Ptolemy, a clear indication
that he is operating on well known grounds, developed long before
his time.

The projection chosen maps the whole sphere onto the plane of the
equator with the south pole as center of projection. In order to determine the center and the radius of the image of a circle, descriptive
methods are again employed. As a simple example might be quoted
the determination of the solstitial circles and of the ecliptic. Let *abgd*
be the equator (Fig. 10); we then consider the same circle also as a
picture of the meridian, turned about the diameter *ag* into the plane
of the equator. Consequently *d* represents the south pole. Let $gh = gn$
$= \epsilon = 23;51°$. The point *h* is therefore a point of the diameter of the
winter solstitial circle, *n* of the summer solstitial circle. Their projections from *d* are *k* and *t* respectively. Because all parallel circles have
their center in *e*, the images of both circles are found. Because the
ecliptic must be represented by a circle touching the solstitial circles

in *t* and *m* respectively, also this circle is known. It is then proved that the straight line *bd*, *b* and *d* being the equinoxes, passes through *e*, a general relation which is often used in the following for intersecting great circles. The conformity of the mapping was apparently unknown.

The main goal of the whole procedure is the determination of the rising times for the zodiacal signs. In order to find the representation of these signs one has only to construct the parallel circles of given declination by the same method which was used for the solstitial circles. The declinations of given ecliptic points are considered known, but it is in principle important to remark that they can also be found by geometric construction, namely from the "daily circle" in the analemma (cf. above p. 1031). Finally it is easy to construct the circle which represents the horizon for a given latitude. The variable positions of the ecliptic with respect to the horizon at different times of the year are in our projection represented by different positions of the horizon circle with respect to the fixed image of the equator-ecliptic system. In order to find the rising times of a given arc of the ecliptic we have only to construct the two positions of the horizon passing through its end points. These two horizon circles intersect the equator in two points whose angular distance is the rising time in question. Because angles on the equator are represented without distortion, our problem is solved by this construction.

The above description shows that the "planisphaerium" could be used for a purely graphical or mechanical solution of problems of spherical astronomy. This was indeed the use made of this method especially by the Arabs whose "astrolabes" are based on the projections described here (Drecker [6] and [7], Michel [23]). In Ptolemy's treatise, however, a different attitude is taken. The geometric constructions are only used for transforming spherical problems into problems of plane geometry which then are solved numerically by means of plane trigonometry. Obviously we have here before us the method used before spherical trigonometry was invented, that is, before the Menelaus theorem was known.

Much speaks in favor of the assumption that the planisphaerium was the tool of Hipparchus (Delambre [4, II p. 453 ff.], Drecker [6, p. 16 ff.]). All computations are based on the latitude of Rhodes, where Hipparchus made his observations. Synesius of Cyrene ascribes the invention of the "astrolabe" to Hipparchus (FitzGerald [10, p. 263]); this statement is certainly to be taken seriously in view of the fact that Synesius was a pupil of Hypathia, who collaborated with her father Theon on the commentaries to the Almagest (Rome [43,

p. LXXXIII]). Finally, Hipparchus determines the positions of stars by a combination of ecliptic and equator coordinates (Vogt [55]); he takes the longitude of the point where the circle of declination through the star meets the ecliptic and then uses the remaining declination as second coordinate. This system finds its direct explanation in the planisphaerium: the first coordinate is given on the image of the ecliptic whereas the circles of declination are mapped into radii.

9. **Map projection.** The work of Mžik and Hopfner [25] has given us a good understanding of Greek mathematical geography as con-

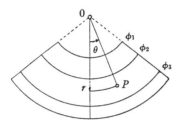

Fig. 11

tained in the first book of Ptolemy's *Geography*. I shall give a short summary, using modern terminology, which will suffice to show how far the Greeks had come in the general problem of mapping a sphere onto a plane. From the earlier development we only know that Ptolemy's predecessor, Marinus (about 100 A.D.), used a cylinder projection which can be described by

$$y = \phi, \qquad x = \lambda \cos 36°$$

where x and y are the Cartesian coordinates of the map, ϕ the geographical latitude, λ the geographical longitude, and $\phi = 36°$ the latitude of Rhodes. Obviously this projection preserves latitudes on all meridians and longitudes for the parallel of Rhodes.

Ptolemy introduced two types of conic projection. In the first type he maps meridians on radii, parallels of latitude on circles with center O (Fig. 11). In order to determine the parameters of this mapping three conditions are imposed. We introduce polar coordinates r and θ, and call $\bar{\phi} = 90 - \phi$ the colatitude.

Condition 1: preservation of length on all meridians

(1) $$r = \phi + c.$$

Condition 2: preservation of length for the latitude $\phi_2 = 36°$ of

Rhodes. Thus

$$r(\bar{\phi}_2) \cdot \theta = \sin \phi_2 \cdot \lambda$$

or from (1)

(2)
$$\theta = \frac{\sin \bar{\phi}_2}{\bar{\phi}_2 + c} \cdot \lambda.$$

Condition 3: preservation of the ratio of lengths on the parallels

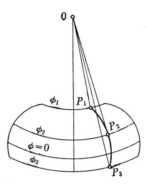

FIG. 12

of Thule ($\phi_1 = 63$) and on the equator ($\phi_3 = 0$). Hence c can be determined from

(3)
$$\frac{\sin \bar{\phi}_1}{\sin \bar{\phi}_3} = \frac{\bar{\phi}_1 + c}{\bar{\phi}_3 + c}.$$

It is obvious that this projection is an improvement of Marinus' simple cylinder projection.

The second type of conic projection, proposed by Ptolemy, assumes that both meridians and parallels of latitude are represented by circles. We again use polar coordinates (Fig. 12) but only the circles $r = \text{const.}$ now have geographical significance as images of circles of constant latitude. The radii, however, are no longer meridians. Nevertheless it is required that length is preserved on all radii. Thus we have

(4)
$$r = \bar{\phi} + c_0$$

where c_0 is an arbitrary parameter which gives the value $r = r_0$ of the image of the equator. In order to determine the circles which represent the meridians we need three points. Thus we can impose the following three conditions: preservation of length on each of the following three parallels $\phi_1 = 63$ (Thule), $\phi_2 = \epsilon = 23;50$ (Syene in Egypt), and $\phi_3 = -16;25$ ($\phi = 16;25$ is the latitude of Meroe on the Upper Nile). Thus

$$(5) \qquad\qquad r(\bar{\phi}_i) \cdot \theta = \sin \bar{\phi}_i.$$

For a given value of λ we can find three values of θ by substituting in (5) the values $\bar{\phi}_1 = 27$, $\bar{\phi}_2 = 66;10$ $\bar{\phi}_3 = 106;25$. The values $r(\bar{\phi}_i)$ are similarly given by (4); thus we are able to construct the circle $P_1P_2P_3$ which represents the meridian λ. If (5) were assumed to hold for all latitudes, the meridians would no longer be circles but we would obtain the Bonne projection, which is preserving length in radial direction and on all parallels.

In concluding, I want to repeat that the topics discussed here were not intended to give a complete picture of mathematical problems which originated from problems of astronomical character. Nevertheless, they may suffice as an illustration of the close relationship between mathematics and astronomy in antiquity, a relationship which thereafter never lost its importance. "Astronomy proceeds to its demonstrations in no uncertain way, for it uses as its servants geometry and arithmetic, which it would not be improper to call a fixed standard of truth" (Synesius of Cyrene in his letter to Paeonius, about 400 A.D.; FitzGerald [10, p. 262]; cf. *Almagest* I, 1; Heiberg [1, p. 6, 17 ff.]).

BIBLIOGRAPHY

1. Apollonius of Perga, *Treatise on conic sections,* ed. in modern notation by T. L. Heath, Cambridge University Press, 1896.

2. Autolyci *de sphera quae movetur liber, de ortibus et occasibus libri duo,* ed. Fr. Hultsch, Leipzig, Teubner, 1885.

3. Paul Boelk, *Darstellung und Prüfung der Mercurtheorie des Claudius Ptolemaeus,* Thesis, Halle, 1911.

4. Delambre, *Histoire de l'astronomie ancienne,* Paris, 1817.

5. J. Drecker, *Das Planisphaerium des Claudius Ptolemaeus,* Isis vol. 9 (1927) pp. 255–278.

6. ———, *Des Johannes Philoponus Schrift über das Astrolab,* Isis vol. 11 (1928) pp. 15–44.

7. ———, *Hermannus Contractus, Über das Astrolab,* Isis, vol. 16 (1931) pp. 200–219.

8. Pierre Duhem, *Le système du monde, histoire des doctrines cosmologiques de Platon à Copernic,* 5 vols., Paris, Hermann, 1913–1917.

9. Euclidis *opera omnia,* vol. 8, *Phaenomena et scripta musica,* ed. H. Menge, Leipzig, Teubner, 1916.

10. Augustine FitzGerald, *The letters of Synesius of Cyrene,* Oxford University Press, 1926.

11. Heiberg, see Ptolemaeus [35].

12. Herons von Alexandria, *Vermessungslehre und Dioptra*, Griechisch und deutsch von Hermann Schöne, Leipzig, Teubner, 1903.

13. Ernst Honigmann, *Die sieben Klimata*, Heidelberg, Winter, 1929.

14. Arthur S. Hunt, *Catalogue of the Greek papyri in the John Rylands Library in Manchester*, vol. I, *Literary texts* (*Nos. 1–61*), Manchester, 1911.

15. Paul Kempf, *Untersuchungen über die Ptolemäische Theorie der Mondbewegung*, Thesis, Berlin, 1878.

16. Jacob Klein, *Die griechische Logistik und die Entstehung der Algebra*. II, Quellen und Studien z. Geschichte der Mathematik B vol. 3 (1934) pp. 122–235.

17. Erik J. Knudtzon and O. Neugebauer, *Zwei astronomische Texte*, K. humanistiska Vetenskapssamfundets i Lund Årsber. 1946–1947 II, pp. 77–88.

18. Max Krause, *Die Sphärik von Menelaos aus Alexandrien in der Verbesserung von Abū Naṣr Manṣūr b. 'Alī b. 'Irāq*, Abhandlungen Gesellschaft der Wiss. zu Göttingen, Philol.-hist. Kl., 3. Folge, vol. 17, 1936, Berlin, Weidmann.

19. Franz Xaver Kugler, *Die babylonische Mondrechnung*, Freiburg, Herder, 1900.

20. ———, *Sternkunde und Sterndienst in Babel*, Münster, Aschendorff, 2 vols., 1907–1924; 3 Ergänzungshefte 1913–1935 (No. 3 by Johann Schaumberger).

21. Albert Lejeune, *Les tables de réfraction de Ptolémée*, Annales de la Société scientifique de Bruxelles (1) vol. 60 (1946) pp. 93–101.

22. P. Luckey, *Das Analemma von Ptolemäus*, Astron. Nachrichten vol. 230 (1937) cols. 17–46.

23. Henri Michel, *Traité de l'astrolabe*, Paris, Gauthier-Villars, 1947.

24. August Ferdinand Möbius, *Gesammelte Werke*, Leipzig, Hirzel, 1887.

25. Hans v. Mžik and Friedrich Hopfner, *Des Klaudios Ptolemaios Einführung in die darstellende Erdkunde*, Klotho vol. 5 (1938).

26. O. Neugebauer, *Vorgriechische Mathematik*, Berlin, Springer, 1934.

27. ———, *Untersuchungen zur antiken Astronomie*, III, Quellen und Studien zur Geschichte der Mathematik, B vol. 4 (1937) pp. 193–346.

28. ———, *Über eine Methode zur Distanzbestimmung Alexandria—Rom bei Heron*, Kgl. Danske Videnskabernes Selskab, Historisk-Filologiske Meddelelser vol. 26, 2 (1938) and vol. 26, 7 (1939).

29. ———, *On some astronomical papyri and related problems of ancient geography.* Transactions of the American Philosophical Society, n.s. vol. 32, 2 (1942) pp. 251–263.

30. ———, *The astronomical origin of the theory of conic sections*, Proceedings of the American Philosophical Society vol. 92 (1948) pp. 136–138.

31. ———, *Astronomical cuneiform texts*, in preparation.

32. O. Neugebauer and A. Sachs, *Mathematical cuneiform texts*, American Oriental Series, vol. 29, New Haven, 1945.

33. J. A. C. Oudemans, *Lösung des sog. Pothenotschen, besser Snellius'schen Problems von Ptolemaeus*, Vierteljahrsschrift d. Astronomischen Gesellschaft. vol. 22 (1887) pp. 345–349.

34. A. Pannekoek, *Calculation of dates in the Babylonian tables of planets*, Akademie van Wetenschappen te Amsterdam, Proceedings vol. 19 (1916) pp. 684–703.

35. Claudii Ptolemaei *Opera quae extant omnia*, vol. I, *Syntaxis mathematica*, ed. J. L. Heiberg, Leipzig, Teubner, 1898, 1903.

36. Des Claudius Ptolemäus *Handbuch der Astronomie*, aus dem Griechischen übersetzt von Karl Manitius, Leipzig, Teubner, 1912, 1913.

37. *Composition mathématique* de Claude Ptolémée, trad. par M. Halma et suivie des notes de of M. Delambre, Paris, 1813, 1816 [reprinted Paris, Hermann, 1927].

38. Claudii Ptolemaei *opera quae extant omnia*, vol. II, *Opera astronomica minora*, ed. J. L. Heiberg, Leipzig, Teubner, 1907.

39. Claudii Ptolemaei *opera quae extant omnia*, vol. III, 1 *Apotelesmatica*, ed. F. Boll-Ae. Boer, Leipzig, Teubner, 1940.

40. Ptolemy, *Tetrabiblos*, transl. by F. E. Robbins, The Loeb Classical Library, 1940.

41. Claudii Ptolemaei *geographia*, ed. C. F. A. Nobbe, Leipzig, 1843.

42. A. Rome, *Le problème de la distance entre deux villes dans la Dioptra de Héron* Annales de la Société scientifique de Bruxelles vol. 42 (1922–1923) Mémoires pp. 234–258.

43. ———, *Commentaires de Pappus et de Théon d'Alexandrie sur l'Almageste*, Roma and Città del Vaticano, 1931, 1936, 1943 (Studi e Testi 54, 72, 106).

44. ———, *Les explications de Théon d'Alexandrie sur le théorème de Ménélas*, Annales de la Société scientifique de Bruxelles, ser. A, vol. 53 (1933) CR pp. 39–50.

45. Schaumberger, see Kugler [20].

46. Olaf Henric Schmidt, *Studies on ancient sphaeric*, Thesis, Brown University, 1942.

47. Paul Schnabel, *Berossos und die babylonisch-hellenistische Literatur*, Leipzig, Teubner, 1923.

48. C. J. Schumacher, *Untersuchungen über die ptolemäische Theorie der unteren Planeten (Merkur und Venus)*, Münster, Aschendorff, 1917.

49. Theodosius Tripolites, *Sphaerica*, ed. J. L. Heiberg, Abhandlung der Gesellschaft der Wissenschaften zu Göttingen, Philologisch-Historische Klasse, NF 19, 3 (1927), Berlin, Weidmann.

50. Theonis Smyrnaei Platonici *liber de astronomia*, ed. Th. H. Martin, Paris, 1849.

51. Johannes Tropfke, *Geschichte der Elementar-Mathematik*, vols. 1 to 4 in 3d ed., 1930–1940, vols. 5 to 7 in 2d ed., 1923, 1924, Berlin-Leipzig, De Gruyter.

52. B. L. van der Waerden, *Zur babylonischen Planetenrechnung*, Eudemus vol. 1 (1941) pp. 23–48.

53. ———, *Egyptian "Eternal Tables,"* Nederlandsch Akademie van Wetenschappen, Proceedings vol. 50 (1947) pp. 536–547, 782–788.

54. Vitruvius, *On Architecture*, transl. by Frank Granger, The Loeb Classical Library, 1945.

55. H. Vogt, *Versuch einer Wiederherstellung von Hipparchs Fixsternverzeichnis*, Astronomische Nachrichten vol. 224 (1925) cols. 17–54.

56. Konrat Ziegler, *Theodosius*, Paulys Real-Encyclopädie der classischen Altertumswissenschaften, vol. 5 A, cols. 1930–1935.

BROWN UNIVERSITY

Reprinted from
Bulletin of the American Mathematical Society
54 (11), 1013–1041 (1948)

THE TRANSMISSION OF PLANETARY
THEORIES IN ANCIENT AND
MEDIEVAL ASTRONOMY

BY O. NEUGEBAUER

1. In Toledo in the year A.D. 1068 Abū'l-Qāsim Ṣā'id ibn Aḥmad, also known as Qāḍī Ṣā'id, wrote a book entitled *The Categories of Nations*.[1] In this work he discusses the people of the world from the viewpoint of their interest in scientific research, stating that "the category of nations which have cultivated the sciences forms the élite and the essential part of the creations of Allah."[2] Eight nations belong to this class: "the Hindus, the Persians, the Chaldeans, the Hebrews, the Greeks, the Romans, the Egyptians, and the Arabs."[3] Taking into account some terminological differences, this list can still be considered fairly complete. What Ṣā'id calls "the Romans" and "the Egyptians" in part coincides with what we would call the Byzantines and the Alexandrian School, while Rome and Egypt in our sense of these words could not compare in importance with India and the Hellenistic or Muslim contributions. But it is wise not to forget that the existence of the Roman Empire was an essential condition for the transmission of Hellenistic science to the Muslim world.

Chronologically the interactions between the leading civilizations are far more complex than Ṣā'id's geographical enumeration would lead us to believe, when he starts with the Hindus in the East and ends with Islamic Spain in the West. We know today that Babylonian astronomy reached a scientific level only a century or two before the beginning of Greek astronomy in the fourth century B.C. The development of Hellenistic astronomy is largely unknown to us except for its last perfection by Ptolemy in the time of Hadrian and Antoninus Pius (probably completed shortly after 141 A.D.). Then, about three centuries later, Indian astronomy appears on the scene, deeply influenced by Greek methods, confronting us with the problem of transmission from West to East, a problem which is made particularly difficult by our ignorance about possible Persian intermediaries. Centuries later, under the Abbasids in Baghdad in the middle of the ninth century, Islamic astronomy begins, influenced from India as well as from

3

Hellenistic sources.   While the Greek component rapidly became dominant in the eastern part of the Muslim world, from Egypt to Persia, the outmost West retained in part methods of Hindu astronomy which left their traces as late as 1475 with Regiomontanus.[4]

It is, of course, impossible adequately to deal with so vast a problem within the framework of one lecture.   All I can attempt to do is to select certain aspects of the history of planetary theories which seem characteristic for some major trends in the development of astronomical ideas and methods during antiquity and the middle ages.

2.   To a modern reader an obvious goal of any "planetary theory" would appear to be the solution of the problem: find, for any given moment t, the position of a planet, that is, its geocentric longitude $\lambda(t)$ and latitude $\beta(t)$ since these are the quantities which are directly accessible to observation.   In fact, however, this formulation is only the result of a considerable sophistication in Hellenistic astronomy. It is preceded by two more "natural" and therefore much more complex problems: (*a*) the Babylonian (very successful) attempts to predict the dates of the first and last visibilities of the planets and (*b*) the Greek search for definite cinematical models which could explain planetary motion.

I shall not be able to describe in any detail either one of these earlier theories but I have to say enough to make their impact on the mediaeval systems intelligible.

We know extremely little about the earlier phases of Babylonian astronomy but it is clear that a mathematical approach to the prediction of lunar and planetary phenomena was not developed before the fifth century B.C., which gives it a very narrow margin of priority over the corresponding Greek steps.   Nevertheless there can be no doubt of the complete independence of Babylonian methods from Greek procedures.

It seems plausible to assume that the Babylonian approach to the planetary theory originated from their lunar theory which was centered around the fundamental problem of the calendar, i. e., the prediction of first and last visibility.   Exactly as the lunar theory tried to answer the question of the length of the lunar month, that is, the time between two consecutive appearances of the new crescent, so did the planetary theory pursue the problem of finding the laws which govern the variations in the length of the intervals in time and longitude between consecutive appearances or disappearances of a planet.[5]

Problems of this type are of great difficulty since they depend not only on the variable velocities of sun and planet in their orbits but also on problems of spherical astronomy (variable inclination between ecliptic and horizon) and on visibility conditions near the horizon which are very difficult to evaluate, even under modern conditions. It is therefore quite astonishing to see to what degree Babylonian astronomers were able to devise effective schemes for the variations as functions of λ of the "synodic arc" $\Delta\lambda$ which separates two consecutive planetary phenomena of the same kind.

The function $\Delta\lambda$ must satisfy several conditions, most of all the exact preservation of the fundamental relation between synodic and sidereal period, combined with the proper amplitudes for the deviations from the mean value of $\Delta\lambda$ and the observed dependence on λ. This was achieved by means of arithmetical devices which can be described, in modern terminology, as representing $\Delta\lambda$ by a step function of λ, or—in a simplified fashion—by a periodic zigzag function. For our present purpose we need not discuss the methods by which the characteristic parameters of these functions were determined. It suffices to emphasize the fact that the Babylonian methods are strictly arithmetical in character and not the consequences of a geometrical model of the planetary motion. It is this feature which suggests ultimate dependence on Babylonian methods whenever it appears later in certain parts of Greek or Hindu and Islamic astronomy.[6]

3. The earliest Greek model designed to account for the appearance of planetary motion was constructed by the famous mathematician Eudoxus of Cnidos (middle of fourth century B.C.). Its basic idea consists in the device to keep a point on the equator of a rotating sphere within a limited area of the sphere by placing its axis as diameter into another concentric sphere which rotates with opposite angular velocity about an inclined axis. By a very simple consideration it can be shown[7] that the resulting curve has the general shape of a lemniscate. This result granted, one only has to add a third uniform rotation in the direction of the axis of the lemniscate to produce a motion which resembles planetary motion: a general "mean" progress combined with periodically recurrent retrograde loops.

It was clear from the outset that such a model could not be more than a qualitative description of the actually observed orbits, since it is an obvious quality of the Eudoxan model that all loops would be congruent and equidistantly spaced. To use the terminology of our discussion of the Babylonian methods: for a planet of the Eudoxan type the synodic arcs $\Delta\lambda$ would be constant, contrary to the empirical fact that $\Delta\lambda$ varies with λ. Notwithstanding such glaring inadequacies this model had a profound impact on planetary theory by demonstrating that superimposition of uniform rotations is capable of producing retrogradations and deviations in latitude. The fact that

Aristotle accepted the principle of Eudoxus' model for his own picture of the world kept it alive with Aristotle's works, resulting in its revival in Spain in the twelfth century by al-Biṭrūjī and thence its spread to the West in Michael Scot's Latin translation "De motibus celorum," completed in Toledo in 1217.[8] In different variants it exercised influence deep into the Renaissance[9] on philosophically minded persons whose convictions about the simplicity or uniformity of the plans of the Creator of the universe were stronger than their desire for numerical agreement between theory and observations.

4. The theory of homocentric spheres does not seem to have reached India, probably because the vehicle of transmission of Greek astronomy to the East seems to have been astrology and not philosophy. Astrology had its main development during the late Hellenistic and early Roman period and makes use both of the Babylonian arithmetical methods and Greek procedures based on eccentric and epicyclic motion. How these cinematic models were developed is known to us only in the most fragmentary outlines. It seems to me a fair conjecture that the mere possibility of a geometric interpretation, opened by Eudoxus, stimulated further attempts. This much is certain: by about 200 B.C. Apollonius had investigated the relationship between eccentric and epicyclic motion and had found the points at which the planet becomes "stationary" between arcs of direct and retrograde motion. About half a century later, Hipparchus aimed for an accurate numerical application of these methods to the solar and lunar motion but he did not think the time ripe for dealing in a definitive fashion with the planetary motion.

Almost 300 years after Hipparchus, about 140 A.D., Ptolemy presented a consistent planetary theory which in its mathematical elegance and efficiency was not superseded until Kepler.[10] But it was not only in the planetary theory that Ptolemy went beyond Hipparchus; also in the theory of the motion of the moon he discovered, by a masterful analysis of intricate observational data, the fact that the Hipparchian theory, which was based on data furnished by lunar eclipses, showed systematic deviations in the parts of the orbit which correspond to the half moons. He succeeded in a modification of the original model of an eccentric epicycle by means of a crank mechanism which changes, with the proper phase, the eccentricity of the orbit in such a way that the variation of the longitudes would be correctly represented. Unfortunately the distances become essentially wrong except for new-moons and full-moons, a deficiency of the model which was eventually corrected by means of a secondary epicycle by the Arabic astronomer Ibn ash-Shāṭir[11] (about 1350). Exactly the same solution was found about 150 years later, by Copernicus and then unluckily extended to his model of planetary motion.

Ptolemy's modification of the lunar theory is of importance for the problem of transmission of Greek astronomy to India. The essentially Greek origin of the Sūrya-Siddhānta and related works cannot be doubted—terminology, use of units and computational methods, epicyclic models as well as the local tradition—all indicate Greek origin. But it was realized at an early date in the investigation of Hindu astronomy[12] that the Indian theories show no influence of the Ptolemaic refinement of the lunar theory. This is confirmed by the planetary theory, which also lacks a characteristic Ptolemaic construction, namely, the "punctum aequans," to use a mediaeval terminology.[13]

Since we know extremely little about the development of Greek astronomy before Ptolemy it is of particular interest to find in Hindu astronomy systems which were influenced by pre-Ptolemaic methods. This would seem to suggest roughly a time of reception near the beginning of our era, a period for which close contact between the Roman empire and India is now well established.[14] On the other hand serious arguments have been brought forth to date the formation of Hindu astronomy closer to the time of Āryabhaṭa and Varāha Mihira, i. e., nearer to 500 A.D.[15] Though one might explain such a delayed reception of pre-Ptolemaic methods by their preservation in astrological treatises it seems to me useful to be aware of the fact that dates obtained by comparison with modernly computed longitudes are not secure as long as there are some doubts possible as to the ancient definition of the longitudes—experience with Babylonian astronomy may serve as a warning.[16]

In early Hindu astronomy, as summarized by Varāha Mihira in the Pañca Siddhāntikā, two distinct methods of approach can be distinguished: the "trigonometric" methods, best known through the Sūrya Siddhānta, and the "arithmetical" methods whose identity with the Babylonian methods of the Hellenistic period has become clear only in recent years.[17] It was possible to show that the South Indian (Tamil) tradition, which was still alive in the seventeenth and eighteenth centuries when India was reached by European colonization, is based on parameters for the lunar motion and the computation of eclipses, ultimately derived from cuneiform texts. Nevertheless we need not assume direct contact between Mesopotamia and India. By a lucky accident we have two Greek papyri from the Roman period[18]— one written about A.D. 90 or somewhat later, the second about A.D. 250—which show that the very same methods were also known in Greek astronomy—a fact of which we would have no indication whatsoever

from Ptolemy's Almagest.   Thus there can be no doubt of a continued
Greek tradition and we can avoid thinking in terms of a direct knowledge of Babylonian astronomy in India.

For the planetary theory equally close parallels between Babylonian
methods and sections of the Pañca Siddhāntikā have come to light,
though in this case direct Greek intermediaries are missing, except for
the occurrence of identical planetary periods in Greek astrological
texts and in Babylonian and Indian treatises.   But there is no reason
to assume for the planetary theory a different route of transmission
than for the theory of the moon.

5.   The fact that Hindu theoretical astronomy was deeply influenced
by the West by no means excludes an independent modification into a
new system.   One very essential practical modification consists in the
replacement of the Greek chord-function by the function Sin α (where
I denote by Sin α the function $R \sin \alpha$ for a radius $R \neq 1$).   Furthermore the Hindus took the reasonable attitude that radial distances
should be measured in the same units in which the length of the circumference is measured, an approach which would have led to the
modern concept of radian,[19] had they not retained the Babylonian
sexagesimal division of the circle into 360 parts which caused them to
make $R = 57;18° = 3438'$ in order to obtain for $2\pi R$ the value $360°$.[20]
Similarly their replacement of a sexagesimal place-value system[21] by a
decimal place-value notation was not applied consistently but kept
alive the sexagesimal division of degrees and hours into minutes and
seconds, thus producing the perverse notation which is still in use today.

How much the planetary theory was modified in India is very difficult to say.   The arithmetical methods still await a detailed investigation of the very difficult text of the Pañca Siddhāntikā and related
Tamil sources.   The trigonometric methods of the Sūrya Siddhānta,
however, can now be fully described in modern terms.[22]   The leading
idea is the following.   The planet moves on an epicycle of radius r
which is carried on a circle of radius R and eccentricity e, the "deferent"
around the observer.   Thus we are dealing with two variables, the
"mean distance" α of the center of the epicycle from the apogee of the
deferent, and the "anomaly" γ which determines the position of the
planet on the epicycle.

The problem now arises to tabulate this rather complicated function
of α and γ.   In order to simplify the procedures two independent
corrections $\mu(\alpha)$ and $\sigma(\gamma)$ were computed which would hold for an
epicycle of radius e or of radius r respectively, but both under the assumption of a distance R of their center from the observer.   For a
position of the planet at mean distance α from the apogee and anomaly

γ, the successive application of the corrections μ and σ to the mean longitude would not lead to the correct true longitude, as can easily be seen from the fact that the sequence in which these corrections are combined affects the result. A close approximation to the correct result can be found, however, by a compromise combining the two corrections such that a preliminary correction is computed in which only $^1/_2\mu$ and $^1/_2\sigma$ enter. Since a detailed discussion of this procedure would lead us too far away from the immediate subject of relationship between planetary theories, I have removed it to an appendix at the end of this paper. Here it will suffice to say that certain variations of this procedure are known from different Hindu treatises which, however, all contain the above-mentioned "halfing of the equation," to use a later expression of Árabic astronomy.[23] This method reappears again in the planetary tables of al-Khwārizmī (about 840) which are preserved for us through the Latin translation (about 1130) of Adelard of Bath of an Arabic version made by al-Majrītī (about 1000). These "methods of the Sind-Hind" survived in Spain to a remarkable degree and thence influenced also the revival of astronomy in non-Muslim Europe.[24] But no careful study has yet been made what in each particular case is meant when an author is said to be following the Sind-Hind.

There are, of course, certain features which are common to several Hindu treatises and which were simply taken over by Islamic works. The zero meridian of Ujjain became the mediaeval "Arin"[25] while the beginning of the Kaliyuga (3102 B.C.) became the "Era of the Flood."[26] Occasionally whole tables were taken over; in the Toledan tables, e. g., the tables of Sines ($R = 150$) and the tables of solar declinations ($\epsilon = 24°$) are identical with the corresponding tables in the Khaṇḍa-Khādyaka of Brahmagupta.[27] Finally, many details in the treatment of problems of spherical astronomy can be traced back to the Sūrya-Siddhānta or the Khaṇḍa-Khādyaka. Not a single text is preserved, however, which could be called "the" Sind-Hind nor is it easy to visualize how a translation of any of the extant Indian treatises could have been used directly without extensive commentaries, numerical examples and tables.

In view of this rather unsatisfactory situation it has become customary to assign to Indian influence all those parts of Muslim tables (called zījes) which cannot be traced directly to Greek predecessors, that is, in practice, either to the Almagest or Theon's "Handy Tables." It is particularly in computational devices (for example, trigonometric interpolations) that it seems plausible to assume Hindu origin. But it is by no means impossible that occasionally a method is called Indian when it actually may come from a Greek source, lost or unpublished, or may be the contribution of a Muslim astronomer.

The popular concept that the decimal Hindu numerals played a decisive role for the progress of Islamic astronomy is completely unfounded since almost all of the countless tables in Arabic zījes are written in alphabetic numerals and with a sexagesimal basis except for integers, thus exactly following the example of the tables of Ptolemy and Theon. For the practice of astronomical computation the existence of the decimal system is totally irrelevant.

6.   That one cannot place too much trust on Islamic reports about Hindu science can be illustrated by the following case. Qāḍī Ṣā'id, whom we have quoted at the very beginning of this paper, describes three Indian systems:[28] The "Sind-Hind" according to which the planets return every 4,320,000,000 years to the beginning of Aries, the followers of "Ārjabhad" who assume a planetary period a thousand times shorter than the previous one, and the system of the "Arkand" of unknown period. But the two periods mentioned are the cosmic periods of the Kalpa and the mahayuga, respectively,[29] not at all characteristic for any system, while Arkand is the technical term ahargana[30] meaning the number of days elapsed since epoch. The whole story about the factor 1/1000 is ridiculed by al-Bīrūnī who explains its origin as due to a misunderstanding of the name Āryabhaṭa: "If the word in this garb wanders back to the Hindus, they will not recognize it," says Bīrūnī.[31]

The famous story about the Indian embassy which came to Baghdad under the second Abbasid Khalif Al-Manṣūr in the year 154 of the hijra (= A.D. 770/1) which had a member from whom al-Fazārī and Ya'qūb ibn Tariq obtained the material for their zīj has been recounted over and over again for at least 900 years without gaining in precision[32] by the process. Similar stories were current about the efforts of the Sasanian ruler Khosro Anōsharwān (531–579) to obtain works of Indian wisdom, particularly the fables of Kalīlah wa Dimnah.[33] The problems concerning the role of the pre-Islamic center at Jundīshāpūr for the contacts with Hindu astronomy have been recently summarized by E. S. Kennedy, who showed that the above-mentioned corrections μ and σ were based on the parameters of the Persian zīj ash-Shāh.[34] Since we have also good evidence for the influx of Greek astrology[35] one may well assume that Sasanian influence on the origin of Muslim astronomy is greater than is recognizable from the extremely fragmentary source material at our disposal.

7.   To establish the connection of the astronomical tables of the late Middle Ages with their Greek and Islamic predecessors would be much

less difficult if more of the existing manuscripts had been published. In fact, however, only two Islamic zījes are published (al-Khwārizmī and al-Battānī) and no Greek Canon except Theon's tables[36] though hundreds of folios of astronomical tables are listed in the catalogues of the major European libraries. If only a modest fraction of the effort which has gone into synthesis and bibliographies would have been spent in making some of the original texts available, our understanding of the scientific trends during the Middle Ages would be on a totally different level.

All that is known about Greek planetary tables except for those in Ptolemy and Theon are two papyrus fragments—one for the year 348/9, the other for 467 A.D.[37] These texts are the prototype of "Almanacs" which begin with the eleventh century in Spain,[38] a development which only in recent years has become better known through the publications of J. M. Millás Vallicrosa. Nevertheless much work remains to be done before we can distinguish between the different influences in the methods of computing planetary longitudes, latitudes, stations, etc.

The complete disregard of modern scholarship for Greek astronomical tables has made it easy to draw a definite historical picture according to which no serious mathematical astronomy exists between the end of the direct Greek tradition in the seventh century and a revival in the fourteenth century, in part under Persian influence, which was revealed in 1645 through the publication by Bullialdus[39] of planetary tables brought to Byzantium by Georgios Chrysokokkes (middle of fourteenth century).

Since no other Greek astronomical table has been published in the last 300 years one must resort to secondary evidence. From references in the "Tribiblos" of Theodoros Meliteniota[40] and from a somewhat different version of the same list,[41] we have references to nine Islamic zījes, evenly distributed in time between Battānī (A.D. 900) and 1300.[42] One may further add to this picture the fact that the astrological writings of Abū Ma'shar (about 880) and of his pupil Shadān are repeatedly quoted in Greek works,[43] thus bringing us back to the Abbasid period. It really does not sound very plausible to assume that four centuries were needed to make the impact of Islamic astronomy felt in Byzantium. In fact it seems much more likely that not only was Arabic and Persian astronomy known in the Byzantium area but that also the direct tradition of the Alexandrian school was not obliterated. The decision whether this hypothesis is correct or wrong can only come from the investigation of the Greek astronomical tables; in its favor can be quoted at least one important set of tables, the famous Cod.

Vat. Gr. 1291 which was first described by Boll[44] and which has often attracted the interest of art historians by its pictorial representation of the zodiac. These tables, which are closely related to Theon's Handy Tables,[45] were written between 813 and 820 and kept in use until 911/-912, as is shown by later additions to its royal canon.[46] It would be remarkable if this were exactly the latest text for centuries to come.

In an informally conducted survey H. Ritter counted some 120,000 unpublished manuscripts in Istanbul libraries.[47] Many hundreds of sheets of Sanskrit planetary tables are easily accessible in the D. E. Smith Collection of Columbia University[48] and the catalogs of European libraries give abundant evidence of the masses of astronomical material which have survived wars and persecutions. But the number of workers is small and it may well turn out that the desire for synthesis and generalization will be the most effective means of obliterating our connection with the past.

APPENDIX. HINDU PLANETARY THEORY

1. Ignoring the theory of latitude the model which forms the basis for the methods followed, e. g., by the Sūrya-Sidhānta or in the Khanda-Khādyaka, is an eccentric epicycle. A model of this type (cf. Fig. 1) is determined by the radius r of the epicycle, the eccentricity e, and the longitude[49] λ_A of the apogee A' of the deferent of radius R. By a well-

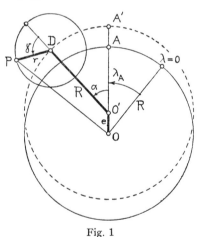

Fig. 1

known theorem[50] an eccenter can be replaced by an epicycle (cf. Fig. 2) of radius e on which the point D moves with the same angular velocity backward from the apogee E as C moves forward away from A. Thus the position of the planet P depends on two independent variables which increase proportionally with time: the *"mean distance"* α of C from A (or the *"mean longitude"* $\bar{\lambda}$ of C counted from $\lambda = 0$) and the *"anomaly"* γ of P on its epicycle. The problem to be solved can then be stated as follows: find, for given α and γ, the correction δ which must be added to the mean longitude $\bar{\lambda}$ of the planet in order to get the *"true longitude"* λ.

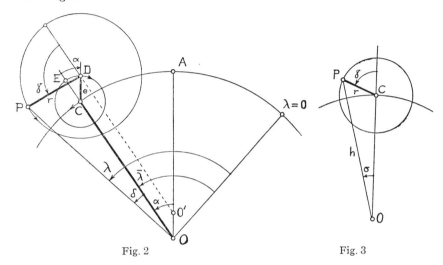

Fig. 2   Fig. 3

In considering α and γ as known quantities for any given moment t, we assume, of course, the knowledge of these quantities for some "epoch" $t=0$ and their rates of increase as functions of time—quantities which follow directly from the observation of the sidereal and synodic periods. What concerns us here is exclusively the transformation of $\bar{\lambda}$ into λ.

The answer to this problem can obviously be given by means of plane trigonometry and therefore was fully within the reach of Greek or Hindu mathematics. It is also evident, however, that the solution is not convenient to handle in numerical respect—as we shall show explicitly by giving the formula for δ as a function of α and γ—and that a tabulation would lead to great difficulties. Not only would such a table be one of double entry but its size would be prohibitive if one would proceed in steps of single degrees (a common requirement in mediaeval tables). It is therefore obvious that the two effects must be separated: the correction caused by the eccentricity and the correction due to the motion of P on its epicycle. If such a separation could be effected we would be dealing with a function of α alone and a second function of γ alone, both of which can be tabulated independently.

Ptolemy of course faced exactly the same problem[51] and solved it by tabulating extremal situations which then are to be modified by interpolation schemes in order to obtain the general case.[52]   The procedure in Hindu astronomy is different.   The tabulated functions are based on the assumption that both epicycles have their center on the deferent of radius 60 and the main problem consists thereafter in evaluating the combined effect of eccentricity and anomaly.   The guiding idea for its solution can easily be formulated if we use modern terminology and consider $e = CD$ and $r = DP$ as vectors; the combined effect of these two displacements is then found by introducing the midpoint of the resultant vector.   We shall show in detail how this leads to the famous "halfing of the equation by the astronomers who follow the Sind-Hind."[53]

  2.   Using the notation of Fig. 2 it is easy to determine $\delta = \lambda - \bar{\lambda}$. One finds

$$\sin \delta = \frac{r \sin \gamma - e \sin \alpha}{\sqrt{(r \sin \gamma - e \sin \alpha)^2 + (R + e \cos \alpha + r \cos \gamma)^2}}.$$

Except for trivial matters of norm[54] all steps which lead to this formula are well known in Hindu astronomy; but it is evident that it is unusable for tabulation.   For this purpose two functions were computed which we call σ ("shīghra") and μ ("manda"), respectively[55] and which isolate the corrections due to the anomaly γ from the effect of the eccentricity or mean distance α.

  We assume from now on that the ecliptic on which longitudes are measured coincides with the deferent.   Let C be the center of an epicycle of radius r (Fig. 3), γ the anomaly.   Then the resulting correction is given by

$$\sin \sigma = \frac{r \sin \gamma}{h} \qquad h = OP = \sqrt{(R + r \cos \gamma)^2 + (r \sin \gamma)^2}. \quad (1)$$

Again except for matters of norm this is the formula derived in S.-S. II 39, 40 and is used for the tabulation of σ, though not in the Sūrya-Siddhānta itself.   In the Khaṇḍa-Khādyaka,[56] however, we find the values of σ listed for γ in irregular intervals from $\gamma = 0$ to 360; in the Almagest[57] this function is called the "prosthaphairesis of anomaly," tabulated in $3°$ or $6°$ intervals; σ is given for single degrees in al-Khwārizmī's tables (called "examinatio argumenti" in Adelard's Latin version[58])—and similarly in many tables which follow the Ptolemaic tradition.   Since formula (1) is employed in all these cases the results are essentially the same, except for the small variations caused by changes in the values of the epicycle radius.   Figure 4 shows the characteristic shape of $\sigma(\gamma)$ in the case of Mars, taken from Kh.-Kh. II, 8, 9.[59]

Formula (1) was derived under the assumption that the center C of the epicycle is located on the deferent of radius R.   If one now makes the same assumption for the epicycle of radius e which describes the influence of an eccentric deferent one would have only to replace in (1)

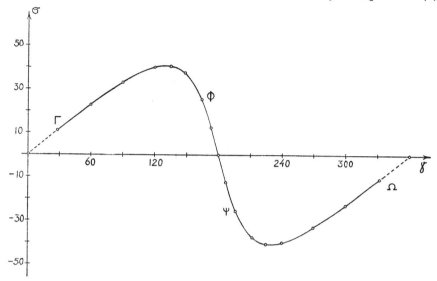

Fig. 4

r by e, and γ by α and would obtain a function $\mu(\alpha)$ of the type of Fig. 4.   This is indeed the procedure followed by Ptolemy whenever he is tabulating an "equation of center."[60]   Hindu astronomy, however, operates in the case of the correction μ for eccentricity with an approximate formula, namely (Fig. 5),

$$R\mu = e \sin \alpha. \tag{2}$$

In other words, a curve of the type Fig. 4 is replaced by a simple sin-curve.

As an example we may refer to Kh.-Kh. I, 19 where the values of μ are listed for the sun in steps of 15° of α:

α	μ	2;14 sin α
0	0°	0
15	0; 35	0; 34,41
30	1; 7	1; 7
45	1; 35	1; 34,45
60	1; 56	1; 56,3
75	2; 9	2; 9,26
90	2; 14	2; 14

Since the values of μ for the single planets are defined[61] as fixed multiples of the corresponding values for the sun the above table demonstrates the use of (2) in all cases.

There is no compelling reason to treat the effect of the eccentricity with so much less accuracy than the effect of the anomaly, except for the fact that usually[62] e is smaller than r. It may be that in the course of the historical development of planetary theory greater emphasis was attached to the phenomena caused by the anomaly than to those due to the eccentricity, but we know so little about the history of planetary theories that we hardly have any choice except to register the facts. For the following discussion it is fortunately sufficient to know that both for shighra- and manda-equations tables were computed from which $\sigma(\gamma)$ and $\mu(\alpha)$ could be obtained either directly or by interpolation.[63]

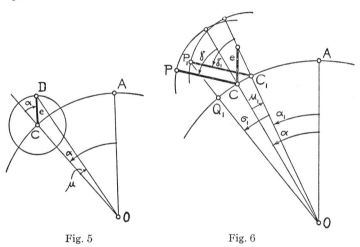

Fig. 5   Fig. 6

3. We must now face the essential difficulty in our problem: how to take into consideration the combined effects of eccentricity and anomaly. We have already mentioned that the Hindu astronomers did not follow Ptolemy's way of using interpolatory devices to bridge the gap between extremal and medium distances of the epicycle. Instead they followed a more arithmetical procedure which must have been founded on roughly the following considerations. At first sight one could argue as follows. Suppose that by its mean motion the center of the epicycle carrying the planet P has reached a point C of the deferent (Fig. 6). We then could take care of the manda correction $\mu(\alpha)$ which would tell us that the distance of the center of the epicycle from the apogee A should not be α but $\alpha_1 = \alpha + \mu_1$ where

$\mu_1 = \mu(\alpha)$. This would move the planet from P to P_1,[64] leading to a shīghra correction $\sigma_1 = \sigma(\gamma_1)$. Thus we would obtain as longitude

$$\lambda = \lambda_A + \alpha + \mu_1 + \sigma_1.$$

With the same right, however, one could also argue as follows (Fig. 7). Corresponding to a mean position C of the center of the epicycle the location of P on the epicycle would give a longitude $\alpha_2 = \alpha + \sigma_2$ where $\sigma_2 = \sigma(\gamma)$. In fact, however, we should also account for the eccentricity which moves CP parallel to itself by the amount e in the direction of OA. This would lead to a corrected longitude

$$\lambda = \lambda_A + \alpha + \sigma_2 + \mu_2$$

where $\mu_2 = \mu(\alpha_2)$.

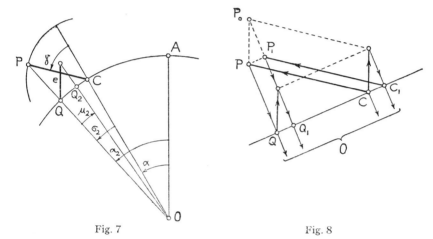

Fig. 7   Fig. 8

Obviously the two procedures would lead to the same result if the arc of the deferent near C could be considered a straight line (Fig. 8). It is furthermore clear that both procedures would give the correct longitude since Q_1 is the projection from O of the point P_0 which represents the actual position of the planet on an eccentric epicycle. In the measure, however, as the circumference of the deferent deviates from a straight line the two processes differ from each other and from the accurate solution.

The actual procedure followed by the treatises considered here can be understood as compromises between the two above-mentioned possibilities but retaining the main advantage of being satisfied with only two tables, that is, the tables of $\sigma(\gamma)$ and $\mu(\alpha)$.

4. The compromise adopted by the S.-S. and by the Kh.-Kh. can be motivated from the preceding discussion. Since it is clear that errors of opposite signs are obtained if one applies the manda-correction

for eccentricity on the opposite ends of the arc which corresponds to the shīghra-correction for anomaly, one may choose the manda-correction which corresponds to the mid-point of the shīghra displacement.   It is essentially this idea which explains the steps which the texts tell us to take in computing the true longitude of a planet.

In order to show this we follow the procedure for an outer planet.[65] For a given moment t we can consider as known

$$\bar{\lambda}_s \ldots \text{mean longitude of the sun}$$
$$\bar{\lambda} \ldots \text{mean longitude of the planet}$$
$$\lambda_A \ldots \text{longitude of apogee A.}$$

From this one forms the anomaly[66]

$$\gamma_1 = \bar{\lambda}_s - \bar{\lambda}$$

since the direction CP′ must always be parallel to the direction from O to the mean sun in order to maintain the fundamental relationship between sidereal and synodic periods (Fig. 9).   To this anomaly γ_1 belongs a shīghra-correction $\sigma_1 = \sigma(\gamma_1)$ which can be found in the table for σ.   One then forms

$$\lambda_1 = \bar{\lambda} + {}^1\!/_2\sigma_1$$

and from it

$$\alpha_1 = \lambda_1 - \lambda_A$$

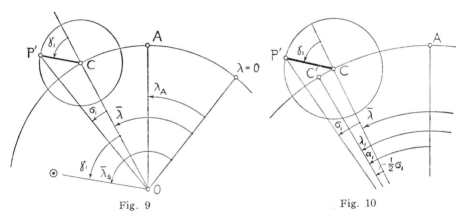

Fig. 9   Fig. 10

which gives the distance of C′ from A (Fig. 10).   With this argument α_1 we find in the table for the manda-correction

$$\mu_1 = \mu(\alpha_1)$$

and then form

$$\lambda_2 = \lambda_1 + {}^1\!/_2\mu_1$$

which means that we have found

$$\lambda_2 = \bar{\lambda} + {}^1\!/_2\sigma_1 + {}^1\!/_2\mu_1.$$

If we again assume for a moment that the segment of the deferent near C′ can be considered a straight line we see that λ_2 is the longitude of the midpoint M of the displacement parallelogram (Fig. 11). For a circular arc this is not correct but the error is less than half of the previous error. In other words we may consider λ_2 as the longitude of a point N which closely represents the longitude of a point M which is the result of a "mean" combination of shīghra and manda correction.

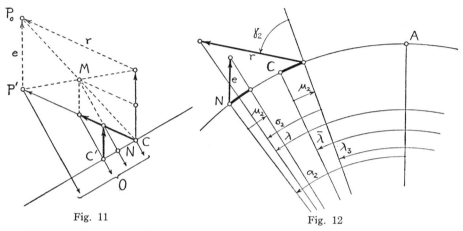

Fig. 11   Fig. 12

The manda-correction of this point N is now considered to be the proper "mean" manda-correction to be used in C. Thus one forms

$$\lambda_2 - \lambda_A = \alpha_2$$

which is the distance of N from the apogee A. To it belongs the manda-correction

$$\mu_2 = \mu(\alpha_2)$$

which is now used in C

$$\lambda_3 = \bar{\lambda} + \mu_2.$$

To this longitude belongs the anomaly

$$\gamma_2 = \bar{\lambda}_s - \lambda_3.$$

Thus we can find

$$\sigma_2 = \sigma(\gamma_2)$$

which, applied to λ_3, gives the true longitude

$$\lambda = \lambda_3 + \sigma_2$$

(Fig. 12). This is the procedure of the Sūrya-Siddhānta.

5. It is now no longer difficult to explain variants of this method which appear in other treatises. For instance we find in the Āryabhaṭīya[67] the following rules for the outer planets.

$$\lambda_1 = \bar{\lambda} + {}^1/_2\mu_1 \qquad \mu_1 = \mu(\alpha) \qquad \alpha = \bar{\lambda} - \lambda_A$$
$$\lambda_2 = \lambda_1 + {}^1/_2\sigma_1 \qquad \sigma_1 = \sigma(\gamma_1) \qquad \gamma_1 = \bar{\lambda}_s - \lambda_1$$
$$\lambda_3 = \bar{\lambda} + \mu_2 \qquad \mu_2 = \mu(\alpha_2) \qquad \alpha_2 = \lambda_2 - \lambda_A$$
$$\lambda = \lambda_3 + \sigma_2 \qquad \sigma_2 = \sigma(\gamma_2) \qquad \gamma_2 = \bar{\lambda}_s - \lambda_3.$$

Note that the only difference between this procedure and the preceding one consists in the arrangement of the first two steps, which aim at the determination of the longitude of the point M (or N) in Fig. 11. Previously we first used $^1/_2\sigma$ and then $^1/_2\mu$ for the result. Now we are using, with equal right, first $^1/_2\mu$ and then $^1/_2\sigma$ and the result would be the same if we could ignore the curvature of the deferent (Fig. 13).

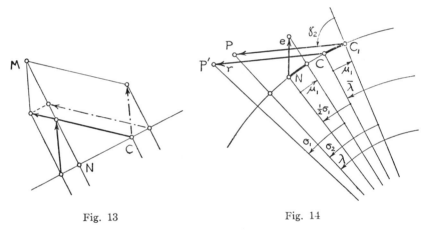

Fig. 13   Fig. 14

The numerical differences between the two procedures are usually very small indeed. For us the real interest lies in the fact that we have here the explicit confirmation of the basic principle that both combinations in the arrangement of manda and shīghra are equally possible.

For the inner planets Āryabhaṭa uses a still more approximative procedure for the determination of N. He now ignores $^1/_2\mu$, in all probability because for Venus and Mercury $^1/_2\mu$ reaches at most 2° and therefore has practically no effect on the subsequent value of $^1/_2\sigma$. In other words Āryabhaṭa uses as longitude for the mean value of the manda-correction no longer the midpoint M of the displacement parallelogram, but instead the midpoint of the longer side CP'.

6. It is interesting to see that exactly the same simplification is applied by al-Khwārizmī to all planets. He furthermore simplified

146

the practice of computation by tabulating not only $\mu(\alpha)$ and $\sigma(\gamma)$ but also the function

$$s(\gamma) = \lambda_A - \tfrac{1}{2}\sigma(\gamma)$$

which in Adelard's Latin version is called "sublimatio examinata."[68] Using these three functions[69] his procedure for an outer planet can be described as follows.

To given mean longitudes $\bar{\lambda}_s$ and $\bar{\lambda}$ of sun and planet we find the anomaly

$$\gamma_1 = \bar{\lambda}_s - \bar{\lambda}$$

and for it the value of

$$s = s(\gamma_1) = \lambda_A - \tfrac{1}{2}\sigma(\gamma_1) = \lambda_A - \tfrac{1}{2}\sigma_1.$$

I might mention at this point that I have so far silently modified the textual rules by assigning to the functions σ and μ positive and negative values as the variables vary between 0 and 360 while the original texts distinguish between addition and subtraction according to the interval to which the variable argument belongs. In the case of the function s al-Khwārizmī tabulates only $s(\gamma) = \lambda_A - \tfrac{1}{2}\sigma$ for γ between 0 and 180 though he gives a second column of arguments $360 - \gamma$ running from 360 to 180. In the same fashion σ is tabulated only for γ between 0 and 180 and one has to remember that these values must be taken negative for γ between 180 and 360. Thus we can say that al-Khwārizmī tabulates actually $|\sigma|$ and $s' = \lambda_A - \tfrac{1}{2}|\sigma|$. For $\gamma > 180$ he therefore gives the rule that one should add s' and $|\sigma|$, that is to say, to form $\lambda_A + \tfrac{1}{2}|\sigma|$. But, for this interval, $\sigma = -|\sigma|$, and thus $\lambda_A + \tfrac{1}{2}|\sigma| = \lambda_A - \tfrac{1}{2}\sigma$. Al-Khwārizmī's rule is therefore the exact equivalent of our statement that he uses $s(\gamma) = \lambda_A - \tfrac{1}{2}\sigma$ for all values of σ.

The next step consists in forming

$$\alpha_1 = \bar{\lambda} - s = \bar{\lambda} - \lambda_A + \tfrac{1}{2}\sigma_1 = \alpha + \tfrac{1}{2}\sigma_1$$

(cf. Figs. 9 and 10). This represents sufficiently accurately the distance from A of the midpoint of CP'. With α_1 as argument one finds

$$\mu_1 = \mu(\alpha_1)$$

and forms with it

$$\alpha_2 = \alpha_1 + \mu_1 \quad \text{and} \quad \gamma_2 = \gamma_1 + \mu_1.$$

With the latter value as argument one finds the shīghra-correction $\sigma_2 = \sigma(\gamma_2)$. This actually solves our problem: having first found the manda-correction μ_1 which corresponds to the midpoint of CP', we use it at C and then apply the corresponding shīghra-correction $\sigma_2 = \sigma(\gamma_2)$ at C_1 (Fig. 14).

All that remains to be done is to compute the longitude of P from the already known quantities. This is done in the following fashion: compute from the previously found α_2 and from σ_2

$$\alpha_3 = \alpha_2 + \sigma_2.$$

Then

$$\lambda = a_3 + s$$

is the desired longitude. The proof can easily be supplied by simple substitution: indeed,

$$\alpha_3 + s = \alpha_2 + \sigma_2 + s = \alpha_1 + \mu_1 + \sigma_2 + s$$

$$= \bar{\lambda} - s + \mu_1 + \sigma_2 + s = \bar{\lambda} + \mu_1 + \sigma_2$$

and Fig. 14 shows that indeed $\lambda = \bar{\lambda} + \mu_1 + \sigma_2$.

7. As is well known the Pañca-Siddhāntika has preserved for us parts of a Sūrya-Siddhānta which are not identical with the existing version of the S.-S. and therefore presumably belong to an earlier version. Also in the procedure for the computation of the planetary longitudes this "old S.-S." is not identical with the procedure in the "modern S.-S." It is now easy to show, however, that the difference is only apparent.

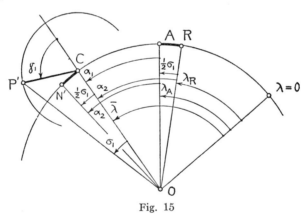

Fig. 15

Pañca-Siddhāntika XVII, 4–11[70] gives a set of rules which can be formulated as follows. First step: for given mean distance α_1 end anomaly γ_1 correct the position of the apogee A to a new one, R, by means of (Fig. 15)

$$\lambda_R = \lambda_A - {}^1\!/_2\sigma_1 \qquad \sigma_1 = \sigma(\gamma_1)$$

and then find

$$\mu_1 = \mu(\alpha_2) \qquad \text{with} \qquad \alpha_2 = \bar{\lambda} - \lambda_R.$$

But since

$$\alpha_2 = \bar{\lambda} - \lambda_R = \bar{\lambda} - \lambda_A + {}^1/_2\sigma_1 = \alpha_1 + {}^1/_2\sigma_1$$

we see that $\mu_1 = \mu(\alpha_2)$ is the manda-correction at the point N' exactly as in the previous cases. In other words, since N'C = AR the manda-correction in C for a vector of length e and of direction OR is the same as the manda-correction in N' for a vector of length e and of direction OA.

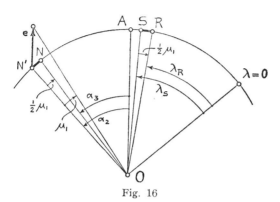

Fig. 16

Second step: correct the apogee R to a new apogee S by means of

$$\lambda_S = \lambda_R + {}^1/_2\mu_1$$

and then find[71]

$$\mu_2 = \mu(\alpha_3) \qquad \text{with} \qquad \alpha_3 = \bar{\lambda} - \lambda_S.$$

Again

$$\alpha_3 = \bar{\lambda} - \lambda_S = \bar{\lambda} - \lambda_R - {}^1/_2\mu_1 = \alpha_2 - {}^1/_2\mu_1$$

which shows us that α_3 is the distance of N from A (Fig. 16), exactly as in the case of the modern Sūrya-Siddhānta.

The remaining steps are not even formally different:

$$\lambda_3 = \bar{\lambda} + \mu_2$$

and with[72]

$$\gamma_2 = \bar{\lambda}_s - \lambda_3$$

find

$$\sigma_2 = \sigma(\gamma_2).$$

Then

$$\lambda = \lambda_3 + \sigma_2$$

is the true longitude of the planet.

8. In the "old" as well as in the "modern" S.-S.[73] additional corrections are made to account for the motion of the planet during the time which corresponds to the interval between mean and true sun. Since no new principle is involved in these corrections we need not discuss them in detail.

There exists, however, one modification of the theory which must be mentioned since it is not yet fully understood. The procedures which we have discussed in the preceding sections do not depend on specific values of eccentricity e and epicycle radius r. All that we really needed was the knowledge of the values of e and r at any given moment. It is therefore perfectly possible to carry out all operations under the assumption that both e and r are functions of their arguments α and γ respectively. Such a periodic variation was indeed assumed by Āryabhaṭa[74] and in the modern Sūrya-Siddhānta[75] where e and r are made variable between two (not greatly different) values which assume their extrema in the apsidal lines and perpendicular to them. For intermediate positions, sinusoidal interpolation is used.

Though it is correct to say that the variability of e and r does not concern the principle of the previously described procedures it must be underlined that the computation of tables for $\mu(\alpha)$ and $\sigma(\gamma)$ becomes somewhat more difficult since e and r are no longer constants. Nevertheless, the main feature of these tables remains intact, namely, to represent functions of one variable only.

Though the mechanism of this modification does not cause any difficulties in principle, I do not know which astronomical elements are reflected in the assumption of variable e and r. One may think of a connection with the theory of the "equant" but if this should be the case the connection does not seem to be evident. It seems to me even possible that we are dealing with a case of learned speculation of no practical significance, comparable to certain secular corrections which could not have been submitted to any empirical test. Future investigations will have to clarify this point.

9. In conclusion a few remarks must be made about the theory of planetary latitudes. It is here that a heliocentric theory has a decided advantage over the geocentric approach since the orbital planes of the planet go through the sun and not through the earth. The analogy with the lunar theory has here misled the ancient astronomers and forced Ptolemy and his followers into a very complicated system of variable inclinations of the planes of the deferents and eccenters.[76]

The Indian theory of planetary latitudes is of a much more primitive type.   It is simply assumed[77] that

$$\beta = \frac{i \, \mathrm{Sin} \, \omega}{h}$$

where ω is the argument of latitude, i. e., the distance from the ascending node of the planetary orbit to the center of the epicycle, i the inclination of the deferent plane to the ecliptic, and h the "variable hypothenuse," i. e. (cf. Fig. 3), the distance from the planet to the observer, thus a function of the anomaly γ.   The underlying geometrical model obviously consists in the assumption that the center of the epicycle moves on an inclined plane, as does the moon in its orbit, while the planet moves on its epicycle, which is kept parallel to the ecliptic.

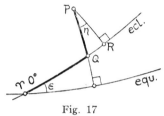

Fig. 17

This generally simple picture of the Hindu treatment of planetary latitudes needs in each specific case of application a special investigation since "latitudes" are often to be understood as arcs of declination between the planet and the ecliptic [78](Fig. 17).   This peculiar combination of ecliptic and equatorial coordinates appears also in Hipparchus' discussion of Aratus' Phenomena.[79]   It would be strange if this parallelism were accidental.

Al-Khwārizmī's rules for the computation of planetary latitudes are again based on a formula of the Hindu type.   The values of $i \, \mathrm{Sin} \, \omega$ $(R = 60)$ are tabulated as "second latitude" as functions of ω.   The latitude of the planet at anomaly γ is then to be found by division of the second latitude by a "first latitude" which is also tabulated and which should correspond to the normed variable hypothenuse h/R. In fact, however, only the general trend of the first latitude agrees with the monotonic decrease of h from $\gamma = 0$ to $\gamma = 180$ but the values are consistently greater than expected;   particularly in the case of Saturn, the values of h/R remain greater than one which would mean that the plane of the epicycle would always be seen below the plane of the deferent.

If so well informed an author as Abū'l Qāsim Sā'id says about al-Khwārizmī,[80] "il découvrit des procédés ingénieux pour l'approximation des calculs astronomiques, mais cela ne compense pas ce que renferme cet ouvrage d'erreurs grossières attestant la faiblesse d'Abū Ja'far [al-Khwārizmī] en géométrie, et son absence de connaissances réelles en astronomie," then it is tempting to apply it to the theory of latitude, particularly when one can only agree with Ṣā'id in the case of al-Khwārizmī's "Algebra," which shows no trace of an original contribution to material known for some two thousand years in the Near East. But as long as we do not control the method of computation which led to the values for the "first latitude" as function of γ, we had better suspend a final judgment.[81]

BIBLIOGRAPHICAL REFERENCES

The Āryabhaṭiya of Āryabhaṭa, trsl. W. E. Clark, Univ. of Chicago Press, Chicago, 1930.

Bīrūnī, The book of instruction in the elements of the art of Astrology, written in Ghaznah, 1029 A.D., trsl. R. Ramsay Wright, London, Luzac, 1934.

Bīrūnī, The Chronology of Ancient Nations, trsl. E. C. Sachau, London, 1879.

Bīrūnī, India, trsl. E. C. Sachau, London, 1910.

Brahmagupta, The Khaṇḍakhādyaka, trsl. Prabodh Chandra Sengupta, Univ. of Calcutta, 1934.

CCAG = Catalogus Codicum Astrologorum Graecorum, Bruxelles, 1898–1953.

Chatterjee, B., Geometrical interpretation of the motion of the sun, moon, and the five planets as found in the Mathematical Syntaxis of Ptolemy and in the Hindu astronomical works. J. Royal Asiatic Soc. of Bengal, Science, Vol. 15 (1949), p. 41–89.

JAOS = Journal of the American Oriental Society.

Kennedy, E. S., A Survey of Islamic Astronomical Tables, Trans. Am. Philos. Soc. N. S., 46, 2 (1956), p. 123–177.

Kh.-Kh. See Brahmagupta.

Khwārizmī: Die astronomischen Tafeln des Muhammed ibn Mūsā al-Khwārizmī in der Bearbeitung des Maslama ibn Aḥmed al-Madjrīṭī und der latein. Uebersetzung des Athelard von Bath auf Grund der Vorarbeiten von A. Bjørnbo und R. Besthorn herausgeg. von H. Suter, Kopenhagen, 1914 [= Kgl. Danske Vidensk. Selsk. Skrifter 7. R., hist.-filol. Afd. III, 1].

Millás Vallicrosa, Estudios sobre Azarquiel, Madrid-Granada 1943–1950.

Millás Vallicrosa, El libro de los fundamentos de las Tablas astronomicos de R. Abraham Ibn Ezra, Madrid-Barcelona, 1947.

Pc.-Sk. See Varāha Mihira.

Reinaud [J. T.], Géographie d'Aboulféda I: Introduction générale à la géographie des orientaux, Paris, 1848.

Reinaud [J. T.], Mémoire géographique, historique et scientifique sur l'Inde. Mémoires de l'Institut National de France, Académie des Inscriptions et Belles-Lettres 18, 2 (1849), p. 1–399, p. 565f.

S.-S. See Sūrya-Siddhānta.

Ṣā'id al-Andalusî, Kitâb Ṭabaḳât al-Umam (Livre des Catégories des Nations). Trad. Régis Blachère, Paris, 1935.

Sūrya-Siddhānta, trsl. Ebenezer Burgess; originally published JAOS, 6 (1860), p. 141–498; reprinted Calcutta 1935 [with introduction by P. Sengupta].

Suter, Khw. See Khwarizmi.

Varâha·Mihira. The Pañchasiddhântikâ. Ed. by G. Thibaut and M. S. Dvivedî, Benares, 1889.

ZDMG = Zeitschrift d. deutschen morgenländischen Gesellschaft.

NOTES

1. Trsl. Blachère; cf. the bibliographical references above s.v. Sâ'id al-Andalusî.
2. Trsl. Blachère, p. 39.
3. Trsl. Blachère, p. 35.
4. Cf. A. Braunmühl, Vorlesungen über Geschichte der Trigonometrie I, Leipzig, 1900, p. 120 and M. Curtze, Urkunden zur Geschichte der Trigonometrie im christlichen

Mittelalter, Bibliotheca Mathematica, 3. Folge 1 (1900), p. 321–416.

5. For details see O. Neugebauer, Astronomical Cuneiform Texts, 3 vols., London, 1955.

6. Also of ultimate Babylonian origin is the sexagesimal division of time and angles as well as the place-value notation for numbers, combined with a symbol for zero, which is in common use in the astronomical cuneiform texts, whence it spread to Greek and Hindu astronomy. It is interesting to see that so competent and authoritative a work as Krumbacher, Geschichte d. byzantinischen Litteratur (2nd ed., Munich, 1897, in Handb. d. klass. Altertumswiss.) contains the statement (p. 624) that the zero symbol reached Byzantium only about A.D. 1300 from India; one need only open the Almagest or any other Greek astronomical treatise to find countless examples of zero from the earliest to the latest periods.

7. Cf. O. Neugebauer, On the "Hippopede" of Eudoxus, SCRIPTA MATHEMATICA, 19 (1953), p. 225–229.

8. Edited by F. J. Carmody, Al Bitrûjî, De motibus celorum, Univ. of California Press, 1952. In this work the reader will find additional information about the earlier and later history of the theory of "homocentric spheres." For an account of the dependence of these theories on Aristotle see L. Gauthier, Journ. Asiatique, 10. sér. 14 (1909), p. 483–510. The essential innovation in al-Biṭrūjī's theory consists in his device to make the inclinations of the axes variable by mounting their endpoints on epicycles. In this way the loops become variable.

9. Cf., e. g., J. L. E. Dreyer, History of the Planetary Systems, Cambridge, 1908 [reprinted under the title "A History of Astronomy," New York, 1953], p. 296ff.

10. Contrary to popular belief the Copernican system is much more complicated than the Ptolemaic. This is again due to adherence to a philosophical (or theological) postulate concerning the Creator's preference for strictly uniform circular motion. This postulate not only caused Copernicus to multiply the number of epicycles and to abandon the concept of "equant" but to assign to the sun a motion on a small circular orbit around the geometrical center of the earth's orbit. Also the planes of the planetary orbits had to pass through this ideal point instead of through the sun, thus destroying one of the main advantages of a heliocentric theory over Ptolemy's model. The question as to which body is "at rest" is of course without any interest, particularly when no such physical body existed in the whole Copernican system.

The real progress of the Copernican theory beyond Ptolemy lies in the insight that in an essentially heliocentric system the size of the epicycle of a planet reflects the size of the earth's orbit and is therefore an indication of the distance of the planet from the sun. In the Almagest all distances (except for sun and moon) are kept relative.

11. This discovery will be published by V. Roberts (Am. Univ. of Beirut) in a paper submitted to Isis on "The solar and lunar theory of Ibn ash-Shāṭir."

12. Cf., e. g., Sūrya-Siddhānta, p. 474/5 (JAOS) = p. 385 (Calcutta).

13. This point is located symmetrically with respect to the center of the orbit—like the second focus in an elliptic orbit—and seen from it (and not from the geometric center) the motion of the center of the epicycle appears of uniform angular velocity.

14. Cf. the excellent survey given by R. E. M. Wheeler, Rome Beyond the Imperial Frontiers, London, 1954.

15. Cf., e. g., the introduction by Sengupta to the Sūrya-Siddhānta, Calcutta, 1935.

16. Cf. my article "Babylonian Planetary Theory" in Proc. Amer. Philos. Soc., 98 (1954), Section 7 (p. 63f).

17. For details see Ch. VI of my book "The Exact Sciences in Antiquity," Copenhagen, 1951, and the article "Tamil Astronomy" in Osiris, 10 (1952), p. 252–276, plus my review of Renou-Filliozat, L'Inde classique II, in Archives Internat. d'Histoire des Sciences, No. 31 (1955), p. 166–173.

18. Knudtzon-Neugebauer, Zwei astronomische Texte, Bull. de la Soc. royale de lettres de Lund 1946–1947, p. 77–88, and O. Neugebauer, The astronomical treatise P. Ryl. 27, Kgl. Danske Vidensk. Selsk., hist.-filol. Meddelelser, 32, 2 (1949).

19. Probably introduced by Euler; cf. Opera, ser. I, vol. 8, p. 134, and note 1 there.

20. This is not the only existing norm; the norm $R = 150$ is, e. g., used for the table which is known as "Kardaga" in late medieval treatises; cf. the references given in note 4. The norm $R = 60$ is used in the Almagest and then frequently in Islamic tables, e. g., al-Khwārizmī (Tab. 58) or al-Battānī (II, p. 55).

21. As mentioned above (note 6) both place-value notation and zero symbol are in ordinary use in Babylonian and Greek astronomy. The Hindu innovation consists only in transferring this method to a number system with decimal order.

22. This is done—to my knowledge, for the first time—in the Appendix to this paper.
23. Cf. Suter, Khw., p. 49.
24. Cf., e. g., O. Neugebauer and O. Schmidt, Hindu astronomy in Newminster in 1428, Annals of Science, 8 (1952), p. 221–228.
25. Arabic Azin misread as Arin (omission of a dot changes Arabic z to r). Ujjain is located at about $\varphi = 23$ (76 East) but was placed at the equator by Muslim geographical lore, then called "the cupola" of the world. Cf. Reinaud, Aboulféda I, p. CCXL f. and Mém. sur l'Inde, p. 373f.
26. Bīrūnī, Chronol., trsl. Sachau, p. 27ff. and p. 133. For the origin of this epoch cf. B. L. van der Waerden, Diophantische Gleichungen und Planetenperioden in der indischen Astronomie, Vierteljahresschrift d. Naturforsch. Ges. in Zürich, 100 (1955), p. 153–170.
27. Kh.-Kh. I, 29, 30 and Millás Vallicrosa, Azarquiel, p. 44.
28. Trsl. Blachère, p. 47.
29. Cf., e. g., S.-S. p. 10f. (Calcutta) = p. 153f. (JAOS).
30. Bīrūnī, India, II, p. 48 and p. 7.
31. India, II, p. 19.
32. Bīrūnī, India, II, p. 15, gives the year 154; p. 67, however, 161. Cf. furthermore Kennedy, p. 124 No. 2, p. 129 No. 28, p. 134 No. 71. Another version is found in al-Qiftī, Ta'rīkh al ḥukamā' ("Chronicle of the Philosophers" about 1200); cf. Reinaud, Mém. sur l'Inde, p. 312ff. and Aboulféda I, p. XLII. Quite embellished (probably influenced by Firdausi, Shāhnāma, written about 1000 A.D.) is the version given in 1160 by Ibn Ezra in the introduction to his Hebrew translation of a work of al-Muṭannā "About the reasons of the tables of al-Khwārizmī" (cf. Millás Vallicrosa, Ibn Ezra, p. 13 and Azarquiel, p. 25). The text of Ibn Ezra's introduction was published by Steinschneider ZDMG 24 (1870), p. 356–359 with a German translation of the main part (p. 353–356); for an English translation of the whole introduction, cf. D. E. Smith and J. Ginsburg in Amer. Math. Monthly, 25 (1918), p. 99–108. The text of the treatise itself is still unpublished (Oxford, Bodl. Libr. MS. Michael 400 fol. 45–76).
33. Cf., e.g., Ṣā'id, trsl. Blachère, p. 47.
34. Kennedy, p. 129f., No. 30, and p. 170ff.—Ṣā'id (trsl. Blachère, p. 102f.) and al-Qiftī (Suter, Khw., p. 33) report that al-Khwārizmī followed the Persians for the "equations," i. e., for our μ and σ.
35. Nallino, Raccolta di Scritti VI, p. 285–303. For the general background of this early period of Islamic science cf. A. Mieli, La science arabe, Leiden, Brill, 1938, p. 66ff.
36. Written about 360 A.D., published by Halma 1822–1825 in a fashion that is not always reliable.
37. P. Heid. Inv. 34, published by the present author in the Meddelelser of the Kgl. Danske Videnskab. Selskab, 1956; and P. Mich. Inv. 1454, published by H. D. Curtis and F. E. Robbins in the Publ. of the Observat. of the Univ. of Michigan, 6 (1938), p. 77–100.
38. Cf., e. g., Millás Vallicrosa, Azarquiel, p. 235ff. For the lost (or unpublished) tables of Ammonios (early sixth century) cf. also CCAG 2, p. 182, 18.
39. Ism. Bullialdus, Astronomia Philolaica, Paris, 1645, p. 211–232. Better MSS seem to be Cod. Vat. 209 and 210; cf. Bybliothecae Apostol. Vaticanae, Codices Vat. Graec., vol. I, p. 258 and p. 261, 3, 5.
40. Cod. Vat. Gr. 1059 F. 350 = CCAG 5, 3 p. 145f.
41. Cod. Laur. Plut., 28, 17 = CCAG 1, p. 89.
42. After Battānī the following works are mentioned which I give in chronological order with references to Kennedy's list in square brackets: Kūshyār (≈ 1010) [7 and 9], al-Fākhir (≈ 1030) [44], Khāzinī, zīj al-Sanjarī (≈ 1120) [27], Abd al-Karīm, zīj al-'Alā'ī (≈ 1150) [84], Athīr ad-Dīn (≈ 1240) [56], Naṣīr ad-Dīn aṭ-Ṭūsī (≈ 1270) [6], Muhi al-Maghribi (≈ 1280) [108], Husām-i Salār (before 1320) [32]. Cod. Vat. Gr. 1058 (unpublished) quotes the zīj al-'Alā'ī[84] and Īlkhānī [6], fol. 261, 10, 11 et passim.
43. Cf., e. g., CCAG 5, 1 p. 142ff. and CCAG 12, p. 96ff. For the Latin version: Thorndike, Isis, 45 (1945), p. 22ff.
44. Franz Boll, Beiträge zur Ueberlieferungsgeschichte der griechischen Astrologie und Astronomie. S. B. d. philos.-philol. u. hist. Cl. d. K. b. Akad. d. Wiss. in München, 1899, Bd. 1, p. 77–140.
45. It is interesting to see with what confidence these tables are assigned sometimes to Theon, sometimes to Ptolemy, without any attempt of proof. In fact there are tables in the Vaticanus which are not in (Halma's) Theon and vice versa. That the whole problem of the "Theon" tables is not so simple is underlined by the existence of a very fragmentary papyrus (in fact probably coming from two texts) which again only partially agrees with

either Theon or Vat. Gr. 1291 (P. Lond. 1278, identified by Dr. T. C. Skeat who kindly put his copy at my disposal). For the date of the miniatures cf. B. L. van der Waerden, Eine byzantinische Sonnentafel, Bayerische Akademie der Wissenschaften, S. B. 1954, p. 159–168, suggesting either A.D. 753/4 or 830/31 as probable date.

46. Boll (cf. note 44), p. 115.

47. H. Ritter, Autographs in Turkish Libraries, Oriens 6 (1953), p. 63.

48. These, and many more, are listed in H. I. Poleman, A census of Indic manuscripts in the U. S. and Canada, Am. Oriental Series 12, New Haven, 1938.

49. If not otherwise specified, "longitudes" are always meant to be sidereal longitudes.

50. Almagest, III, 3.

51. It is of no principal interest for the problem of tabulation that his model contains the added refinement of an "equant" since it is the fact of *two* independent variables which causes the trouble.

52. Cf. for details Almagest, XI, 10f.

53. Al-Bīrūnī, Chronology, ed. Sachau p. XXXXVII, and Suter, Al-Khwārizmī, p. 49.

54. I mean the use of the functions Sin $\alpha = R \sin \alpha$ and Cos $\alpha = R \cos \alpha$ (instead of our sin α and cos α, with $R \neq 1$) and the use of circumferences $c = 2\pi r$ instead of radii.

55. From *shīghra* "fast" and *manda* "slow." Combined with *ucca* = extremum, apex one calls *shīghrocca* the apogee of the epicycle, where the motion of the planets is fastest (and where it is in conjunction with the sun, thus the translation "conjunction"), and *mandocca* the apogee of the excenter. Cf. E. Burgess, S.-S., p. 54f., Calcutta = p. 192, JAOS. From *ucca* is derived Arabic *awj* (cf. al-Bīrūnī, Astrol. §171) and from it Latin *aux* for apogee.

56. II, 8 to 17.

57. XI, 11 referring to a position of the epicycle at mean distance from O.

58. Suter, Table 27 or p. 11.

59. The Greek letters indicate the values of γ or σ for which the planet becomes invisible, reappears, or is stationary.

60. Cf., e. g., Almagest, III, 6.

61. Kh.-Kh., II, 6, 7.

62. This does not hold for Saturn.

63. The numerical examples given for the year A.D. 864 by Prithūdaka in his commentary to the Kh.-Kh. show that strictly linear interpolation was used in tables with 15° steps. Obviously the accuracy of the results is purely fictitious—an observation which can be made very often in all areas of mediaeval astronomy.

64. $C_1P_1 \parallel CP$.

65. Cf., e. g., Kh.-Kh., p. 59ff.

66. For an inner planet, $\lambda = \lambda_s$ and γ_1 is independently given for given t.

67. III, 22–24.

68. Suter, Tables 27–56; μ is the "examinatio centri," σ the "examinatio argumenti."

69. Suter, who did not have the Khanda-khādyaka at his disposal, tried to determine from al-Khwārizmī's tables 27–44 their law of computation. He found correctly (p. 235) the equivalent of our formula (1), p. 14 for $\sigma(\gamma)$. For $\mu(\alpha)$, however, he suggested sin $\mu = e \sin \alpha$ instead of $\mu = e \sin \alpha$. He overlooked the fact that in all tables the value of μ for $\alpha = 30°$ is exactly half of the extremal value (at $\alpha = 90°$). His values for the eccentricities are therefore inaccurate.

70. Thibaut, p. 93f.

71. Pc.-Sk., XVII, 7 seems to say $\lambda_8 = \lambda_R + \mu_1$ but only $1/2\mu_1$ makes sense. Cf. also the formulation given by B. Chatterjee, p. 69.

72. For an inner planet $\gamma_2 = \gamma_1 + \lambda_1 - \lambda_3$.

73. Pc.-Sk., XVII, 10f. and S.-S. II, 46, respectively.

74. I, 8, 9.

75. S.-S., II, 38.

76. Almagest, XIII, 1–6.

77. Kh.-Kh., VIII, 1, 2, 5 = VI, 1–3; S.-S., II, 56, 57.

78. Cf., e. g., S.-S., II, 58. Another example is the computation of QR (Fig. 17) in Kh.-Kh., VI, 4. If we call β the arc of declination PQ and if we treat the triangle PQR as plane, then QR $= \beta \sin \eta$. For $\lambda = 0$ the angle η has its maximum value ϵ; for $\lambda = 90$ it is zero. Using, in typical Hindu fashion, a sinusoidal interpolation for intermediate values of λ, one has

$$QR = \beta \sin \epsilon \cos \lambda = \beta \operatorname{Sin} (\lambda + 90)/\frac{R^2}{\operatorname{Sin} \epsilon}$$

as given in Kh.-Kh. VI, 4.

79.   Instead of the longitude of a star Hipparchus gives the longitude of the simultaneously culminating point of the ecliptic; cf. H. Vogt, Astronomische Nachrichten, 224 (1925), cols. 17–54.

80.   Trsl. Blachère, p. 103; cf. also p. 130 and p. 135 for criticism of the Sind-Hind.

81.   An obvious error is the definition of ω as the sum, instead of the difference, of planetary longitude and ascending node (chapter 17, Suter, p. 14).   But this error is so trivial that it may well be the fault of the Latin version only.   Kennedy in his "Survey" mentions (p. 128, No. 21) a criticism of al-Khwārizmī's theory of planetary equations by al-Bīrūnī.

BROWN UNIVERSITY AND INSTITUTE FOR ADVANCED STUDY

Reprinted from
Scripta Mathematica
22, 165–192 (1956)

THE SURVIVAL OF BABYLONIAN METHODS IN THE EXACT SCIENCES
OF ANTIQUITY AND MIDDLE AGES

O. NEUGEBAUER

Professor of the History of Mathematics, Brown University

(*Read April 19, 1963, in the Symposium on Cuneiform Studies and the History of Civilization*)

Non omnis sapientia penes Chaldaeos et Orientem fuit. Etiam Occi-
dentis aut Septentrionis homines fuerunt λογικὰ ζᾶα.
Scaliger, De emend. temp. (1629) p. 171.

AMONG the many parallels between our own times and the Roman imperial period could be mentioned the readiness to ascribe to the "Chaldeans" discoveries whenever their actual origin was no longer known. The basis for such assignments is usually the same: ignorance of the original cuneiform sources, excusable in antiquity but less so in modern times. Given this situation, it seems to me equally important to establish what we can say today about knowledge which the Babylonians *did* have and to distinguish this clearly from methods and procedures which they did *not* have. In other words, it seems to me that it is high time that an effort is made to eliminate historical clichés, both for the Mesopotamian civilizations and their heirs, and to apply common sense to the fragmentary but solid information obtained from the study of the original sources during the last hundred years.

My approach to these problems will earn the displeasure of many scholars. Classicists who are still fighting the Persian Wars and see only barbarians in the Orientals, scholars who discover Iranian influences wherever they look, Orientalists who are convinced that "ex oriente lux," and philosophers who think that science originates from preconceived doctrines will join in disagreeing with every one of my conclusions, however different their points of view may be in all other respects. All that I can say in my defense is expressed in a sentence by Louis de la Vallée-Poussin: "je ne suis qu'un lecteur de textes."

How strongly historical interpretations can be influenced by generally accepted clichés may be illustrated by the following incident. During the excavations by the University of Pennsylvania at Nippur (1887 to 1900) for the first time a substantial number of mathematical texts came to light. Except for a few simple ideograms, they contain nothing but columns of numbers, written in the sexagesimal system. Obviously, concluded Hilprecht in editing these texts, one had here the oriental source on which ultimately rested Plato's number mysticism, contained in the *Republic* VIII and in the *Laws* V, since Plato was following Pythagoras who "derived his mathematical science and doctrines from the East." Thus Hilprecht discovered Plato's "Nuptial Number" 12960000 = 60⁴ in his tablets which he then transcribed, e.g., as

$$4 \ldots \ldots 3240000$$
$$5 \ldots \ldots 2592000$$
$$6 \ldots \ldots 2160000.$$

In fact the numbers in the right-hand column all contain the factor 216000 = 60³, arbitrarily chosen by Hilprecht, while the text actually reads

4	15
5	12
6	10.

Since the product of left and right is always 60, one has here a simple table of reciprocals, telling us that ¼ of any unit is 15 minutes, ⅕ equals 12 minutes, etc.—and so with all the remaining texts which are nothing but elementary aids for sexagesimal computing.

Since 1929, when W. Struwe and I succeeded in bringing sense also into nontrivial (i.e., algebraic and geometric) mathematical cuneiform texts, our available sources have expanded to constitute probably the largest body of scientific original documents from a pre-Hellenistic civilization. No trace of number mysticism has ever been found in these often highly sophisticated but perfectly rational mathematical texts which range from the twentieth to the first century B.C. Nevertheless, the "Babylonian" origin of whatever is irrational or mystical (in fact or reconstructed) remains the inexhaustible resource for synthetic histories of science and philosophy, whether it concerns the Pythagoreans or the Ionians, Plato or Eudoxus, Nicomachus, Proclus, etc., etc.

This tendency is by no means new; in fact it is inherited from antiquity. The Babylonians or the Egyptians (e.g., in the Hermetic literature) were held responsible whenever one needed authority with a high reputation. For less inspiring subjects, like mathematics, simple nursery stories (e.g., the origin of geometry from the inundation of the fields in Egypt) gained general currency because of their simple finality. It is very illuminating to collect these ancient stories about "origins"[1] from Herodotus to the Church Fathers. Their uniformity and simplemindedness is most impressive; for our knowledge of pre-Hellenistic science these sources are not only practically valueless but seriously misleading. In order not to remain in the realm of generalities, let me quote only one example, the statements from antiquity concerning the reckoning of the beginning of the "day" in Mesopotamia. Relying exclusively on sources from classical antiquity—e.g., Pliny who claims (N.H. II, 188) that the Babylonians reckon their days from sunrise—modern scholars have come to the most contradictory results, whereas the astronomical cuneiform texts from the Seleucid period show in their date columns that sunset represents the civil epoch but that midnight was used for astronomical reasons in the computations of the lunar theory of the most advanced type ("System B"). This shows how little of contemporary Babylonian science had become common knowledge. I should not wish to extrapolate this experience to other areas of ancient cultural history and literary criticism where we do not have at our disposal original sources of unquestionable precision and authority.

That the Babylonian priest Berossos, dedicating his "Babyloniaca" to Antiochus I, transmitted Babylonian astronomy to his Greek pupils in Kos is common knowledge. Schnabel, who edited the fragments of the *Babyloniaca*,[2] added to his edition gratuitously (and with many errors) sections from two ephemerides from cuneiform texts which have nothing to do with Berossos. The little, however, that is preserved of astronomical character

in the fragments[3] suffices to demonstrate that Berossos was totally ignorant of the contemporary Babylonian astronomy when he was teaching that the lunar phases were the result of a rotation of the moon which he supposed to be half luminous, half dark. The mathematical theory of the lunar phases constitutes the best developed and most sophisticated section of Babylonian astronomy in the Seleucid period, leaving no room for such primitive doctrines. They were proper meat for Greek philosophers; for the transmission of Babylonian astronomy, however, Berossos can be safely ignored.

If we now turn to positive evidence, provided by cuneiform texts, we stand on safe chronological grounds. Mathematics is fully developed in the Old Babylonian period, while mathematical astronomy originated probably in the Persian period. Cuneiform texts of both classes still exist from the last century or two B.C. Of the origin of Babylonian mathematics, we know nothing; my guess would be that it developed fairly rapidly without a Sumerian antecedent. Whatever the case may be, we know today that it had reached by the nineteenth century B.C. a full command of sexagesimal techniques based on a place value notation (though without a symbol for zero), including higher exponents and their inverses, and a great deal of insight into algebraic and plane geometric relations, among which "Thales' Theorem" about the right triangle in a semicircle and in particular the "Pythagorean Theorem" for the right triangle take a common place. The famous Plimpton Tablet[4] reveals full understanding of the mathematical laws which govern "Pythagorean" triples of integers, i.e., solutions of $a^2 + b^2 = c^2$ under the condition that a, b, and c be integers. To quote only the first three solutions on which our text is based:

a	b	c
2,0 (= 120)	1,59 (= 119)	2,49 (= 169)
57,36 (= 3456)	56,7 (= 3367)	3,12,1 (= 11521)*
1,20,0 (= 4800)	1,16,41 (= 4601)	1,50,49 (= 6649)

This goes far beyond such trivialities as the "discovery" that $3^2 + 4^2 = 5^2$.

Since we have mathematical cuneiform texts from the Seleucid period and since Greek and Demotic papyri from the Greco-Roman period in Egypt show knowledge of essentially the same

---

[1] *Cf.* for this whole problem A. Kleingünther, ΠΡΩΤΟΣ ΕΥΡΕΤΗΣ, *Philologus*, Suppl. 26, 1 (1933).

[2] *Berossos und die babylonisch-hellenistische Literatur* (Leipzig, 1923).

[3] Schnabel, Nos. 16 to 26.

[4] O. Neugebauer and A. Sachs, *Mathematical Cuneiform Texts* (American Oriental Series 29, New Haven, 1945), pp. 38–41.

* Error in the text for 1,20,25 (=4825).

basic material, one can no longer doubt that the discoveries of the Old Babylonian period had long since become common mathematical knowledge all over the ancient Near East. The whole tradition of mathematical works under the authorship of Heron (first century A.D.), Diophantus (date unknown), down to the beginning Islamic science (al-Kwârizmî, ninth century) is part of the same stream which has its ultimate sources in Babylonia.

Probably also in the Persian period, perhaps in connection with beginning mathematical astronomy, the sexagesimal place value notation was perfected by the introduction of a symbol for "zero" (a separation mark). Thus, when Greek astronomy began its own development in the early second century B.C., using sexagesimally written Babylonian parameters, the use of a separation mark for zero was also adopted. With Greek astronomy the place value notation, including zero, came to India, where this system was finally extended also to decimally written numbers, whence our "Hindu (Arabic) numerals" originated. Again, the basic idea is undoubtedly Babylonian in origin.

For Greek mathematics the picture now becomes quite clear. It hardly needs emphasis that one can forget about Pythagoras and his carefully kept secret discoveries. It is also clear that a large part of the basic geometrical, algebraic, and arithmetical knowledge collected in Euclid's *Elements* had been known for a millennium and more. But a fundamentally new aspect was added to this material, namely the idea of general mathematical proof. It is only then that mathematics in the modern sense came into existence. Parallel with the development of modern axiomatic mathematics it became clear that the discovery of "irrationals" by Theaetetus and Eudoxus caused the transformation of intuitively evident arithmetical and algebraic relations into a strictly logical geometric system. From this moment the theoretical branch of Greek mathematics severed all relations with the ultimately Mesopotamian origins of mathematical knowledge.

It is very illuminating to compare Euclid's *Elements* with another work by the same author, his spherical astronomy (the *Phaenomena*).[4a] In the

[4a] Unfortunately I did not overlook a recent paper on "Greek Astronomy and its Debt to the Babylonians" by Leonard W. Clarke in *The British Journal for the History of Science* 1 (1962): 65–77. May it suffice to characterize the author's competence by his opinion "that Euclidean geometry pre-supposes a flat earth" (p. 75).

Elements everything is subject to a perfectly logical structure (though notoriously based on different earlier presentations, in part of the fifth century). In the *Phaenomena* we find only fumbling attempts to obtain some quantitative and some geometric insight into the mutual relations of equator, ecliptic, and parallel circles in relation to the yearly solar motion. The *Elements* were unsurpassed for two millennia to come; the *Phaenomena* try to settle elementary astronomical questions which had to wait for four more centuries of efforts until Menelaos' spherical geometry. I think the reason for such a marked difference is clear. In the *Elements* a well-known body of material was readily at hand and made subject to the newly discovered logical principles—and it is worth repeating that this constitutes a scientific achievement of the very first order but adds very little to the factual material. For spherical geometry, however, the situation was quite different. All that we know of Babylonian astronomy (and mathematics) speaks against the existence of any spherical geometry. Crude arithmetical schemes mark the beginning of astronomical endeavors—e.g., concerning the variation of the length of daylight during the year—eventually to be highly refined by means of arithmetical sequences of different order. Here the Greeks had to begin from scratch. Eudoxus' "Homocentric Spheres" were a flash of genius but astronomically valueless, and the subsequent entanglement with philosophical principles did not help matters. Euclid and Aristarchus drastically demonstrate the inadequacy of the traditional mathematics to cope with spherical astronomy and trigonometry at the end of the fourth century. Only with Apollonius (around the end of the third century) does real progress begin by returning to plane geometry, masterfully applied in Archimedean fashion to the problems of eccentric and epicyclic motion. We do not know whether Apollonius had Babylonian data at his disposal; from what we know about his work, there is no compelling reason to assume it. With his successor Hipparchus, however, it is clear that the empirical data of the contemporary Babylonian astronomy were available to him while in all probability the theory of stereographic projection (perhaps influenced by Apollonius' theory of conic sections) made it possible to solve spherical trigonometric problems in the plane. From now on, the full impact of the Babylonian sexagesimal place value notation is felt and remains the backbone of astronomical computations to the present day such that the sexagesimal division of time and angles still encumbers instruments and tables. Only in the first century of our era did Menelaos find the proper generalization of plane geometry to the sphere by operating with great circles only and thus create spherical trigonometry. With Ptolemy, in the next generation, astronomy reaches the same perfection as mathematics in the *Elements* almost five centuries earlier. He eliminated the remnants of Babylonian-Hipparchian parameters by the systematic refinement of all empirical data. This was made possible by comparison of his own observations and improved cinematical models with older data. Except for the sexagesimal computational procedures the role of Babylonian astronomy is now ended for mathematical astronomy.[5] Exactly as Babylonian mathematics lingers on for many more centuries in the Heronic-Diophantine literature which is continued into the high Middle Ages, so too we shall find residues of Babylonian astronomy in branches untouched by the Almagest tradition, in particular in the Hindu-Arabic tables and treatises. But for the scientific main stream, the Babylonian influence ends for Greek mathematics in the fifth century B.C. and for astronomy with Menelaos and Ptolemy in the first and second century A.D.

Astral religion has as little to do with the origin of astronomy as Genesis with astrophysics. Nor does the interest in celestial omens—as one class of omens among many—lead to astronomy. As long as one takes extraordinary events as a means of communications between the gods and men one will devote every effort to deciphering this complicated divine language. One will record the repetitions of phenomena in order to find their significance but one will not try to predict the recurrence of the ominous phenomena themselves. Astrology, which operates with the use of mathematical astronomy, is the very antithesis of omen-lore. Its basis is causality, not communication of arbitrary decisions of a divine will. This causal root of

astrology is still clearly recognizable in Greek astrology in its relation to weather prognostication. Just as sun and moon visibly influence the seasons, the winds, and climatic conditions, so may also the stars and planets be influential. Now it makes sense to predict; but Hesiod and Aratus, the parapegmata and related treatises, show how qualitative and how crude all such attempts had to remain before the necessary mathematical tools existed. The situation in Mesopotamia was perhaps not very much different. It is probably not accidental that the first calendaric cycles, in particular the "Metonic" nineteen-year cycle, appear simultaneously in Babylon and in Greece. In both areas simple arithmetical schemes gained wide currency: in Babylonia the schemes for the variable length of daylight and night, in Greece similar schemes for the determination of hours by means of the length of the human shadow. Very little is known of an "astrology" in the Hellenistic sense of the word from this early time in Mesopotamia. But the really decisive difference lies in the fact that the Babylonians made a serious—and in the end extremely successful—attempt to control their lunar calendar mathematically by unravelling step by step the different periodic variations which cause the intricate pattern in the variation of the time between consecutive new crescents. The resulting mathematical procedures represent undoubtedly one of the most outstanding scientific achievements of antiquity. Its direct effect was, however, very small indeed as its scientific content is concerned; but peripheral elements had an enormous spread, geographically as well as chronologically, and provide for the historian who is familiar with the original context one of the most powerful tools for establishing the routes of transmission of scientific methods.

The popular belief of antiquity that astrology originated with the "Chaldeans" certainly contains an element of truth, although the textual evidence for Babylonian astrology (as distinct from celestial omens) is rather meager and belongs only to the latest periods. This fact has caused W. Gundel to go to the other extreme and to seek the origin of astrology in Egyptian civilization. In fact, however, one can be sure on the basis of all Egyptian documentation that the only component of ultimately Egyptian origin in astrology is the "decans" which were assimilated to the zodiac of Babylonian origin in the early Ptolemaic period. But the enormous expansion of astrology into an all-encompassing doctrine is undoubtedly a purely

---

[5] As another example of historical clichés in which practically every sentence is wrong may be quoted (remarks in [ ] are mine): "The final synthesis by Ptolemy in the Almagest . . . shows in fact great mathematical resourcefulness, a new use of Babylonian techniques, but no change except a yielding of principles: the uniformity of circular motion . . . had to be abandoned at last [why?] in favor of more complicated hypotheses. But by that time no one thought of forcing his way back [to where?]. The tool [which one?] controlled the men." (Giorgio de Santillana, *The Origin of Scientific Thought* (Chicago, 1961), p. 250).

Hellenistic product, developed from a comparatively modest Babylonian nucleus.[5a] It is perhaps useful occasionally to remember that the so-called Greek mind not only produced works of the highest artistic and intellectual level but also could indulge in the development of the most absurd doctrines of a pseudo-rational superstition which contributed heavily to the "darkness" of later ages.

For the spread of certain scientific methods, however elementary, the ultimate Babylonian origin of astrology is of great importance. Most of the methods which spread with Hellenistic astrology to India and (directly or indirectly) to the West belong to the elementary level of Babylonian astronomy. Thus early Babylonian-Hellenistic scientific methodology, at that level, remained the main tool of astrological practice and the refinements neither of Babylonian mathematical astronomy nor of the Almagest had an essential influence on the formation of astrological methodology for the determination of the position of the celestial bodies and for questions of spherical astronomy. There are strong indications, however, that much which we find of Hellenistic material in Hindu astronomy reflects the situation of astronomical knowledge in the time of Hipparchus. That direct Babylonian influence reached India seems to me not very likely—e.g., all the foreign technical terms are Greek—such that Babylonian components in Hindu astronomy can be taken as evidence for a corresponding influence on Hellenistic astronomy for the early period. This view is supported by the discovery, in Greek papyri of the early Roman imperial period, of methods which are characteristic for a certain class of sources from India (Tamil).[5b] Only in passing may be remarked that there is no evidence whatever for a Babylonian origin of the concept of "lunar mansions," however frequent such an origin has been assumed in the literature.

Beside the use of the sexagesimal number system and the zodiacal division of the ecliptic, the employment of the "lunar days" of exactly $\frac{1}{30}$ of a mean synodic month goes back to Babylonian astronomy. The consistent use of a strict lunar

---

[5a] For clear evidence of Babylonian astrology in Hellenistic Egyptian texts see R. A. Parker, *A Vienna Demotic Papyrus on Eclipse- and Lunar-Omina* (Providence, Brown University Press, 1959).
[5b] The whole problem of Babylonian influences on astronomy and astrology in India is now discussed in a masterful article by David Pingree, *Isis* 54 (1963) : 229–246.

calendar made it necessary to introduce smaller units in terms of the lunar month as fundamental unit of time measurement. This clear and convenient definition was perverted in later Indian astronomy and astrology to the use of thirtieths (*tithi*) of true lunar months, i.e., to units variable in a very complicated fashion from month to month. In other words, the Hindus reintroduced into the definition of *tithis* (as they are still used in India today) exactly the complication which it was the purpose of the Babylonian invention to avoid.

Since from early times the Babylonian calendar (not the Assyrian one) had the tendency to coordinate the lunar calendar more or less with the seasons of the solar year, the mathematization of astronomy in the fifth century B.C. is also reflected in a definite intercalation cycle, usually called the "Metonic cycle," which intercalates seven additional months in nineteen lunar years. This quite accurate and convenient cycle in combination with the continued counting of the regnal years of Seleucus I, beginning at 312 B.C., constitutes one of the greatest advances in practical chronology. Here we have for the first time a precise era in which dates can be accurately established according to simple computational rules. It is not surprising that Islamic astronomers made much use of this era (or a modification, the era of Philip) and it should be mentioned that the chronology of the modern historians for the Hellenistic age is based on Father Epping's decipherment of the terminology of astronomical cuneiform texts in their relation to the Seleucid era. Another aspect of the later history of the nineteen-year cycle is contained in the stormy history of the Easter cycles—analogous to the history of the *tithis* in so far as a simple and practical solution of one problem was contaminated by additional conditions (e.g., Easter limits) which deprived the original solution of its main value, simplicity. Indeed, that simplicity was the element that recommended the nineteen-year cycle to the Babylonian astronomers is demonstrated by the fact that the mathematical astronomical texts of the whole Seleucid-Parthian period maintained the use of the nineteen-year cycle as the chronological skeleton of their computations in spite of the fact that they used, for their lunar ephemerides themselves, relations of higher accuracy than those reflected in the calendaric cycle.

That much of the astronomical elementary

knowledge of later times originated in Babylonian astronomy is not difficult to show. For example relations for planetary phenomena, fundamental for the Babylonian "Goal-year-texts" [6] (but refined in the mathematical-astronomical texts), reappear in Greek and mediaeval astrological treatises. Similarly, certain patterns for the anomalous motion of the moon and of the planets have been found in Greek as well as in Demotic papyri [7] while their use in Indian and in Islamic sources is equally attested.[8]

A particular modification of an early scheme in Babylonian astronomy has greatly influenced ancient geography: I refer to their arithmetical patterns for relating the variable length of daylight to the position of the sun in the ecliptic. This simple scheme (existing in two variants, "System A," strictly linear, and "System B," with double the ordinary difference in the middle of the increasing and decreasing branches) was adapted, probably in the second century B.C.,[9] to the latitude of Alexandria and subsequently to other geographical latitudes. There is no trace anywhere in Babylonian astronomy for the concept of "geographical latitude" and consequently for provisions necessary for the adaptation of their procedures for other localities. Had astronomy originated during the Assyrian empire the situation might have been different; but the astronomers of the fourth century could not feel the need to see their computation applied outside a narrow area from Uruk to Babylon.

The world of the Greeks in Alexandria was of different dimensions; it extended from the Far East and India to Spain and from the Upper Nile to the Crimea and beyond. Obviously the Babylonian scheme for the variation of the length of days and nights would not do for such an area. But the mathematical device in itself was simple and easy to modify—again in a typical Babylonian fashion by a linear variation of the extremal length of daylight but otherwise unchanged pattern. According to this scheme one distinguishes "climates" of equal length of daylight, arranged in the simple pattern of half-hour increment of the longest day.

Again, as everywhere else, the strictly Babylonian procedure was eventually eliminated (probably shortly before Ptolemy, if not by himself) when Greek spherical trigonometry replaced the cruder arithmetical patterns. But, as a concept, the sequence of the climates of linearly increasing length of the longest day remained unchanged and dominated geographical lore from antiquity through Islam and the western Middle Ages. Simultaneously the original Babylonian scheme (even in such details as the definition of the eighth degree of Aries as the solar position at the vernal equinox) remained in use in the astrological literature (e.g., among many others, in the *Anthology* of Vettius Valens of the second century A.D.) and spread with it to India, where we find the unchanged Babylonian System A in the writings of Varaha Mihira (sixth century A.D.) applied for latitudes entirely different from Babylon. Intelligent modifications of the Babylonian arithmetical scheme for the latitude of Persia are described by al-Bīrūni (around A.D. 1000) as used by "the people of Babylon."[10]

The absence of recognition in Babylonian astronomy of any influence of geographical coordinates on astronomical procedures makes it obvious that at no time of its existence could Babylonian astronomy predict that the path of a certain solar eclipse—whether visible at Babylon or not—would cross Asia Minor. In other words it is clear that even the methods of the Seleucid period would not explain the alleged approximate prediction by Thales of a solar eclipse for Ionia. Since for a given region no simple periodicity of solar eclipses exists, the only way to predict a solar eclipse would have been to investigate mathematically every conjunction (or at least every sixth conjunction) for the given place and moment. This is indeed the procedure still followed by Ptolemy and by Islamic and Byzantine astronomers—not before the Renaissance could one compute eclipse paths, because this requires a much better knowledge of the solar parallax than the observational methods of antiquity, restricted to naked-eye techniques, could provide. That Thales had even the faintest idea of the problems involved is out of the question, quite aside from the fact that Mesopotamian "astronomy" of the early sixth century B.C. made at best the first stumbling attempts to construct quantitative schemes to describe the variation of

---

[6] *Cf.*, for this concept, A. Sachs, "A Classification of Babylonian Astronomical Tablets of the Seleucid Period," *Journal of Cuneiform Studies* **2** (1950) : 271–290.

[7] E.g., R. A. Parker, "Two Demotic Astronomical Papyri in the Carlsberg Collection," *Acta Orientalia* **26** (1962) : 143–147.

[8] I am here unpublished material recently uncovered by E. S. Kennedy.

[9] Hypsicles' *Anaphorikos* is our earliest source.

[10] Mark Lesley, "Bīrūni on Rising Times and Daylight Lengths," *Centaurus* **5** (1956–1958) : 138.

the length of daylight or to measure the shadow lengths.

While the enthusiasts for Ionian philosophy will not desist from allowing Thales to borrow non-existing methods from Babylonian astronomers, another fabulous achievement, the discovery of the precession of the equinoxes, has a better chance of disappearing from the literature because this amounts to restoring a pearl to the crown of the Greeks. Since Schnabel's theory of the Babylonian discovery of precession was based on taking seriously a scribal error (interchange of cuneiform 4 and 7—as common as A and Δ in Greek), it sufficed to find the other half of the text, extant in Chicago and unknown to Schnabel, containing other scribal errors more than outbalancing the first one. In fact before Newton's understanding of the relation between the precession of the equinoxes and the flattening of the earth's globe, the discovery of a slow gradual change in the longitudes of stars is neither a great achievement nor of theoretical interest for ancient astronomers—it only requires the preservation of records sufficiently old to establish beyond doubt the existence and amount of the effect.

The really significant contribution of Babylonian astronomy to Greek astronomy, in particular to Hipparchus' astronomy, lies in the establishment of very accurate values for the characteristic parameters of lunar and planetary theory [11] and in particular in the careful separation of the components of the lunar motion—longitude, anomaly, latitude, and nodal motion. The value of one of these parameters, the evaluation of the length of the mean synodic month as 29;31,50,8,20 days is not only fundamental for Hipparchus' theory of the moon but still appears in the Toledan Tables of the eleventh century. Islamic astronomy of the ninth century received its first impulses from Persia and India and assimilated the Ptolemaic refinements only somewhat later (in particular through al-Battâni, around 900). The earlier phase is mainly represented by al-Khwârizmi and influenced Spanish Islamic astronomy which remained somewhat outside the development in the Near East, Persia and Byzantium. But the European revival of astronomy beginning in the twelfth and thirteenth centuries took place mainly in Spain and Southern France and thus reflects again the Hindu, ultimately Babylonian, component of Is-

lamic astronomy. Not until the full recovery of the Ptolemaic methods in the Renaissance of the fifteenth century did the influences directly traceable to Babylonia of the Seleucid-Parthian period disappear.

For the modern historian there remains as the greatest unsolved question the problem of the transmission of astronomical knowledge from the temple schools in Uruk and Babylon to men like Hipparchus or Apollonius. The situation for astronomy is very different from the parallel situation in mathematics. This difference is very outspoken even in modern historical research. Ancient pre-Greek mathematics is easy to understand since it concerns only the elementary facts of arithmetic, geometry, and algebra. This material must have been accessible in countless elementary treatises at all periods and in all areas of the Near East. The astronomy of the Hellenistic period is a quite different matter. Not that the mathematical methods in themselves are more advanced than in the ordinary contemporary mathematics. The real difficulty lies in the astronomical motivation for the complex interplay of difference sequences which represent the different components of lunar and planetary motions. The determination of the characteristic parameters of periodic difference sequences of different order as well as the design of interpolation methods applicable to these difference sequences requires arguments totally outside the framework of ordinary ancient mathematics—in fact often strangely similar to the numerical methods of the latest Islamic period and to the beginning of modern "applied" mathematics.

It is obvious that the transmission of this type of material cannot be ascribed to any latent knowledge contained in easily accessible treatises. Even if we completely disregard the very serious practical difficulty of utilizing cuneiform material, we must assume a careful and extended training by competent Babylonian scribes and computers in order to account for the profitable use of any of the Babylonian ephemerides. We have at our disposal enough cuneiform instructions for the computation of lunar and planetary ephemerides to be able to say that they require a great deal of study and additional instruction and astronomical knowledge before it is possible to use them properly. And of the arguments upon which this mathematical theory was constructed these "Procedure Texts" contain nothing. It is therefore not surprising that the Greek astronomical literature does not contain a trace of factual information concerning the theo-

[11] Cf. Asger Aaboe, "On the Babylonian Origin of some Hipparchian Parameters," *Centaurus* **4** (1955–1956): 122–125.

retical foundations of Babylonian astronomy. Perhaps this silence is not accidental but reflects the fact that Greek astronomy did not know too much about the details of the Babylonian techniques and their theoretical and historical foundations. This might have forced the Greeks to look for methods of their own to solve the problems which arise in the mathematical description of the motion of the celestial bodies, a task which Apollonius and Hipparchus began and Ptolemy completed by bringing the planetary theory to the same degree of perfection as the lunar theory of old, but on the basis of cinematic models and spherical trigonometry, both unknown to his Babylonian predecessors.

Zur Transkription mathematischer und astronomischer Keilschrifttexte.

Von O. N e u g e b a u e r (Göttingen).

Für die Transkription der Keilschriftzeichen ist durch die Listen Thureau-Dangins ein einheitliches System geschaffen, das gestattet, in eindeutiger Weise aus der Transkription auf die Zeichen des Textes zu schliessen. Eine ähnliche Uebereinkunft für die Zahlzeichen fehlt jedoch, und so möchte ich mir erlauben, einige dieses spezielle Gebiet betreffende Vorschläge zu machen. Für mathematische Texte ist dies kein nebensächlicher Punkt, denn eine ungeschickte Umschrift der Zahlen kann die Transkription unverständlich machen und den Sinn einer Rechnung vollkommen entstellen.

I. Z a h l z e i c h e n.

Oberste Forderung: d i e T r a n s k r i p t i o n m u s s d i e V e r h ä l t n i s s e d e s T e x t e s s o g e t r e u w i e m ö g l i c h w i e d e r s p i e - g e l n. Selbstverständlich reicht es aus, irgend eine e i n d e u t i g e Beziehung Text — Transkription zu schaffen; es ist also überflüssig, ◁⟨⫯⫯ durch XII zu umschreiben, zumal ja bereits IV oder V nicht mehr den keilschriftlichen Bild entspricht. Da die Schreibung der Zahlen (abgesehen von graphischen Varianten wie ⟨⟨⟨ und ⟨⟨⟨) feststeht, so reichen die üblichen arabischen Ziffern völlig aus, um erkennen zu lassen, welche Zeichengruppen im Text stehen. Treten mehrstellige Sexagesimalzahlen auf, z. B. ⟨⟨, so wäre also 1 40 die sinngemässe Transkription. Aus drucktechnischen Gründen ist aber vorzuziehen, die Zeichen voneinander zu trennen. Dafür einen P u n k t zu verwenden, scheint mir aus zwei Gründen u n - g ü n s t i g zu sein: 1) ist der Punkt bei mathematischen Dingen schon mit der Operation der Multiplikation verbunden, 2) spielt ja das Trenn-

zeichen ⟨ die Rolle unseres Schlusspunktes, so dass man (z. B. bei seleukidischen Texten) 1.40 für ⟨⟨ halten könnte. Ich schlage also als T r e n n u n g s z e i c h e n d e r S e x a g e s i - m a l s t e l l e n d a s K o m m a vor, da dieses weder ein mathematisches Symbol ist, noch für ein Keilschrift-Aequivalent gehalten werden kann. Steht dagegen ⟨⟨ im Text (ist also etwa 3640 im Gegensatz zu 100 gemeint), so ist dies durch 1,.,40 sinngemäss zu umschreiben. Da Keilschriftzahlen irgendwelche S t e l l e n w e r t - b e z e i c h n u n g grundsätzlich nicht zukommt, so ist sie auch i n d e r T r a n s k r i p t i o n eines Textes absolut zu v e r m e i d e n.

Erst in der U e b e r s e t z u n g eines Textes scheinen mir positionelle Angaben gerechtfertigt. Hier gibt es eine ganze Musterkarte von Schreibweisen: Hochgestellte römische Ziffern, dann Abkürzungen m(inuten), s(ecunden), t(erzen) usw., ° ′ ″ usw. Neuerdings hat Thureau-Dangin sogar ‵‵ ‵ ° ′ ″ in Anwendung gebracht. Alle diese Methoden haben viele Mängel: drucktechnisch kompliziert, undeutlich und unbequem und nicht einheitlich zu brauchen. Man denke sich z. B. einen Mondrechnungstext mit den besonders beliebten Zeichen ° ′ ″ : In einigen Spalten bedeuten die Zahlen wirklich Winkel-Grade usw. In anderen Winkelgeschwindigkeiten, Geschwindigkeitsänderungen, Zeiten usw. Welches Chaos wird aus allen astronomischen Texten, wenn man sich auf ‵‵ ‵ ° ′ ″ einigte! Zum Ausdruck eines Stellenwertes bedarf es statt dessen bloss der einzigen Uebereinkunft: h i n t e r d e r G r u n d e i n h e i t setze man ein S e m i - k o l o n statt eines Kommas (das wegbleiben kann, wenn die letzte Ziffer Einer bedeutet). Dann ist 1;40 = 1 + 2/3, 1,40; = 1,40 = 100. Ist aber keine der vorhandenen Ziffern den Grundeinheiten entnommen, so hat man entsprechend Nullen einzuschalten, z. B. 0;40 = 2/3, 4,0 = 240, 1,0,40 = 3640. Die letzte Schreibung würde besagen, dass im Text nur ⟨⟨ steht, dass aber sachliche Gründe dazu zwingen, diese Zahl gerade als 3640 zu interpretieren (der Abstand der Zahlzeichen sagt im allgemeinen — z. B. bei Tabellentexten! — über den Stellenwert gar nichts aus, kann also bei der Transkription ignoriert werden). Dagegen würde 1,.; 40 bedeuten, dass ein ⟨⟨ des Textes aus irgendwelchen sachlichen Gründen als 60 2/3 zu interpretieren ist. Dieses Transkriptionsverfahren scheint mir äusserst übersichtlich, absolut eindeutig und so beschaffen, dass man noch beliebige Mass-

---

[6]) Heron, *Stereometr.* I, 38.

[7]) Pap. Vind., Aufg. 13.

[8]) S. hierzu in dem Anm. 3 genannten Aufsatz S. 246 ff. Ueber Stumpfaufgaben in der indischen Mathematik siehe B. Datta, *On the supposed indebtedness of Brahmagupta to Chiu-chang Suan-shu*: *Bull. Calc. Math. Soc.* 22, 1930, S. 39—51, bes. S. 45 ff.

[9]) Vergleiche auch den sonstigen Text, z. B. *Stereom.* I, 33, ed. Heiberg, S. 34, 15 ff.: σύνθες κορυφὴν καὶ βάσιν· γίνονται λβ̄· ὧν τὸ L'· γίνονται ις̄· ἐφ' ἑαυτὰ γίνονται σνς und BM 84194, V, Z. 44 ff: sà-sùm ù mu-ḫa kumur 17 ta-mar [mišil 17 ḫe-pé 8.]30 tamar šutamḫir 1.12.15 ta-mar.

und sonstige Bezeichnungen hinzufügen kann.

Schliesslich möchte ich mir noch einen Vorschlag für die Zahlzeichen in astronomischen Texten erlauben. In der Transkription eines Textes ist das eben geschilderte Verfahren natürlich ohne weiteres anwendbar. In der Übersetzung und beim Kommentar bedarf man aber im wesentlichen dreier Einheiten: Bogengrade °, babylonische Stunden H (im Wert von 4 unserer Stunden) und unsere Stunden h. Alle anderen Symbole für die sexagesimalen Teile dieser Einheiten sind überflüssig, insbesondere die „Minuten". So kann man etwa schreiben: Die Länge des synodischen Monats beträgt $29^d + 12;44,3,20^h = 29^d + 3;11,0,50^H = 29;31,50,8,20^d$. Bei Winkelangaben kann man die lästigen „Zeitgrade" ohne weiteres entbehren, da man entweder Angaben in H oder h machen kann oder in Bogengraden. Ich würde vorschlagen, auch für „Bogenminuten" grundsätzlich nicht ' zu schreiben, sondern etwa 0;17°. Damit wäre eine wirklich einheitliche und auch drucktechnisch sparsame Umschriftweise aller Sexagesimalzahlen erreicht.

II. Termini.

Während die Transkriptionsweise der Zahlen eine reine Angelegenheit der Konvention ist, greift die Frage nach der Transkription von Ideogrammen mathematischer Termini in ein rein philologisches Gebiet ein, in dem sich schon eine feste Tradition gebildet hat. Ich bin mir bewusst, dass ich mit dem Folgenden gegen die opinio communis verstosse, möchte aber doch einiges zur Diskussion stellen, das mir sachlich wichtig zu sein scheint.

Ein Beispiel. Thureau-Dangin transkribiert (a) *GAM* (b) *DU*-ma mit (a) *adi* (b) *tubal-ma*. Dazu ist zu bemerken: warum zweite Person und nicht Imperativ? Beides kommt in mathematischen Texten vor, die Auswahl bei *DU*-ma ist reine Willkür (ebenso die Stämme, I oder II und dgl.). Warum *DU* = *abâlu* und nicht z. B. = *šakânu* (letzteres kommt oft in mathematischen Texten vor)? Und ähnliche Fragen mehr in anderen Fällen.

Es ist keine blosse Nebensächlichkeit, die hier gestreift wird, sondern ein für mathematisch-astronomische Texte ganz wesentlicher Punkt. Zunächst: man kann unter Umständen aus der Transkription nicht mehr erkennen, welches Zeichen im Text stand (was z. B. für die Frage der Zusammengehörigkeit gewisser Textgruppen durchaus wesentlich sein kann). Dann: wie hat man sich in Fällen zu verhalten, wo ganze su-

merische Phrasen in akkadischen Texten stehen? Warum muss man sie akkadisieren? Mit demselben Recht müsste man das Lateinische oder Französische aus im Prinzip deutschen Schriften von Gelehrten des 17. Jahrhunderts ausmerzen. Und schliesslich: Man zerstört durch das absolute Akkadisieren eine fundamentale sachliche Rolle der Ideogramme: dass sie nämlich vollkommen wie mathematische Symbole wirken und man wenn man e⁻x² durch „e-hoch-minus-ix-Quadrat" ersetzte oder (weil man ja im allgemeinen gar nicht weiss, wie ein solches Symbol ausgesprochen wird) noch einen „richtigen" grammatischen Satz daraus fabrizierte. Dass die Ideogramme an dem so auffallend stark algebraischen Charakter der babylonischen Mathematik einen sehr wesentlichen Anteil haben (ob ursprünglich gewollt oder nicht, ist hier gleichgültig), ist mir bei der Beschäftigung mit dieser Textgattung immer deutlicher geworden. Charakteristischer Weise ist diese Verwendung der Ideogramme für Operationen und Grössen in den späteren Texten viel ausgeprägter als in den älteren; das heisst aber, dass ihr Wert für den mathematischen Formalismus immer mehr erkannt und ausgenutzt worden ist. Für den Aussenstehenden sei aber etwa an die reine Symbolik der astronomischen Texte erinnert. Hat es Sinn, sich bei

$$25, 9,30 \quad lal$$
$$11, 4,30 \quad lal$$
$$9,48 \quad tab$$
$$24, 8 \quad tab$$

usw. den Kopf zu zerbrechen, welche akkadischen Aequivalente man für die „Vorzeichen" tab und lal einsetzen soll? Wenn man es tut (was doch offenbar weitgehende Willkürlichkeiten verlangt), so hat man nur ein präzises mathematisches Symbol durch unsichere Wortäquivalente ersetzt. Wer würde ahnen, dass *rêš sikkati* gerade „Dreieck" bedeutet? Statt dessen ist nicht mehr erkennbar, welche Zeichen der Text hat, während *SAG-DÛ* mindestens dieses ohne weiteres leistet. Ich bitte, das mathematische Symbol arc cos zu übersetzen!

Ich bin mir bewusst, dass es für die Umschreibung der Termini keine festen Regeln geben kann, wenn man nicht wieder ins andere Extrem verfallen will; aber es scheint mir der Beachtung wert, dass durch das grundsätzliche Akkadisieren aller Ideogramme geradezu die innere mathematische Struktur eines Textes zerstört werden kann. Dies zu vermeiden, scheint mir wichtiger als jeglicher Purismus.

Reprinted from
Archiv für Orientforschung
8, (1932/3)

2. Egyptian

Die Bedeutungslosigkeit der ‚Sothisperiode' für die älteste ägyptische Chronologie.

Von

O. Neugebauer, Kopenhagen.[1]

Lasciate ogni speranza

1. Vorbemerkungen. Es wäre ein leichtes gewesen, die Einfachheit der Überlegungen dieser Arbeit zu verbergen, einerseits durch eine Auseinandersetzung mit der umfangreichen Literatur zur ältesten ägyptischen Chronologie, andererseits durch ein Eingehen auch auf andere Fragen des altägyptischen Kalenderwesens, die nur mehr oder minder indirekt mit der ‚Sothisperiode' zu tun haben. Ich habe beides vermieden, denn das wesentliche Resultat der folgenden Betrachtungen ist gerade das, daß es mit den elementarsten Argumenten möglich ist zu zeigen, daß es kein Mittel gibt und geben kann, die Einführung des ägyptischen Kalenders auf den Tag genau zu datieren (ob man nun dafür nach alter Manier den 19. Juli des Jahres 4241 v. Chr. ansieht oder unter Berücksichtigung genauerer astronomischer Rechnung ein anderes ähnliches Datum des fünften Jahrtausends v. Chr. wählt), daß vielmehr die Unsicherheit in der Bestimmung dieses Anfangspunktes *Jahrhunderte* beträgt und die Diskussion um Tage oder einige Jahre etwa dieselbe Bedeutung hat wie ein Streit um die Zentimeter und Millimeter einer Längenmessung, bei der die Meter ungewiß sind.

Daß wir uns ein ausführliches Eingehen auf die Literatur ersparen können, verdanken wir der Existenz eines erschöpfenden Werkes, in dem alle älteren und neueren Untersuchungen zur ägyptischen Chronologie berücksichtigt und referiert sind, nämlich

---

[1] Ich möchte hervorheben, daß der Kern dieser Arbeit in gemeinsamer Diskussion mit Dr. W. Feller, Stockholm, entstanden ist. Ebenso habe ich Herrn Dr. Feller für die Literaturnachweise über den Nil sowie für Durchsicht von Manuskript und Korrekturen zu danken.

Acta orientalia XVII.

12

L. Borchardt ‚Die Mittel zur zeitlichen Festlegung von Punkten der ägyptischen Geschichte und ihre Anwendung‘.[1] Wenn im folgenden Borchardt besonders häufig zitiert wird, auch bei Argumenten, die wir nicht zu den unseren machen können, so darf dies nicht als eine gerade gegen diesen Autor gerichtete Polemik aufgefaßt werden, sondern bedeutet nur, daß jeder, der sich über die Vorgeschichte dieser Schlußweisen orientieren will, alle Angaben am einfachsten bei Borchardt findet.

Was andere Fragen des ägyptischen Kalenderwesens betrifft (so z. B. die verschiedenen Formen eines Mondjahres), so konnten wir auch sie ruhig beiseitelassen, weil diese Fragen zwar oft mit der ‚Sothisperiode‘ gemeinsam behandelt worden sind, aber mit dem Kern der Sache nichts zu tun haben. Wir legen im Gegenteil gerade Wert darauf zu betonen, daß diese äußerst komplizierten historischen Probleme ganz unabhängig sind von der Diskussion über den Beginn der ersten Sothisperiode; vermutlich hat aber die gleichzeitige Behandlung von ernsten Fragen des ägyptischen Kalenderwesens und der nach dem Entstehungsdatum des Kalenders an Hand der ‚Sothisperiode‘ dazu beigetragen, die letztere Frage nicht unmittelbar als Scheinproblem erkennen zu lassen. In der Tat besteht alles, was im folgenden zu tun sein wird, nur darin, einmal die Voraussetzungen klar zu formulieren, auf denen die Sothisberechnungen beruhen. Dies allein ist ausreichend, um zu zeigen, daß die älteste Chronologie Ägyptens genau dieselbe Größenordnung von Unsicherheit aufweist wie jede andere Chronologie einer Epoche weit vor der Existenz einer schriftlichen Überlieferung.

2. Definitionen. Da wir wünschen, daß unsere Bemerkungen auch einem Historiker voll verständlich und nachprüfbar sind, dessen Spezialgebiet nicht gerade Chronologie ist, so stellen wir hier die wenigen Begriffe zusammen, die wir benutzen müssen. Irgend etwas Neues enthält dieser Abschnitt nicht.

---

[1] Quellen und Forschungen zur Zeitbestimmung der ägyptischen Geschichte, Bd. 2, Kairo, Selbstverlag, 1935. Wir zitieren dieses Werk im folgenden kurz als ‚Borchardt‘. Für andere Abkürzungen vgl. das Literaturverzeichnis am Schluß.

Ägyptisches Jahr oder *Wandeljahr*: ein Zeitraum von 365 Tagen.

Julianisches Jahr: ein Zeitraum von $365 + {}^{1}/_{4}$ Tagen.

Diese beiden Zeiträume sind astronomisch bedeutungslos und reine Rechnungseinheiten (worüber man sich selbstverständlich bei ihrer Einführung keineswegs im klaren war). Dagegen ist das *tropische Jahr* oder *Sonnenjahr* (oder, wenn kein Mißverständnis zu befürchten ist, *Jahr* schlechthin) eine Periode, die durch den tatsächlichen Sonnenlauf bestimmt wird, definiert etwa durch den Zeitraum zwischen zwei benachbarten Augenblicken, in denen sich die Sonne im Frühlingspunkt befindet. Die Länge des tropischen Jahres ist streng genommen nicht genau konstant; es genügt aber, sich zu merken, daß es etwas (ca. 11 Minuten) kürzer ist als das julianische Jahr und daß sich unser Kalender (der ‚*Gregorianische*‘) mit Hilfe gewisser Schaltregeln in historischer Zeit nie erheblich von dem tropischen Jahr entfernt,[1] so daß ein *gregorianisches Datum* stets die Jahreszeit richtig erkennen läßt, da sich das Datum des Frühlingspunktes nie wesentlich vom 21. März greg. entfernt. Der *Julianische Kalender* kennt nicht die etwas kompliziertere gregorianische Schaltregel, sondern nur die Schaltung eines 29. Februar jedes vierte Jahr, während alle übrigen Monate die heute übliche Anzahl von Tagen erhalten. Zwischen 200 und 300 n. Chr. sind julianische und gregorianische Daten gleichwertig, um — 3000 sind dagegen von julianischen Daten bereits 24 Tage abzuziehen, um gregorianische zu erhalten, um — 4000 bereits 31 Tage.[2]

Heliakischer Aufgang. Damit ein Stern sichtbar wird, muß es ausreichend dunkel sein, d. h. die Sonne muß tief genug unter dem Horizont stehen (der genaue Betrag hängt von der Helligkeit des Sternes sowie von der Lage des Beobachtungsortes ab). In Fig. 1

---

[1] Da Kalenderdaten notwendigerweise ganzzahlig sind, so müssen die Abweichungen gegen den genauen Moment eines astronomischen Phänomens in jedem Kalender gelegentlich Fehler von der Größenordnung eines Tages aufweisen. Was allein ausgeschlossen werden kann, ist, daß sich diese Fehler auf die Dauer summieren.

[2] Bequeme Umrechnungstafeln z. B. bei P. V. Neugebauer, Hilfstafeln, Tafel 15.

12*

bedeute H den Ost-Horizont, ⊙ die Sonne in jenem Abstand unter
dem Horizont, bei dem ein bestimmter Stern S gerade noch sichtbar wäre, wenn er über dem Horizont stände. Dies soll durch den
punktierten Kreis um die Sonne angedeutet sein, der ganz unter
dem Horizont liegen muß, damit es für die Sichtbarkeit von S
noch dunkel genug ist. Fig. 1a gibt eine Situation, in der S nicht
sichtbar wird, denn wenn die tägliche Drehung des Himmels S über
den Horizont hebt (großer Pfeil), so ist es gleichzeitig bereits zu
hell geworden, um S noch zu erkennen. Fig. 1b zeigt die Stellung
um einen Tag später; da die Bewegung der Sonne relativ zu den
Fixsternen (kleiner Pfeil) gegen die Richtung der täglichen Drehung

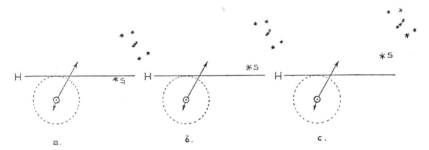

Fig. 1. Heliakischer Aufgang.

erfolgt, ist der Abstand Sonne—Stern bereits so groß, daß S gerade
noch über den Horizont gelangt, bevor die Helligkeit zu groß wird;
dieser erstmalige sichtbare Aufgang von S nach einem Zeitraum
der Unsichtbarkeit heißt der *heliakische Aufgang* von S. In Fig. 1c
ist der Abstand Sonne—Stern neuerlich größer geworden, so daß
auch die Zeitdifferenz zwischen Aufgang von S und Unsichtbar-
werden von S wegen des folgenden Sonnenaufgangs zunimmt, was
nun von Tag zu Tag deutlicher wird, bis die Sonne fast die ganze
Ekliptik durchwandert hat und S wieder von der anderen Seite her
in dem Schein der Sonne verschwindet und dann bis zum nächsten
heliakischen Aufgang ganz unsichtbar bleibt.

 Sothis = Hundsstern = Sirius ein sehr auffallender Fixstern,
wenig östlich vom Orion (s. Fig. 1).

Präzession, Eigenbewegung von Fixsternen, arcus visionis spielen (obwohl gelegentlich zitiert) für die Betrachtungen dieser Arbeit nur die Rolle von Statisten, brauchen also nicht näher definiert zu werden.

3. Die klassische Schlußweise. Wir wollen nun angeben, in welcher Weise man das Datum der Einführung des ägyptischen Kalenders zu bestimmen pflegt, und zwar zunächst in der älteren Fassung und dann mit einer kleinen rein astronomisch bedingten Korrektur, die noch an der einfachen Rechnung der ersten Formulierung anzubringen ist.

Als absolut gesicherter Ausgangspunkt dient die Tatsache, daß wir das Verhältnis des ägyptischen Kalenders zum Julianischen und damit auch zum Gregorianischen Kalender z. B. dadurch genauestens kennen, daß Ptolemäus im Almagest (ca. 150 n. Chr.) seine Beobachtungsdaten stets auf den ägyptischen Kalender reduziert, weil dieser auf Grund seiner Unabhängigkeit von allen lokalen Ären und Schaltregeln besonders für astronomische Zeitangaben geeignet war. Da man andererseits imstande ist, die Ptolemäischen Angaben (meist Finsternisse) mit den modernen Rechnungen in Beziehung zu setzen und da die ägyptischen Jahre stets unverändert 365 Tage enthielten, so ist also der ganze ägyptische Kalender astronomisch genau fixiert.

Aus einer Angabe von Censorinus[1] folgt ferner, daß im Jahre 139 n. Chr. der Neujahrstag des ägyptischen Kalenders (d. h. der erste Tag des ersten Monats der sog. ‚Überschwemmungsjahreszeit‘) mit dem heliakischen Siriusaufgang zusammenfiel und daß das Datum dieses Tages der 19., bzw. 20. Juli jul.[2] ist.

Nun ist das ägyptische Jahr um einen Vierteltag kürzer als das julianische Jahr; die beiden Jahresanfänge entfernen sich also

---

[1] Borchardt S. 15 f. Der Text ist publiziert: Censorini de die natali liber, cap. 21, ed. Hultsch (Leipzig, Teubner, 1867) p. 46, 13 ff., bzw. cap. 18 (p. 38, 22 ff.), ebenso Meyer, Chron. S. 23 f. Diese Schrift Censorins ist 238 n. Chr. abgefaßt (vgl. RE **3**, Sp. 1909).

[2] P. V. Neugebauer, Hilfstafeln, Vorbemerkungen zu Tafel 21 (bzw. Borchardt S. 15 f.).

von Jahr zu Jahr um einen Vierteltag voneinander, um erst nach
$4 \times 365 = 1460$ Jahren wieder zusammenzufallen.[1] Ferner: ‚infolge
seiner Eigenbewegung und der Präzession der Nachtgleichen hat der
(heliakische) Siriusaufgang Jahrtausende hindurch mit dem julianischen Jahr gleichen Schritt gehalten‘,[2] so daß die Periode von
1460 Jahren auch die Periode zwischen Koinzidenz von ägyptischem
Neujahrstag und heliakischem Siriusaufgang ist; man bezeichnet sie
daher als ‚Sothisperiode‘. Da schließlich der heliakische Aufgang
der Sothis ‚im fünften und vierten Jahrtausend mit dem ersten Beginn der Nilschwelle zusammenfiel‘,[2] so erhält man in dieser Zeit
noch eine weitere Koinzidenz, nämlich daß die erste Jahreszeit des
ägyptischen Kalenders, die ‚Überschwemmungszeit‘ heißt, auch tatsächlich mit der Überschwemmung zusammenfällt.[3] Rechnet man
also vom 19. Juli jul. 140 n. Chr.[4] um je 1460 Jahre zurück, so
ergeben sich die Koinzidenzen: 1321 v. Chr., 2781 v. Chr. und
4241 v. Chr. 19. Juli jul., und da der ägyptische Kalender um
3000 v. Chr. bereits in Gebrauch war, muß das zuletzt genannte
Jahr das der Einführung des Kalenders gewesen sein. Eduard Meyer
formuliert dies in dem lapidaren Satz:[5] ‚Auf Grund der gewonnenen

---

[1] Dabei ist der Vierteltag des Koinzidenzjahres ignoriert; die Koinzidenz
bestimmt daher genau genommen nicht *ein* Jahr, sondern *vier* Jahre, bis sich die
Daten wieder mit Sicherheit um einen vollen Tag unterscheiden. Diese kleine
Unsicherheit pflegt man natürlich als historisch belanglos beiseite zu lassen.

[2] Meyer GA I, 2 § 159 (S. 29).

[3] Nach der Bemerkung von S. 171 entspricht nämlich dem 19. Juli jul. der
Zeit um — 4000 ein um etwa einen Monat früheres gregorianisches Datum. Wie die
Zusammenstellung von modernen Daten von minimalen Wasserständen des Nils
zeigt (vgl. Fig. 4 S. 190), kann man aber Mitte oder Ende Juni ganz gut als die
Zeit der beginnenden Nilschwelle ansehen, für weiter nördlich liegende Orte natür-
lich entsprechend später.

[4] Daß die Rechnung mit 140 n. Chr. statt wie oben angegeben mit 139
begonnen wird, beruht auf einer Korrektur, die Ed. Meyer an dem Censorin-
Datum anbringt (Meyer, Chronol. S. 28). Wir können uns ein Eingehen auf diese
Korrektur ersparen, da wir doch sogleich eine Korrektur erwähnen müssen, die
aus astronomischen Gründen an der Länge der ‚Sothisperiode‘ angebracht werden
muß, die alle Daten um einige Jahre verschiebt.

[5] Meyer, Chronol. S. 45.

Ergebnisse stehe ich nicht an, die Einführung des ägyptischen Kalenders
am 19. Juli 4241 v. Chr. als das erste sichere Datum der Weltgeschichte
zu bezeichnen.‘

Damit ist die klassische Schlußweise zur Berechnung des
Datums der Einführung des ägyptischen Kalenders angegeben. Es
bedeutet nur eine ganz unwesentliche Modifikation, daß man die
Behauptung einer praktisch vollkommenen Identität zwischen juliani-
schem Jahr und Zeitdifferenz zwischen zwei heliakischen Siriusaufgängen durch eine astronomisch etwas korrektere ersetzt, wonach
die Koinzidenzen zwischen ägyptischem Jahr und heliakischem
Siriusaufgang nicht genau durch 1460 Jahre voneinander getrennt
sind, sondern durch Zwischenräume von 1455, 1456, bzw. 1458 Jahren,
was (von 139 n. Chr. nach rückwärts gezählt) 1317 v. Chr., 2773 v. Chr.
und 4231 v. Chr. als die Jahreszahlen ergibt, in denen der heliakische Siriusaufgang auf den ersten Tag des ägyptischen Kalenders
fällt. Auch die Tagesdaten verschieben sich etwas, so daß schließlich
der 16./17. Juli jul. 4231 v. Chr. als Geburtstag des ägyptischen
Kalenders zu gelten hätte.[1]

Diese zuletzt angegebene rein astronomische Korrektur ist
natürlich für das historische Bild gänzlich belanglos und ändert
auch nichts an dem Prinzip der ganzen Schlußweise. Sie hat aber
zweifellos den psychologischen Effekt, den Glauben an die Zu-
verlässigkeit der ganzen Methode zu erhöhen, denn wenn es sich
lohnt, um einige Jahre oder gar Tage zu diskutieren, so hat dies
doch nur Sinn, wenn die Datierungsmethode als solche in der Haupt-
sache völlig gesichert ist.

4. Die ‚Astronomen‘ von Heliopolis — eine reductio ad
absurdum. Als Resultat der bisher referierten Schlüsse ergibt sich
also das Jahr 4231 v. Chr. als das der Einführung des ägyptischen
Kalenders. Wir wollen noch ein kleines Stück in unserem Bericht
über die übliche Darstellungsweise dieser Begebenheit fortfahren
und dann versuchen, uns bei den dabei gebrauchten Worten etwas

---

[1] P. V. Neugebauer, Hilfstafeln, Vorbemerkungen zu Tafel 21.

Konkretes vorzustellen; dies wird genügen, um uns die innere Un-
haltbarkeit des ganzen Bildes klar zu machen und zugleich auch
eine unbewiesene Voraussetzung deutlich hervortreten zu lassen
(Nr. 5), die man nur zu streichen hat, um eine ganz natürliche Ent-
wicklungsgeschichte des ägyptischen Kalenders skizzieren zu können
(Nr. 6).

Das ägyptische Jahr der geschichtlichen Zeit kennt bekanntlich nur ganz formale Monate von stets 30 Tagen Länge, die zu je
vieren auf die drei Jahreszeiten ‚Überschwemmungs-‘, ‚Wachstums-‘
und ‚Erntezeit‘[1] verteilt sind, zu denen noch die fünf ‚Zusatztage‘ hinzu-
kommen, um die Zahl 365 der Tage voll zu machen. Diese ‚Monate‘
haben also mit dem Mondlauf nicht das geringste mehr zu tun; es
gibt aber eine Reihe von Anzeichen, aus denen hervorgeht, daß es
auch echte Mondmonate von bald 29, bald 30 Tagen gegeben hat.[2]
Da das gänzlich schaltungslose 365tägige Jahr als großer systemati-
scher Fortschritt gegen jede Art von Monatsschaltungen anzusehen
ist, ging, so schließt man, dem Kalender der Wandeljahre fester Länge
eine Periode anderer Kalenderformen voraus. So faßt man auch ganz
konsequent die Einführung des neuen Kalenders als scharfen Bruch
mit der Vergangenheit auf, ‚bei der es also nicht ohne Gewalt-
maßnahmen abging‘.[3] Aber zum Glück kennt man auch die großen
Autoritäten, die diese Reform durchgesetzt haben: es war in erster
Linie der Oberpriester von Heliopolis, der ‚Oberastronom‘ (so wird
sein Titel ‚Großer der Schauenden‘ interpretiert), der an einer ‚Sternwarte‘ wirkte, ‚die in ganz Ägypten anerkannten Ruf hatte‘,[3] an
der ‚Landessternwarte‘ in Heliopolis.[4] Dort geschah es dann, daß
‚ein hervorragender Astronom oder gar eine Mehrheit von solchen
eine Kalenderänderung erdachten und durchführen konnten, die den

---

[1] Vgl. z. B. Sethe, Zeitr. S. 294 ff. Danach ist die ursprüngliche Wortbe-
deutung der dritten Jahreszeit vermutlich ‚Wassermangel‘, die der zweiten auch
‚Hervorkommen (der Äcker aus dem Hochwasser)‘.

[2] Vgl. z. B. Sethe, Zeitr. S. 296 ff., oder Borchardt S. 5 ff., S. 19 ff., S. 36 ff.
und auch sonst passim.

[3] Borchardt S. 7.

[4] Borchardt S. 8. Ähnliche Termini passim in der Literatur.

Zweck hatte, die Schwankungen des Jahresanfangs „um den längsten Tag" oder besser „um den Hundssternfrühaufgang" herum und die verschiedenen Längen der Monate zu vermeiden, sowie den Kalender mit dem Sonnenlauf und seinen Jahreszeiten in Einklang zu bringen‘.[1]

Wir wollen uns nun gestatten, dieses Bild von der ägyptischen Astronomie des Jahres 4231 v. Chr. etwas genauer zu analysieren. Dabei müssen wir leider sogleich auf den ‚Oberastronomen‘ verzichten, denn daß man den Titel ‚Großer der Schauenden‘ (*wr mꜣ-w*) gerade auf astronomisches Schauen, also Beobachten, bezieht, ist eine reine petitio principii. Man wird uns vielleicht die Tatsache eines heliopolitanischen Sonnenkultus entgegenhalten; aber eine noch so hervortretende Verehrung der göttlichen Kraft des Tagesgestirns ist noch lange keine Astronomie — das zeigt z. B. die Amarnazeit ebensogut wie die babylonische Religion oder die Sonnenkulte der römischen Kaiserzeit.

Aber die Frage der Titulatur unserer Astronomen ist nicht sehr belangreich; sehen wir also besser nur auf die konkrete Seite ihrer Beschäftigung. Wir hören gleich von mehreren hervorragenden Astronomen, sogar von verschiedenen mehr oder weniger angesehenen ‚Sternwarten‘ (in einem Lande der Größe Dänemarks), kurz von einem geradezu erstaunlichen Zustand zu einer Zeit, die weit vor der Entstehung der Schrift liegt, wo man die Töpferscheibe noch nicht kannte,[2] in Lehm- und Schilfhütten wohnte und die Toten, bloß mit einer Matte bedeckt, in bescheidene Gräber am Wüstenrand verscharrte.

In diesem Milieu wirkten also unsere Astronomen an ihren Sternwarten. Womit beschäftigten sie sich? Irgend ein theoretisch-astronomisches Gebäude haben die Ägypter in ihrer ganzen langen Geschichte nicht errichtet.[3] Ernstliche Rechnungen auszuführen hatten sie weder Anlaß noch Möglichkeit: noch reichlich 2000 Jahre

---

[1] Borchardt S. 7.

[2] Vgl. z. B. Scharff, GÄV S. 19.

[3] Auch in Babylonien entstand ein solches erst in den letzten Jahrhunderten v. Chr.

später wäre die Mathematik des Mittleren Reiches gänzlich außer-
stande gewesen, eine einzige der Aufgaben zu lösen, wie sie auch
nur die einfachsten Fragen des Periodenausgleichs der Bewegungen
der Gestirne stellt.[1] Wie sollten die Astronomen von Heliopolis
vom Jahre 4231 da rechnen, wo sie ihre Zahlen, wie die späteren
hieroglyphischen Zahlzeichen noch deutlich genug zeigen, höchstens
in Kerbhölzer schnitten, Einer nach Einer, und dann zu Gruppen
nach der Zahl ihrer Finger zusammenfaßten. Zu den schwierigen
Prüfungen, die der Tote in der Unterwelt zu bestehen hat, gehört
das Zählen seiner Finger, denn ,dieser herrliche Gott wird sagen:
Hast du mir einen Mann übergefahren, der seine Finger nicht
zählen kann?‘, und der Tote muß den Fährmann durch das Vor-
zählen seiner Finger von seinem überlegenen Wissen überzeugen.[2]
So war also das mathematische Niveau der gelehrten Priester beschaffen, aus deren Kreis die Reformatoren des prä-prädynastischen
Kalenders hervorgingen.

Mit einer ,theoretischen Astronomie‘ ist es also nichts. Aber vielleicht waren unsere Astronomen doch wenigstens ,große Schauende‘?
Ein argumentum ex silentio ist nicht schön, aber ist es reiner Zufall, daß uns aus den 4000 Jahren ägyptischer Geschichte nicht
eine einzige Beobachtung einer Sonnen- oder Mondfinsternis erhalten ist,[3] obwohl wir gleichzeitig aus einer unendlichen Fülle von
Texten aller Art die minutiösesten Einzelheiten des ägyptischen
Lebens, des Kultus, der Verwaltung und der politischen Geschichte
überliefert bekommen haben? Ist es reiner Zufall, daß wir aus
Babylonien über Unmengen kalendarischer und astronomischer Be-

---

[1] Für den rein additiven Charakter der ägyptischen Mathematik vgl.
O. Neugebauer, Die Grundlagen der ägyptischen Bruchrechnung, Berlin, Springer,
1926, S. 16 f. u. S. 41 f., oder meine ,Vorlesungen‘ Kap. IV.

[2] Sethe [1] S. 16 sowie Gunn [1]. Vgl. dazu auch H. Kees, Totenglauben
und Jenseitsvorstellungen der alten Ägypter (Leipzig, Hinrichs, 1926) S. 112 ff.
Der zitierte Text stammt aus dem ,Totenbuch‘ des Mittleren Reiches, geht aber
zweifellos auf sehr viel ältere Vorbilder zurück (vgl. Gunn [1]).

[3] Borchardt S. 3 ff.

richte bis weit zurück in die Zeit von Ur III (ca. 2300)[1] oder auch nur bis in die Hammurapizeit (ca. 2000 v. Chr.)[2] verfügen,. und daß noch Ptolemäus (ca. 150 n. Chr.) praktisch geschlossene Reihen babylonischer Beobachtungen seit ca. 750 v. Chr. kennt,[3] aber nicht eine einzige ägyptische Beobachtung irgendeiner Art zitiert, obwohl er doch in Alexandrien lebte? Um aber bei positiven Tatsachen zu bleiben: so viel steht fest, daß unsere großen heliopolitanischen Astronomen die astronomischen Jahreszeiten konsequent ignoriert haben; sie teilen ja ihr Jahr in *drei* ‚Jahreszeiten', deren Namen jeden Zweifel daran ausschließen, daß sie nur auf die durch den Nil bedingten landwirtschaftlichen Ereignisse Bezug nehmen, und der Ausgangstag, der 16. Juli jul., liegt um etwa 12 Tage von der Sommersonnenwende entfernt.[4] Nicht einmal der erste Schritt zu einer systematischen Astronomie ist also getan worden, nämlich die Feststellung der Tag- und Nachtgleichen und der Sonnenwenden. Und in voller Übereinstimmung damit ist auch die Aufgabe einer Ausgleichung von Sonnen- und Mondlauf nicht angegriffen worden: man lese z. B. Borchardts Referat über die chaotische Lage der Mondfeste[5] und vergleiche damit die babylonischen Bemühungen zur Schaffung eines kombinierten Mond- und Sonnenkalenders, um sich klar zu machen, daß auch hier die Leistung der Astronomen von Heliopolis gleich Null gewesen ist.

Man wird gegen die letzte Bemerkung vielleicht einwenden wollen, daß diese Astronomen gerade durch Einführung des 365tägigen Jahres die Rücksicht auf den alten Mondkalender ab-

---

[1] Vgl. etwa die detaillierten Untersuchungen von N. Schneider, Die Zeitbestimmungen der Wirtschaftsurkunden von Ur III, Analecta Orientalia **13** (Rom 1936).

[2] Vgl. etwa Langdon-Fotheringham-Schoch, The Venus tablets of Ammizaduga, Oxford 1928.

[3] Almagest Buch III Kap. 7.

[4] Borchardt S. 5 Anm. 7. Mit Recht lehnt Borchardt (S. 6 Anm. 9) Sethes Versuch ab, eine Berücksichtigung der Winterwende im ägyptischen Kalender nachzuweisen.

[5] Vgl. o. S. 176 Anm. 2.

schaffen wollten. Aber auch hier stimmt etwas nicht. Unsere ‚Astronomen‘ waren ja doch ‚Priester‘,[1] und gerade im kultischen Kalender hat sich die Rücksicht auf den Mond allein erhalten. Als Astronomen haben die Priester von Heliopolis den Mond radikal aus dem Kalender entfernt, als Priester haben sie ihn zäh beibehalten. ‚Zwei Seelen wohnen, ach! in meiner Brust.‘

Es wäre leicht, in derselben Weise fortzufahren;[2] das Resultat bleibt dasselbe: von dieser ganzen prähistorischen Astronomie bleibt nichts Faßbares zurück als *zwei* Dinge: eine astronomisch recht ungenaue Annahme über die Jahreslänge, nämlich von 365 Tagen, und als einziges Glanzstück die Festlegung des Jahresanfangs durch den heliakischen Siriusaufgang.

Auf die Frage der Jahreslänge kommen wir etwas weiter unten nochmals zurück. Wir wollen jetzt nur noch auf die *Kombination* der beiden einzigen Begriffe, die die Astronomie von Heliopolis produziert hat, unser Augenmerk richten. Haben wir es nicht hier wenigstens mit einer wirklich hervorragenden Leistung zu tun? Statt einfach und primitiv den Anfang der Überschwemmungszeit mit dem Anfang der wirklichen Überschwemmung im Jahre der Kalenderbegründung zusammenfallen zu lassen und sich dann der bequemen Überzeugung zu überlassen, daß man ja die Jahreslänge genau genug bestimmt habe, so daß auch in Hinkunft alles in Ordnung bleiben würde, legte man eine wahrhaft wissenschaftliche Voraussicht und Sorgfalt an den Tag und gab noch eine zweite, von der ersten ganz unabhängige Definition des Jahresanfangs, nämlich die durch den heliakischen Siriusaufgang.

Aber warum eigentlich diese Duplizität? Traute man den kommenden Astronomen-Generationen nicht zu, daß sie bis 365 zählen

---

[1] Vgl. z. B. Borchardt S. 7 und passim. — Das Wort ‚Priester‘ impliziert übrigens unwillkürlich die Vorstellung von einer wohlorganisierten Klasse von Theologen. Vielleicht wäre ‚Medizinmänner‘ in dieser Kulturstufe besser.

[2] Auch eine *Astrologie* existiert nicht im alten Ägypten (sie ist in der modernen Bedeutung dieses Wortes erst ein Produkt der hellenistisch-römischen Zeit). Vgl. z. B. H. Kees, Handbuch der Altertumswissenschaften III, 1 Bd. III, 1 S. 305 f.

konnten? Oder wußten bereits die Väter des Kalenders, daß ihre
Jahreslänge von 365 Tagen recht ungenau war und verübten sie einen
bewußten Schwindel? Das letztere ist doch kaum anzunehmen, nicht
etwa aus moralischen Gründen,[1] sondern aus einer Ursache, die
schon Borchardt klar formuliert hat:[2] ,Das Kalenderjahr sollte mit
dem Tage des Hundssternfrühaufganges beginnen. Da es aber um
rund $1/_4$ Tag zu kurz war, so tat es das schon im 5. Jahr seines
Bestehens nicht mehr. Da fiel der Anfang des Kalenderjahres schon
auf den Tag vor dem Hundssternfrühaufgang, im 9. Jahr schon
zwei Tage davor und so fort. Wenn man auch zuerst vielleicht ungenaue Beobachtung des Hundssternfrühaufganges als Grund für diese
vielleicht nur scheinbare Verschiebung vorschützen konnte, zehn Jahre
nach seiner Einsetzung wird man sich doch darüber klar gewesen
sein, daß das neue Kalenderjahr kein Sternjahr war, sondern wan-
derte, wenn man auch die schweren Folgen dieses Wanderns erst
viel später in ihrer ganzen Bedeutung erkannt haben wird, nämlich
daß die Kalendermonate, die in den drei Jahreszeiten gezählt wurden,
gar nicht mehr in die Jahreszeiten fielen, in denen sie eigentlich
gezählt wurden.' Dieser Schluß ist völlig zwingend; nur am Ende
scheint mir Borchardt die Astronomen von Heliopolis doch zu schlecht
zu beurteilen; als Reformatoren eines durch wiederholte Schaltver-
suche unbrauchbar gewordenen älteren Kalenders mußten sie sich
doch mindestens darüber klar sein, welches die Folgen zweier offenbar unvereinbarer Definitionen des Jahresanfanges sein mußten und
sie konnten sich buchstäblich an ihren Fingern abzählen, in welcher
Zeit der Wirrwarr noch größer sein mußte als je zuvor, wenn der
Fehler in zehn Jahren schon mehr als zwei Tage ausmachte.

Nun weiß man zwar, daß die Ägypter konservativer waren als
alle anderen Völker der Geschichte, aber nachdem sie doch gerade
eine revolutionäre Änderung ihres Kalenders vorgenommen hatten,
ist doch nicht gut verständlich, warum sie bereits zehn Jahre nach-

---

[1] Mangels jeglicher Texte sind wir über den Moralkodex der ägyptischen
Stein- und Bronzezeit nicht ausreichend orientiert. Vgl. auch Anm. 1 S. 182.

[2] Borchardt S. 9.

her nicht einen der beiden Jahresanfänge wieder aufgegeben haben
sollten. Aber dazu reichte die Entschlußkraft der Astronomen von
Heliopolis nicht mehr aus. Und zum Schluß kann man ihnen den
Vorwurf gröbsten Dilettantismus nicht ersparen: hätten sie ihre
Kalenderreform auch nur auf ein zehnjähriges *älteres* Beobachtungs-
material aufgebaut, so hätten sie doch wissen müssen, daß zwischen
zwei heliakischen Siriusaufgängen mehr als 365 Tage lagen. Aber
keine Rede davon; ohne eine Spur von Vorarbeit kam der große
Seher von Heliopolis plötzlich auf die Idee, gerade den heliakischen
Siriusaufgang als Beginn eines neuen 365tägigen Jahres zu wählen,
und dabei blieb es ohne Rücksicht auf Vergangenheit und Zukunft.
So erweist sich die erste Kalenderreform der Geschichte als ein
Werk der größten Hast und Unüberlegtheit, und erst die weitere
Geschichte des Kalenders bis in unsere Tage zeigt jenen Hang zum
Beibehalten des Ererbten, der für die Ägypter so charakteristisch
sein soll.

5. Das Sirius-Dogma. Unser Resultat ist also einfach das:
*was überhaupt noch von der heliopolitanischen Astronomie des Jahres
4231 übrigbleibt, ist absurd.* Oder vorsichtiger ausgedrückt: die
konsequente Verfolgung der Schlußweise, mit deren Hilfe ‚das erste
sichere Datum der Weltgeschichte‘ berechnet wurde, zwingt zu ab-
surden Vorstellungen von den Qualitäten der Begründer des Kalenders.[1] Also wird es gut sein, die Voraussetzungen dieser Schluß-

---

[1] Diesen Widerspruch haben natürlich auch andere empfunden, so Ed. Meyer
und Sethe. Es ist interessant zu sehen, zu welchen Konsequenzen sie dadurch
geführt worden sind. Sethe sagt (Zeitr. S. 309 f.): ‚So ist nicht daran zu zweifeln,
daß die Ägypter die wahre Länge des einzelnen Jahres = 365 $\frac{1}{4}$ Tage sehr früh
erkannt haben müssen. Um so merkwürdiger ist es, daß sie nicht zur Korrektur
des Fehlers ihres bürgerlichen Jahres geschritten sind. . . .‘ Sethe nimmt also an,
daß eine klare Erkenntnis bewußt nicht benutzt worden ist. — Meyers Schlußweise zeigt den inneren Widerspruch womöglich noch deutlicher. Er sagt (Chronol.
S. 11): ‚Schon nach wenigen Jahren mußten die Sternkundigen beobachten . . .‘,
daß eigentlich ‚die Einfügung eines Schalttages in jedem vierten Jahre‘ notwendig
gewesen wäre, was aber bekanntlich nicht geschehen ist. ‚Der Grund dafür‘, fährt
er fort, ‚wird zunächst in der Scheu gelegen haben, durch eine neue Schaltung
von neuem ein schwankendes Jahr zu bekommen und dadurch wieder in eine

weise nochmals zu überprüfen. An der 365tägigkeit des ägyptischen Jahres und seiner Einteilung ist, soweit unsere Texte zurückreichen, nicht zu rühren. Auch nicht daran, daß der heliakische Siriusaufgang als anderer Jahresanfang (im Gegensatz zum Neujahrstag des Wandeljahres[1]) und als ‚Bringer des Nils‘ gilt.[2] Unsere Rechnung hat aber mehr benutzt, nämlich daß dieser Zusammenhang schon weit in vorgeschichtlicher Zeit bestanden hat. Aber dafür gibt es keinen Beweis und kann es keinen geben, denn es handelt sich dabei um eine Zeit, die um viele Jahrhunderte der Entstehung der Schrift vorausliegt. *Daß bereits die Begründer des ägyptischen Kalenders den heliakischen Siriusaufgang als Jahresanfang definiert haben, ist eine reine Hypothese, die mit gleichem Recht durch jede andere Annahme ersetzt werden kann.* Wir wollen zeigen, daß sich ohne Sothis ganz leicht ein Bild von der Entstehungsgeschichte des ägyptischen Kalenders skizzieren läßt, in das sich zwanglos alles einordnen läßt, was wir an realen Tatsachen kennen, wobei wir allerdings darauf verzichten müssen, präzise Daten zu berechnen oder den Ort von Sternwarten zu entdecken.

6. Der Nilkalender. Daß der Rhythmus von Steigen und Sinken des Nils das Leben in Ägypten beherrscht hat, seit die klimatischen Verhältnisse Nordafrikas im wesentlichen die heutigen Formen angenommen haben, ist nicht zu bezweifeln. So ist es nur selbstverständlich, daß die allmählich zum Ackerbau übergehende Bevölkerung ihr Leben nach dem Stande des Nils geordnet hat;

Kalenderverwirrung hineinzugeraten, der man eben entronnen war; alsbald kam dann die Macht der Tradition hinzu.‘ Also noch ältere Kalendersysteme (d. b doch bis weit ins 5. Jahrtausend zurück), dann revolutionäre Änderung und ‚alsbald‘ Macht der Tradition bei eben den Leuten, die gerade das Alte radikal verändert hatten.

[1] Vgl. z. B. Meyer, GA I § 159 (S. 29).

[2] Damit soll nicht behauptet werden, daß ich überzeugt bin, daß ein derartiger Ausdruck *ohne nähere Untersuchung des ganzen Textzusammenhanges* genau auf den heliakischen Siriusaufgang bezogen werden muß und nicht etwa so unbestimmt gemeint sein kann wie unsere Redensart, daß die Schwalben den Sommer bringen. Vgl. auch S. 192.

Überschwemmung, Wachstum und Ernte sind die naturgemäßen Jahreszeiten dieses Landes, in dem die Veränderlichkeit der Tageslängen nur noch eine recht bescheidene Rolle spielt.[1] Wie viele Tage genau diese Perioden dauern, kümmert den Bauern im Grunde herzlich wenig. Er weiß aus langer Erfahrung, wie er seine Arbeit nach dem Stande des Nils einzuteilen hat. Die allmähliche Entstehung größerer Siedlungen, die Entstehung kultischer und staatlicher Zentren bringt aber auch in diese primitivste Zeitrechnung allmählich neue Züge. Es entwickelt sich in den größeren Kultstätten eine Art von Theologie mit bestimmten Vorstellungen von Göttern und Ereignissen, mit festen Riten und bestimmten Festlichkeiten.[2] So erhalten auch die kosmischen Erscheinungen allmählich ihre mehr oder minder geregelten Kulte neben der Verehrung der Baum- und Steinfetische, der Tiere der Wüste, der Sümpfe und des Nils. Der Mond wird dabei der naturgemäße erste etwas genauere Zeitmesser. ‚Astronomie‘ und ‚Kalender‘ ist das noch lange nicht, aber ein Anlaß zu einer Art von Zeitrechnung, ein Hilfsmittel zur genaueren Fixierung auch anderer Feste. Neben diesen lokal verschiedenen, zeitlich noch sehr variablen Kulten läuft nach wie vor der alles beherrschende Rhythmus der Jahreszeiten des Nils. Aber der bewegtere Ablauf des Lebens in den mehr städtischen Siedlungen lehrt auch allmählich die große Periode des Niljahres an den kleinen Zeitintervallen messen. Auch die Veränderlichkeit des Nils selbst wird genauer verfolgt. Nahe Heliopolis am Beginn des Deltas fand sich schon in vordynastischer Zeit ein Nilmesser auf der Insel Roda.[3] An Hand der Reihenfolge der Feste, an Hand der Mondmonate ist die Zahl der Tage leicht auszuzählen,[4] die

---

[1] Vgl. Sethe, Zeitr. S. 295.

[2] Wieviel sich noch aus der späteren Literatur über das staatliche und kultische Leben dieser Frühzeit erkennen läßt, zeigt Sethes meisterhaftes Werk ‚Urgeschichte und älteste Religion der Ägypter‘ (1930).

[3] Vgl. z. B. Sethe, Urgesch. § 109 ff.

[4] Dazu bedarf es wieder keiner besonderen Mathematik. Kerbmarken, Knotenschnüre (vgl. die Hieroglyphen für die Einer, bzw. für 100) gibt es in den primitivsten Kulturen.

zwischen zwei ungefähr gleichen Pegelständen verflossen sind, z. B. zwischen den tiefsten Wasserständen, von denen an das Steigen des Nils wieder beginnt.

Wir wollen diese Zeitdifferenz, etwa zwischen zwei Niedrigwassern, kurz ein ,Niljahr' nennen[1]; es hat eine recht schwankende Länge, wie die moderne Statistik zeigt. In Fig. 2 sind die Zeitdifferenzen in Tagen zwischen den minimalen Wasserständen während

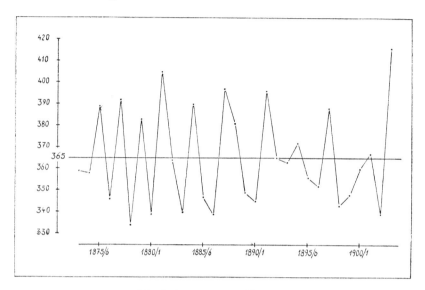

Fig. 2. Länge von ,Niljahren'.

der Jahre 1873 bis 1904 angegeben.[2] Man sieht, daß die Schwankungen mehrere Wochen ausmachen können (vgl. auch Fig. 4 S. 190) und daß in den von uns hier herausgegriffenen 31 Jahren nur ein einziges Mal (zwischen 1892 und 1893) genau 365 Tage zwischen zwei Minima liegen. Und trotzdem ermöglichen auch diese stark schwankenden ,Niljahre' ohne weiteres eine Bestimmung der Jahreslänge. Nehmen wir nur die uns vorliegenden Zahlen. Das erste ,Nil-

---

[1] Damit soll nicht behauptet werden, daß dieser Begriff als solcher je bestanden hat.

[2] Diese Zeitdifferenzen sind entnommen Willcocks, Nile, Appendix K, Table 41. Daß sie sich auf Assuan beziehen, spielt für unsere Zwecke gar keine Rolle.

jahr' hat die Länge von 359 Tagen, das zweite die von 358. Im Mittel könnte man also zunächst $358\,^1/_2$ für die Jahreslänge halten. Das dritte Niljahr dauert aber 389 Tage; eine dreijährige Verbuchung der Tageszahlen lieferte also $^1/_3\,(359 + 358 + 389) = 368\,^2/_3$ Tage. Nach zehn Jahren würde sich ein Mittelwert von $364\,^3/_4$ Tagen ergeben usw. Eine ganz primitive Statistik über die Anzahl der verflossenen Tage und ihre Gruppierung in Abschnitte gleicher Länge (eine wirkliche ‚Rechnung‘ ist auch dazu gar nicht nötig, es genügt ja die Verteilung von Strichmarken in gleich große Gruppen) führt

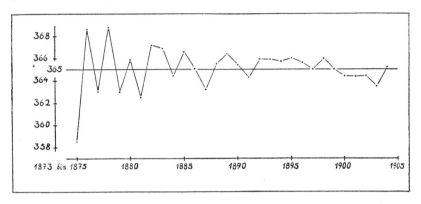

Fig. 3. Mittlere Länge von ‚Niljahren‘.

also zu Durchschnittszahlen, die sich bis auf Bruchteile ganz von selbst dem Wert 365 immer mehr nähern müssen, einfach aus dem Grunde, weil in jedem Jahr genau *eine* Nilschwelle eintritt und die Schwankungen des Datums sich im Mittel immer mehr ausgleichen, je länger man sie beobachtet. Dies zeigt das Beispiel von Fig. 3 ganz deutlich: bereits nach 15 Jahren überschreitet die Abweichung der mittleren Länge des Niljahres von 365 nur noch ein einziges Mal einen vollen Tag, und dies auch nur um ganz wenig, obwohl die entsprechende tatsächliche Schwankung des Datums besonders extrem war (sie betrug sieben Wochen; vgl. Fig. 2 S. 185, bzw. Fig. 4 S. 190 am rechten Ende der Kurve). Also: *Bereits die allerelementarste Aufzeichnung über die Nilstände, durch etwa 50 Jahre fortgeführt, liefert ganz von selbst als Durchschnittslänge eines ‚Jahres‘*

365 Tage und selbst genaue Rechnung würde nur Bruchteile eines Tages als Abweichung ergeben. Dieses Resultat ist ohne jede Astronomie am primitivsten Nilmesser und mit dem primitivsten Markensystem zur Notierung der vergangenen Tage zu gewinnen, also genau mit den Hilfsmitteln, deren Existenz allein durch unser archäologisches und inschriftliches Material[1] wirklich gesichert ist.[2]

Wir haben also gesehen: Eine Jahreslänge von 365 Tagen erscheint als Mittelwert von ‚Niljahren' zwar durchaus nicht sofort, sondern erst nach einer längeren Reihe von Jahren, über die man mittelt. Dann prägt sie sich aber immer deutlicher aus, je länger man fortschreitet, und nur sie. Denn die Bruchteile schwanken als Folge der Unregelmäßigkeit der Nilstände immer etwas hin und her, wenn auch mit immer geringeren Beträgen. Würde man die Mittelbildungen streng weiterführen, so müßte man (das wissen wir heute) auch bei den Bruchteilen immer deutlicher $1/4$ bevorzugt finden. Aber solche wirkliche Beobachtungssystematik anzunehmen ist historisch sinnlos. Was die primitive Auszählung der Tage und ihre Verteilung auf die Anzahl der verflossenen Jahre mit scheinbar größter Deutlichkeit nach einiger Zeit ergeben muß, ist, wie auch die Nilstände im einzelnen schwanken mögen, ganz klar und deutlich die Tageszahl 365. *Ein oder zwei Generationen Beobachtung an einem Nilpegel führt also mit völliger Notwendigkeit gerade zur ägyptischen Jahreslänge.* Die einzige Voraussetzung, die wir bei dieser Entstehungsgeschichte des ägyptischen Jahres machen mußten, ist die einer gewissen Dauer des Probierens und Beachtens. Dies scheint mir gerade eine besondere Stütze für diese Betrachtungsweise zu sein, im Gegensatz zu der Plötzlichkeit, die man bei einer Bestimmung dieser Jahreslänge durch die heliakischen Siriusaufgänge annehmen müßte, um sie nicht sogleich durch die elementarsten Beobachtungen widerlegt zu sehen.

---

[1] Bekanntlich enthält z. B. auch der Palermostein, der die Annalen Ägyptens bis gegen das Ende der 5. Dynastie verzeichnet (Meyer, GA I Nachtr. S. 41 f.), Angaben über Nilhöhen (Sethe, Beitr. S. 103 ff.).

[2] Damit werden natürlich auch alle Spekulationen über ein ursprünglich 360tägiges Jahr (z. B. Sethe, Zeitr. S. 302 ff.) gänzlich überflüssig.

13*

Archäologisches Material (Nilmesser bei Heliopolis aus vordynastischer Zeit), *Aufbau und Namengebung des ägyptischen Kalenders* (reine Nil-Jahreszeiten und Jahreslänge von 365 Tagen) *sowie
historische Plausibilität* (kulturelles Niveau ohne alle Astronomie
oder Mathematik) *führen also übereinstimmend zu dem Resultat, daß
das ägyptische Jahr dem Nil und dem Nil allein seine Entstehung
verdankt* und daß diese Entstehung eine allmähliche gewesen ist,
ohne alle ‚Reform'-Gedanken, ‚Reichsgründungen' u. dgl. m.[1] Das
ägyptische Jahr ist einfach erklärbar als der ganz allmählich präzi-
sierte Kalender eines Bauernvolkes, das auf Gedeih und Verderb an
Steigen und Fallen des Nils interessiert war. Daneben mögen zahl-
reiche lokale Mond- und sonstige Festkalender existiert haben, die
sich allmählich enger oder lockerer an den Nilkalender mit seinen
365 Tagen angeschlossen haben, wie es noch das spätere ägyptische
Kalenderwesen erkennen läßt. Aber nichts zwingt uns, irgendwelche
Gegensätze zwischen Festkalender und Bauernkalender, zwischen
‚Priestern' und ‚Astronomen' anzunehmen.

7. Die ‚Sothis'. Wie wir gesehen haben, ist das ägyptische
Jahr von 365 Tagen nur verständlich, wenn man es sich entstanden
denkt *ohne* jede Kontrolle durch astronomische Beobachtungen, die
seine Ungenauigkeit bereits in ganz kurzer Zeit deutlich hätten
zeigen müssen. Ebenso haben wir schon bemerkt, daß seine Einteilung in drei Jahreszeiten in keiner Relation zu den astronomischen Jahreszeiten steht. Auch die feinere Jahresteilung des geschichtlichen Kalenders in stets 30tägige Monate zeigt die gänzliche
Bedeutungslosigkeit der Astronomie für diesen Kalender.[2] Daß

---

[1] So sagt Sethe ÄZ **44** (1907) S. 26 Anm. 1: ‚Nur in einem Einheitsstaat,
wie dem von Heliopolis, war wohl auch eine solche Kalenderregulierung möglich.'
Einer solchen Schlußweise ist natürlich jetzt jeder Boden entzogen.

[2] Diese Einteilung in 30tägige Monate ist gewiß eine Art von Rücksicht-
nahme auf die wirklichen Mondmonate, von deren Existenz wir schon mehrfach
gesprochen haben. Aber der ganz undurchsichtige Wechsel zwischen 29- und
30tägigen Monaten war viel zu kompliziert, um ihn ernstlich zu berücksichtigen
und außerdem war klar, daß die Mondmonate doch nicht zu dem paßten, was
allein interessierte, nämlich zur mittleren Dauer des Niljahres. Nachdem man

gerade er dereinst der bevorzugte Kalender der großen Astronomen der hellenistisch-römischen Zeit werden sollte, haben sich seine (vielen und unschuldigen) Väter gewiß nicht träumen lassen.[1]

Wir sahen auch, daß es einige Zeit gedauert haben muß, bis man aus den schwankenden Nilständen zu dem festen Wert von 365 Tagen des Niljahres kommen konnte. Etwa 50 Jahre oder vielleicht das Doppelte muß man für diesen Prozeß in Anschlag bringen — historisch ist dies ja auch nur natürlich. Als man sich endlich von der Konstanz der Zahl 365 zur Genüge überzeugt hatte, konnte man auch mit diesem Kalender wohl zufrieden sein. Hatte man einmal den Neujahrstag, den ersten Tag der ‚Überschwemmungsjahreszeit‘, auf einen Tag gelegt, an dem ein gewisser Pegelstand den Beginn der Überschwemmung anzeigte, so war man tatsächlich sicher, daß man nach 365 Tagen wieder in der Nähe dieses Phänomens war. Einmal begann der Nil etwas früher dieses Niveau zu überschreiten, einmal etwas später, aber im großen und ganzen stimmte die Sache Jahr für Jahr, und mehr war für diesen Bauernkalender nicht nötig. Nun, wir wissen, daß man doch zu früh zufrieden war. Jedes Jahr blieb der ägyptische Neujahrstag um fast genau einen Vierteltag gegen das Sonnenjahr zurück. Für unsere weiteren Betrachtungen können wir ruhig sagen: genau einen Vierteltag. Wir wollen uns wieder an der modernen Statistik[2] der Nil-Minima aus den Jahren 1873 bis 1904 überlegen, was geschehen mußte. Fig. 4 zeigt die Daten dieser Minima. Wir wollen als Beispiel einmal annehmen, daß der Neujahrstag zu Anfang auf dem ersten Minimum[3] unserer Statistik lag. In vier Jahren ist

---

sich schließlich überzeugt hatte, daß dieses 365 Tage betrüge, hat man eben den Begriff des ‚Monats‘ so gut und so simpel es ging in dieses Jahr übertragen — eine Art des Vorgehens, die uns in der ägyptischen (und jeder anderen) Kulturgeschichte ja so oft begegnet.

[1] Es ging ihm wie einer großen Bibliothek: auch die raffinierteste Aufstellung nach sachlichen Gesichtspunkten wird nach kurzer Zeit zur bloßen Plage, während das triviale Alphabet für alle Zeiten brauchbar bleibt.

[2] Vgl. oben S. 185 Anm. 2.

[3] Damit soll nicht behauptet werden, daß genau das Minimum als Anfang des Steigens des Nils angesehen wurde (vgl. dazu auch u. S. 191).

dann das Datum dieses Neujahrstages um einen vollen Tag geringer
geworden (vgl. die stark ausgezogene Linie[1] in Fig. 4) und so geht
es weiter. Man sieht: für die Dauer unserer Statistik (über 30 Jahre)
hat man nicht den geringsten Grund, an der Richtigkeit des Kalenders
zu zweifeln: immer schwanken die Nilminima um den Neujahrstag
herum. Und dies gilt noch viel länger; wie unsere Statistik zeigt,
ist die Variationsbreite der Minima, auch wenn man besonders extreme Fälle beiseite läßt, reichlich 40 Tage.[2] Es dauert also min-

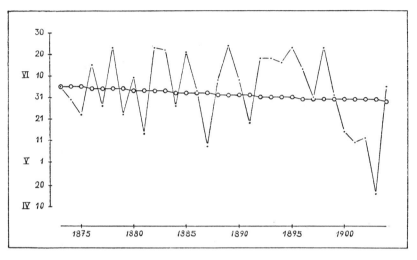

Fig. 4. Gregor. Daten von Nil-Minima und Tage gleichen ägyptischen Datums.

destens $4 \times 40 = 160$ Jahre, bis eine Zone dieser Breite vom Neujahrstag durchschritten ist, und auch dann noch werden immer
wieder vereinzelte Minima auf den Neujahrstag fallen. Für eine
Periode von mindestens 200 Jahren hat man also nicht den geringsten
Grund, das Niljahr von 365 Tagen für unrichtig zu halten. Auch

---

[1] Da nach gregorianischer Schaltregel 1900 kein Schaltjahr war, tritt für
diese Zeit einmal keine Datumsänderung auf. Unsere Linie zeigt also sogar genauer
das Verhalten gegen das wahre Sonnenjahr, als wir es im Text der Einfachheit
halber annahmen.

[2] Die Statistik der Daten der Maxima ergibt das gleiche Bild (vgl. Willcocks,
Nile, Appendix K, Table 41).

noch weiterhin kann man sich noch lange damit trösten, daß erfahrungsgemäß wochenlange Verspätungen des Nils gegen den Neujahrstag vorkommen.

Ein weiteres Moment kommt hinzu, das auf die Verschleierung des wahren Sachverhaltes hinwirken mußte. Die Nilstandsangaben der prädynastischen Zeit waren ja keine modernen Präzisionsmessungen; welchen Wasserstand man als ‚Beginn‘ des Steigens des Nils oder

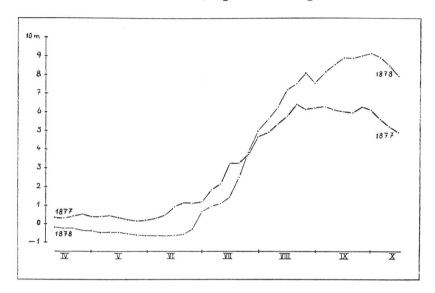

Fig. 5. Nilwasserstände.

als ‚Beginn‘ der Überschwemmung ansehen sollte, ist mit primitiven Mitteln gar nicht scharf zu definieren. Dies läßt sich unmittelbar an beliebigen Beispielen über die wechselnden Nilwasserstände erkennen, wie sie Fig. 5 zeigt. Dort sind nach modernen Messungen die Nilhöhen für zwei aufeinanderfolgende Sommer angegeben.[1] Wir haben nicht den geringsten Grund für die Annahme, daß das Anschwellen des Nils im Altertum regelmäßiger erfolgt ist. Man konnte also wieder durch viele Jahre die Nichtübereinstimmung von Neujahr

---

[1] Nach Willcocks, Nile, Appendix K, Table 45. Die Daten beziehen sich auf Assuan.

und Nilschwelle den Zufälligkeiten in der Art des allmählichen
Anstiegs zuschreiben.

Es ist eine reine Ansichtssache, wie groß wir glauben sollen,
daß die Verschiebung des Anfangs der kalendarischen ‚Überschwemmungszeit‘ gegen die wirkliche Überschwemmung geworden
sein muß, bis man sich darüber klar wurde, daß der Kalender *nicht*
ein Niljahr, sondern ein *Wandeljahr* definierte. Aber das eine ist
jedenfalls klar: als dies geschah, hat der Kalender mindestens schon
zwei bis drei Jahrhunderte bestanden und damit ist es auch schon
ohne Zuhilfenahme einer speziell ägyptischen Psychologie völlig verständlich, daß man ihn längst als etwas Altgewohntes betrachtete
und an keine Reform dachte.

Aber man sah sich nun doch nach einem Kriterium um, das
gestattete, das Steigen des Nils mit einiger Sicherheit vorherzusagen.
Auch hier wird eine alte Erfahrung vorgelegen haben: wenn der
Sirius wieder am Morgenhimmel sichtbar wird (nach etwa $2^1/_2$ monati-
ger Unsichtbarkeit), dann ist das ungefähr die Zeit der Nilschwelle
— so unbestimmt und allgemein, wie wenn wir sagen: wenn die
Tage länger werden, kommen die Frühlingsstürme. Auch dazu sind
weder Astronomie noch Astronomen nötig; es genügt eine alte Bauernregel: der Sirius ist ‚der Bringer des Nils‘.

Wann dieses Wiedererscheinen der Sothis zum *ersten* Wiedererscheinen präzisiert worden ist, also genau zum ‚heliakischen Aufgang‘ von einer bestimmten Stelle aus gesehen, kann man wohl
kaum mehr entscheiden und bedürfte einer speziellen Untersuchung
an Hand aller erreichbaren Texte. Die Frage ist aber wenig interessant; für die Chronologie ist sie jedenfalls ganz bedeutungslos und
selbst für die spezielle Geschichte der ägyptischen Astronomie nicht
belangreich, denn es handelt sich doch nicht um etwas, was man
als astronomische Entdeckung bezeichnen kann, sondern höchstens
als Kanonisierung einer Bauernregel.

8. Chronologische Folgerungen. Wir sind am Ende unserer
Betrachtungen angekommen. Wir können sie sehr einfach dahin
zusammenfassen, daß sich die Entstehung des ägyptischen Jahres

ganz ungezwungen aus der alleinigen Rücksicht auf den Nil erklären läßt, daß es sich dabei um einen Prozeß handelt, der sich über Jahrhunderte erstreckt und daß schließlich, als die Differenz zwischen ‚Kalender' und Nil all zu groß und offenbar irreparabel geworden war, der Aufgang der Sothis als anderer Jahresanfang eingeführt worden ist, an den sich der Nil diesmal besser hielt. Kurz: *die Rücksicht auf die Sothis und damit die Duplizität der Jahresformen wird überhaupt erst verständlich, wenn sie nach der Loslösung des 365 tägigen Jahres vom Nil begründet wurde.*

Diese, wie mir scheint, historisch ungemein einfache und zwanglose Entstehungsgeschichte zu widerlegen dürfte nicht leicht sein, denn eine solche Widerlegung müßte sich, sollte man sie ernst nehmen, nicht auf Dogmen über eine Astronomie der grauesten Vorzeit stützen, sondern auf Texte, und diese werden sich aus der Zeit vor der Erfindung der Schrift schwer beibringen lassen.

Daß es damit auch mit der Berechnung des Datums des Kalenderanfanges vorbei ist, ist klar. Von dieser ganzen Schlußweise bleibt nur übrig, daß zu Anfang die ‚Überschwemmungsjahreszeit' zur tatsächlichen Überschwemmung gepaßt haben muß, und dies kann etwa um 4200 der Fall gewesen sein. Da aber jede Verschiebung des Anfangs der Überschwemmung um *einen* Tag für die Chronologie vier Jahre ausmacht und da diese Daten damals gewiß ebenso wie heute um Wochen variiert haben, so beträgt die Unsicherheit der Chronologie *mindestens* ein oder zwei Jahrhunderte nach oben und unten. Es ist daher prinzipiell ausgeschlossen, mehr zu sagen als: um etwa 4200 v. Chr. $\pm$ etwa 200 Jahre hat man begonnen, ein 365 tägiges Jahr zu benutzen. Nach weiteren ein oder zwei Jahrhunderten wurde dann vielleicht der Sirius als ‚Bringer des Nils' als anderer Jahresanfang hinzugefügt. Und schließlich: für die Kulturgeschichte ist aus diesen Dingen nicht mehr zu lernen, als man längst aus archäologischen und religionsgeschichtlichen Quellen erschlossen hat, nämlich daß die Kultur dieser Periode bereits zu einer gewissen Ordnung des staatlichen Lebens fortgeschritten war. Auch für die Geschichte der Astronomie verliert die Geschichte der ägyptischen Kalenderbegründung jede Besonderheit und ordnet sich zwanglos dem Bilde unter, das man sich auch sonst von den ersten kalendarischen Versuchen einer sich entwickelnden Kultur zu machen pflegt.

͵ Ich bin mir darüber klar, daß durch diese Betrachtungen ein Bild zerstört wird, das vielen Historikern und Chronologen lieb und wert geworden ist. Ich hoffe aber, daß es auch Historiker und Prähistoriker geben wird, die erleichtert aufatmen werden, wenn sie von einem Gespenst befreit werden, das, gepanzert mit Präzession und Eigenbewegung, arcus visionis und heliakischem Aufgang, jedem Versuch zu spotten schien, an der Chronologie und Kulturgeschichte des Reiches von Heliopolis zu rühren.[1] — Die Frage nach dem ersten sicheren Datum der Weltgeschichte hat genau das gleiche historische Interesse wie die nach dem Familiennamen des ersten Menschen.

---

[1] A. Scharff (GÄV S. 54 ff.) hat trotzdem einen solchen Versuch gewagt, indem er das ganze Gespenst als solches um eine volle Sothisperiode zurückversetzen wollte. Dies ist zweifellos etwas zu gewaltsam, zeigt aber deutlich, *wie* unbequem das feste Datum allmählich geworden ist. Borchardt (S. 33 Anm. 3) begnügt sich mit einer kurzen Abfertigung: ‚Die dort hervorgesuchten Scheingründe … bedürfen für Einsichtige keiner besonderen Widerlegung.‘ Auch ich glaube, daß Scharffs Argumente unzureichend sind; historisch unwahrscheinlicher als die üblichen Vorstellungen von der Astronomie des Jahres —4240 sind sie aber gewiß nicht.

Literaturverzeichnis.

ÄZ   Zeitschrift für ägyptische Sprache und Altertumskunde.

Borchardt   s. S. 170 Anm. 1.

Gunn [1]   B. Gunn, ‚Finger-numbering' in the Pyramid texts. ÄZ **57** (1922)
                  S. 71 f.

Meyer, Chronol.   Ed. Meyer, Ägyptische Chronologie, Abh. Berl. Akad. **1904**,
                  phil.-hist. Kl.

Meyer, GA I, 2   Ed. Meyer, Geschichte des Altertums, Bd. I, 2. Hälfte, Die ältesten
                  geschichtlichen Völker und Kulturen bis zum 16. Jahrhundert.
                  4. Aufl. (Omnitypiedruck), Stuttgart und Berlin, Cotta, 1921.

Meyer, GA I, Nachtr. Ed. Meyer, Die ältere Chronologie Babyloniens, Assyriens
                  und Ägyptens, Nachtrag zum ersten Bande der Geschichte des
                  Altertums, Stuttgart und Berlin, Cotta, 1925.

O. Neugebauer, Vorlesungen   O. Neugebauer, Vorlesungen über Geschichte der
                  antiken mathematischen Wissenschaften, Bd. I, Vorgriechische
                  Mathematik. Berlin, Springer, 1934.

P. V. Neugebauer, Hilfstafeln   P. V. Neugebauer, Hilfstafeln zur technischen Chrono-
                  logie, Astronomische Nachrichten Bd. **261**, Nr. 6250, 6261 und
                  6264 (1937). [Auch als Sonderdruck erschienen, Kiel 1937.]

RE   Paulys Real-Enzyklopädie der klassischen Altertumswissenschaft,
                  Neue Bearb., Herausgeg. von Wissowa, Kroll, Mittelhaus; Stutt-
                  gart, Metzler, 1893 ff.

Scharff, GÄV   Alex. Scharff, Grundzüge der ägyptischen Vorgeschichte, Morgen-
                  land **12**, Leipzig, Hinrichs, 1927.

Sethe [1]   K. Sethe, Ein altägyptischer Fingerzählreim, ÄZ **54** (1918) S. 16 ff.

Sethe, Beitr.   K. Sethe, Beiträge zur ältesten Geschichte Ägyptens (= Unter-
                  suchungen zur Geschichte und Altertumskunde Ägyptens **3**),
                  Leipzig, Hinrichs, 1905.

Sethe, Urgesch.   K. Sethe, Urgeschichte und älteste Religion der Ägypter (= Abh.
                  f. d. Kunde d. Morgenlandes **18**, 4) Leipzig, Brockhaus, 1930.

Sethe, Zeitr.   K. Sethe, Die Zeitrechnung der alten Ägypter im Verhältnis
                  zu der der anderen Völker, Nachr. d. K. Ges. d. Wiss. zu Göttingen,
                  phil.-hist. Kl. **1919**, 287 ff.; **1920**, 28 ff.; **1920**, 97 ff.

Willcocks, Nile   Sir William Willcocks, The Nile in 1904, London, Spon, 1904.

Reprinted from
Acta Orientalia
17, 169−195 (1938)

THE ORIGIN OF THE EGYPTIAN CALENDAR

O. NEUGEBAUER

Probably no calendaric institution has continued over a longer period than the Egyptian calendar. After its uninterrupted use during all of Egyptian history, the Hellenistic astronomers adopted the Egyptian year for their calculations. Ptolemy based all his tables in the *Almagest* on Egyptian years; even as late as A.D. 1543 Copernicus in *De revolutionibus orbium cœlestium* used Egyptian years. The explanation of this fact is very simple: astronomers are practical-minded people who do not connect more or less mystical feelings with the calendar, as the layman frequently does, but who consider calendaric units such as years, months, and days as nothing but conventional units for measuring time. And because *the* main requirement of every measuring unit is, of course, its constancy, the Egyptian calendar is an ideal tool: twelve months of thirty days each, five additional days at the end, and no intercalation whatsoever. It is no wonder that the Hellenistic astronomers preferred this system to the Babylonian lunar calendar with its very irregularly changing months of twenty-nine and thirty days combined with a complicated cyclic intercalation—not to mention the chaos of Greek and Roman calendars.[1]

The ideal simplicity of the Egyptian calendar, however, raises serious problems for the historian. Should we assume that astronomers, for the sake of their own calculations, imposed on the rest of the population a calendar with no respect for sun and moon? No scholar will accept this viewpoint, even if he does not hesitate to speak (in cases of no consequence) of Egyptian "astronomers" or Egyptian *Kalendermacher*. Only one other solution seems to remain: the simplicity of the Egyptian calendar is a sign of its primitivity; it is the remainder of prehistoric crudeness, preserved without change by the

---

[1] It may be remarked that for the same reason modern astronomy does not use the Gregorian calendar for computations but Julian years instead (the continued use of Egyptian years would be inconvenient because of the discrepancy of about one and a half years with the adopted historical chronology of our times).

Egyptians, who are considered to be the most conservative race known in human history.

Even this second solution, however, is by no means satisfactory. I do not have in mind the sophisticated argument that one of the strongest foundations for the belief in the extreme Egyptian conservatism is the very maintenance of the calendar and should therefore not be used as an *explanation* of the calendar. What I mean is the fact that there is no astronomical phenomenon which possibly could impress on the mind of a primitive observer that a lunar month lasts 30 days and a solar year contains 365 days. Observation during one year is sufficient to convince anybody that in about six cases out of twelve the moon repeats all its phases in only 29 days and never in more than 30; and forty years' observation of the sun (e.g., of the dates of the equinoxes) must make it obvious that the years fell short by 10 days! The inevitable consequence of these facts is, it seems to me, that *every theory of the origin of the Egyptian calendar which assumes an astronomical foundation is doomed to failure.*

Four years ago I tried to develop the consequences of this conviction as far as the Egyptian *years* are concerned. I showed[2] that a simple recording of the extremely variable dates of the inundations leads necessarily to an average interval of 365 days. Only after two or three centuries could this "Nile calendar" no longer be considered as correct, and consequently one was forced to adopt a new criterion for the flood, which happened to be the reappearance of the star Sothis. I do not want to repeat the discussion here, but I should like to state that I still think that this theory is in perfect agreement with the structure of the Egyptian calendar, which has only three seasons, admittedly agricultural and not astronomical, and which has no reference to Sothis at all.[3] I did not see, at that time, any satisfactory

[2] O. Neugebauer, "Die Bedeutungslosigkeit der Sothisperiode für die ältere aegyptische Chronologie," *Acta orientalia*, XVII (1938), 169–95.

[3] I wish to take this opportunity to make some remarks about an interesting paper by H. E. Winlock, "The Origin of the Egyptian Calendar" (*Proceedings of the American Philosophical Society*, LXXXIII [1940], 447–64), where the problem of the Egyptian year is treated independently of my paper.

The most important point seems to me that Winlock reached the same conclusion, namely: the classical theory that *both* Nile and Sothis are responsible for the beginning of the years must be abandoned. The old story of the "creation" of the Egyptian calendar in 4231 B.C. can now be considered as definitely liquidated. An objection has been raised against my theory of a "Nile-year" resulting from averaging the strongly fluctuating

explanation of the second characteristic element of the Egyptian calendar—the *months* of invariably 30 days' length. How are we to explain these artificial months, seemingly so contradictory to all our experience with ancient calendaric systems?

The solution which I finally believe to have found for this problem is nothing but the radical abandoning of the concept that the 30-day months should be explained by some kind of primitive astronomy and the clear insight into the fact that the 30-day months are by no means peculiar to Egypt but play a very important role also in Mesopotamia, the classical country of the strictly lunar calendar.

I can best start by quoting two sentences from Sethe's *Zeitrechnung:* "Bei den Aegyptern haben sowohl Lepsius als Ed. Meyer und Andere die Existenz eines Mondjahres für die Urzeit als a priori selbstverständlich vorausgesetzt . . . . und Brugsch wollte sogar das Fortbestehen eines solchen Mondjahres in geschichtlicher Zeit neben dem Siriuswandeljahr aus zahlreichen Angaben über Mondstände

---

intervals between the inundations. This objection is that there is no proof of the existence of "Nilometers" at so early a period (*ibid.*, p. 450, n. 11). However, no precise Nilometer is required for my theory. The sole requirement is that somebody recorded the date when the Nile was clearly rising. As a matter of fact, *every* phenomenon which occurs only once a year leads to the same average, no matter how inaccurately the date of the phenomenon might be defined. The averaging process of a few years will automatically eliminate all individual fluctuations and inaccuracies and result in a year of 365 days. Fractions, however, would be obtained only by much more extensive recording and by accurate calculation. The actual averaging must, however, be imagined as a very simple process based on the primitive counting methods as reflected in the Egyptian number signs: the elapse of one, two, or three days recorded by one, two, or three strokes. After ten strokes are accumulated, they are replaced by a ten-sign, thereafter ten ten-signs by a hundred symbol, etc. This is the well-known method of all Egyptian calculations. This method finally reduces the process of averaging to the equal distribution of the few marks which are beyond, say, three hundred-signs and five ten-signs; in other words, there is no "calculation" at all involved in determining the average length of the Nile-years. Of course, we need not even assume the process of counting all the single days every year: the averaging of the excess number of days over any interval of constant length (say twelve lunar months) gives the same result. This equal distribution of counting-marks finally makes it clear that no fractions will be the result of the process.

Winlock's own theory assumes the prediction of the flood at an early epoch according to lunar months (pp. 454 ff.). Thereafter, Menes is credited with having begun to determine the beginning of the years by observing the Sothis star (pp. 457–58), the seasons still being of variable length because of their composition by lunar months (p. 459). Finally, Djoser around 2773 b.c. is supposed to have dropped the actual New Year's observations by installing the year of "12 times 30 +5 days" because "experience of centuries by now had seemed to show that the year should contain 365 days" (p. 462). I cannot see how *experience* from observing Sothis could have created this assumption of the length of the year because Sothis after one hundred years of 365 days each rises 25 days too late! This obvious *contradiction* between the year of 365 days and any astronomical observations seems to me just the most striking argument in favor of looking for another phenomenon which leads to a 365-day year—the flood of the Nile.

.... schliessen."[4] But, he goes on, "schwer liesse sich von einem solchen alten Mondjahre die Brücke zu dem geschichtlichen Wandeljahre .... schlagen." This conclusion of Sethe is obviously the generally accepted viewpoint. However, how can one justify the total ignoring of the textual evidence amply collected, for example, by Brugsch in his *Thesaurus*,[5] which shows clearly a great interest in the real lunar months? Indeed, Brugsch's assumption of the existence of real lunar months has only been confirmed since his time.[6] I admit, of course, that Borchardt[7] overemphasized the importance of the full-moon festivals for the coronation ceremonies and that his chronological construction, based on this theory, requires checking. The fact remains, however, that at all periods of Egyptian history the real lunar months had their well-defined religious significance. One need only recall the countless passages where we are told about the loss and restitution of the moon's eye, of its magical importance, etc. Indeed, one should be surprised that the behavior of the real moon should have been totally disregarded and have been replaced by meaningless intervals of 30 days. Moreover, we now know that the "short" and "long" years mentioned in the list of offerings at Benihasan[8] (Twelfth Dynasty) are the years containing either twelve or thirteen lunar festivals (say, new moons), respectively; this is shown by a Demotic papyrus in which a simple cycle of twenty-five years is developed according to which one can tell whether a certain year contains twelve or thirteen new moons and on what dates in the civil calendar they can be expected.[9] In other words, we have to admit the *coexistence* of real lunar months and of the civil calendar with its 30-day months. Sethe's contradiction then disappears, and we no longer need astronomy to explain the 30-day months: all "astro-

[4] K. Sethe, *Die Zeitrechnung der alten Aegypter* ("Nachr. Ges. Wiss. Göttingen, Phil.-hist. Kl.," 1919, pp. 287–320, and 1920, pp. 28–55, 97–141), pp. 300 and 301.

[5] H. Brugsch, *Thesaurus inscriptionum aegyptiacarum*, Vol. I: *Astronomische und astrologische Inschriften altaegyptischer Denkmäler* (Leipzig, 1883).

[6] Winlock, *op. cit.*, pp. 454 f.

[7] L. Borchardt, *Die Mittel zur zeitlichen Festlegung von Punkten der ägyptischen Geschichte und ihre Anwendung* (Cairo, 1935).

[8] *Urkunden d. aeg. Altertums*, VII, 29, 18 = P. E. Newberry, *Beni Hasan I*, p. 25, ll. 90 f.

[9] Neugebauer-Volten, "Untersuchungen z. antiken Astronomie. IV: Ein demotischer astronomischer Papyrus (Pap. Carlsberg 9)," *Quellen u. Studien z. Gesch. d. Mathematik*, Abtl. B, IV (1938), 383–406.

nomical" interest is restricted to the actual observation of the real moon with no resultant influence on the civil calendar.

But how are we to explain the coexistence of the schematic 30-day months side by side with the real lunar months? The answer sounds paradoxical at first but is actually very simple: *schematic months are the natural consequence of a real lunar calendar.*

Here the analogy with the situation in Mesopotamia enters the picture. The actual behavior of the moon is so complicated that not before the very last centuries of Babylonian history was a satisfactory treatment of the movement of the sun and the moon developed sufficiently accurate to predict the length of the lunar months for an appreciable time in the future. In other words, only a highly developed theoretical astronomy (today we would say "only celestial mechanics") is able to determine the further course of a lunar calendar. Private and public economy require the possibility of determining future dates regardless of the irregularity of the moon and the inability of the astronomers to predict the outcome. A simplified calendar is equally useful also for the past because it eliminates the necessity of keeping exact records of the actual length of each month. It is amply testified from Babylonian sources how this natural demand was met: beside the real lunar calendar there was a schematic calendar of twelve months of 30 days each, regardless of the real moon. A few well-known examples are sufficient to prove this statement: contracts for future delivery were dated in this schematic calendar, regardless of the actual outcome in the particular year,[10] past expenses[11] and rents are calculated according to a 360-day business year and to 30-day months,[12] etc. But it is interesting to see that this schematic year was also in use in astronomical texts. Solstices and equinoxes are listed as falling on the fifteenth of the Months I, IV, VII, and X, although everybody knew that the dates in the real lunar calendar would be totally different in almost all cases. The same holds with

[10] Thureau-Dangin, *RA*, XXIV (1927), 188 ff. These examples belong to the Old Babylonian, Persian, and Neo-Babylonian periods.

[11] Kugler, *ZA*, XXII (1908), 74 f.

[12] Neugebauer, *Mathem. Keilschrift-Texte*, III, 63.

the lengths of day and night,[13] the shadow length,[14] rising and setting
of fixed stars,[15] etc.[16] This use of the schematic calendar in an astronomical context is especially important; it demonstrates clearly that
the schematic dates do not represent an attempt to approximate as
closely as possible the real facts but merely constitute a way of
expressing future dates in round numbers according to a general
scheme whose exact relation to the real lunar calendar remains to be
established later on when actually needed.

It is evident that the analogous situation in Egypt is sufficient to
explain analogous consequences. No one was able to predict exactly
the moon's behavior, and a schematic calendar was therefore quite
necessary wherever economic life demanded regularity and simplicity. "The" Egyptian calendar is therefore in all respects the
result of practical needs alone, and "astronomy" is restricted to the
simple fact that the real lunar festivals were regulated by direct
observation, with no attempt to influence the civil calendar, and vice
versa. It is only a slight difference in emphasis which brought about
the almost total eclipse of the schematic calendar in Babylonia and of
the lunar calendar in Egypt. The deeper reasons for this difference
in emphasis can perhaps be found in the difference of social and economic structure of the two countries. In unified Egypt with its cen-
tralized administrative system the schematic calendar naturally had a
much higher importance for the life of the whole country[17] than in the

---

[13] E.g., Weissbach, "Bab. Miscellen," *Wiss. Veröff. DOG*, IV (1903), 50 f., and Kugler, *Sternkunde*, Ergänzungsheft, 88 ff.

[14] Weidner, *AJSL*, XL (1924), 186 ff.

[15] CT, XXXIII, 1–8.

[16] It is very possible that many dates in cuneiform sources are actually meant in the schematic calendar, but we have no means to prove it. It would be, however, equally diffi-cult to prove that the real lunar calendar is meant.

[17] When I reviewed the content of this paper at the meeting of the American Oriental Society in Boston, Professor H. Frankfort asked whether the institution of the schematic calendar could be assumed to belong to the reign of Djoser. I think that no serious objec-tion can be raised against such an assumption, because the only condition for the creation of the schematic calendar is a sufficiently well-organized and developed economic life. On the other hand, means to determine such a date by *astronomical* considerations do not exist.

The problem of the invention of the schematic months must not be confused with the problem of the period at which the 365-day year was introduced. The two institutions are absolutely independent—at least in principle. The 365-day year must have been created

city-states of early Mesopotamia, where each community enjoyed the right of having a calendar of its own.[18]

It is worth noticing that the parallelism between the Babylonian and Egyptian situation also holds for the astronomical documents which we possess from the Twelfth Dynasty and from the New Kingdom. The decanal lists in the coffins from Asyut[19] represent the same type of schematic astronomical calendars as do the Babylonian texts,[20] and the same holds for the star calendars around the figure of Nut in the cenotaph of Seti I and in the tomb of Ramesses IV.[21] Here again, as in Babylonia, we see that astronomy in its earlier stages of development makes no attempt to give exact dates but applies simple schemes which strongly idealize the real facts.[22]

To summarize, both the Egyptian and the Babylonian calendaric concepts display a higher complexity than usually admitted by modern scholars. One point needs special stressing: this complexity must not be considered as the struggle of two or three competing calendaric systems in the modern sense of the word but represents the peaceful coexistence of different methods of defining time moments and time intervals in different ways on different occasions. The situation is here very much the same as in ancient metrology: no need is felt to measure, e.g., grain and silver and fishes by the same units of

---

at a period when the inundation coincided roughly with the season called "inundation." Such a coincidence held for the centuries around 4200 and again in the centuries around 2800. The latter date (i.e., the time of Djoser) has been considered by Winlock (*op. cit.* p. 462) as the date of the definite establishment of the Egyptian year. The analysis of all available evidence for the use of the 365-day year by A. Scharff (e.g., *Historische Zeitschrift*, CLXI [1939], 3–32) also shows that there is no reason to maintain the earlier date (as I was still inclined to do in my paper in *Acta orientalia*).

[18] Cf., e.g., N. Schneider, "Die Zeitbestimmungen der Wirtschaftsurkunden von Ur III," *Anal. Or.*, Vol. XIII (1936).

[19] Cf., e.g., Pogo, *Isis*, XVII (1932), 16–24, and *Osiris*, I (1935), 500–509.

[20] Of course, only as far as the *method* is concerned; the content is totally different.

[21] For the astronomical and mythological interpretation of these texts see Lange-Neugebauer, "Papyrus Carlsberg I", *Kgl. Danske Vidensk. Selsk. Hist.-fil. Skrifter*, Vol. I, No. 2 (1940). It is a methodical mistake to use these documents as astronomically precise and to calculate their date under this assumption—not to mention the fact that there does not yet exist a satisfactory explanation of essential features of the "diagonal calendars" on the coffin lids.

[22] The same can be observed in early Greek astronomy, e.g., in Autolycus (*ca.* 300 B.C.), *De ortibus et occ. II*, theorem 6 (ed. Hultsch, p. 118).

weight, nor is an attempt made to establish well-defined relations between these measures. Exactly in the same sense all modern talk
about ancient "luni-solar calendars" constitutes an anachronism:
some elements of ancient life are regulated according to the seasons;
others, according to the moon (and in Egypt also according to the
Nile and Sothis). But no Egyptian thought about a Sothis-lunar
calendar or any analogous construction. The key to understanding
the origin of the Egyptian calendar seems to me to be the insight into
the *independence* of all its elements which we still see in existence in
historical times: the Nile, the Sothis star, the fiscal calendar, and
the moon.

BROWN UNIVERSITY

Reprinted from
Journal of Near Eastern Studies
1, 397–403 (1942)

The Egyptian "Decans"

O. Neugebauer

Brown University, Providence, R.I., U.S.A.

Summary

It is shown that the 10° sections of the ecliptic, called decans by the Greeks, were originally constellations rising heliacally 10 days apart, and invisible for 70 days. Such stars belong to a zone south of the ecliptic and include Sirius and Orion. The use of the decans for time measurement at night leads to a twelve-division of the period of complete darkness. From this is eventually derived the twenty-four division of day and night.

1. THREE different systems of astronomical reference were independently developed in early antiquity: the "zodiac" in Mesopotamia, the "lunar mansions" in India, and the "decans" in Egypt. The first system alone has survived to the present day because it was the only system which at an early date (probably in the fifth century B.C.) was associated with an accurate numerical scheme, the 360-division of the ecliptic. The lunar mansions, *i.e.* the twenty-seven or twenty-eight places occupied by the Moon during one sidereal rotation, were later absorbed into the zodiacal system which the Hindus adopted through Greek astronomy and astrology. With Islamic astronomy the mansions returned to the west but mainly as an astrological concept. A similar fate befell the decans. When Egypt became part of the Hellenistic world the zodiacal signs soon show a division into three decans of 10° each. As "drekkana" they appear again prominently in Indian astrology, and return in oriental disguise to the west, forming an important element in the iconography of the late Middle Ages and the Renaissance.

2. We shall be concerned not with these wanderings of early astronomical concepts but with the much discussed problem of the localization of the decans. I think we now can satisfactorily solve this problem and simultaneously gain an insight into the origin of the twelve-division of night and day in Egypt, from which eventually our twenty-four-division was derived (Sethe, 1920). This progress has been made possible by utilizing information contained in a Demotic papyrus of the Roman period, purchased by the Carlsberg Fund about twenty years ago for the Egyptological Institute of the University of Copenhagen. The late H. O. Lange recognized the importance of this text, now called "P. Carlsberg 1", which was then published by him and the present writer (Lange-Neugebauer, 1940). In recent years, my colleague, Professor R. A. Parker, and I assumed the study of this text in connection with our plans for a comprehensive publication of all available astronomical texts from Egypt. It was from our discussions that it became clear that the Egyptian texts of the Middle and New Kingdom contain all the information required for determining, at least qualitatively, the position of the decans and their use for time measurement.

3. The "decans" make their appearance in drawings and texts on the inner side of coffin lids of the tenth Dynasty (around 2100 B.C.). Here we find thirty-six constellations arranged in thirty-six columns of twelve lines each in a diagonal pattern of which the following scheme represents the right upper corner (the columns proceed

47

from right to left, as is customary in Egyptian inscriptions):

day 21	day 11	day 1	
S_3	S_2	S_1	hour 1
	S_3	S_2	hour 2
		S_3	hour 3

The constellations "S" are our thirty-six "decans". Among them figure Sirius and Orion, which, except for the Big Dipper, are the only two identifiable asterisms of the Egyptian sky.

The use of these "diagonal calendars" was first explained by POGO (1932). Each vertical column serves as a star clock during the particular decade the first day of which is quoted at the top of the column. For example, the rising of decan S_3 indicates during the first decade the third hour of the night; in the second decade, the second, *etc.* When a decan rose in the first hour of the night, it was obviously near its acronychal setting ten days later. (Fig. 1 may serve as an interesting illustration of the decans.)

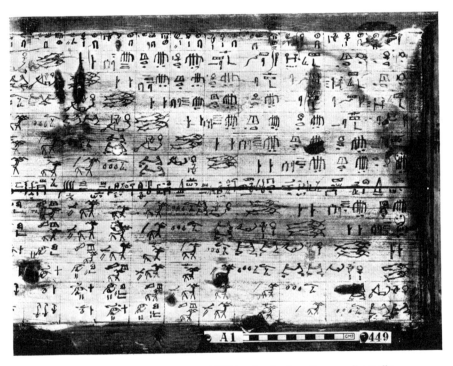

Fig. 1. The Decans on the Coffin of H K ꜣ·t, showing how these are diagonally
arranged. (Cairo Mus. 28127)

206

We shall not discuss the details of these texts, particularly those modifications of the scheme which became necessary because of the five epagomenal days at the end of the Egyptian year. On the contrary, we shall simplify our discussion by being quite unhistorical and replacing ten-day intervals by 10°-segments of the ecliptic. The error thus committed has no influence on the interpretation of such crude schemes and can be remedied at once if the need should arise.

Using the rising of stars as indications of the beginning of "hours" means that all stars which rise simultaneously—synanatellonta in Greek terminology—are in principle equally serviceable. Thus the diagonal calendars tell us only that the decans were stars located at thirty-six positions of the eastern horizon, such that these horizons intersect the ecliptic at points that are about 10° apart.

4. The next bit of information comes from monuments of Seti I and Ramses IV (about 1300 and 1170 B.C.), where the decans are represented on the body of the sky goddess Nut, and from P. Carlsberg 1, which is an extensive commentary to the often very cryptic inscriptions on the monuments. Again we shall not describe details but utilize only one fact which became clear only through the ancient commentator: all decans are invisible for 70 days between acronychal setting and heliacal rising. In other words, all the decans have (at least ideally) the same duration of invisibility as their leader, Sirius. This shows clearly the origin of the whole concept of the decans. The heliacal rising of Sirius marks, in principle, the beginning of the year, and similarly one chose other stars whose rising indicated the beginning of the consecutive decades of the Egyptian civil calendar. The gradual removal of each such constellation away from heliacal rising was used to mark intervals of the night—we shall call them the "decanal hours". And in order to make the whole scheme as uniform as possible the selection of the decans from among all simultaneously rising stars was made such that they had been seen for the last time 70 days earlier, just as Sirius had spent 70 days in the nether world before rising again at the end of the last hour of the night.

5. Before discussing the resulting decanal hours we shall combine what we know from the diagonal calendars and from P. Carlsberg 1. The diagonal calendars told us that the decans were located on thirty-six horizons which meet the ecliptic at intervals of 10°. But a similar condition is imposed on the thirty-six positions of the horizon with respect to the ecliptic when the stars set. The two groups must be in the relation to each other such that 70 days elapse between corresponding settings and risings. The thirty-six intersections between such pairs give the places of the thirty-six decanal stars.

It is easy to carry out this construction graphically. We make, say, a cylinder projection of the celestial sphere with the equator as circle of contact (see Fig. 2). Let B and A be two positions of the Sun, 70 days apart. Let EH be the position of the horizon when a star S rises heliacally, WH the position of the western horizon when S sets acronychally. For a star of given brightness, it is known how far the Sun in A and B must be distant from the horizon. Thus for given A and B, 70 days apart, we can find S at the intersection of the two proper horizons. Moving A and B 10 days ahead, we get a new pair of horizons and, at their intersection, the next decan. And because Sirius is one of the decans, we are given an initial position AB from which to start. Thus all the other decans can be found, at least in principle.

5

Of course, this presupposes that we know the brightness of the stars—which is obviously not the case except for Sirius. Nevertheless it is clear that less brightness must remove the two horizons from A and B and thus bring S closer to the ecliptic. And the opposite holds for bright stars. Thus we obtain for the decans a zone, instead of a curve, following by and large the ecliptic toward the south, with Sirius located at its farthest boundary.

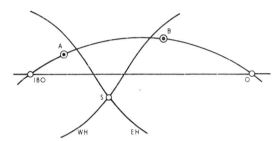

Fig. 2. Graphical determination of the Decans

6. There would be no point in trying to push the identification of the decans any further. What we now know definitely is that both hypotheses which have found support among scholars, namely, that the decans are either ecliptical or equatorial stars (considering the only certain identifications, Sirius and Orion, as "exceptions" to confirm the rule), are equally wrong. It is clear that the 70-day invisibility cannot be taken absolutely literally, not to mention the idealizations made by the Egyptians in order to maintain the relationship between stars and rotating calendar. All these effects soon enough rendered unusable the whole device of the diagonal calendars, and it was probably already obsolete in practice when we meet it as the traditional time instrument for the use of the dead person on his coffin. Indeed, as PARKER has recognized, the star tables of the temples of the Ramessides abandon decanal risings in favour of transits of quite different constellations. But the decans maintain to the end of Egyptian history their rôle as representatives of the consecutive decades of the year and as such they were readily absorbed into the Hellenistic zodiacs.

7. We must now come back to the "decanal hours". Obviously these "hours", determined by the rising of stars on horizons with constant longitudinal difference, are neither constant nor even approximately 60 min in length. They vary as the oblique ascensions of the corresponding sections of the ecliptic and are about 45 min long because during each night eighteen decans rise and set. Thus twelve decanal hours seem too short to measure time at night. In so arguing, one forgets, however, that each decan has to serve for ten days as indicator of its hour. If we furthermore require that these "hours" of the night never should be part of twilight, we get a satisfactory covering of the time of total darkness by means of only twelve decans during ten consecutive days, especially for the shorter summer nights.* Again for the sake of consistency a simple scheme which held true for Sirius near the shortest summer nights was extended to all decans and all nights, thus making the number of

---

* This can be checked, e.g. by using the tables of oblique ascensions in the Almagest II, 8.

"hours" twelve for all seasons of the year. And finally the symmetry of night and day, of upper and nether world, suggested a similar division for the day. This parallelism between day and night is still visible in the "seasonal hours" of classical antiquity. It is only within theoretical astronomy of the Hellenistic period that the Babylonian time-reckoning with its strictly sexagesimal division, combined with the Egyptian norm of 2×12 hours led to the twenty-four "equinoctial hours" of 60 min each and of constant length.

REFERENCES

LANGE, H. O. and NEUGEBAUER, O. 1940   Papyrus Carlsberg No. 1, ein hieratisch-demotischer kosmologischer Text, Danske Videnskabernes Selskab Hist.-filol. Skrifter 1, No. 2.

POGO, A. 1932   Calendars on coffin lids from Asyut, *Isis* 17, pp. 6–24.

SETHE, K. 1920   Die Zeitrechnung der alten Aegypter, Nachr. d. Ges. d. Wiss. zu Göttingen, Philol.-hist. Kl., p. 28–55; p. 97–141.

Reprinted from
Vistas in Astronomy (Ed., A Beer), Vol. 1
Pergamon Press, London and New York, 1955

On the Orientation of Pyramids

by

O. NEUGEBAUER*

The remarkable accuracy of the orientation of the Great Pyramid has led to a number of suggestions concerning the methods on which the determination of the cardinal directions could have been achieved. All astronomically based theories, however, face serious difficulties, e.g. that there is no bright star exactly at the celestial pole, or that rising and setting amplitudes suffer from the poorly defined position of the observer (as well as other practical difficulties). It is therefore perhaps permissible to suggest as a possible method a procedure which combines greatest simplicity with high accuracy, without astronomical theory whatsoever beyond the primitive experience of symmetry of

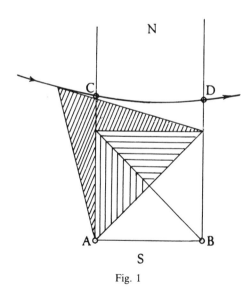

Fig. 1

* History of Mathematics Department, Brown University, Box 1900, Providence, Rhode Isl. 02 912, U.S.A.

Centaurus 1980: vol. 24: pp. 1–3

shadows in the course of one day. In short, one can use the shadow of a pyramid as an excellent instrument for orientation.

All one has to do is to place an accurately shaped pyramidal block (e.g. the capstone of the pyramid under construction) on the accurately levelled ground which will eventually carry the monument. Let its square base be oriented according to a reasonably accurate estimate of the SN/EW directions.[1] Then one observes the path of the shadow cast by the apex of the pyramid from some time before noon to some time after noon. This path describes a curve (which we now know to be a branch of a hyperbola, concave toward North in the winter half of the year, concave toward South in the summer, a straight line at the equinoxes), which will intersect first the western,

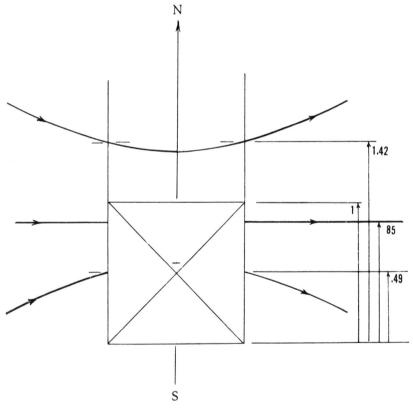

Fig. 2

then the eastern base of the pyramid or a straight continuation in a northerly direction (cf. Fig. 1). If these points of intersection are at different distances from, e.g., the south corners (AC and BD respectively), then the orientation is not yet correct. A slight turn of the base and repeated observation on the next day will improve the situation. Not only can this process be repeated many times until high stability is reached but by waiting some weeks one utilizes different tracks and thus in effect averages small errors of individual observations. For example, observations scattered over half a year would lead to a neat set of mid-points between the two parallel base-sides providing the desired SN direction, valid for any structure erected on the same ground. Figure 2 gives a scale drawing of the range of shadow curves available at Memphis, assuming a slope of 52° for the faces of the pyramid. But the method in no way depends on these specific data. For example, any accurately shaped pyramidal model can be used as "gnomon".

1. This preliminary orientation toward north may well be carried out ceremoniously by the king, looking at the northern constellations. Such rituals imply nothing for the technicalities of the actual construction, not to mention that the accurate determination of the north celestial pole by means of naked eye observations of the Big Dipper is a problem at least as difficult as the orientation on the ground.

3. Babylonian

Über eine Untersuchungsmethode astronomischer Keilschrifttexte

Von O. Neugebauer-Kopenhagen

Ich möchte mit diesen Bemerkungen die Aufmerksamkeit derer, die sich mit der Geschichte der babylonischen Astronomie beschäftigen, auf ein Verfahren hinlenken, das geeignet ist, manche überflüssige Rechenarbeit zu ersparen und in wenigen Zeilen Schlüsse abzuleiten, zu deren Begründung manchmal umständliche astronomische Betrachtungen verwandt worden sind, die man de facto ganz aus dem Spiel hätte lassen können. Mathematisch ist das Verfahren ein altbekannter Kunstgriff, den man stets in einer oder der andern Form anzuwenden pflegt, wenn es sich um die Untersuchung periodischer Vorgänge handelt.

Bekanntlich besteht eines der wichtigsten Verfahren zur Beschreibung periodischer Vorgänge (etwa der Bewegung des Mondes) in der babylonischen Astronomie in der Bildung von Zahlenreihen konstanter Differenz, die bis zu einem bestimmten Maximum M ansteigen, um dann wieder bis zu einem Minimum m zu fallen, dann wieder bis M zu steigen usw. Die einzelnen Zahlen entsprechen etwa den aufeinanderfolgenden Monaten oder Tagen u. dgl. — jedenfalls gehören sie zu Zeitpunkten gleichen Abstandes, d. h. kurz gesagt, sie sind die Ordinaten einer linearen Funktion zu aequidistanten Abszissen (vgl. Abb. 1). Wir nennen einen solchen Funktionsverlauf eine *„lineare Zackenfunktion"*. Die einzelnen Abszissen-Punkte, zu denen die Texte die Ordinaten angeben, numerieren wir fortlaufend mit den ganzen Zahlen, z. B. indem wir in einem bestimmten Text irgendeine Zeile mit 0 bezeichnen, die folgenden Zeilen mit 1, 2, 3, 4, ..., die vorangehenden rückwärts gehend mit —1, —2, —3, ... Allgemein bezeichnen wir die Zeilennummern mit k, den zugehörigen Funktionswert mit y_k (s. Abb. 1). Als „Abstand" zweier Abszissenpunkte nehmen wir einfach die Differenz ihrer Nummern, d. h. wie bezeichnen ein Zeilen-Intervall als Ein-

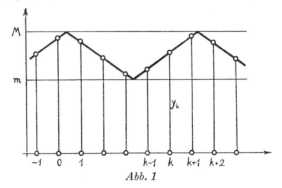

Abb. 1

heit, gleichgültig, welche astronomische Bedeutung ihm zukommt. Das Intervall, in dem die Zackenfunktion hin und her geht, bezeichnen wir als „*Schwingungsbreite*" $\triangle = M - m$, ferner die konstante Differenz zwischen zwei benachbarten Ordinaten als $|\,d\,|$. Bezüglich der Zacken ist zu bemerken, daß sie im allgemeinen nicht mit einem der angegebenen

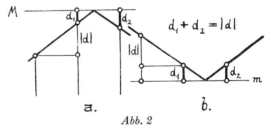

Abb. 2

Ordinaten-Endpunkten zusammenfallen werden, sondern daß die Extrema zwischen zwei Punkte fallen. Dann gilt einfach die Regel, daß die Abstände d_1 und d_2 der beiden Nachbarpunkte eines Extremums vom Extremwert zusammen wieder $|\,d\,|$ ausmachen (vgl. Abb. 2). Diese letzte Regel erlaubt uns nun, den ganzen Sachverhalt sehr viel einfacher darzustellen, und gerade darauf möchte ich hier hinweisen. Wir können nämlich z. B. im Falle von Abb. 2a den abstei-

genden Ast an der Oberkante unseres Streifens „spiegeln",
d. h. den Punkt nach dem Maximum um ebensoviel, d. h.
um d_2, über die Oberkante legen, als er vorher unter ihr
lag. Da $d_1 + d_2$ gleich der üblichen Differenz $|d|$ sein sollte,

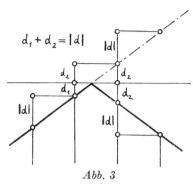

Abb. 3

so bedeutet das, daß der gespiegelte absteigende Ast die
geradlinige Fortsetzung des aufsteigenden ist (vgl. Abb. 3)[1]).
Wir gehen nun von Punkt zu Punkt weiter, bis wir an dem
Punkt vor dem nun folgenden „Minimum" ankommen, das

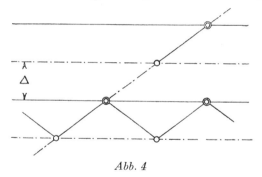

Abb. 4

durch die Spiegelung als ein „Maximum" im zweiten Streifen
erscheint (Abb. 4). Statt zu knicken, spiegeln wir neuerlich usw. Auf analoge Weise gehen wir von der ursprüng-
lichen Zacke nach links, nur spiegeln wir jetzt immer an der

---

1) Dies gilt selbstverständlich auch für den Fall, daß die Zacke
zufällig genau mit einem der Punkte zusammenfällt, denn die Differenzen im absteigenden Ast sind dieselben wie im aufsteigenden.

Unterkante nach unten zu. *Durch dieses Verfahren ersetzen wir also die in dem Streifen der Breite △ hin und her schwingende Zackenfunktion durch eine einzige Gerade, die ein System von Parallelstreifen der Breite △ unter der Steigung | d |: 1 durchsetzt.* Zeichnen wir einen solchen Streifen als „Ausgangsstreifen" aus, und nennen das in ihm verlaufende Geradenstück einen „aufsteigenden Ast", so entspricht dem Stück im nächsten Streifen ein absteigender Ast, dann kommt wieder ein aufsteigender, und so abwechselnd weiter (vgl. Abb. 4); entsprechend sind den Schnittpunkten der Geraden mit den Streifenrändern abwechselnd Maxima und Minima zugeordnet. — Zu beachten ist bei dieser Darstellungsweise einer Zackenfunktion nur noch, daß sie nur das Verhalten der Funktion in nerhalb des Streifens wiedergibt, während dieser Streifen in Wirklichkeit um den Abstand m von der Abszissenachse entfernt ist (vgl. Abb. 1). Man hat also bei allen Rechnungen zu bedenken, daß alle Ordinaten stets um den Wert m verringert erscheinen, der eventuell für das Schlußresultat wieder hinzuzufügen ist.

Die Darstellungsmethode einer Zackenfunktion gemäß Abb. 4 bedeutet eine große Vereinfachung aller Überlegungen mit solchen Funktionen: wird doch die umständliche Beschreibung der immerwährenden Knickungen ersetzt durch die einfachste Funktion überhaupt: durch eine einzige lineare Funktion. Einige Beispiele mögen dies erläutern.

Beispiel 1. Bekannt seien die ersten 5 Zeilen der Kolonne H des Textes BM 34580[1]):

Zeile 1	20,20,0
2	14,52,30
3	8,5,0
4	1,17,30
5	5,30,0

Man soll den Wert in Zeile 39 berechnen.

Die Zeilen 1 bis 5 reichen aus, um $| d |$, m, M (also auch $△ = M - m$) zu berechnen, denn zwischen Zeile 1 und 2 liegt

---

1) „SH 272" bei Kugler, Mondrechnung S. 13, oder in Kugler-Schaumberger, Sternkunde, Ergänzungen S. 392, Kol. VIII.

ein Maximum, zwischen 4 und 5 ein Minimum. Man erhält nach dem üblichen Verfahren

$$|d| = 6{,}47{,}30 \qquad M = 21{,}0{,}0 \qquad m = 0 \qquad \text{also} \qquad \triangle = 21{,}0{,}0.$$

Wir kennen also einen absteigenden Ast der Funktion vollständig. Wir denken sie uns nun nach dem Schema von Abb. 4 geradlinig fortgesetzt (vgl. Abb. 5). Um zu Zeile 39 zu gelangen, müssen wir zu Zeile 1 noch 38 Zeilen addieren,

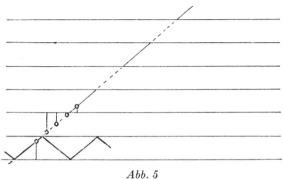

Abb. 5

d. h. die Funktion steigt von Zeile 1 an um 38 Stufen der Höhe $|d|$, d. h. um

$$38 \cdot |d| = 38 \cdot 6{,}47{,}30 = 4{,}18{,}5{,}0.$$

Die Anzahl der Streifen der Breite $\triangle$, die sie dabei durchsetzt hat, ist zu finden durch

$$\frac{38 \cdot |d|}{\triangle} = \frac{4{,}18{,}5{,}0}{21{,}0{,}0} = 12 + \frac{6{,}5{,}0}{21{,}0{,}0}.$$

Das heißt also, daß die Gerade zwischen Zeile 1 und Zeile 39 12 volle Intervalle der Breite $\triangle$ und noch einen Rest gestiegen ist, d. h. sich im 14. Intervall befindet, wenn man das Intervall, in dem der erste Punkt liegt, als erstes zählt. Da 14 eine gerade Zahl ist, entspricht das betreffende Geradenstück einem absteigenden Ast, denn der zweite, vierte, ..., $2n$te Streifen entspricht einem absteigenden Ast. Wie weit ist die Gerade bereits in diesen 14. Streifen eingedrungen? Die Antwort ist leicht zu geben: wir wissen ja schon, daß sie insgesamt um $38 \cdot |d| = 12 \cdot 21{,}0{,}0 + 6{,}5{,}0$ gestiegen ist. Ein Teil des Restbetrages $6{,}5{,}0$ wird für den Punkt 1 benötigt, der um $M - y_1 = 21,0,0 - 20,20,0 = 40,0$ unter der Oberkante des ersten Streifens liegt (vgl. Abb. 5). Also sind für den 14. Streifen noch $6,5,0 - 40,0 = 5,25,0$ an Eindringtiefe verfügbar. Der Funktionswert, der uns interessiert, ist der Ergänzungswert über der Geraden, denn wir befinden uns auf einem absteigenden Ast (vgl. Abb. 6), also $y_{39} = \triangle - 5,25,0 = 21,0,0 - 5,25,0 = 15,35,0$, und dies ist in der Tat die Zahl, die im Text steht[1]).

Die Überbrückung dieser 38 Zeilen hätte man durch direkte Weiterrechnung natürlich auch erreichen können. Wollte man aber, statt um 38 Zeilen, um 2000 Zeilen weitergehen (da der Zeilenabstand 1 Monat bedeutet, so wären dies ungefähr 160 Jahre), so wäre die direkte Rechnung eine

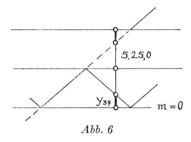

Abb. 6

ungeheure Arbeit, während wir in unserer Rechnung nur an die Stelle des Faktors 38 den Faktor 2000 zu setzen hätten, um den Wert an einer um 2000 Zeilen weiteren Stelle direkt zu berechnen. Allgemein: soll man die gesetzmäßige Fortsetzung einer Zackenfunktion an irgendeiner beliebigen Stelle vom Zeilenabstand n berechnen (hier = Abstand in synodischen Monaten), so bestimmt man zunächst durch Division von $n \cdot |d|$ durch $\triangle$ die Anzahl der vollen Streifen und verteilt den Rest auf Ausgangswert und neuen Wert, wobei man daraus, ob die Streifenanzahl gerade oder ungerade ist, ablesen kann, welcher Art der Ast ist, auf dem der neue Wert liegt.

---

1) Man beachte, daß wir eigentlich nur das Stück $y_{39} - m$ berechnet haben (Abb. 6), so daß wir jetzt noch m zu addieren hätten, wenn nicht hier $m = 0$ wäre.

Beispiel 2. Gegeben zwei Texte mit je einer Zacken-funktion, die in den Zahlen $|d|$, M und m übereinstimmen. Es soll entschieden werden, ob der eine Text die Fortsetzung des andern sein kann und wenn ja, wie groß ihr Abstand ist.

Lösung. Man wähle z. B. in jedem der beiden Texte einen Wert gerade vor einem Maximum. Angenommen, die Texte ließen sich aneinanderfügen und der Abstand der beiden ausgewählten Zeilen wäre n Zeilen (positives n bedeutet dann, der Text mit der Zeile n liegt rechts von dem mit der Zeile 0, negatives n das Gegenteil). Dann muß es

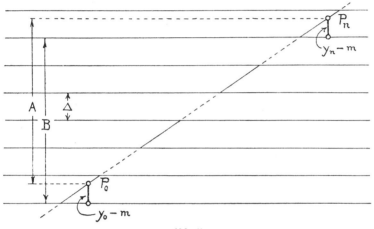

Abb. 7

möglich sein, die beiden Texte gemäß Abb. 7 aneinander-zufügen. Der Vertikalabstand A zwischen den Punkten P_0 und P_n beträgt einerseits $A = n\,|d|$ ($=$ Zeilenzahl mal Diffe-renz); andererseits können wir diesen Abstand A auch zu-sammensetzen aus einer Anzahl a von Streifenbreiten $\triangle$ (ge-rechnet von der Basis des Streifens, der P_0 enthält bis zur Basis des Streifens P_n d. h. Strecke B in Abb. 7) wozu wir am oberen Ende noch das Stück $y_n - m$ hinzufügen, am unteren Ende aber $y_0 - m$ wieder abziehen müssen. Somit haben wir für die Strecke A den zweiten Ausdruck $A = a\,\triangle + (y_n - m) - (y_0 - m) = a\,\triangle + y_n - y_0$ gefunden. Da P_0 und P_n beide in einem aufsteigenden Ast liegen, muß unsere Anzahl a

von Streifen eine gerade Anzahl sein. Also haben wir das Resultat: die beiden Texte lassen sich dann und nur dann ineinander fortsetzen, wenn gilt:

$$n\,|\,d\,| = a\,\triangle + y_n - y_0 \qquad a \text{ eine gerade ganze Zahl.}$$

Man vergegenwärtige sich nochmals, was dies besagt: die Texte aneinanderschließen heißt 1. eine ganze Zahl n von Differenzen $|\,d\,|$ und 2. eine gerade Anzahl a von Intervallen $\triangle$ muß so gefunden werden, daß

$$(1) \qquad n\,|\,d\,| - a\,\triangle = y_n - y_0 \qquad (a \text{ gerade})$$

ist, wo y_0 bzw. y_n die Zahlenwerte sind, die im ersten bzw. zweiten Text stehen. Oder: *die beiden Texte sind dann und nur dann aneinanderschließbar, wenn die „diophantische Gleichung" (1) bei gegebenen $|\,d\,|$, $\triangle$, y_n, y_0 in ganzen Zahlen n und a lösbar ist, wobei außerdem a noch gerade sein muß.*

Man kann sofort sagen, wann die Texte nicht aneinanderschließbar sind. Es sind $|\,d\,|$, $\triangle$, $y_n - y_0$ gegebene Zahlen, also auch ihre Primfaktoren. Angenommen, $|\,d\,|$ und $\triangle$ enthielten einen gemeinsamen Primfaktor. Dann ist dieser Primfaktor auch in $n\,|\,d\,|$ und auch in $a\,\triangle$ enthalten, also auch in $n\,|\,d\,| - a\,\triangle$ (man kann ihn ja einfach ausklammern), also wegen (1) auch in $y_n - y_0$. Daraus folgt: *enthalten $|\,d\,|$ und $\triangle$ einen gemeinsamen Primfaktor* (sind sie nicht „relativ prim", wie man sagt), *der nicht auch in $y_n - y_0$ enthalten ist, so ist (1) unerfüllbar, also der erste Text nicht in den zweiten fortsetzbar.*

Andererseits besagt ein klassischer Satz der Zahlentheorie: *Enthalten $|\,d\,|$ und $\triangle$ keine gemeinsamen Primfaktoren, so ist (1) immer lösbar*, was auch $y_n - y_0$ für einen Wert haben möge. Die Frage, ob zwei Texte in einer gewissen Kolonne ineinander fortsetzbar sind oder nicht, ist also immer ohne weiteres entscheidbar. *Sind $|\,d\,|$ und $\triangle$ relativ prim, so kann man die Texte immer aneinanderfügen, andernfalls nur dann, wenn die gemeinsamen Primfaktoren von $|\,d\,|$ und $\triangle$ auch in $y_n - y_0$ stecken.* Dabei ist zu beachten, daß alle Textzahlen hier stets als ganze Zahlen gelten (unabhängig von ihrer astronomischen Bedeutung), so daß die letzte über-

haupt berücksichtigte Sexagesimalstelle als Stelle der Einer
zählt.

Beispiel 3. Die Kolonne G von BM 34580 (Text 1) ist
nicht in die von BM 34580 (Text 2) fortsetzbar[1]). Beweis:
die beiden Texten gemeinsamen Konstanten sind $|d| =$
22,30,0, $M = 4,29,27,5$, $m = 1,52,34,35$, also $\triangle =$
2,36,52,30. Als y_0 eignet sich Vs. VII, 16 in Nr. 1, als y_n
Vs. VIII, 2 in Nr. 2 (beide unmittelbar vor einem Maximum).
Damit hat man

$$y_n - y_0 = 4,11,19,35 - 4,23,37,30 = -12,17,55.$$

Es sollte also sein

$$n \cdot 22,30,0 - a \cdot 2,36,52,30 = -12,17,55$$

oder unter Weglassung gemeinsamer Faktoren[2]):

$$n \cdot 54,0 - a \cdot 6,16,30 = -2,29,31,$$

was unmöglich ist, denn die linke Seite ist immer gerade, die
rechte aber ungerade[3]).

Beispiel 4. Schnabel sagt ZA **37**, 47: „Den Beweis (daß
sich die Kolonnen H und J von BM 34580[4]) in den Text
VAT 7844 fortsetzen lassen) könnte ich allerdings nur durch
Abdruck der Koll. H und J für Neumond von 110—210 S.-Ä.
liefern, was mindestens 17 Seiten Raum von ZA in Anspruch
nehmen würde."

Kürzer und zugleich allgemeiner läßt sich ein solcher Be-
weis folgendermaßen führen, wobei ich allerdings nicht die
Kol. H oder J, sondern G aneinanderschließe[5]). Für diese
Kolonne ist beiden Texten gemeinsam

---

1) Bezüglich Text 1 siehe die Zitate von Anm. 1 S. 124. Text 2 = „99,
81—7—6" bei Kugler, Mondr. S. 42 (allerdings mit mehreren Tran-
skriptionsfehlern).

2) Vgl. unten S. 130 Anm. 1.

3) Man kann nicht weiter verkürzen, da 2,29,31 eine Primzahl ist.

4) Vgl. Anm. 1 S. 124.

5) H ist nämlich in VAT 7844 („Text 3") nicht direkt enthalten,
und ich müßte daher etwas über ihre Berechnung sagen, was hier ganz
nebensächlich wäre. Bei J kommen, wie Schnabel gezeigt hat, noch
zusätzliche, regelmäßige Korrekturen hinzu. Am Prinzip ändert sich
nichts.

$$|d| = 22,30,0 \quad M = 4,29,27,5 \quad m = 1,52,34,35$$
$$\text{also} \quad \triangle = 2,36,52,30.$$

Als Funktionswerte kann man z. B. wählen

Text 3 (VAT 7844, Vs. II, 4)
$$y_0 = 4,19,52,30$$
Text 1 (BM 34580 Vs. VII, 2)
$$y_n^{\varepsilon} = 4,22,22,30$$

$\left.\right\}$ also $y_n - y_0 = 2,30,0$.

So erhält man die Bedingung

$$n \cdot 22,30,0 - a \cdot 2,36,52,30 = 2,30,0 \qquad a \text{ gerade}.$$

Befreit man sie von überflüssigen gemeinsamen Faktoren[1]), so erhält man

(2) $\qquad n \cdot 36 - a \cdot 4,11 = 4 \qquad a \text{ gerade}.$

Eine Lösung liefert ein Blick in eine Multiplikationstabelle[2]): es ist nämlich $7 \cdot 36 = 4,12$, also

$$7 \cdot 36 - 1 \cdot 4,11 = 1.$$

Um rechts 4 zu erhalten statt 1, hat man also nur beide Seiten mit 4 zu multiplizieren, d. h. es ist

(3) $\qquad 28 \cdot 36 - 4 \cdot 4,11 = 4$

oder $n = 28$, $a = 4$ eine Lösung. Wir gewinnen aus ihr leicht beliebig viele weitere, indem wir irgendeine ganze Zahl g nehmen und damit die Zahl $g \cdot 4,11 \cdot 36$ bilden, die wir im ersten Glied links addieren, vom zweiten Glied links aber subtrahieren, so daß man keinen Fehler begeht. So ergibt sich:

---

1) Man macht dies natürlich nicht durch Primzahlzerlegung, sondern durch Multiplikation mit geeigneten Faktoren: Multiplikation mit 2 gibt $n \cdot 45,0,0 - a \cdot 5,13,45,0 = 5,0$, wobei man schon eine Stelle weglassen kann. Dann durch 5 kürzen: $n \cdot 9,0 - a \cdot 1,2,45 = 1,0$. Hier kann man offenbar überall durch 15 kürzen, was man durch Multiplikation mit 4 leistet (und eine Stelle wegläßt) $n \cdot 36 - a \cdot 4,11 = 4$ und $4,11$ ist Primzahl.

2) Eine solche ist z. B. in meinen Math. Keilschrift-Texten (Quellen und Studien zur Geschichte der Mathematik A 3) S. 84 angegeben.

$$(28 + g \cdot 4,11) \cdot 36 - (4 + g \cdot 36) \cdot 4,11 = 4$$

woraus folgt, daß

$$n = 28 + g \cdot 4,11 \qquad a = 4 + g \cdot 36$$

(g irgendeine positive oder negative ganze Zahl oder 0) ebenfalls Lösungen von (2) sind und die Zahlentheorie lehrt, daß damit alle möglichen Lösungen gefunden sind. Also: die beiden Texte sind in der Kol. G aneinanderschließbar und die beiden ausgewählten Zeilen haben einen der folgenden Abstände

$$(28 - g \cdot 4,11) \quad \ldots \quad -7,54 \quad -3,43 \quad 28 \quad 4,39 \quad 8,50$$
$$\ldots \quad (28 + g \cdot 4,11).$$

Will man den Abstand der ersten Zeilen der beiden Texte feststellen, so hat man nur alle Zahlen um 2 zu verändern und erhält für den Abstand der beiden Texte Vs. 1:

$$30 + g \cdot 4,11 \qquad g \gtreqless 0 \text{ ganz.}$$

Damit sind alle Möglichkeiten, die beiden Texte in Kol. G aneinanderzuschließen, vollständig angegeben. Im vorliegenden Fall ist es leicht, daraus eine auszuwählen, denn der Text 1 ist genau datiert, und der Text 3 trägt wenigstens auf der Rs. ein Datum, das die Größenordnung seines Abstandes von Text 1 sichert. Dazu paßt nur $g = 5$, d. h. ein Abstand von $30 + 5 \cdot 4,11 = 21,25$. Da hier Zeile = Monat ist, erhält man für VAT 7844 Vs. 1 den Nisan 104 SÄ. Nachdem so der Abstand der beiden Texte bekannt ist, ist es nach der Methode von Beispiel 1 ein leichtes, auch in andern Kolonnen den Abstand zwischen beiden Texten zu überbrücken und zu prüfen, ob auch andere Kolonnen konsequent weiter gerechnet sind oder, wenn nicht, wie groß die Abweichungen in ihnen sind.

Ich hoffe, daß durch die vorangehenden Beispiele klar geworden ist, wie man mit wenigen Zeilen Rechnung die verschiedensten Fragen in voller Allgemeinheit entscheiden kann, die bei expliziter Nachrechnung oft Tausende von Zeilen erfordern würden. Das Schlimmste, was bei der hier geschilder-

9*

ten Methode passieren kann, ist, daß eine diophantische Gleichung der Form (1) nicht ohne jede Rechnung gelöst werden kann, wie in unserm Beispiel 3. Dann genügt aber jedenfalls die Kettenbruchentwicklung von $\dfrac{|d|}{\triangle}$, um die Lösung zu gewinnen[1]).

Zum Schluß sei noch auf eine wichtige Anwendung der geschilderten Methode verwiesen, die häufig dazu führen kann, einen Text absolut zu datieren. Dazu ist allerdings nötig, daß er sich in zwei unabhängige Kolonnen in einen bereits datierten Text fortsetzen läßt. Man hat dann das Verfahren von Beispiel 4 auf jede der beiden Kolonnen für sich anzuwenden. In jedem Fall erhält man eine ganze Schar von Möglichkeiten für den Abstand zweier bestimmter Zeilen. Für die erste Kolonne möge die Lösung lauten

$$(4) \qquad A_1 = a + g_1 b$$

für die zweite

$$(5) \qquad A_2 = c + g_2 d.$$

Darin sind also a, b, c, d gewisse bestimmte ganze Zahlen (im vorangehenden Beispiel wäre $a = 30$, $b = 4{,}11$), während g_1 und g_2 irgendwelche ganze Zahlen bedeuten. Sie besagen, daß in der ersten Kolonne die verschiedenen Möglichkeiten um Vielfache von b, in der zweiten um Vielfache von d voneinander entfernt sind. Soll nun der erste Text in den zweiten hinsichtlich beider Kolonnen fortsetzbar sein, so bedeutet das, daß für beide Kolonnen ein gemeinsamer Abstand existieren muß. Es muß also $A_1 = A_2$ gemacht werden können, d. h. g_1 und g_2 so bestimmt werden, daß

$$a + g_1 b = c + g_2 d$$

ist. Nennen wir $c - a$ zur Abkürzung e, so heißt dies: Damit die Texte in beiden Kolonnen übereinstimmen, muß die diophantische Gleichung

$$(6) \qquad g_1 b - g_2 d = e$$

---

1) Man vgl. z. B. Weber-Wellstein, Enzyklopädie der Elementarmathematik (5), Bd. 1, § 59.

lösbar sein. Setzt man die Lösung[1]) g_1 in (4) ein (bzw. g_2 in (5)), so ist damit ein bestimmter Abstand gewonnen, für den auch die zweite Kolonne richtig von einem Text in den andern übergeht. Die Lösungszahl g_1 (oder g_2) von (6) ergibt sich selbstverständlich wieder mit der charakteristischen Unbestimmtheit von Lösungen diophantischer Gleichungen, nämlich so, daß man irgendwelche Multipla einer gewissen Zahl noch hinzufügen oder abziehen darf. Diese Periodizität bei g_1 hat aber bereits eine solche Größe, daß sie gewöhnlich weit über alle historisch zulässigen Möglichkeiten hinausführt, so daß von selbst nur der kleinste Wert von g_1 überhaupt zulässig ist. Sobald aber g_1 festliegt, ist der Abstand der beiden Texte eindeutig bestimmt.

Das Wesentliche dieser ganzen Methode liegt darin, daß sie gestattet, durch ganz einfache Rechnungen eine Reihe von Fragen zu entscheiden, *ohne irgendwelche Hypothesen über die sachliche Bedeutung der Kolonnen zu benutzen oder astronomisches Vergleichsmaterial auf Grund moderner Rechnungen zu benötigen.* Auch werden alle Fragen, wie weit man Vergleichsrechnungen ausdehnen soll, überflüssig, da man mit einem Schlag immer a l l e überhaupt möglichen Lösungen erhält. Das einzige, was man benutzt, ist die innere Gesetzmäßigkeit der einzelnen Kolonnen, d. h. man verläßt de facto an keiner Stelle das aus den Texten selbst gegebene Material. Was das moderne Verfahren von dem der Texte unterscheidet, ist nur die Einsparung der Rechenarbeit in den Intervallen, in denen man die Kenntnis der Einzelzahlen gar nicht benötigt.

Es möge noch bemerkt werden, daß das geschilderte Verfahren auch noch auf eine Reihe anderer Fälle ausdehnbar ist, also z. B. keineswegs verlangt, daß es sich um Zackenfunktionen des einfachsten Typus handelt. So kann man etwa durch eine leichte Modifikation auch gesetzmäßige Abänderung der Differenzen mit berücksichtigen, wie sie z. B.

---

1) Sie existiert immer, falls b und d relativ prim sind. Andernfalls sind die beiden Texte nur dann aneinanderschließbar, falls auch e die gemeinsamen Primfaktoren von b und d enthält.

in der Breitenkolonne E des Mondrechnungssystems des Kidinnu auftreten u. dgl. m. Das Entscheidende ist nur die Existenz einer strengen Periodizität, damit man das Wellenbild durch das Streifenbild ersetzen kann. Bei dem traurigen Zustand der meisten Texte wird man trotz alledem oft nicht vermeiden können, Zusatzannahmen zu erproben, wo man lieber mit den bloßen Textangaben auskommen möchte. Eine solche Zusatzannahme ist z. B. die von Kugler[1]) aufgestellte Schaltungsregel der Seleukidenära. Da auch sie einen rein periodischen Vorgang darstellt, kann man sie ebenfalls nach unserm Schema behandeln und z. B. Schaltregel und Fortsetzung einer Zackenfunktion zur Aufstellung einer diophantischen Gleichung für die Periodennummer g_1 des Textabstandes benutzen (s. o. S. 132) und so wieder mit wenigen Zeilen Rechnung zu einer absoluten Datierung kommen. Der Vorteil bleibt auf alle Fälle bestehen, daß man alle Fragen mit einem Minimum an Rechenarbeit klären kann und die unlösbaren Fälle von Anfang an praktisch ohne jede Rechnung auszuschalten imstande ist (s. o. S. 128).

---

1) Sternkunde I S. 212.

Reprinted from
Zeitschrift der Deutschen Morgenländischen Gesellschaft
90 (1), 121−134 (1936)

Reprinted from ISIS, Vol. XXXVI : Pt. 1 : No. 103 : 1945

STUDIES IN ANCIENT ASTRONOMY. VII.*
MAGNITUDES OF LUNAR ECLIPSES IN BABYLONIAN MATHEMATICAL ASTRONOMY

By O. Neugebauer

1. When F. X. Kugler's monumental work "Die Babylonische Mondrechnung" appeared in 1900, it presented for the first time an insight into the methods and achievements of Babylonian astronomy of the Hellenistic age. Continued research made it increasingly clear that the Babylonian lunar theory is equalled only in the best works of Greek mathematical astronomy. The results reached in the present article point in exactly the same direction. It is the purpose of this study to show that several features in the Babylonian theory of eclipses are the results of a common methodological idea which in itself is of great historical interest.

It is not my intention to give an account here of all the steps leading from Kugler's initial discoveries to the present results. It need only be recalled that Kugler already recognized the existence of two different methods for the computation of lunar ephemerides: an older one, here called "System A," and a more recent "System B." [1] Kugler's investigations concentrated for the most part on System B, because he had at his disposal an almost complete ephemeris for the new moons of two years computed according to System B. As to System A, the available ephemerides were in such a bad state of preservation that he was forced to restore the main features of System A from a text for lunar eclipses which naturally gave only the elements at intervals of six or five months. In spite of the difficulties caused by the missing intermediate values, Kugler reached the following results. He identified a column (denoted by E) representing the moon's latitude, established the main rules for its computation, and found by comparison with Oppolzer's "Canon" that one column (called Ψ here) in the eclipse text was closely related to the magnitude of the eclipse. [2]

The investigation of System A was given a fresh

---

* The first five articles of this series were published in the "Quellen und Studien zur Geschichte der Mathematik," Ser. B, Vol. 4 (1937–1938), under the title "Untersuchungen zur antiken Astronomie." The sixth study will be published in the anniversary volume dedicated to G. Sarton.

[1] Kugler's Systems II and I respectively.

[2] Kugler BMR p. 155.

---

impetus by the appearance of new textual material and particularly by the consistent application of the idea that Diophantine equations should be the tool to answer the question whether or not a given fragment of an ephemeris could be the result of continued computation starting from another ephemeris of the same system. [3] It could be shown by this method that all known fragments of System A were parts of the same undisturbed process of computation of all columns with the single exception of the column for the moon's latitude in a single text. [4] As to the column Ψ, it became clear that the relationship between the latitude E of the moon and the magnitude Ψ of an eclipse was given by

$$(1) \qquad \Psi = c + \text{sign } K \cdot \frac{E}{6}$$

where $c = 17,24,0$ is a certain constant and sign $K = +1$ for an ascending node but sign $K = -1$ for a descending node, and where E is the latitude of the moon at the moment of opposition. In addition, it was possible to determine for both systems the exact shape of the curve which led to the specific rules for the computation of column E in the ephemerides. This is to be understood as follows. Assume that we are given E as a *continuous* function of time (called the "true function"); we then obtain the ephemerides from it by tabulating the values of this function for the moments of the syzygies. The above statement is then equivalent to saying that it was possible to discover the true functions from which the ephemerides were derived. Their graph is given in Fig. 1, where the points N and $N+1$ indicate the moments of opposition in the months N and $N+1$, respectively; the values of E at these points are the values found in the ephemerides.

By the consideration of variable points on the surface of the moon, I made an attempt [5] to explain the peculiar shape of these curves and to deduce the value of the angular diameter of the moon and of the

---

[3] Neugebauer UAA II.

[4] Neugebauer UAA II p. 89.

[5] Neugebauer UAA III.

shadow of the earth. PANNEKOEK recognized [6] that I had committed a trivial mistake in this discussion which made untenable my results concerning these diameters. PANNEKOEK returned to KUGLER's initial hypothesis [7] that the greater inclination of E in System A in the nodal zone is the result of an attempt to describe better the actual variation of the latitude of the center of the moon. This interpretation will be fully confirmed in the following discussion.

2. In order to describe the progress made beyond the state of affairs explained above, I must return to the exceptional text mentioned before. The text [8] in question contains, instead of a latitude function as described in Fig. 1a, a function of the same period but with unaltered slope in the nodal zone, which I had interpreted as the latitude of the moon's center and therefore called the "center-function" E' (dotted line in Fig. 2). The fact that all values were only one-third of the values corresponding to the dotted line in Fig. 2 could be understood, at least in principle, as a change of metrological units. More serious, however, appeared the fact that this new function had not the simple relation to the latitude function indicated by the dotted line in Fig. 2 and Fig. 3 but showed nodes in advance of the corresponding nodes of E (thick line in Fig. 3). At the time of my publication, it seemed possible to explain this in the following way. The text in question is the latest text (49 B.C.) of System A, whereas all the other texts belong to the period from 175 B.C. to 59 B.C.; one could assume that the uninterrupted use of E for more than 150 years eventually led to an accumulated error which was corrected between 59 and 49 B.C. This explanation, which was not very plausible in itself because of the great accuracy of the period of E, became untenable when, by means of the Diophantine method, I recently dated two fragments which contain the same function as the latest text but belong to the years 102 B.C. and 163 B.C. respectively.[9] All columns of these texts were connectible with all the remaining texts with the sole exception of the "center-function" E' which, in turn, led in unbroken continuation to the corresponding column of the latest text. This result

[6] PANNEKOEK [1] p. 12.
[7] KUGLER BMR p. 132, PANNEKOEK [1] p. 13.
[8] Incompletely published by SCHNABEL Ber. p. 244/5. The relevant part is transcribed in NEUGEBAUER UAA II p. 52 (cf. also UAA III p. 340).
[9] The first text, which is very badly preserved, was published by KUGLER BMR plate XI (Sp. II, 74 = B.M. 34600). Only traces of three numbers are visible in the critical column, to be restored as [1,24,4,4]8 [43,28,5]4 and [3,53,4]0. The second text (to be published in NEUGEBAUER ACT), however, is just well enough preserved to leave no doubts concerning the column in question.

seemed to confront us with the absurd situation that for more than 150 years two latitude functions existed side by side with exactly the same period but with different nodes. All the other columns, however, were computed exactly alike.

This riddle was completely solved when I recognized that the very same numerical value which occurred in the function E' in the ephemeris of 163 B.C. also occurred in the eclipse text, mentioned above, but in the column Ψ. In other words, the alleged center-function E' is nothing but the continuation of the eclipse magnitude Ψ. Consequently, all the texts of System A form, without exception, a uniform group of ephemerides computed with no modification whatsoever for more than 150 years, — with the sole restriction that some ephemerides contain merely the moon's latitude (E), others give E and Ψ, and still others Ψ alone.

Certain details must be presented in order to make these statements fully clear. First of all, it must be said that the "magnitude" of an eclipse, as defined by (1), p. 10, is measured in a way which is slightly different from the way it is done today. We would say that the greatest total lunar eclipse corresponds to latitude zero (central eclipse). The corresponding value of Ψ according to (1) would be $\Psi = c$ because $E = 0$. For eclipses smaller than the central eclipse, the modern definition and (1) would agree; that is to say, $\Psi \leqq 0$ indicates that no eclipse will take place and that the magnitude increases from $\Psi = 0$ to $\Psi = c$ with decreasing latitude. If the immersion of the moon increases beyond the central position, we would ascribe to the eclipse a value decreasing with increasing distance of the center of the moon from the ecliptic, whereas (1) indicates that Ψ further increases from c to $2c$. Fig. 4 will illustrate this situation; it is evident that it is only a difference of terminology which distinguishes (1) from the modern definition. For the comparison of Ψ with modern tables for eclipse magnitudes, it is convenient to modify Ψ according to the modern definition (dotted line in Fig. 4).

The second remark is much more essential. To this point we have tacitly assumed that Ψ is computed for only such oppositions where the latitude of the moon is so small as to make an eclipse, at least in principle, possible. In other words, Ψ has until now been defined only for every sixth, or perhaps fifth, full moon. Indeed, we have ephemerides in which E is computed for all syzygies, but Ψ only for those months when E is closest to zero. These values are then selected to build up an eclipse table. We said above, however, that Ψ was also computed for all *intermediate* oppositions where eclipses were obvi-

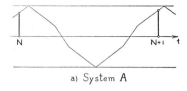

a) System A

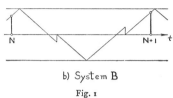

b) System B

Fig. 1

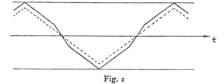

Fig. 2

ously excluded. This statement is not fully correct without a minor modification in our definitions. We call Ψ' a function defined by

$$(2) \qquad \Psi' = \text{sign } K \cdot c + \frac{E}{6}$$

for values of E sufficiently close to zero (in a sense which will be defined accurately in (3)). A comparison of (1) and (2) shows that Ψ and Ψ' are identical for ascending nodes but have opposite signs for descending nodes. This change of sign is necessary if Ψ should be extended to a *continuous* function with the same period as E. If (2) were used for all values of E, the function Ψ would have the same period as E as well as the same change in slope as E (Fig. 1) for values $|E| \geqq \kappa$. There is no reason, however, to introduce this complication into the computation of the values which are in any case of no significance outside the nodal zone. We obtain uniform slope for Ψ' if we define Ψ' as follows for all values of E:

$$\begin{aligned} &\Psi' = \text{sign } K \cdot c + \frac{E}{6} \quad \text{if} \quad |E| \leqq \kappa \\ (3) \\ &\Psi' = \text{sign } K \cdot c + \frac{2E \mp \kappa}{6} \quad \text{if} \quad \begin{cases} E \geqq \kappa \\ E \leqq -\kappa \end{cases} \end{aligned}$$

where κ is the value of E where the slope of E changes to half of its value in the nodal zone.

It is this function which we find in the texts and which we denoted previously by E' (dotted line in Fig. 2). It is identical with Ψ for ascending nodes and with $-\Psi$ for descending nodes. The ephemerides contain either E and Ψ in separate columns, Ψ being computed only for every sixth or fifth month, or Ψ' alone, because E can always be obtained very easily from Ψ'.

3. Before turning to the discussion of System B, we must clarify the metrological units which are used in Systems A and B in connection with our problem. It is well known that angular distances were expressible either in degrees or in certain other units, the most important of which is the cubit (*ammatu*, abbreviated in the following by a). Unfortunately, the relation between cubits and degrees seems to be ambiguous, the cubit corresponding [10] either to $2\frac{1}{2}°$ or to $2°$. This ambiguity can be eliminated by introducing the smaller unit, "finger," denoted in the following by f. In the case of $1^a = 2°$, KUGLER found that $1^a = 24^f$; but in the case of $1^a = 2;30°$, it is $1^a = 30^f$. In both cases,

$$(4) \qquad 1^f = 0;5°$$

holds regardless of whether 24 or 30 fingers are called one cubit.

We must now determine the units employed to measure the latitude of the moon E and the magnitude of eclipses Ψ. The values of the characteristic parameters of E and Ψ' in System A are

$$\text{for } E \qquad \begin{aligned} M_E &= 7,12,0,0 \\ \kappa &= 2,24,0,0 \end{aligned}$$

$$\text{for } \Psi' \qquad \begin{aligned} M_{\Psi'} &= 2,0,0,0 \\ c &= 17,24,0 \end{aligned}$$

In my previous investigation of the latitude function, I found no other way to explain these numbers satisfactorily except to assume that they are measured in thirds of cubits. Unfortunately, such units are not attested in independent use elsewhere, but they at least gave reasonable values for the magnitudes in question. On the other hand, serious difficulties were encountered in extending this assumption to System B. The whole problem of the units employed took a new turn when I found the units of "finger" mentioned in connection with eclipse magnitudes in one of the so called "procedure texts," which give rules for the computation of the ephemerides.[11] It became increasingly clear that the assumption of thirds of cubits as units for E and Ψ was untenable. Indeed,

[10] KUGLER BMR p. 127, SSB I p. 25 and SSB II p. 547 ff.
[11] To be published in NEUGEBAUER ACT.

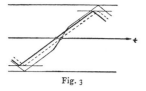

Fig. 3

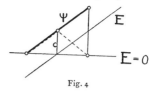

Fig. 4

all difficulties disappeared when I realized that the units in E are "barleycorn" (še), units well attested in Babylonian metrology. Ordinarily, the barleycorn (denoted in the following by $š$) is $\frac{1}{180}$ of the shekel (gín). There exists, however, a small text from Assur,[12] which shows that the barleycorn was also used to denote $\frac{1}{180}$ of the cubit. Using the cubit of $2\frac{1}{2}°$, we have with (4)

(5) $1^š = 0;0,20^a = 0;0,50° = 0;10^f$

and for E we obtain

$$M_E = 7,12^š = 6°   \kappa = 2,24^š = 2°.$$

The value of 6° for the extremal latitude (affected, of course, by parallax) is also independently attested by an unpublished procedure text.

In order to find the units of Ψ', we need only notice that according to (5)

$$1^f = 6^š.$$

The factor $\frac{1}{6}$ in the formula (3) defining Ψ' by

$$\Psi' = \text{sign } K \cdot c + \frac{E}{6}$$

can therefore be interpreted as changing barleycorn into fingers. In other words, $\frac{E}{6} = E^f$ is the latitude of the moon expressed in fingers, and therefore c and Ψ' must also be measured in fingers, as is explicitly confirmed by the procedure text mentioned above and by observational texts.[13] We can thus write instead of (3)

(6a) $\Psi' = \text{sign } K \cdot c + E^f$ if $|E^f| \leq \kappa^f$

(6b) $\Psi' = \text{sign } K \cdot c + 2E^f \mp \kappa^f$ if $\begin{cases} E^f \geq \kappa^f \\ E^f \leq -\kappa^f \end{cases}$

where

(6c) $c = 17;24^f = 1;27°$ $\kappa^f = 24^f = 2°.$

[12] Published by THUREAU-DANGIN [1] p. 33. Cf. also NEUGEBAUER UAA III p. 280.
[13] KUGLER SSB I p. 276 b s.v. SI.

These formulae have a very simple interpretation. Formula (6a) shows that for small latitudes (less than $\pm 2°$) the magnitude of an eclipse increases with decreasing latitude, whereas (6b) tells us that outside the nodal zone Ψ' has a slope twice that of the curve for E. But outside the nodal zone, the function E has only half the slope it had inside. Thus (6b) shows that Ψ' does not change its slope but continues with the slope of the inner strip. The reason is clear: it is important to describe the latitude of the moon by a function which represents closely enough the fact that the latitude changes more rapidly near the ecliptic than near the extrema. The eclipse magnitude Ψ', however, has physical significance only for small latitudes, and the way it is continued from node to node is therefore irrelevant if only its period is correct and if the relation (6a) is restored whenever eclipses are possible. The result is illustrated by Fig. 5.

I think one is correct in saying that this construction of the function Ψ' shows an amazing highly developed "modern" attitude toward the whole problem. For reasons of purely mathematical expediency, the initial definition (1) of the magnitude Ψ of an eclipse is modified by the construction of a continuous function Ψ' which has no astronomical significance whatsoever in general but gives the right numerical results whenever the conditions for an eclipse are satisfied. It is precisely this trend to a mathematical formulation of all methods which we find wherever we reach a complete understanding of the basic ideas of System A.

4. If we compare our above results with the material available for ephemerides of System B, it becomes immediately evident that the alleged columns for latitude of the type represented by Fig. 1b are actually columns for the magnitude of eclipses,[14] extended to a function Ψ'. The discontinuous jumps after each crossing of the zero line can be explained as follows. As in the case of System A, the "magnitude" Ψ of an eclipse ranges from 0 to $2c$, where the value c corresponds to the exactly central

[14] This confirms a conjecture, mentioned to me by VAN DER WAERDEN several years ago; at that time it seemed impossible to prove it, with the hopelessly entangled metrological situation and with no known parallel in System A.

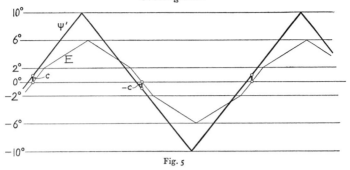

Fig. 5

eclipse (latitude $E = 0$). Outside of the interval $0 \leqq \Psi \leqq 2c$ no eclipses are possible and it is therefore irrelevant how we pass from one node to the next by means of a function Ψ'. Fig. 6 shows how Ψ' was chosen with respect to the nodes of E.

Three groups of units are known[15] to have been used in texts of System B for Ψ'. In the first group, we have

$$c = 18,0 \qquad\qquad M = 1,58,27$$

where M denotes the maximum of Ψ'. Comparison with System A shows immediately that the above quantities must be measured in fingers because in System A we have $c = 17;24^f$ and $M = 2,0^f$. We thus obtain for this first group from (4) p. 12

$$(7a) \qquad c = 18^f = 1;30° \qquad M = 9;52,15°$$

This result is directly confirmed by the next group of units which are given by

$$c = 1,30,0 \qquad\qquad M = 9,52,15$$

thus being identical with (7a). In the last group we have

$$c = 1,0,0 \qquad\qquad M = 6,34,50.$$

The values of c and M in this group are two-thirds of the corresponding values in the second group. Their significance becomes evident if we remember that $\Psi = c$ is the magnitude of the greatest total eclipse. If we call this value 1^c, then all other eclipses are measured in units of the greatest possible eclipse. Thus we have

$$(7b) \qquad c = 1^c \qquad\qquad M = 6;34,50^c.$$

These results are very satisfactory from the point of view of metrology. The only units which occur in the lunar ephemerides for the measurement of latitude and eclipse magnitude are the following:

barleycorns, fingers, and degrees. In the column Ψ, indicating eclipse magnitudes, a measurement in units of the maximal eclipse is added in some texts of System B, but this is actually not a new metrological unit but merely a convenient expression of eclipse magnitudes. *Complete metrological uniformity is thus established for all lunar ephemerides.*

5. The discovery that the function which we once considered as the latitude of the moon really represents the eclipse magnitude has many implications, the detailed development of which must be postponed for later studies. Only a few points must be mentioned.

It is evident that we must assume that the latitude function E of System B must have a shape of the type indicated in Fig. 6, analogous to the shape of E in System A (Fig. 5). The question arises at what value κ the slope of E has to be changed and by what amount. Assuming that κ is again 2° and the slope for $|E| > \kappa$ half of the slope for $|E| < \kappa$ as in System A, we obtain for the extremal latitude of the moon $6;1,7,30°$ instead of 6° in System A. This speaks in favor of our assumption but can only be proved by new texts.

The fact that we have texts of System B which contain two functions of type Ψ' side by side, essentially proportional but with quite different nodes, remains unexplained.[16] This difficulty is not due to our new interpretation, because it was equally unintelligible why two latitude columns were given side by side.

Another unsolved problem is the question of the angular diameter of the moon. It is, of course, very tempting to compare the eclipse magnitudes given by Ψ in "fingers" with magnitudes measured in "digits." This is further supported by the fact that $c = 17;24^f$ (System A) and $c = 18^f$ (System B) would be a fair approximation of the actual value of

---

[15] Cf. the list given in NEUGEBAUER UAA III p. 271.

[16] Cf. NEUGEBAUER UAA III p. 310.

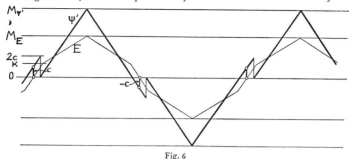

Fig. 6

about 22 digits for the largest lunar eclipse. This correspondence would imply that the lunar diameter was assumed to be $d = 12^I$ or, because of (4), $d = 1°$ which is almost exactly twice the real value. Additional difficulties occur if one tries to determine the radius of the shadow cone. It seems to me that no decision can be made at present because of the interference of the lunar parallax which must have distorted the ancient measurements in a still unknown fashion. Only new texts or a better understanding of the terminology of the procedure texts will bring the solution of this question.

Finally, it must be mentioned that the clear separation of latitude and eclipse magnitude requires a reinvestigation of all previous attempts to use alleged latitudes for the comparison of results of Babylonian ephemerides of different systems with each other and with modern computations.

Bibliography

KUGLER BMR
F. X. KUGLER, Die Babylonische Mondrechnung. Freiburg 1900.

KUGLER SSB
F. X. KUGLER, Sternkunde und Sterndienst in Babel. 2 vols. Münster 1907/1924.

NEUGEBAUER ACT
O. NEUGEBAUER, Astronomical Cuneiform Texts. Babylonian Ephemerides for the Movement of the Sun, the Moon and the Planets from the Seleucid Period and their Computation. In preparation.

NEUGEBAUER UAA II
O. NEUGEBAUER, Untersuchungen zur antiken Astronomie. II. Datierung und Rekonstruktion von Texten des Systems II der Mondtheorie. QS B 4 p. 34-91 (1937).

NEUGEBAUER UAA III
O. NEUGEBAUER, Untersuchungen zur antiken Astronomie. III. Die babylonische Theorie der Breitenbewegung des Mondes. QS B 4 p. 193-346 (1937).

OPPOLZER, Canon
TH. OPPOLZER, Canon der Finsternisse. Akad. d. Wiss. (Wien) Math.-nat. Klasse, Denkschriften 52 (1887) [Reprinted by Stechert, New York, 1921].

PANNEKOEK [1]
A. PANNEKOEK, Some Remarks on the Moon's Diameter and the Eclipse Tables in Babylonian Astronomy. *Eudemus* 1, p. 9-22 (1941).

QS
Quellen und Studien zur Geschichte der Mathematik, Astronomie und Physik.

RA
Revue d'Assyriologie.

SCHNABEL Ber.
P. SCHNABEL, Berossos und die babylonisch-hellenistische Literatur. Leipzig-Berlin, 1923.

THUREAU-DANGIN [1]
F. THUREAU-DANGIN, Notes assyriologiques 49. RA 23 (1926) p. 33 f.

Reprinted from ISIS, Vol. 37: Pts. 1 & 2: Nos. 107 & 108: 1947

STUDIES IN ANCIENT ASTRONOMY. VIII.
THE WATER CLOCK IN BABYLONIAN ASTRONOMY

By O. Neugebauer

Two clear-cut types of Babylonian astronomy can be distinguished: one, a highly developed mathematical astronomy which flourished the last three centuries before our era; the other, a rather primitive and crude astronomy which we find in texts of which the archetypes, though there is no direct evidence, possibly go back as far as Old-Babylonian times (i.e., about 1600 B.C.). The first-mentioned mathematical astronomy has as its goal the computation of ephemerides for the moon and the planets; this is accomplished by means of the ingeniously devised use of periodic arithmetic progressions of various orders and related interpolatory methods. The high accuracy of these ephemerides, however, is not based on very precise observations but merely on an extensive use of relations between the periods of the phenomena in question.[1] The reason for this tendency is obvious: ancient observational instruments are so inaccurate that they cannot compete with the accuracy obtainable by mathematical means from the comparison of few but sufficiently distant observations. This not only holds for Babylonian astronomy but is also the admitted principle on which Greek mathematical astronomy is based, e.g., in the Almagest.

The strictly mathematical character of late Babylonian astronomy permits us to get a very detailed picture of its methods and results. Many obstacles, however, lie in the way of our understanding the astronomy of the preceding periods. One of the major difficulties lies in the character of the source material. Our main information must be obtained from two "series" of texts, named after their opening words "Enuma-Anu-Enlil" and "Mul-Apin." Enuma Anu-Enlil is a large collection of tablets containing astronomical omina from all periods.[2] Mul-Apin only contains, so far as we know, two tablets,[3] but is much more "astronomical" in content. None of these texts, however, contains practi-

cal examples of astronomical computations and leaves to us the interpretation of the usually difficult context. Of a more concrete character is a collection of rules for lunar and planetary phenomena, written down in Seleucid times but going back, at least in part, to about 800 B.C.[4] We shall quote this text by its number of publication,[5] TU 11.

The driving force in the development of Babylonian astronomy was calendaric problems. The largest and most highly developed part of the theoretical astronomy of the Seleucid period is devoted to the computation of the new moons. This interest was due to the fact that the civil calendar never abandoned the use of real lunar months, and it is therefore not surprising that attempts to solve questions in connection with the lunar calendar already appear in the earlier material, e.g., the question of the interval between the setting of the moon and sunset. So far as the solar year is concerned, we know that no serious attempt was made in earlier periods to establish fixed relations between the solar and the lunar year; the intercalation of a 13th month was simply regulated according to the actual agricultural conditions of each individual year. For formal purposes, however, there existed a schematic calendar which assumed a year of 12 months of 30 days each.[6] This schematic calendar was the basis for economic transactions, payment of interest, etc. — in short, it was a convenient business calendar. The texts which we will discuss here are all based on this calendar. Dates given in this schematic calendar were obviously understood to be of only qualitative correctness and subject to adjustments if comparison with the actual lunar calendar showed that this was required. In order to characterize the solar year within this scheme, four points were selected: the equinoxes and the solstices, arbitrarily[7]

---

[1] It should be mentioned in this connection that the general assumption of hoary observational records as a requirement for such period relations is greatly exaggerated. Observations of a few decades or, at the most, a century or two are enough to provide the basis for the methods which we find applied in late Babylonian astronomy.

[2] Main edition: Virolleaud ACh.

[3] Weidner [1].

[4] This has been shown by Schaumberger in an unpublished study of this extremely difficult text. I wish to acknowledge my indebtedness to the results obtained by Prof. Schaumberger and communicated to me shortly before the war.

[5] Thureau-Dangin, TU, Plates 22 and 23; inventory No. AO 6455.

[6] Cf. Neugebauer [2] p. 400 f.

[7] "Arbitrarily" means without deeper astronomical significance.

placed on the 15th day of the months I, IV, VII, X of the schematic calendar. The question arises how these four points were determined.

Ever since WEIDNER's publication [8] of the second tablet of Mul-Apin, we know that one way to find the change in the length of daylight during the seasons consisted in the use of the sundial.[9] We shall not discuss here the scheme which was used to describe the variation in the lengths of the shadow,[10] but we must mention that it is based on the assumption of a ratio 3:2 for the lengths of the longest (M) and the shortest (m) daylight. The very same ratio $M:m=3:2$ occurs explicitly in another text closely related to Mul-Apin and establishes at the same time the relation of this ratio to the schematic calendar.

The passage in question [11] is as follows: "$1\frac{1}{3}$ longest daylights are one day, [50 longest daylights are] one month, 600 longest daylights are one year." In other words, $\frac{5}{3}M=1$ day; hence $M=\frac{3}{5}$ of one day and [12] $m=\frac{2}{5}$ of one day, i.e., $M:m=3:2$. Moreover, 50 $M=30$ days $=1$ month and 600 $M=360$ days $=1$ year.

The ratio $M:m=3:2$ was first discovered by KUGLER in the ephemerides of the Seleucid period;[13] this revealed the basis for the classical tradition which characterized the "climate" of Babylon by this ratio of the longest to the shortest daylight.[14] The above-quoted passage shows that the same ratio was already assumed in the oldest astronomical texts accessible to us. This will be of importance for the discussion of our main problem, the measurement of time by water clocks. Before turning to this subject, we must briefly investigate the significance of the ratio $M:m=3:2$.

When KUGLER first discovered the value $M=\frac{3}{5}$ day (i.e., 14 hours and 24 minutes) for the longest daylight in the ephemerides of the Seleucid period, he concluded that this value could not refer to the latitude of Babylon ($32\frac{1}{2}°$) but to a place at $35°$

latitude. Later,[15] he revised his opinion and found satisfactory agreement with the latitude of Babylon by taking, e.g., atmospheric refraction into account. Actually, the values in question are not the result of accurate observations but are nothing more than round numbers close enough to the truth to be useful in practice. This is evident not only from the simplicity of the ratio 3:2 itself, but is also underlined by the fact that the same ratio was used for the ratio of the longest to the shortest night. The longest night, however, is actually shorter than the longest day, so that at least two of the four values cannot be the result of exact measurements, even if we assume that the other pair was observed. In other words, it is methodologically incorrect to treat one of these values as an accurate astronomical constant instead of realizing that all of them are parts of a simple scheme devised in some early phase of Babylonian culture to describe the variability of daylight and night.

For the measurement of time by water clocks, our main source of information is again Mul-Apin. The first tablet of this series was published in 1912 by L. W. KING in vol. 33 of the "Cuneiform Texts from Babylonian Tablets etc. in the British Museum" (BM 86378) and soon became the subject of several studies, the most important of which is BEZOLD's "Zenit- und Aequatorialgestirne."[16] KING himself gave a short description of the text; from this I quote:[17] "An interesting indication of the practical character of the treatise may be seen in the facts that notes are given as to the payments made to the day and night watch respectively: during the six months, from the 15th Tammuz to the 15th Tebet, the day-watch was paid four manehs and the night-watch two, but during the remainder of the year the payments were reversed, the night-watch receiving twice the pay of the day-watch. The observers who were on duty during the longer and colder nights of winter and the long scorching days of the summer months were naturally more highly recompensed." BEZOLD paid scant attention to this passage,[18] but it follows from his translation and commentary [19] that he interpreted the contents astronomically. "(It is) daylight 4 (sixths) of the 'watch'; (it is) night 2 (sixths) of the 'watch'." KUGLER objected to this

---

[8] WEIDNER [1].
[9] This sundial apparently had the simple form of a vertical pole on a horizontal plane, called "gnomon" by the Greeks.
[10] I intend to do this in a separate paper.
[11] K 2164+2195+3510 obv. 26 f., published WEIDNER [3] and correctly explained by KUGLER SSB Erg. p. 89/90 note 1. Text: 1,40 ud-da-zal-e u_4-mu [50 ud-da-zal-e] itu 10 ud-da-zal-la-e mu-an-na.
[12] Assuming, as usual, $M+m=1$ day.
[13] KUGLER BMR p. 75 ff.
[14] Cf. NEUGEBAUER [1].

[15] KUGLER SSB I p. 174 f., SSB II p. 588, SSB Erg. p. 89 and p. 377.
[16] BEZOLD [1]. Cf. also KUGLER SSB Erg. I p. 21 ff., WEIDNER Hdb. p. 35 ff.
[17] CT 33 p. 5, elaborated in KING [1] p. 41 f.
[18] Cf. in the "Appendix" given below the passage No. 1 and its variants Nos. 2 and 3.
[19] BEZOLD [1] p. 27 and p. 54.

interpretation, stating [20] that "the ratio 2:1 of the longest daylight to the shortest night is absolutely excluded for Babylonia." He maintained the astronomical interpretation, however, and added the insight [20a] that the weight "mana" (the Greek μνᾶ, about one pound) must be understood as the weight of water in a water clock. As to the ratio 2:1, he proposed a solution by considering the time difference between sunset and the rising and setting of the moon which he found discussed in other texts. Also, WEIDNER found in the section of the second tablet of Mul-Apin which deals with the sundial, mentioned above, that again the 15th of month IV and month X were characterized by 4 mana and 2 mana, respectively.[20b] WEIDNER, too, declared the ratio 2:1 to be impossible but reached no clear decision as to how to explain it.[21]

The main cause of trouble obviously lies in the fact that the ratio 2:1 is not only extremely bad for the latitude of Mesopotamian cities [22] but contradicts the ratio 3:2 attested in the very same texts. The key to the solution of this dilemma lies in recognizing that, since KUGLER's interpretation of mana as the weight of water in a clepsydra is undoubtedly correct, the outflow of water must not be taken as simply proportional to time. All that we have to do is to look for simple models of water clocks and to compute the weight of water as a function of time, excluding all cases where a time ratio 3:2 does not lead to an increased ratio of water weight. The simplest possibility is certainly the outflow from a cylindrical [23] vessel of height h, emptied through a hole at the bottom. The time t needed to empty such a vessel completely is then given by $t = c\sqrt{h}$, where c is a constant depending upon the outlet and the area of the cross section of the cylinder. The weight w of water contained in the cylinder is proportional to h. If M and m denote two time intervals needed to empty the vessel filled with water of

[20] KUGLER SSB Erg. p. 89.
[20a] KUGLER SSB Erg. p. 95 f., p. 188. In SSB II p. 542 f. KUGLER thought he had found an explicit mention of the water clock in a procedure text (BM 32651 obv. II, 16–19). A close study of this text shows, however, that KUGLER was misled by taking ki in *ki-ṣir* as the ideogram for *qaqqaru*. Actually, this passage explains how to compute the lunar velocity after it reached its greatest or smallest value.
[20b] Text given in Appendix Nos. 4 and 5.
[21] In [2] p. 68 he considered it the result "of the sufficiently known tendency toward schematization (Schematisierungswut) of the Babylonians," whereas in [1] p. 201 he concluded that the Babylonians assumed an invariable(!) length of daylight during the year.
[22] The ratio 2:1 would suit a latitude of about 48°, e.g., Stalingrad.
[23] A prismatic vessel, of course, gives the same results.

weight W and w respectively, we therefore obtain

$$M:m = \sqrt{W}:\sqrt{w}$$

or a ratio $W:w = 9:4$ if $M:m = 3:2$. We thus reach within narrow limits the ratio 2:1 mentioned in the texts as the ratio between weights; and we see that this ratio necessarily follows from the measurement of time by the weight of water flowing from a cylindrical clepsydra.

Our result can be completed in various directions. First, it can be shown that one must assume that the weights indicated mean the weights corresponding to complete emptying of the vessel, because an outflow beginning from a constant level would result in ratios closer to 1 instead of 2. Secondly, it is easy to see that conical vessels would reduce the ratio $W:w$ which corresponds to a given $M:m$, but the reduction from 9:4 to 8:4 is so small that the obtained cone would only deviate very slightly from a cylinder which, therefore, remains the most plausible container. Our conclusion in favor of a cylindrical (or prismatic) shape for the clepsydra finds direct confirmation from Old-Babylonian mathematical texts. THUREAU-DANGIN discovered [24] that some examples in a British Museum mathematical tablet concern a water clock. The calculations assume that the volume of the vessel is proportional to the height,[25] thus excluding conical vessels. In these examples, the volume is measured in capacity units called qa and not in weights. These qa-units are in turn brought into relation with prismatic containers, showing us that $1 \ qa = \frac{1}{120}$ cubic cubits,[26] but this of course does not exclude a cylindrical shape for the clepsydra.

Finally, we must say a few words about the procedure followed in the interval between the two solstices. Experience with various types of primitive astronomy, e.g., early Greek astronomy, would make it probable that linear interpolation was used.[27] This is indeed confirmed by our texts,[28] which give linearly varying values for the manas corresponding to one watch on the 1st and 15th days of each month. Linear increase or decrease of the weights does not correspond to linear change in time, but this is certainly not significant. Having once decided that

[24] THUREAU-DANGIN [1].
[25] NEUGEBAUER MKT I p. 173 ff.
[26] This relation, tentatively proposed in NEUGEBAUER MKT I p. 181, is now confirmed by new textual evidence. (NEUGEBAUER-SACHS MCT p. 96).
[27] Cf., e.g., the linear increase of the length of daylight assumed in the "Eudoxos papyrus" (BLASS [1] p. 13 f., transl. TANNERY HAA p. 284 No. 5).
[28] Appendix Nos. 6, 7 and 8.

the extremal values are adequately approximated by 4 and 2 mana, respectively, one was satisfied with the simplest possible scheme for the interval between.

We thus have reached the following simple result. In order, e.g., to define the length of a "night watch" at the summer solstice, one had to pour 2 mana of water into a cylindrical clepsydra; its emptying indicated the end of the watch. One-sixth of a mana had to be added each succeeding half-month. At equinox, 3 mana had to be emptied in order to correspond to one watch, and 4 mana were emptied for each watch of the winter solstitial night. The ratio 2:1 between the extremal weights is a convenient, rounded-off value which corresponds in practice closely enough to the corresponding time ratio 3:2. Here again one was satisfied with assuming perfect symmetry between summer and winter, and daylight and night. We see once more that very simple but natural schemes characterize the methods of early Babylonian astronomy.

With the above explanation we have removed one very disturbing obstacle to the explanation of time measurement in the early phase of Babylonian astronomy. Many questions, however, remain unanswered. The most disturbing one is the problem of metrological units. KUGLER, seeking to avoid the interpretation of our passages as referring to the variable length of daylight and night, reached the conclusion that 1 mana corresponds to 16 minutes of time.[29] In view of our above discussion, this result can no longer be upheld. I am not convinced, however, that we can again return to the assumption made by BEZOLD[30] and WEIDNER,[31] namely that one day corresponds to 6 mana, 1 mana thus representing 4 equinoctial hours. WEIDNER, e.g., interprets the passage "3 mana is the watch of the day, 3 mana is the watch of the night" (at equinox) as a division of the day into six parts, but it seems to me more natural to think of the length of a *single* watch,

three of which constitute daytime and three night. This would lead to the correspondence 18 mana = 1 day or 1 mana = 80 minutes. Support for this interpretation might be found in the division of the day into 18 parts which occurs in the Book of Enoch, a fact to which my attention was drawn by Dr. A. J. SACHS.[32] The relationship to our problem is evident from the ratio 2:1 which is here assumed for the ratio of longest to shortest daylight. In favor of BEZOLD and WEIDNER, however, might be quoted TU 11 (Appendix No. 14), e.g., the passage rev. 12 "4 daylight 2 night" (at summer solstice), though this might be only an abbreviated statement. Another type of units appears in an ivory prism which contains, among other things, information about shadow lengths.[33] Here we find $M:m=8:4$ which indicates that we are dealing with units of weight. They must contain 30 subunits because "$\frac{2}{3}$" and "10" (not 20 !) constitute one of the higher units. This might explain the numbers 8 and 4 instead of 4 and 2, but I do not know of units satisfying this condition.

With our problem is connected a large complex of questions concerning first and last visibility of the new moon and related phenomena at full moon. In the latest period of Babylonian astronomy a clear understanding of the variable causes for the duration of the visibility of the moon was reached.[34] In the preceding periods, however, an attempt had been made to find a simple scheme, exclusively depending on the seasons.[35] It would lead us far beyond the scope of this article to analyze all the relevant passages of the texts, but the translation of Section 19 of TU 11, given in Appendix No. 14, will show the general direction. Future studies will have to investigate these methods which appear already in Mul-Apin II (Appendix Nos. 4–10) and similar texts.

Appendix. Extracts from the Texts

A. Mul-Apin I

1. **BM 86378** (CT 33, pl. 1–8. Photo: KING [1] facing p. 42)

   Obv. II, *ina* itu šu u₄-15-kam . . . 4 ma-na en-nun
   42/43: u₄-me 2 ma-na en-nun ge₆

   Rev. III, *ina* itu ab u₄-15-kam . . . 2 ma-na en-nun
   7/9: u₄-me 4 ma-na en-nun ge₆

2. **Rm. IV, 337** (KUGLER SSB I pl. 23, No. 26)

   Obv. *ina* itu šu u₄-[15-kam . . .] 4 ma-na en-
   1/2: nun u₄ [. . .

[29] KUGLER SSB Erg. p. 94. On p. 78 and p. 96 he concluded that there existed also another mana, equivalent to 24 minutes of time.
[30] BEZOLD [1] p. 54.
[31] WEIDNER Hdb. p. 43 f. and [2] p. 66 f.
[32] Chapter 72, 10 ff. Cf. CHARLES, Enoch p. 153 ff. and WEIDNER [4].
[33] Appendix No. 13. LENORMANT, Choix 86, interpreted the text as a "règles d'un jeu," SAYCE [1] as an "augural staff"; LANGDON (Men. p. 55) recognized its astronomical character but did not discover the relation with the gnomon tables.
[34] Cf. SCHAUMBERGER Erg. p. 380 ff.
[35] A typical example is the text BE 13918, published WEISSBACH BM pl. 15 No. 4 and p. 50 f., explained by KUGLER SSB Erg. p. 96 ff.

Obv.   *ina* itu du$_6$ u$_4$-15-kam . . . 3 ma-na^{36} en-
8/9:   nun [. . .

3. AO 7540 (Weidner [1] p. 190/191)

Obv. II,   *ina* itu du$_6$ u$_4$-15-kam . . . 3 ma-na en-
2/3:   nun u$_4$-*mi* 3 ma-na en-nun ge$_6$

Obv. II,   *ina* itu ab u$_4$-15-kam . . . 2 ma-na en-nun
8/10:   u$_4$-*mi* 4 ma-na^{37} en-nun ge$_6$

B. Mul-Apin II.

4. VAT 9412. (Unpublished).^{38}

Obv. II,   *ina* itu bár u$_4$-15-kam 3 ma-na ^{39} u$_4$-*mi* 3
21:    ma-na en-nun ge$_6$

Obv. II,   *ina* itu šu u$_4$-15-kam 4 ma-na en-nun u$_4$-*mi*
25:    [2 ^{40} ma-na en-n]un ge$_6$

Obv. II,   *ina* itu du$_6$ u$_4$-15-kam 3 ma-na en-nun u$_4$-*mi*
31:    3 ma-na en-nun ge$_6$

Obv. II,   *ina* itu ab u$_4$-15-kam 2 ma-na en-nun u$_4$-*mi*
35:    4 ma-na en-nun ge$_6$

5. AO 7540. (Cf. No. 3).

Rev. I, 3: *ina* itu bár u$_4$-15[-kam . . .
Rev. I, 7: *ina* itu šu u$_4$-15[-kam . . .

6. VAT 9412. (Cf. No. 4).

Obv. II,   *ina* itu bár u$_4$-1-kam 3 ma-na 10 gín en-nun
43:    ge$_6$ 12 uš 40 GAR šú šá [*sin*]

44:    *ina* itu bár u$_4$-15-kam 3 ma-na en-nun^{41} 12
uš kur šá [*sin*]

45:    *ina* itu gu$_4$ u$_4$-1-kam 2 ⅝ ma-na en-nun ge$_6$
11 uš 20 GAR šú š[*á sin*]

ina itu gu$_4$ u$_4$-15-kam 2 ⅔ ma-na en-nun ge$_6$
10 uš 40 GAR kur [*šá sin*]

etc.

Rev. III,   *ina* itu še u$_4$-1-kam 3 ½ ma-na en-nun ge$_6$
11:    14 uš šú šá šá *sin*

12:    *ina* itu še u$_4$-15-kam 3 ⅔ gín^{42} en-nun ge$_6$
13 uš 20 GAR kur šá *sin*

13:    šá igi-gub-be-e igi-du$_8$-a šá *sin* 3 ma-na en-nun ge$_6$

14:    aná 4 íl-*ma* 12 igi-du$_8$-a šá *sin* igi

15:    40 *nap-pal-ti* u$_4$ *u* ge$_6$ aná 4 íl-*ma* 2,40
nap-pal-ti igi-du$_8$-a igi

Translation of the last section:

13:   Concerning the coefficients of visibility of
the moon; 3 mana, the watch of the night,

14:   multiply by 4 and you will see 12, the visibility of the moon.

^{36} Kugler (pl. 23 and p. 230) gives e-na, which is cer-
tainly a misreading of ma-na.
^{37} Text: giš-na instead of ma-na.
^{38} With the exception of four lines (obv. I, 68–71) in
AfO 7 (1923) p. 269.
^{39} en-nun omitted.
^{40} One wedge preserved.
^{41} ge$_6$ omitted.
^{42} Sic, instead of 3 ⅓ ma-na.

15:   40, the difference of day and night, multiply
by 4 and you will see 2,40, the difference
of the visibility.

These three lines give a summary of the whole table.
A duration of 12° visibility corresponds to 3 mana (i.e.
equinox). The daily increase or decrease of the length
of the watches is 0;0,40 mana. To it corresponds a daily
variation of the visibility by 0;2,40°. Both quantities
are obtained by linear interpolation.

7. VAT 8619 (AfO 12 (1937/39) pl. XII and p. 147
note 23)

Obv. 1:   [*ina* itu b]ár u$_4$-1-kam 3 ma-na 10 gín en-
nun ge$_6$ 12 uš 40 GAR šú šá *sin*

2:   [*ina* itu bá]r u$_4$-15-kam 3 ma-na en-nun ge$_6$
12 uš kur šá *sin*

etc.

Rev. 8:   [*ina* itu še] u$_4$-1-kam 3 ma-na en-nun ge$_6$
14 uš šú šá *sin*

9:   [*ina* itu še u$_4$]-15-kam 3 ⅓ ma-na en-nun ge$_6$
13 uš 20 GAR kur šá *sin*

8. K 2164+2195+3510 (Weidner [3] pl. II)
Beginning destroyed

Rev. 2:   *ina* [itu ap]in u$_4$-1-kam 3,10 *ù* 3,10 a-rá 4
12,40 *aná* šú šá *sin*

3:   *ù* u$_4$-15-kam 3,20^{43} *ù* 3,20^{43} a-rá 4 13,20
aná kur šá *sin*

etc.

10:   *ina* itu še u$_4$-1-kam [3,]30 *ù* 3,30 a-rá 4
14 *aná* šú šá *sin*

11:   *ù* u$_4$-15-kam [3,20] *ù* 3,20 a-rá 4 13,20^{44}
aná kur šá *sin*

9. Rm. II, 174 (Virolleaud ACh. Suppl. II No. 67,
p. 95)

Rev. I,   Duplicate of No. 8. Only beginnings of lines
1–10:   preserved.

10. VAT 9415 (Unpublished; cf. Weidner [1] p.
187).

C. Enuma-Anu-Enlil XIV

11. K 6427 (Craig AAT 17 and Virolleaud Sin 30)
Beginning destroyed

Rev. 1:   [*ina* itu zíz u$_4$-30-kam . . .] 3 m[a . . .

2:   [*ina*] itu še [u$_4$-15-kam] 3 [ma(-na)] ge$_6$
aná(?)^{45} gín(?) [. . .

3:   [*ina*] itu še u$_4$-30-kam 3 ma(-na) gín en-
nun g[e$_6$ . . .

4:   [24]^{46} *ni-ip-lu* [. . .]

5:   [*ina*] itu bár u$_4$-1-kam 11 uš 40 GAR igi-
du$_8$-a šá ^{d}*sin* . *ina* itu bár [u$_4$-15-kam . . .

^{43} Text: 3, 10.
^{44} Text: 13, 30.
^{45} One would expect an expression for ⅓.
^{46} Traces of 4 visible.

6:   [*ina* itu gu₄ u₄-1-]kam 10 uš igi-du₈-a[47] *šá*
     ᵈ*sin . ina* itu gu₄ u₄-15-kam[48] 9 u[š . . .
     etc.
14: [*ina* itu ab u₄-1-]kam 1 [4][49] uš igi-du₈-a
     šá ᵈ*sin ina* itu zíz u₄-15-kam [. . .
15: *ina* itu zíz u₄-1-kam 12 uš 40 GAR igi-du₈-a
     šá ᵈ*sin . ina* itu še u₄-15-kam[48] [. . .
16: 24 igi-du₈-a-meš *ù* kur [. . .]

This text is in great disorder because of the omission of
one line.[50] See SCHAUMBERGER Erg. p. 280 f. for the
restoration. The correct end is given in the following
fragment.

12. British Museum 80–7–19, 273 (CRAIG AAT 16 and VIROLLEAUD Sin 30)

Rev.   Beginning destroyed
1:   *ina* it[u ab
2:   *ina* itu zíz u₄[-1-kam . . .
3:   *ina* itu zíz u₄-15[-kam . . .
4:   *ina* itu še u₄-1-kam [. . .
5:   *ina* itu še u₄-15-kam [. . .
6:   igi-du₈-a-meš *u* kur [. . .]

D. Varia

13. British Museum, Ivory prism (LENORMANT Choix No. 86 p. 224 f., SAYCE [1] p. 336, LANGDON Men. p. 55)

Face C and D, lines 3 to 5.

	C		D	
3.	u₄ gu₄ kin	ge₆ gu₄ kin	u₄ šu	ge₆ šu
4.	ge₆ apin še	u₄ apin še	ge₆ ab	u₄ ab
5.	6⅓ u₄	5 10 ge₆	8 u₄	4 ge₆

14. AO 6455 (THUREAU-DANGIN TU 11)

Section 19 (Rev. 8–15)

8   *gaba-ri* u₄-ná-a *aná epēšika* [dù-*ka*] *šumma* [BE-*ma*]
     bar *šá šatti*(mu)-*ka eš-še-tú* 27 25 KUR 3,20
     ME 2,40 ge₆ 3,20 a-rá 4
9   13,20   13,20 *ultu*(ta) 25 zi-*ma* 11,40 *uḫ-ḫur*
     ūmu(u₄) 28 11,40 *aná muḫḫi*(muḫ) *šamáš ri-ḫi*
     13,20 *ultu*(ta) 11,40 zi-*ma*
10   *ūmu*(u₄) 29 1,40 *sin aná šamáš etiq*(dib-*iq*) 13,20
     aná muḫ-ḫi 1,40 tab-*ma* 15 *ūmu*(u₄) 30 *ūmu*

---

[47] VIROLLEAUD igi-du₆-a, CRAIG igi-du₈-*šá*.
[48] VIROLLEAUD u₄-15-kam, CRAIG u₄-14-kam.
[49] Upper part of 4 visible. VIROLLEAUD erroneously restored 17.
[50] Consequently, all numbers are one month early. I am
convinced that this has no astronomical significance (against
KUGLER SSB Erg. p. 94).

(u₄) 15 *sin aná šamáš etiq*(dib-*iq*) *šumma* (BE-
     ma) 27 15 KUR
11   3,20 a-rá 4 13,20 13,20 *ultu*(ta) 15 zi-*ma* 1,40
     uḫ-ḫur 28 1,40 *aná šamáš ri-ḫi* 1,40 *ulta*(ta)
     13,20
12   zi-*ma* 11,40 *ūmu*(u₄) 29 *ani' šamáš etiq*(dib-*iq*)
     šumma(BE-*ma*) 27 24 KUR 4 ME 2 ge₆ 4 a-rá
     4 16 16 *ultu*(ta) 24 zi-*ma*
13   8 *uḫ-ḫur* 28 8 *aná muḫḫi*(muḫ) *šamáš ri-ḫi* 16
     ultu(ta) 8 zi-*ma* 29 8 *aná šamáš etiq*(dib-*iq*)
     meš-lu šá 16 8
14   *aná muḫḫi*(muḫ) 8 tab-*ma* 16 29 *ina* ŠÚ *šamáš* 16
     NA *šumma*(BE-*ma*) ME *aná* ge₆ dirig u₄-*mu*
     a-rá 4 *tal-lak šumma*(BE-*ma*) ge₆ *aná* u₄-*mu*
15   dirig ge₆ a-rá 4 *tal-lak* ki-lal ME *u* ge₆ igi-*ma itti*(ki)
     dirig kaskal *tal-lak*

TRANSLATION

8   in order for you to find the time of invisi-
     bility (of the moon). (A.) If in the month I
     of your new year (on the) 27th the last visibility
     (is) 25°, 3;20 (mana) the daylight, 2;40 (mana)
     the night, (multiply) 3;20 by 4; (the result is)
9   13;20. Subtract 13;20 from 25, and 11;40 re-
     mains; on the 28th day (the moon) remains
     11;40° behind the sun. Subtract 13;20 from
     11;40;
10   on the 29th day the moon passed the sun 1;40°.
     Add 13;20 to 1;40, and 15 (is the result); on
     the 30th day the moon passed the sun 15°. (B.)
     If on the 27th the last visibility is 15°
11   (multiply) 3;20 (mana) by 4; (the result is) 13;20.
     Subtract 13;20 from 15, and 1;40 remains; on
     the 28th (the moon) remains 1;40° behind the
     sun. Subtract 1;40 from 13;20,
12   and 11;40 (is the result); on the 29th day (the
     moon) passed the sun (11;40°). (C.) If on the
     27th the last visibility is 24°, 4 (mana) daylight,
     2 (mana) night, (multiply) by 4; (the result
     is) 16. Subtract 16 from 24, and
13   8 remains; on the 28th (the moon) remains 8° be-
     hind the sun. Subtract 16 from 8; on the 29th,
     (the moon) passed the sun 8°. One-half of 16,
     namely, 8,
14   add to (the preceding) 8, and (the result is) 16;
     on the 29th at sunset the first visibility is 16°.
     (D.) If the daylight exceeds the night, (multi-
     ply) the daylight by 4 and you shall proceed
     (with this amount). If the night exceeds the
     daylight,
15   (multiply) the night by 4, and you shall proceed
     (with this amount). Consider (finally) equinox
     and proceed with the difference of the path (of
     sun and moon).

Bibliography and Abbreviations.

AfO
Archiv für Orientforschung.

BEZOLD [1]
C. BEZOLD — F. BOLL — A. KOPFF, Zenit- und Aequatorialgestirne am babylonischen Fixsternhimmel. Sitzungsber. d. Heidelberger Akad. d. Wissenschaften, Philos.-hist. Klasse 1913 No. 11.

BLASS [1]
F. BLASS, Eudoxi ars astronomica. Programm, Kiel 1887.

CHARLES, ENOCH
R. H. CHARLES, The book of Enoch. Oxford, Clarendon Press, 1912.

CRAIG AAT
J. A. CRAIG, Astrological-astronomical texts copied from the original tablets in the British Museum. Leipzig, 1899.

CT
Cuneiform Texts from Babylonian Tablets, etc., in the British Museum.

KING [1]
L. W. KING, A Neo-Babylonian astronomical treatise in the British Museum, and its bearing on the age of Babylonian astronomy. Proceedings of the Soc. of Biblical Archaeology 35 (1913) p. 41–46.

KUGLER BMR
F. X. KUGLER, Die babylonische Mondrechnung. Freiburg 1900.

KUGLER SSB
F. X. KUGLER, Sternkunde und Sterndienst in Babel. 2 vols. Münster 1907/1924.

KUGLER SSB Erg.
F. X. KUGLER, Sternkunde und Sterndienst in Babel. Ergänzungen. I, Münster 1913, II, 1914 (III see SCHAUMBERGER).

LANGDON Men.
S. LANGDON, Babylonian Menologies and the Semitic Calendar. London 1935.

LENORMANT Choix
F. LENORMANT, Choix de textes cunéiformes inédits ou incomplètement publiés. Paris 1873.

NEUGEBAUER [1]
O. NEUGEBAUER, On some astronomical papyri and related problems of ancient geography. *Trans. of the Am. Philos. Soc.*, N.S.32 (1942) p. 251–263.

NEUGEBAUER [2]
O. NEUGEBAUER, The origin of the Egyptian calendar. *J. of the Near Eastern Studies* 1 (1942) p. 396–403.

NEUGEBAUER MKT
O. NEUGEBAUER, Mathematische Keilschrift-Texte. Quellen und Studien zur Geschichte der Mathematik A 3 (1935/38) 3 vols.

NEUGEBAUER — SACHS MCT
O. NEUGEBAUER — A. J. SACHS, Mathematical Cuneiform Texts, American Oriental Series vol. 29 (1945).

SAYCE [1]
A. H. SAYCE, Miscellaneous notes 18. An Assyrian augural staff. *Zeitschrift für Assyriologie* 2 (1887) p. 335–337.

SCHAUMBERGER Erg.
J. SCHAUMBERGER, part III of KUGLER SSB Erg. Münster 1935.

TANNERY HAA
P. TANNERY, Recherches sur l'histoire de l'astronomie ancienne. Paris 1893.

THUREAU-DANGIN [1]
FR. THUREAU-DANGIN, La clepsydre chez les Babyloniens. *Revue d'Assyriologie* 29 (1932) p. 133–136.

THUREAU-DANGIN TU
FR. THUREAU-DANGIN, Tablettes d'Uruk. Musée du Louvre, Textes cunéiformes 6, Paris 1922.

VIROLLEAUD ACh.
CH. VIROLLEAUD, L'astrologie chaldéenne. Paris 1908/1912.

WEIDNER [1]
E. F. WEIDNER, Ein babylonisches Kompendium der Himmelskunde. *Am. J. of Semitic Languages and Literatures* 40 (1924) p. 186–208.

WEIDNER [2]
E. F. WEIDNER, Alter und Bedeutung der babylonischen Astronomie und Astrallehre (= Im Kampfe um den Alten Orient 4) Leipzig 1914.

WEIDNER [3]
E. F. WEIDNER, Zur babylonischen Astronomie. III. Mondlauf, Kalender und Zahlenwissenschaft. *Babylonica* 6 (1912) p. 8–28.

WEIDNER [4]
E. F. WEIDNER, Babylonisches im Buche Henoch. *Orientalistische Literaturzeitung* 19 (1916) col. 74 f.

WEIDNER Hdb.
E. F. WEIDNER, Handbuch der babylonischen Astronomie. I. (= Assyriologische Bibliothek 23), Leipzig 1915.

WEISSBACH BM
F. H. WEISSBACH, Babylonische Miscellen. Wissensch. Veröff. d. deutschen Orient-Ges. 4. Leipzig 1903.

THE ALLEGED BABYLONIAN DISCOVERY OF THE PRECESSION OF THE EQUINOXES

O. NEUGEBAUER

BROWN UNIVERSITY

> Rien n'égale la rapidité avec laquelle se
> répand l'erreur historique si ce n'est la
> ténacité qu'elle oppose aux tentatives de
> réfutation.
>
> Duhem, *Le Système du Monde*, I, 21.

1. HISTORIANS constantly face two closely related problems: to make new textual material available and to destroy generally accepted theories. The present paper is concerned with the latter aspect, in the case of the more and more frequently quoted statement that the Babylonian astronomer Kidinnu was the discoverer of the precession of the equinoxes and that this event can be dated in 379 B.C., thus antedating Hipparchus by about two and one-half centuries. It may seem as if we were dealing here with one of those questions of priority which are of very little significance. Actually the problem has wider implications. It is closely related to the problem of the date of origin of Babylonian mathematical astronomy, which exercised a deep influence on Greek astronomy and its continuation in the Middle Ages. It is furthermore of importance for the evaluation of Babylonian astronomy and the mutual role of observation versus theory during the Seleucid period. Consequently it seems to me worth while to discuss in some detail the above-mentioned theory (which was developed by Schnabel in 1923 and 1927) and to demonstrate that none of the arguments on which it was based can be upheld. I hasten to add that I do not have the slightest interest in questions of personal or national glory and that I see no special merit in restoring to Hipparchus the priority which he held before Schnabel's publications; nor do I pretend to know now more about the history of precession than one knew 50 years ago. As a matter of fact, I shall mention at the end a short notice from Theon Alexandrinus which makes the early history more difficult than ever.

I have tried to make the following discussion as non-technical as possible. The reader who is familiar with the methods of ancient astronomy will excuse me for explaining well-known concepts and for being not very concise in the discussion of mathematical details which can easily be supplied by the specialist.

2. The 'seasons' can be defined by the variable length of daylight and night. The decrease in the length of the nights hails the approaching summer. Simultaneously another observation can be made. During the winter nights totally different constellations are visible than during summer. Thus it seems as if one could characterize the seasons also by means of the constellations. To realize that this is not the case means to recognize 'precession.' In about 13,000 years constellations which began as winter constellations move into the summer and vice versa. And in 26,000 years a constellation has travelled once through all four seasons.

To make this statement a little more precise we may focus our attention on one specific moment during the year, e. g. the spring equinox, when day and night are of exactly equal length. Projecting the sun at this moment onto the background of the stars we may mark this point as the 'vernal point.' If this point remained fixed through the years with respect to the surrounding fixed stars we could define the beginning of spring equally well by means of equinox or by means of the return of the sun to the same star. Again, it is the 'precession' of the vernal point which excludes this possibility. The sun requires about 11 minutes less than $365\frac{1}{4}$ days to return again to the equinox but it takes it about 9 minutes more than $365\frac{1}{4}$ days to return to the same star. The first period is called the 'tropical,' the second the 'sidereal' year.

We need still one more twist in describing 'precession.' The projection of all the positions of the sun onto the background of the fixed stars is called 'ecliptic.' This circle which the sun travels in one year is divided into 360 degrees. Suppose that we begin the count at some arbitrarily chosen point, which might be marked by a star. Call this star 'Aries 0°.' Suppose we observe the position of the

1

sun at equinox of a particular year. This vernal point may be found to be 10° distant from Aries 0°. We know already that precession will slowly change the distance of the vernal point from the fixed star which we called Aries 0.° Indeed, the vernal point will be only 9° distant after 72 years, 8° distant after 144 years, etc. This amount of $\frac{1}{72}$ degree per year is called the 'constant of precession.' Ptolemy, from the comparison of his own observations with the observations of his predecessors, especially Hipparchus, concluded that the vernal point moves only 1° in a hundred years, and this remained the accepted value deep into Byzantine astronomy until new Arabic observations corrected (or rather over-compensated) Ptolemy's value.

3. Kugler recognized in his Babylonische Mondrechnung (1900) that two different methods existed in Seleucid Mesopotamia for the prediction of the lunar movement. He also realized that one of these systems showed definite improvements over the other and that the more highly developed (and consequently later) system was utilized by Hipparchus. In a text from this second system, henceforth called 'System B,' he found in the colophon the name of Kidinnu, while Weidner discovered the name of Nabu-rimannu in a text of the more primitive system ('A'). Since both names occur also in classical sources (e. g. Strabo and Pliny [1]) it has become customary to consider Nabu-rimannu and Kidinnu the inventors of the lunar theories A and B respectively. Only in passing it might be said that the basis for this assumption is exceedingly slim. The names occur only in three tablets of the latest period, the reading and translation of the colophons is full of difficulties, the classical sources say nothing about the authorship of these men nor is their relation known to the scribal families to which the owners and scribes of the tablets of the Uruk archive belong. Still less is known about the texts from Babylon and no material is available from any other site in spite of many statements to the contrary in the literature.[2]

[1] The details were given in a famous paper by Cumont, *Florilegium . . . Melchior de Vogüé* (Paris 1910), p. 159-165.
[2] These problems were investigated by A. J. Sachs and myself and will be presented in the introductory chapter to my forthcoming edition of *Astronomical Cuneiform Texts*.

What we really know about the two 'Systems' is exclusively deduced from their mathematical context. The chronological arrangement from A to B is based on the greater refinement of B but the actually preserved tablets of both systems are contemporaneous. For our specific problem one fact established by Kugler is of great importance. Both systems contain schemes for the variable length of daylight depending upon the position of the sun in the ecliptic. In these schemes equinox corresponds in System A to the solar position Aries 10°, in System B to Aries 8°. This difference in the position of the vernal point plays a central role in the discussion about the Babylonian discovery of the precession.

4. In 1923 P. Schnabel, in his book Berossos und die babylonisch-hellenistische Literatur, developed the theory that Kidinnu, supposedly the founder of System B, discovered precession about 315 B. C. Kugler (who himself had considered in 1900 the possibility of a Babylonian discovery of the precession [3]) in the meantime reached the conviction that this was not the case and sharply contradicted [4] Schnabel (1924). The latter answered in an often-quoted article, Kidenas, Hipparch und die Entdeckung der Präzession (1927), which he concluded with the sentence that Kidinnu's discovery of precession (now in 379 B. C.) was 'endgültig festgestellt.'

At this point the discussion has rested ever since, obviously because nobody who was sufficiently familiar with the methods of Babylonian astronomy investigated carefully Schnabel's arguments. In my own study of System B, and especially of the theory of eclipses, I reached the result ten years ago that Schnabel was hopelessly wrong, and I stated this on several occasions, but only in passing, because the full discussion of my arguments would have required a detailed explanation of rather complicated sections of the theory of System B. In the meantime, however, several new facts have come to light and make it possible to disprove Schnabel's theory directly without being forced to be familiar with the details of the Babylonian lunar theory.

5. I shall quote Schnabel's arguments in greater detail during my subsequent discussion. To sim-

[3] Kugler [BMR] p. 103 ff.
[4] Kugler [SSB] II p. 582 ff.

plify the reading, however, I state his main arguments now in a rather summary fashion.

A. The shift from Aries 10 to Aries 8 for the vernal point in System A and System B respectively can be explained as the correction for precession between Naburianu and Kidinnu.

B. The fall equinoxes were accurately observed, thus guaranteeing the correct value for the tropical year.

C. The planetary entries into zodiacal signs, listed in observational records, show the influence of precession.

D. The periods of two columns of the lunar theory (called H and J) should be equal, but actually deviate by a small amount which corresponds to precession.

E. A tablet in Berlin shows a sudden correction which can be explained as a correction for precession.

F. Ptolemy says only that 'also Hipparchus' was concerned with precession, but not that he was the discoverer.

G. Kidinnu being a Babylonian, he could not have deviated from his predecessors without serious reasons, because the conservatism of the Babylonians is notorious.

6. In discussing these arguments I shall follow the inverse arrangement, because this corresponds more or less to the seriousness of the argument. Indeed I have quoted G only as an example of the intrinsically absurd concept that alleged national characteristics can explain individual steps in the development of scientific theories (or, for that matter, of anything else).

7. Argument F is taken from the Almagest, where Ptolemy, in discussing the length of the year in the introduction to Book VIII, states that much uncertainty existed about this point among the 'old' astronomers, as can be seen from their writings and especially from those of Hipparchus because also (καὶ) he was concerned about the difference between the time of return of the sun to the next equinox and to the same fixed star. Consequently he, Hipparchus, reached the conclusion that all fixed stars participate in a common motion like a very slowly moving planet.

Much more information, not quoted by Schnabel, can be obtained from the Almagest about Hip-

parchus's theory of precession; e. g. from the introduction to Book VIII we learn that he had a preliminary theory according to which only the stars of the zodiac were involved in this slow motion. From this and from similar passages it is evident that Ptolemy considers Hipparchus as the first astronomer who consciously tried to bring order into the contradictory results of observations about the length of the year. The passage quoted by Schnabel only underlines what we would know without Ptolemy's explicit statement, namely, that many astronomers were concerned about the exact length of the year. Against this unbiased interpretation of the whole material offered in the Almagest one cannot consider a simple connecting καὶ as a statement of intended chronological significance for the discovery of precession.

8. Argument E would have required, some years ago, a lengthy astronomical discussion. Schnabel found a text in the Berlin collection, VAT 7821, in the middle of which the day-by-day positions of the sun are changed from the expected value $3°27'15''$ to $3°24'15''$. By means of an argument which I was never able to understand, he derived from this 'attested empirical correction' a value for the length of the year which proved the consideration of precession.

Yet, a trivial remark should have been made from the very beginning. Cuneiform 7 and 4 are two signs often erroneously interchanged because of the similarity of their appearance. Thus the theories of Schnabel rested on the basic assumption that 24 was not a simple scribal error for 27. A few years later I found the missing half of the Berlin tablet in Chicago; two additional scribal errors are found in this section of the text, more than outweighing the alleged 'empirical correction' of the first part. Many pages of very learned discussion can be crossed out for good.

9. Argument D is the only one which implies some mathematical concepts. Schnabel has correctly shown that the mean period of column J in System B of the lunar theory is 12;22,8 months, whereas the column of the differences of J, called H, has a slightly different period. Schnabel argued that there was no reason why column J should be a sequence of second order unless H has an independent astronomical significance. Thus he took the difference between the periods seriously and combined it with another 'empirical correction' (actually again a simple scribal error) in order to find for the period of H the tropical year while he correctly understood that the period of J should correspond to the anomalistic year, which was, however, not distinguished by the ancients from the sidereal year.

This argument can be refuted on many grounds. First of all, it would be a rather absurd procedure to take the slow-moving effect of precession into account by introducing a rapidly oscillating function of only slightly different period. Positively, Schnabel's argument can be answered by explaining the need for a second order sequence by considerations which are closely similar to my explanation of the use of second order sequences for the lunar latitude.[5] Thus H and J are tied together on arithmetical grounds and cannot represent independent astronomical phenomena.

It is furthermore easy to explain why the periods of H and J are slightly different. First of all it is clear that only a very small difference is permissible, because the zeros of H must in the mean coincide with the extrema of J, because otherwise J obtains sharp maxima and minima instead of flat ones.[6] Thus the ideal case would be exact equality. But the period P of H is given by the quotient of $2M$ and d where M is the maximum, d the difference of the linear zigzag function H. On the other hand the arithmetical relationship between H and J requires that M is the quotient of 4Δ and P where Δ is the amplitude of J. To satisfy these relations exactly would require that all divisions can be carried out without remainders. Furthermore it is of primary importance for the practical computation of the lunar tables that the numbers involved have only a small number of places and are not too unhandy for all the arithmetical processes which are needed. It is easy to see from the texts that the latter requirement was given priority over the absolute exactness of the period relation. It is a mere accident that the deviation is approximately of the same amount as the constant of precession and thus could be mistaken for the difference between tropical and sidereal year.

It might be remarked that it is easy to develop numerous slightly different 'years' from Babylonian astronomy, all of which are only different

because of the small deviation from the exact theory caused by the practical requirements of not too complicated computation. It can be shown furthermore that the value 12;22,8 months, which Schnabel discovered, lies at the basis of the theory in both System A and B, thus establishing a direct relationship between the two theories which would be hard to explain if one system admitted precession while the other did not.

10. Arguments B and C are both based on the assumption that the records which are contained in a class of texts from the Seleucid period, often called 'observation texts,' are records of observations. Recently, however, it has been shown by Sachs and myself[7] that almost all of the entries found in these texts are computed and not observed. This holds especially for the solstices and equinoxes and for the entry of planets into zodiacal signs, referred to by Schnabel. Hence these texts are completely eliminated from any discussion of Babylonian accuracy of observation of these (and many other) phenomena.

Because the above statements have far-reaching implications for our whole outlook on Babylonian astronomy I shall mention a few additional details, though they are not directly necessary for the discussion of the discovery of precession. Since Epping and Kugler succeeded in deciphering the lunar and planetary computations of the Seleucid period, it has become clear to all who seriously studied these texts that they were based on an exceedingly small number of observations, the majority of which consisted in the establishment of relations between periods. These relations do not require a very high accuracy of individual observation but only the counting of gross variations. For example, in order to find that m synodic months correspond to n draconitic months one has only to count how often the moon crossed the ecliptic from one side to the other between two lunar eclipses of similar magnitude and direction. Kugler's discoveries have shown how such simple facts were in a really ingenious way utilized for the development of mathematical methods for the prediction of lunar and planetary phenomena. It seems to me one of the most admirable features of ancient astronomy that all efforts were concentrated upon reducing to a minimum the influence

---

[5] Neugebauer [1] § 5 and § 8.
[6] Cf. Neugebauer [2] p. 1023, fig. 4.

[7] Sachs [1] and Neugebauer [3].

of the inaccuracy of individual observations with crude instruments by developing to the farthest possible limits the mathematical consequences of very few basic elements.

Already Epping found that equinoxes and solstices, recorded in the 'observation texts' of the Seleucid period, divide the year simply into four equal parts. The question arose as to which one of the four characteristic points in the year was actually observed while the three other ones were schematically derived from it. After some discussion it was generally accepted that the autumn equinoxes agreed best with the facts and had thus to be considered to be the basic observation. The summer solstice seemed to be out of the question because it showed least agreement. In 1948, however, I found [8] in a tablet from Uruk a simple arithmetical scheme which gave the dates of consecutive summer solstices such that the same dates would appear in each 19-year cycle of the civil calendar. The dates are given as months and 'days' with fractions but the 'days' are 'lunar days' i. e. thirtieths of mean synodic months. Hence the real calendaric dates might deviate by one unit from the 'days' given by the scheme. After this discovery, Sachs and I investigated the dates of solstices and equinoxes recorded in the 'observation texts' of the Seleucid period. We found that all these dates without exception were derived from the Uruk scheme by simply identifying the dates as given by the scheme with dates in the civil calendar, ignoring fractions, however close they might come to the next unit. Thus it is evident that not a single one of these dates is observed and that a possible error of at least two days was considered as irrelevant. It is also clear that it is a mere accident that the autumn equinoxes agree well with modern computations. The inaccuracy of the 19-year cycle and the systematic errors in the Uruk scheme accidentally compensate each other in one series of dates. Once more it has become evident that a comparison with modern calculation is of no value whatsoever unless one knows exactly how the ancient results were obtained.

What was shown for the solstices and equinoxes holds for a still larger range of records. Sachs could demonstrate that also the dates for the rising of Sirius are derived by a scheme similar to that of the summer solstices. He could furthermore show

[8] Neugebauer [4].

that also the lunar and planetary phenomena, risings and settings, entry into zodiacal signs, etc. are computed and not observational records. All that remains of real observations are positions of the moon and of the planets with reference to nearby 'normal stars' i. e. certain bright stars conveniently selected along the zodiac. Finally it can be made plausible that the computation of the planetary phenomena was based, aside from period relations, on an absolute minimum of empirical facts, namely a single heliacal rising.

These results are indeed bound to change radically the traditional picture of Babylonian astronomy during the Seleucid period, the only period for which we have a reasonably complete amount of textual information. Several layers of mathematical methods can be distinguished while the role of observations as well as their accuracy is reduced to an extremely modest amount. It is obvious that also the earlier phases of Babylonian astronomy will appear in a different light since the situation in the latest period has become clearer. And it is evident that all attempts to combine the alleged 'observations' with modern computations are without significance because they reflect only the inherent inaccuracies of the ancient mathematical methods, without revealing anything about actual observations.

11. To return to Schnabel's arguments about the discovery of precession we have only left the explanation of the shift of the vernal point from Aries 10 to Aries 8. It is clear that a correction for precession might be the cause of this shift and that this deviation reflects the difference in time of origin of the two systems. But this argument is certainly not generally applicable. We know, e. g. that Eudoxos called the vernal point Aries 15 while Hipparchus normed it as Aries 0. Since we know that only two centuries separate Eudoxos (about 370 B. C.) from Hipparchus it is clear that precession cannot be responsible for this shift or for the difference between the norm. Similarly the two Babylonian norms are simultaneously applied during the whole Seleucid period, and even still later in Greek astrological texts. If precession were recognized in System B but not in System A, one should expect an increasing deviation between the two systems. There is no trace of such an effect in our texts. The most one could possibly admit would be that precession produced the necessity for a correction in the position of the vernal point from 10° to 8°. But even if this were the case (I personally doubt it), we could only say that the correction was made without realizing that no single correction could solve the problem.

With this question is connected the problem of the date of origin of the two lunar theories in Mesopotamia. Kugler [9] concluded as follows. An ephemeris for the new moons of the years from 104 to 101 B. C. would show good agreement with modern computations if one would diminish all longitudes given in the text for the conjunctions by about 3°. Because the vernal point in this system is given the longitude [10] Aries 8° we should conclude that its longitude was actually Aries 5° instead of Aries 0°. For this error Kugler saw two causes, working in the same direction. First, precession would result in an error of $1/72$ of a degree per year; secondly, he found that the solar velocity assumed in the text was slightly too small, resulting in an error of $1/321$ degree per year. Thus the total error per year would be $1/72 + 1/321$ degree, from which it follows that it takes about 295 years to accumulate an error of 5 degrees; thus System B was invented about 295 years earlier, i. e. about 400 B. C. Similarly he found 500 B. C. for System A.

Serious objections can be raised against this method. It assumes that originally, when the lunar theory was invented, the vernal point was determined correctly. We have seen before that there is no basis for this assumption. But a deviation of only one degree implies a corresponding deviation of 60 years in the chronology. Furthermore, the correction for the inaccuracy in the solar velocity implies that the solar longitudes in System B were computed without changes for three centuries. This is by no means borne out by our texts, which show many interruptions in System B. Indeed, Kugler's method leads to very different dates if applied to different texts at our disposal. Kugler himself considered his computation only as a rough estimate and finally abandoned it completely.

Schnabel, however, not only maintained its validity but tried to improve on it by varying slightly the numbers involved. From this resulted finally

the often-quoted dates 508 B. C. for Naburianu and 379 B. C. for Kidinnu. Of course this increased accuracy is purely fictitious because the basic assumptions are the same as Kugler's—which means that not even a full century up and down could really be guaranteed in view of our complete ignorance of the accuracy of the initial observations. But Schnabel overlooked another point. The whole method is based on the assumption that the accumulated error is caused by ignoring precession. Thus he computes the date of origin of Kidinnu's system under the assumption that these texts do not show any consideration for precession, though he had just tried to demonstrate that Kidinnu had discovered precession. He even determined [11] the value which Kidinnu had found, namely one degree in 45.8 years. In other words, according to Schnabel, Kidinnu had even overcompensated the amount of precession. This has rather catastrophic consequences for the dating of the invention. If precession actually amounts to $1/72$ degree per year and if one corrects it with $1/46$ degree per year, then the total error is obviously $\frac{1}{72} - \frac{1}{46} \approx -\frac{1}{130}$ degree per year i. e. a negative quantity. In other words, one degree of error found in a text indicates that the system was invented 130 years after (!) the text was written. Thus Kidinnu should have lived many centuries after the beginning of our era if he had lived when his observations were correct.

12. I hardly need underline that even without drawing this particularly absurd consequence of Schnabel's theory, there is no basis visible anywhere in the cuneiform texts of the Seleucid period to support the assumption of a conscious recognition of precession. Similarly I see no way to determine exactly the date of origin of the methods of mathematical astronomy. The basic idea of all these attempts, namely the assumption of accurate agreement between facts and initial observations, is utterly naive. The inaccuracy of individual ancient observations is notorious, wherever we can check them. The 'doctoring' of numbers for the sake of easier computation is evident in innumerable examples of Greek and Babylonian astronomy. Rounding-off in partial results as well as in important parameters can be observed frequently, often depriving us of any hope of reconstructing

---

[9] BMR p. 100 ff.

[10] Kugler actually uses 8;15° but this value is obtained by an incorrect argument. The value 8° is secured through the scheme for the length of daylight (cf. Kugler BMR p. 99).

[11] Berossos p. 236.

the original data accurately. Thus nothing remains but the dating on purely historical evidence, i. e. from the dates available in preserved texts. At present it is open to anyone to guess whether it is significant or purely accidental that we have no texts from the time before 300 B. c. in contrast to the very complete coverage for the next two and one-half centuries.

13. Also the early history of the theory of precession is far from well known. There is, of course, no room for doubt of the authenticity of the facts concerning Hipparchus as revealed by the Almagest. But Ptolemy did not intend to give a historically complete report. Delambre, in his famous Histoire de l'astronomie ancienne already in 1817 drew the attention of historians to a passage of Theon's ' small ' commentary [12] to Ptolemy's ' Handy Tables.' There Theon tells us [13] that Ptolemy considered as wrong a theory of ' the old astrologers ' (οἱ παλαιοὶ τῶν ἀποτελεσματικῶν) according to which the vernal point oscillates between Aries 8° and Aries 0° at a rate of one degree in 80 years. He furthermore tells us that the extremal point, Aries 8°, was reached 128 years before Augustus (158 B. c.[14]). This remark of Theon is reflected in the Arabic literature, e. g. in Al-Battānī (900 A. D.) and Al-Bīrūnī [15]. The latter quotes Ptolemy in his Chronology (written 1000 A. D.) and Theon in his ' Astrology ' (written 1029 A. D.). In both works he substituted the Babylonians for the ' old ' astrologers. He thus inaugurated a tradition which is still dear to many modern scholars, namely, to consider every reference to any predecessor called ' old ' (παλαιός) as a reference to oriental science. Yet Synesius of Cyrene, a younger contemporary of Theon of Alexandria, in speaking [16] about ' old ' astronomers calls Hipparchus παμπάλαιος ' very old ' thus clearly indicating that the ' old ' ones are much nearer to his own time, hence certainly some Greek astronomers.

The theory of alternating precession and recession of the vernal point, later called ' trepidation,' is of great historical interest because of the influence it exercised in various forms during the Middle Ages, in Hindu astronomy, in Arabic astronomy, and in Europe, where it found followers until Copernicus. Theon's remark raises great difficulties for the early history of this theory. Aries 8° is exactly the position of the vernal point in System B of the Babylonian lunar theory. The same norm is very common in Hellenistic astrology. While Hipparchus called the vernal point Aries 0° it has been known since Kugler's discoveries that he was familiar with the basic elements of System B of the Babylonian theory. Finally, the date given by Theon falls just within the years which Ptolemy quotes [17] for Hipparchus's observations of solstices in order to determine the length of the year.

Several possibilities are open for discussion. One might think of a Babylonian origin; no basis for such a theory can be found in the existing contemporary texts, which use Aries 8° as vernal point unchanged through the whole following century. One might assume that we are dealing with a preliminary theory of Hipparchus, eventually discarded like his hypotheses of a precession of ecliptic stars only. No support can be found for this possibility in Ptolemy or Theon. Closest to Theon's words remains the assumption that the theory of trepidation is the product of early Hellenistic astrologers, who based their theory on Hipparchian observations of equinoxes but without accepting his norm for the vernal point. Completely in the dark remain the reasons for assuming an oscillation of the vernal point and for its amplitude. We are certainly very far from a real insight into the development of Hellenistic astronomy before Ptolemy.

---

[12] Written in the last quarter of the fourth century A. D. For the intricate relationship between the different commentaries to Ptolemy's works by Pappus, Theon, and Hypathia see Rome [1] p. 213.

[13] Halma [CT] I p. 53, or Delambre II p. 625.

[14] Augustus —127 Thoth 1 = —157 Oct. 2.

[15] Al-Battānī: Nallino I p. 126 f. and p. 298 ff.; Al-Bīrūnī, Chronology, trsl. Sachau p. 322; Astrol., trsl. Wright, No. 191. The theory of ' trepidation ' is also mentioned by Proclus, Hypotyposis III, 54, Manitius p. 68/69.

[16] Migne, Patrol. Graeca 66, 1584 = Terzaghi, Sym. opuscula 138, 20.

[17] Almagest III, 1 (Heiberg 195, 16): Autumn equinox —157 Sept. 27. This is accidentally exactly the same date obtained by using the Babylonian scheme, which gives Sel. Era 154 Ulul 13 and converting it to Julian dates by means of the tables of Parker-Dubberstein, which give Ulul 12 but the two other observations are exactly the same: —161 Sept. 27 = S. E. 150 VI 28; —158 Sept. 27 = S. E. 153 VII 2.

REFERENCES

Halma [CT], *Commentaire de Théon d'Alexandrie sur les tables manuelles astronomiques de Ptolemée*, I, Paris, 1822.

F. X. Kugler [BMR], *Babylonische Mondrechnung*, Freiburg, 1900.

F. X. Kugler [SSB], *Sternkunde u. Sterndienst in Babel*, 2 vols., Münster, 1907-1924.

O. Neugebauer [1], Die babylonische Theorie der Breitenbewegung des Mondes, *Quellen u. Studien z. Gesch. d. Math.*, Ser. B 4 (1938), p. 193-346.

O. Neugebauer [2], Mathematical methods in ancient astronomy, *Bull. Am. Math. Soc.*, 54 (1948), p. 1013-1041.

O. Neugebauer [3], Solstices and equinoxes in Babylonian astronomy during the Seleucid period, *J. Cuneiform Studies*, 2 (1949), p. 209-222.

O. Neugebauer [4], A table of solstices from Uruk, *J. Cuneiform Studies*, 1 (1948), p. 143-148.

A. Rome [1], Le problème de l'équation du temps chez Ptolémée, *Ann. Soc. Sci. de Bruxelles*, sér. I, 59 (1939), p. 211-224.

A. J. Sachs [1], A Classification of the Babylonian Astronomical Tablets of the Seleucid Period, to appear in the *Journal of Cuneiform Studies*, vol. 2.

P. Schnabel, *Berossos und die babylonisch-hellenistische Literatur*, Leipzig-Berlin, Teubner, 1923.

P. Schnabel [1], Kidenas, Hipparch und die Entdeckung der Praezession, *Zeitschr. f. Assyriol.*, 37 (1927), p. 1-60.

Reprinted from
Journal of the American Oriental Society
70 (1), 1-8 (1950)

From the ASTRONOMICAL JOURNAL
72, No. 8, 1967, October—No. 1353
Printed in U. S. A.

Problems and Methods in Babylonian Mathematical Astronomy
Henry Norris Russell Lecture, 1967

O. NEUGEBAUER
Brown University, Providence, Rhode Island
(Received 17 July 1967)

An account is given of the methods used in ancient Babylonia for predicting the positions of the planets.

IT is certainly for the first time, and in all probability also for the last time, that important astronomical achievements gain recognition by a Russell Lecture with a delay of some 22 centuries. I consider it, of course, as a great honor to have been chosen to report on the work of our earliest colleagues but I must emphasize from the outset that I am heavily indebted in the study presented here to predecessors and contemporaries. It is, however, not only because of requirements of fairness that I mention in the following some details in the history of our studies but also because some insight into the accidents of exploration will make it clearer what types of problems confront us.

1. Around 1875 one had reached the point where one could read, with a fair understanding, cuneiform script. Unfortunately at an early phase texts with obvious relevancy to biblical stories had been discovered, with the result that the new field of "assyriology" became involved in a bitter controversy about the historicity of the Bible, a controversy in which German scholars played a dominant role. The Jesuits, at that time expelled from Germany in the so-called "Kulturkampf," realized that they could only take a successful stand in the struggle if they had the necessary competence in discussing the primary source material. It was against this background that Father J. N. Strassmaier (1846–1920) became one of the leading assyriologists of his time. He made a lasting impact on the new field by his insistence on making as many texts as possible available from the enormous quantities of clay tablets which reached the British Museum by the tens of thousands. Working in particular with late Babylonian material he realized that a great number of fragments were of clearly astronomical content, as was evident from the columns of numbers combined with year numbers and month names and ideograms for the zodiacal constellations and planets. But otherwise these texts remained a mystery to Strassmaier, and he therefore asked a friend for help, Father J. Epping (1835–1894), who was teaching mathematics and astronomy in Falkenburg, Holland. After many fruitless attempts, in 1881 Epping found the key to the understanding of a lunar ephemeris and soon uncovered also the underlying chronological patterns, explaining for the first time correctly the ideograms for the planets and their phases and many other details of great historical interest, finally published in a remarkable little book called *Astronomisches aus Babylon* (1889). Few modern historians know that their chronological framework for the history of the "Hellenistic" period (from Alexander to the Roman imperial

period) rests on astronomical data established by Epping.

All the sources for these investigations came from Strassmaier's hand copies of more or less fragmentary tablets in the British Museum, and the same material provided the basis for the monumental volumes on *Mondrechnung* (1900) and *Sternkunde* (1907–1924, 1935) by Father F. X. Kugler (1862–1929). We know now, mainly through investigations by A. Sachs, that these texts come from an archive in Babylon. Since the beginning of the present century, material from another source appeared in the antiquities market, ending up in the collections in Paris, Berlin, Chicago, Istanbul, etc., but originally coming from the city of Uruk in southern Mesopotamia (cf. Fig. 1 for the chronological and local distribution of our sources).

This increase of scattered material and the huge amount of still unpublished texts from Strassmaier's copies impressed on me the need for a comprehensive edition of the whole of the mathematical *Astronomical Cuneiform Texts* available to me, a work which was completed in 1955, henceforth referred to as "ACT." This term will now also be used to denote a certain type of texts, mostly ephemerides for the moon and the planets plus procedure texts which contain rules for computing the ephemerides. In contrast one speaks often of "observational" texts, a term only in part justifiable (as we shall see in one specific case) but which will be used also in the following as a convenient abbreviation. Roughly we have about 300 texts of the ACT type and about five times as many of the observational type.

In making such a sharp distinction we already utilize the more recent phase of our studies. In 1948 my colleague A. Sachs published a systematic classification of the non-ACT texts which will be the skeleton to an edition of these sources. The study of the ACT material has been refined by B. L. van Waerden and P. Huber in Zürich and in recent years most successfully by A. Aaboe of Yale University. While Epping and Kugler were reaching their most definitive results in the understanding of the techniques for the computation of lunar ephemerides (mainly what we call "System B"), it is today possible to understand some of the basic features of the planetary theory, mainly thanks to Aaboe's work. It is here that we for the first time reach some insight into the strategy of Babylonian mathematical astronomy beyond the mere tactics of computational devices.

2. One short lecture cannot possible pretend to give

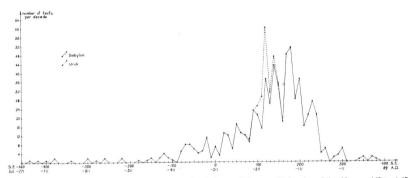

Fig. 1. Number of known astronomical cuneiform texts per decade from −771 to A.D. 84, i.e., from Seleucid era −460 to +400. The texts from Uruk are added to the texts from Babylon.

an adequate summary of Babylonian astronomy. Consequently I restrict myself to two topics which are well enough defined to convey, I hope, a general impression of the methods applied and of the consequences for our historical analysis. One concerns the length of the year, the other some basic ideas in the description of planetary motion.

What I am bypassing in this survey is by far the greatest bulk of the ACT material. The lunar ephemerides alone outnumber all planetary texts. But their structure is so complex (the determination of the visibility of the new crescent month by month can require, e.g., up to 18 parallel columns) that an adequate description would have to introduce far too many historical and technical details. I am also bypassing any attempt to describe what we know about the history of mathematical astronomy. But in this case the reason for ignoring this problem is different. In fact we know extremely little about it and for our present purposes it will suffice to say that the overwhelming mass of our material belongs to the last three centuries B.C. (cf. Fig. 1) and to admit that we cannot outline any consistent sequence of steps as we can, e.g. (to a modest extent), describe the methods which lead to the Ptolemaic system. But I may perhaps

say that it is my impression that none of the essential methods were developed before, say, 500 B.C., and that an even more rapid progress toward the level which we know from the Hellenistic period would not surprise me. This does not preclude, e.g., that the recording and cyclic classification of eclipses began as early as the middle of the eighth century.

Before turning to the two specific topics mentioned above, a few remarks are necessary to characterize the mathematical methods of Babylonian astronomy. First of all, and most striking, is the total absence of any geometric or cinematic model which seems to us so indispensible for the description of celestial motions. The Babylonian methods are strictly arithmetical in character, based on numerical sequences, e.g., of constant differences. Furthermore the independent variable usually operates with mean synodic months and their thirtieths [which we call "tithis" (τ), a term borrowed from Indian astronomy], not with solar days. The reason for this lies in the use of a real lunar calendar without any fixed pattern for the arrangement of full or hollow months. Hence mean lunations of 30^r permit the computation of dates without reference to the actual lunar civil date. The consistent use of mean lunations avoids accumulative errors and in final results one simply identifies tithis with calendar dates, a procedure sufficiently accurate, e.g., for planetary

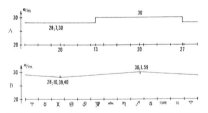

Fig. 2. Solar velocity in degrees per mean synodic month as function of the solar longitude, according to System A (top) and System B (bottom), respectively.

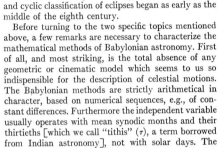

Fig. 3. Linear zigzag function. Symmetric reflection takes place at M and m such that the total change is always $|d|$.

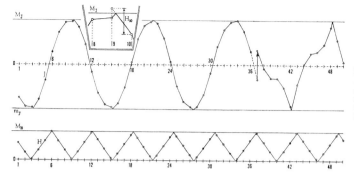

FIG. 4. The linear zigzag function H (bottom) gives the differences for the function J (top). Inset: example of reflection at the maximum M_J of J. At point 36: consequence of a computing error (wrong sign).

phases. The lunar theory itself is, of course, far more refined in its time reckoning.

As an example of typical mathematical methods we use a (modern) graph, Fig. 2, for the solar velocity, expressed in degrees per mean synodic month (i.e., 30^r). The "System A" variety assigns the sun only two constant velocities, one of $30°$ per month from $\lambda = 167$ to $\lambda = 357$, one of $28;7,30°/^m$ on the remaining arc of the ecliptic, with discontinuous changes at the two points mentioned. It is easy to see that this model corresponds to a norm in which

$$1 \text{ year} = 12;22,8 \text{ mean synodic months.} \quad (1)$$

This parameter plays a fundamental role not only in both "systems" of the lunar theory but also underlies many parts of the planetary theory. If we had an accurate value for the length of the mean synodic month, reckoned in days, we would have the length of the year. But an equally famous parameter

$$1 \text{ month} = 29;31,50,8,20 \text{ days} \quad (2)$$

can probably not be combined with (1) without a slight change in the last two digits, because (2) seems to be restricted to System B only. It would also be misleading to deduce from Fig. 2 an "apsidal line" for the sun. We have no trace of such geometrical concepts anywhere in Babylonian astronomy, and we have similar planetary models where the velocity zones have no line of symmetry.

System B of the lunar theory describes the solar motion through a "linear zigzag function" (cf. Fig. 2, lower half). This is one of the most commonly used

FIG. 5. Tabulation of a function with the draconitic month as period at points which are one mean synodic point apart.

tools of Babylonian astronomy, obviously developed from the simple concept of linear variation and interpolation. Figure 3 shows that to a constant difference $\pm d$ and an amplitude $\Delta = M - m$ belongs a period $P = 2\Delta/d$. Again the representation by a continuous curve is unhistorical. What we really have is only sequences of numbers tabulated for equidistant intervals of time. The changes if M or m would be transgressed are made in such a fashion that the effective amount remains $|d|$ while M and m need not occur ever in the sequence. Since such sequences are strictly arithmetical in character, with no roundings or approximations, we may consider all numbers as integers, subject to the same arithmetical law. Such a sequence must be periodic after a number π of steps, called the "number period." Obviously π can be found from $P = \pi/Z$, where π and Z are the smallest relatively prime integers in the quotient $2\Delta/d$.

From a linear zigzag function (H) can be derived a smoother periodic function by summation (cf. Fig. 4) which we would describe as composed from parabolic arcs (J) obtained by integration of H. In fact, however, this description is not applicable to arithmetical functions which obey a rule of linear reflection at M_J and m_J as determined by the values in H (cf. the inset in Fig. 4). It is only by summation over twice the whole number period π of H that the proper parameters of J are obtained. The consideration of arithmetical progressions in their behavior over the whole number period, usually a very large number, is a very typical methodological idea of Babylonian astronomy. In practice it means that variations in the small will have no accumulative effect since the initial situation will be exactly restored after π events.

Figure 4 shows also the effect of an error. At No. 37 the function J should have been given a negative value while the computer still kept the positive sign. From then on he proceeded correctly with $\Delta J = H$ but the sinusoidal character of J is now totally destroyed.

Another aspect of Babylonian mathematical techniques is illustrated in Fig. 5, which is taken from lunar ephemerides. The points 1, 2, 3, $\cdots$ of the time axis are spaced one mean synodic month apart; the continuous function which varies between $+M$ and $-M$ has a shorter period and it is easy to show that it is the period of the draconitic month. But it is clearly not the lunar latitude which has discontinuities at each crossing of the zero line. One can, however, show that the quantity represented gives the eclipse magnitude whenever an eclipse takes place at the node in question; otherwise the zigzag function serves only as a mathematical tool to pass from one eclipse to the next. In fact the tabulated numbers can also serve to show whether an eclipse is excluded or possible (and then deserves further computation). To the student of ancient astronomy this implies an important warning: a sequence of tabulated numbers must by no means represent a sequence of actual events. Interpretations are secure only when the underlying theory is fully understood as well.

3. For the modern astronomer it is a natural tendency to look into the ancient Babylonian material with the expectation of finding from it data for secular variations, e.g., concerning the earth's rotational velocity. Again I must bypass such a discussion because of the complexity of the relevant historical problems. Nevertheless the very simple case described in this section illustrates some aspects of the unexpected difficulties one meets in the utilization of Babylonian sources.

It is one of the important historical contributions of Kugler's work that he securely established the arrangement of a 19-yr calendaric cycle which is not only valid for the Hellenistic period but remained in use far into Islamic and western mediaeval astronomy. This is the so-called "Metonic" cycle with 12 ordinary and 7 intercalary (13 months) lunar years, i.e., a cycle of 235 lunations (cf. Fig. 6). These years are usually counted in the "Seleucid Era" (S.E.) which begins its first year with the new crescent of -310 April 3. We know now that before -500 intercalations were irregular, that the fixed rule is strictly followed from -380 in Babylonia and that Meton's announcement of such a pattern was made in Athens in -431 (without any practical effect).

Kugler, investigating "observational" texts from Strassmaier's material, found many references to dates of equinoxes and solstices. He soon established the fact that the intervals divided the year into four equal seasons (with an inevitable variance of one day) and in comparing the dates with modern computations he concluded that the vernal equinoxes constituted the empirical backbone of these data since they agreed best with facts. If therefore somebody would like to obtain, say, empirical data for precession he would naturally resort to the use of Babylonian vernal equinoxes.

But it is an extremely treacherous conclusion to assume that an ancient method which probably will

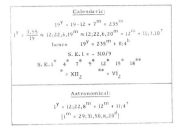

FIG. 6. Metonic intercalation cycle (top) and length of year in mean synodic months (bottom).

contain some systematic errors must have been initiated at the moment of best agreement with modern theory. This will become clear enough for the case in point.

Among the many small fragmentary tablets from Uruk, now in Istanbul, two pieces turned up (cf. Fig. 7) which could be joined together, giving a simple arithmetic sequence of dates. The first column gives the year numbers from S.E. 143 to 157 (-168 to -154); the second, the lunar months; the third, day numbers and fractions, obviously "tithis" since these numbers show a constant difference of $11;3,10^r$ when reckoned modulo 30. The names of the months, either III or IV depending on the character of the year, suggest that we are dealing with summer solstices, and this conjecture is indeed fully confirmed in the following. The asterisks mark the intercalary years. Figure 6 shows that this difference of $12^m + 11;3,10^r$ is an excellent approximation for the 19-yr cycle, being based on a year of $12;22,6,20^m$ in contrast to the astronomical year of $12;22,8^m$ which we have found before [cf. Eq. (1)]. This again is a healthy warning that, for different purposes, different parameters may be used, although representing in principle the same astronomical entity.

With this "Uruk-scheme" of summer solstice dates known, A. Sachs and I combed through the whole

U 107 + 124, ACT 199

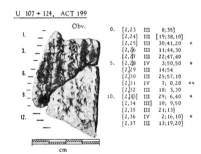

FIG. 7. Fragments of text (in Istanbul) from Uruk with moments of consecutive summer solstices. E.g., 2,23 = S.E. 143 = -168, month III day 8+35/60 (27 June).

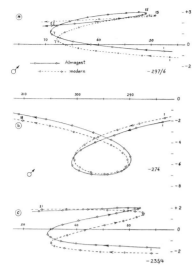

FIG. 8. Apparent motion of Mars in longitude and latitude, computed with the tables of the Almagest (about A.D. 150) in comparison with modern results (tables by B. Tuckerman).

available textual material, now considerably larger than in Kugler's time. The following facts became apparent (Neugebauer 1948): The dates which are given in the "observational" texts are not observations at all but obtainable from the Uruk scheme by simple truncation of the fractional parts (and taking tithis as day numbers). Hence, without exception, all solstice and equinox dates are derived from the sequence of summer solstice dates of the Uruk-scheme which is constructed to fit the 19-yr calendaric cycle. All dates are then arithmetically derived from this backbone. In other words we are in fact dealing with a matrix of 4 times 19 elements, where all elements are known as soon as one of them is given. Obviously it is futile to single out any one of these numbers as based on observations.

But things went even further. Sachs (1952) also extended these investigations to the equally frequent references to the rising, setting, and acronychal rising of Sirius, records which one had taken as observations without any question. It turned out that also these dates were in fixed arithmetical relation to the Uruk pattern of solstices and equinoxes. Hence the matrix mentioned before contains 7 times 19 elements, each one of which could define all the others. It is completely beyond our control to reconstruct or date the ultimate empirical elements upon which this pattern was founded.

It may be mentioned, in passing, that the situation is in fact still more complicated. The lunar ephemerides contain, of course, columns for the variable length of daylight, required for the computation of the visibility of the first crescent or of eclipses. The equinoxes are then related to a certain solar longitude which is labeled Aries 10° in System A, but Aries 8° in System B of contemporary ephemerides. In general longitudes are reckoned as sidereal—but no trace of a recognition of precession has been established—and therefore longitudes of stars and planets can be reduced to modern longitudes only approximately by subtracting about 5° from Babylonian longitudes around −100 (Huber, 1958). I think this will suffice to show how much caution is needed in the use of ancient data as empirically established elements.

4. We now turn our attention to the planetary theory. The principles of the "Ptolemaic system" are familiar enough: the planet moves on an epicycle of an eccentrically viewed deferent with an angular velocity that is constant with respect to a point located symmetrically to the observer on the apsidal line. Latitudes are obtained through a proper tilting of the plane of the deferent while the plane of the epicycle remains parallel to the ecliptic. This cinematic model can be immediately utilized for the computation of ephemerides, and the results are very satisfactory indeed, as is shown in the examples of the motion of Mars depicted in Fig. 8 in comparison with the actual geocentric path (the latter from Tuckerman's tables, 1962). [The ancient data are purposely not corrected for precession (which is too small in the Ptolemaic theory).]

The Babylonian theory operates on a completely different basis. First of all, latitudes are ignored (in marked contrast to the lunar theory). Secondly, the basis of the whole theory is not a cinematic model for the motion $\lambda(t)$ of the planet itself but a description of the displacement of isolated characteristic phases, e.g., first and second stationary points ϕ and ψ, opposition Θ, first and last visibility Γ and Ω near conjunction. If we call the longitudinal difference $\Delta\lambda$ between, e.g., consecutive first stations the "synodic arc," we can say that the main part of the planetary theory consists in establishing methods of computing the synodic arcs in their dependence on the section of the ecliptic where these phases occur. In other words, each phase is treated as if it were an independent celestial object (much as we speak about the motion of a node), regardless of the motion of the planet itself. This approach has, among other things, the advantage that the synodic arcs are always positive. The problem of retrogradations is thus removed to a secondary position which involves only the question of the relative arrangement of the above-mentioned points. Let us assume that we know for each of the phases their synodic arcs; then we can compute, e.g., the consecutive longitudes and moments for the first appearances $\Gamma_1, \Gamma_2, \cdots$ of a planet. If furthermore a rule is given how such a synodic arc (and the corresponding synodic time) $\Gamma_1\Gamma_2$ is subdivided by the other phases, $\phi, \Theta, \psi, \Omega$,

259

then we have again a matrix of data:

$$
\begin{array}{ccccc}
\Gamma_1 \rightarrow & \phi_1 \rightarrow & \Theta_1 \rightarrow & \psi_1 \rightarrow & \Omega_1 \\
\downarrow & \downarrow & \downarrow & \downarrow & \downarrow \\
\Gamma_2 & \phi_2 & \Theta_2 & \psi_2 & \Omega_2 \text{ etc.,} \\
\downarrow & \downarrow & \downarrow & \downarrow & \downarrow
\end{array}
\tag{3}
$$

here of 5 columns. The number of lines depends, of course, on the planet and is the "number period" π of events needed to bring the phase exactly back to its starting point. Since all data are again strictly arithmetically controlled, we have here once more a matrix in which any one element determines all the others. In principle, Babylonian planetary theory is the most deterministic theory ever conceived, but based on an arithmetical pattern without any underlying cinematic or even mechanistic model.

In practice many short cuts of such a rigorous pattern are described in procedure texts and applied in ephemerides, but the ideal scheme of the ACT material is undoubtedly the matrix structure for all planetary phases.

In the description of the schemes that determine the synodic arcs and times of any individual phase one must distinguish between a System A type which operates with step functions and a System B based on linear zigzag functions. Of particular methodological interest is the procedure of System A. The ecliptic is divided into k arcs of fixed length $\alpha_1, \cdots \alpha_k$ and on each of these arcs a synodic arc $w_1, \cdots w_k$ is prescribed. For example Mars has six arcs of 60° length each, on which the following synodic arcs are given:

$$
w_1 = 30°, \quad w_2 = 40°, \quad w_3 = 60°,
$$
$$
w_4 = 90°, \quad w_5 = 67;30°, \quad w_6 = 45°. \tag{4}
$$

If, e.g., Γ_1 takes place at a longitude λ near the beginning of α_1 the event Γ_2 will take place at $\lambda + w_1$, also in α_1. But Γ_2 is now so near to the end point of α_1 that $(\lambda + w_1) + w_1$ reaches into α_2. This excess is then to be changed by a factor $w_2/w_1 = 4/3$, etc. Hence we see that the progress of Γ smoothens the discontinuities of the underlying step function into a "velocity" distribution as represented in Fig. 9. We return to the question that one will naturally ask: How was it possible to derive from some empirical data a pattern like the above-given synodic arcs w_i which then produce resulting synodic arcs as shown in Fig. 9. But this much is clear: We are again far removed from simple empirical data when we investigate the planetary theory of the Hellenistic age.

5. We can now consider the result of these procedures to be given. The longitudes and times of the characteristic phases can be established from any initial configuration. There remains the question about the behavior of the planet itself during the time between the phases. In other words, at the end one might wish to find $\lambda(t)$ of the planet, e.g., for consecutive days.

There is, of course, no longer room left for a cinematic theory of the daily motion of the planet and the answer

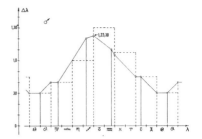

FIG. 9. Synodic arcs for Mars, constant on 60° sections of the ecliptic (dotted step function) and resulting synodic arcs taking the effect of consecutive sections into account (solid line).

to this final question can only be found by interpolation between the phases that have been determined beforehand. Linear interpolation will, of course, give a fair estimate of the planet's position for certain sections of its orbit, but we also have evidence for very refined arithmetical methods to interpolate with sequences of second or even third order. Investigations of progressions of different order are well established in Babylonian mathematical texts, and these methods

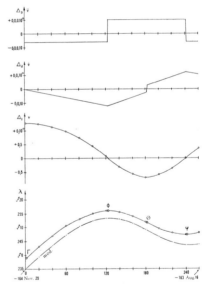

FIG. 10. Day by day motion in longitude of Jupiter (lowest graph) according to Babylonian interpolation methods (upper curve) compared with actual motion (lower curve). A longitudinal difference of about 5° is only due to a difference in zero points. The three upper graphs show the corresponding first, second, and third differences.

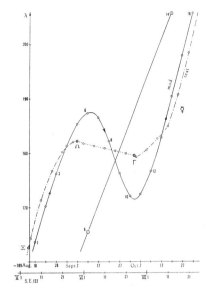

FIG. 11. Day by day motion in longitude of Mercury according to Babylonian crude interpolation methods (dotted curve) compared with actual motion (solid curve).

were applied to the construction of sequences which could bridge the gaps between given numbers that represent the longitudes of the phases in their natural arrangement. A text which covered in about 400 lines one complete synodic period of Jupiter has been step by step recovered by Kugler, myself, and Huber (1957). Figure 10 shows the different sequences in greatly different scale, resulting in a smooth day-to-day motion which compares favorably with the actual motion. (The longitudinal gap corresponds to the zero-correction mentioned above in Sec. 3.)

It is probably due only to the accidents of preservation that we have very few texts with such refined interpolations for the construction of ephemerides in the proper sense of the word. But, even so, we have an example of a much less skillful handling of second-order sequences in the case of an ephemeris of Mercury. Figure 11 gives the velocity diagram of the actual motion (solid curve) for the sun and for Mercury. The dotted line is taken from the text (again not corrected for the difference in zero longitude; probably Ω should be lowered to a level a little above point 3) and shows a rather poor approximation in the retrograde part. It must be remembered, however, that the planet is invisible between Ω and Γ or, actually, between about points 5 and 11.

Visibility questions of Mercury are a topic that

has caused much misunderstanding. Schoch accused Ptolemy of gross incompetence as an observer because he chose the parameters of his theory of planetary phases in such a fashion that Mercury is morning star at the beginning of Taurus and as evening star at the beginning of Scorpio does not reach sufficient elongation from the sun to become visible at the latitude of Alexandria (Almagest XIII, 8). Schoch insisted that these phases were observable in Babylon since they are listed in the texts. But we meet here the same situation that we mentioned in the case of the eclipse magnitudes (above Sec. 2), where one cannot conclude from the mere fact of tabulation that the Babylonian astronomers considered eclipses possible month after month. Also the Mercury "appearances" are only computed for arithmetical requirements for all transitions between morning and evening phases, even if other criteria exclude actual visibility. In this case, however, the "appearance" was often expressly marked by an ideogram meaning "passed by," a fact not recognized in Schoch's time (Neugebauer 1951). This incident shows once more how careful one must be before one deduces "empirical parameters" from ancient sources as long as one has not reached a complete understanding of the terminology as well as the purely mathematical devices.

6. The most trivial and hence most commonly made mistake in the interpretation of Babylonian astronomy is the concept of enormously far extended observational records serving as the basis for the latest development. Indeed, if one looks at the period relations on which the ephemerides are constructed (Fig. 12) one sees that many centuries would be required to establish these data. Fortunately we have textual evidence as to how such period relations were obtained (cf. Fig. 13). The most trivial approximate period of Jupiter of 12 yr is supposed to bring the planet back with an excess of 5° in longitude, whereas the 71-yr period results in a deficit of 6°. The subsequent list given by the text shows linear combinations of these two basic periods intended to result in smaller deviations until eventually 427 yr eliminate the errors completely. Similarly the 284-yr period of Mars is designed to compensate opposite errors accumulated during 47 and 79 yr. Hence the huge periods are no better and no worse than the round estimates established for much shorter intervals. A number like the 1151 yr for Venus is totally fictitious as far as the extension of observational records is concerned.

In spite of the fact that the large number periods on

	occur.	sid. rot.	years
Saturn	256	9	265
Jupiter	391	36	427
Mars	133	151	284
Venus	720		1151
Mercury	1513		480

FIG. 12. Periods used in the computation of Babylonian ephemerides.

which all the arithmetical procedures of Babylonian astronomy are built are not the result of direct observations extended over many centuries, it nevertheless remains certain that the requirements of exact periodicity form the cornerstone of the planetary and lunar theory. [This holds in particular also for the theory of eclipses in their relation to the so-called "Saros" of 223 lunations (cf. Neugebauer 1957). Newcomb (1879) in his study of the solar eclipse cycles did not suspect that he had Babylonian predecessors who had fully grasped the significance of the periodicity also of the lunar anomaly.] With this fact must be combined the methodological idea to concentrate all attention on the sequence of isolated characteristic phenomena, the syzygies for the moon, the phases for the planets, and to deal with all questions of the intermediary motions of the celestial bodies only in a secondary fashion after the "motion" of the phenomena had become known.

Starting with these principles one can now reconstruct the main steps in the mathematical development. We consider given the number period π of a certain phenomenon; for example after $\pi = 391$ oppositions (or stations, etc.) Jupiter will be exactly back at the same (sidereal) point of the ecliptic, having traversed the ecliptic $Z = 36$ times (cf. Fig. 14). It is natural to first consider the mean situation. Obviously the mean synodic arc $\langle \Delta\lambda \rangle$ that separates two consecutive events of the same type is given by $360 \cdot 36/391 \approx 33;$ $8{,}45°$. Beginning at point 0 of the ecliptic the point 11 will be about $4\frac{1}{2}°$ beyond point 0, point 22 about $9°$ etc. The final result will be a set of points which form a regular polygon of $\pi = 391$ vertices which are separated by an arc of $\bar{\delta} = 360°/391 \approx 55'$. According to hypothesis these 391 points represent all possible points of the ecliptic at which Jupiter can ever come to opposition. [In Newcomb's theory of solar eclipses (1879) these points are called "conjunction points".] Let us renumber them $0, 1, 2, \cdots, \pi$ as they follow each other at the distance $\bar{\delta}$. Then we also know that, if the planet is at any point k, it will next be in the same phase at the point $k + Z$. This gives us a complete control over the motion of all phases starting from a given initial position under the assumption of mean motions. In

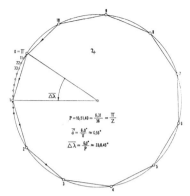

FIG. 14. Positions of phases of Jupiter (e.g., oppositions) in the ecliptic under the assumption of mean motions. The resulting regular polygon has $6{,}31 = 391$ sides (cf. Fig. 12). Consecutive phases are 36 steps apart, corresponding to a longitudinal difference $\langle \Delta\lambda \rangle$ ("mean synodic arc").

fact, however, neither sun nor planets always proceed with mean motions and consequently the distribution of the π points at which a given phase may be expected will not be represented by a regular polygon. Nevertheless the two parameters π and Z must be preserved: After π events, everything repeats itself, and two consecutive events are separated from one another by Z points at which sooner or later a phase takes place. That this simple rule suffices to derive all arithmetical rules which characterize a System A model has been seen only recently by Aaboe (1965), and by Aaboe and Sachs (1966), after van der Waerden (1957) had reached similar results for Mars. All that is required is an arithmetical idealization of an empirically recognized variation of density of the points at which the phases occur. For Mars one may have found that about 12 stationary points can be fitted into the two signs Capricorn and Aquarius, whereas three times as many occur in Cancer and Leo. While the mean density corresponds on the whole ecliptic to the same distance $\bar{\delta}$, one now introduces arcs α_i and densities δ_i such that on the arc α_i the phases again have constant distance δ_i. Hence we have $\Sigma\alpha_i = 360°$ and each α_i contains a certain number ν_i of intervals of length δ_i, hence $\nu_i\delta_i = \alpha_i$. Finally $w_i = Z\delta_i$ always separates, in principle, consecutive phases. For Mars, e.g., all six arcs α_i are of equal length $60°$. Then the ν_i are chosen to be

$$\nu_1 = 24, \quad \nu_2 = 36, \quad \nu_3 = 27, \quad \nu_4 = 18, \quad \nu_5 = 12, \quad \nu_6 = 16,$$

hence

$$\delta_1 = 2{;}30°, \quad \delta_2 = 1{;}40, \quad \delta_3 = 2{;}13{,}20,$$
$$\delta_4 = 3{;}20, \quad \delta_5 = 5, \quad \delta_6 = 3{;}45.$$

Since for Mars $\pi = 133$ and $Z = 18$ (cf. Fig. 12), we

Jupiter	
12 years = a	deviation: $+ 5°$
71 years = b	$- 6$
83 years = b + a	$- 6 + 5 = - 1$
95 years = b + 2a = c	$- 6 + 10 = + 4$
166 years = b + c	$- 6 + 4 = - 2$
261 years = b + 2c	$- 6 + 8 = + 2$
427 years = 2b + 3c	$- 12 + 12 = 0$

Mars	
284 years = 47 + 3·79	

FIG. 13. Construction of larger periods from cruder smaller ones by means of linear combinations which reduce the deviations to zero.

have for the synodic arcs

$$w_1=45°,\ w_2=30,\ w_3=40,\ w_4=60,\ w_5=90,\ w_6=67;30,$$

as listed before in Eq. (4). The final synodic arcs as represented in Fig. 9 are then the outcome of this redistribution of the density of the phases, always combined in groups of $Z=18$ steps. All arithmetical rules which govern System A (and in a similar fashion also System B) can be derived from a proper distribution of π points, either constant on given arcs, or in arithmetic progression.

We have reached the point where we have gained insight into the mathematical ideas that determined the computational methods of an important section of Babylonian astronomy. But the empirical data or observational techniques which must somewhere have provided the basic experiences have vanished for us even farther away than one would have suspected in Kugler's time. This does not mean that one must assume a long chronological development from empirical data to mathematical description. On the contrary, I think that it was in rather rapid succession that the final mathematical methods were reached after the main principles had become clear. But for us these preliminary steps, however rapidly or slowly they were made, remain hidden and will remain hidden as long as sources from a different period do not become available to us.

7. It might be tempting to conclude this summary with a eulogy of the importance of Babylonian astronomy for the development in later periods—indeed our use of degrees and minutes is a still present reminder of Babylonian influence. In fact, however, this influence is far less dominant than one might expect. As soon as Greek astronomers decided upon cinematic models, most of the Babylonian methodology became inapplicable. What remains is the fundamental period relations and similar parameters and the sexagesimal place-value notation for numerical computations. And

indeed these are exactly the elements which we find taken over by Hipparchus in the second century B.C. But in the Greek geometric approach the characteristic phases are a secondary consequence, not a starting point for the description of the actual motion of the planets. Hence it is only in side branches of astronomical techniques, in part still recognizable in certain Indian and Islamic methods, that typically Babylonian procedures did survive (Neugebauer 1952; Pingree 1959). But for the main stream of astronomical development, Babylonian influence was of only secondary importance. In conclusion, I think, one should not view the historical significance of Babylonian astronomy in the same light as the influence of the Ptolemaic system on the Copernican, but see it in a quite different direction. It is here that for the first time in human history purely mathematical methods were shown to open the road to a most successful description, and hence prediction, of natural phenomena, free from all philosophical principles which, more than anything else, have been the main obstacle to scientific development.

REFERENCES

Aaboe, A. 1965, *Centaurus* 10, 213.
Aaboe, A., and Sachs, A. 1966, *J. Cuneiform Studies* 20, 1.
Epping, J. 1889, *Astronomisches aus Babylon*. Ergänzungshefte zu den Stimmen aus Maria-Laach 44, Freiburg.
Huber, P. 1957, *Z. Assyriologie* 18, 265.
——. 1958, *Centaurus* 5, 192.
Neugebauer, O. 1948, *J. Cuneiform Studies* 2, 209.
——. 1951, *Proc. Am. Phil. Soc.* 95, 110.
——. 1952, *Osiris* 10, 252.
——. 1955, *Astronomical Cuneiform Texts* (Lund Humphries, London), 3 vols.
——. 1957 *Kgl. Danske Videnskab. Selskab, Mat.-Fys. Medd.* 41, 4.
Newcomb, S. 1897 *Astron. Papers* 1, 1.
Pingree, D. 1959, *J. Am. Oriental Soc.* 79, 282.
Sachs, A. 1948, *J. Cuneiform Studies* 2, 271.
——. 1952, *ibid.* 6, 105.
Tuckerman, B. 1962, "Planetary, Lunar, and Solar Positions 601 B.C. to A.D. 1," *Am. Phil. Soc. Mem.* 56.
Van der Waerden, B. L. 1957, *Vierteljahrsschrift Naturforsch. Ges. Zürich* 102, 2, 39.

The Origin of »System B« of Babylonian Astronomy

O. NEUGEBAUER*

TO W. K. FELLER AS A TOKEN OF LIFELONG FRIENDSHIP

1. Our understanding of the Babylonian mathematical astronomy of the Seleucid-Parthian period progressed in several rather distinct steps. In its first phase, forever associated with the names of Epping, Kugler, and Schaumberger, the main efforts were concentrated on clarifying the astronomical meaning of the computational devices and the terminology, a truly pioneering work of decipherment and interpretation. Building upon these results my own efforts were directed toward a uniform mathematical classification of the textual material and its systematic arrangement in a corpus, at least so far as the strictly mathematical-astronomical texts were concerned. During the last decade these studies have entered a new phase which one may perhaps characterize as the "historical" phase in which one endeavors to penetrate into the Babylonian way of constructing a methodology, the results of which are now comparatively clear before us. Perhaps the most important result in this line of research is the demonstration by A. Aaboe[1] that the computational rules for ephemerides of the "System A" type can be derived from very simple assumptions concerning the distribution of the events under consideration on the ecliptic. It is the purpose of this paper to show that exactly the same methodological idea leads also to the rules of "System B". In the past I repeatedly had occasion to point out that the commonly accepted evolutionary sequence from A to B is by no means securely established. Obviously the existence of a common methodological basis makes a chronological distinction between the two systems still more difficult.

2. Aaboe's results can be summarized in the following fashion. Consider the points of occurrence on the ecliptic of some phenomenon (e.g.

---

* Institute for Advanced Study, Princeton, New Jersey.

Centaurus 1968: vol. 12: no. 4: pp. 209–214

15 CENTAURUS, vol. XII

a planetary stationary point or a full moon) which comes back to the same place with the period Π, having in the meantime traversed the ecliptic Z times. If these phenomena would depend exclusively on mean motions the points in question would define a regular Π-gon on the ecliptic, and consecutive phenomena would always be separated from one another by Z intervals (we assume, of course, Π and Z to be relatively prime integers), hence again being equidistant. In fact, however, such a regularity does not exist. Because of the anomalies in the motion of the celestial bodies the actually observable phenomena are more densely packed in some regions of the ecliptic than in others (for the sake of simplicity of expression we assume all anomalies to be sidereally fixed). What Aaboe has shown is the following: in order to cope with this situation we divide the ecliptic into a suitable number of fixed arcs each of which carries a definite number of "basic intervals" of constant length on each arc[2], different from arc to arc, but such that their total remains Π. The intervals between consecutive events, the so-called "synodic arcs", are again the total of Z intervals, obviously constant as long as all the Z constituent basic intervals belong to the same arc of the ecliptic. But the synodic arc will change at the transition to another arc of the ecliptic. The importance of this simple model lies in the fact that from it all the rules can be derived which characterize "System A" in which the synodic arcs can be represented as being regulated by a discontinuous step function on the previously mentioned arcs of the ecliptic.

3. What I shall show in the following is a similar derivation of "System B" which operates with linear zigzag functions. We divide the ecliptic into two halves (for the sake of simplicity I assume Π even, hence Z odd) and mark on each half, in a strictly symmetric fashion, the points at which the phenomenon in question was supposed to occur, such that the lengths of the intervals form an arithmetic progression of $\dfrac{\Pi}{2}$ elements.

If we then take always Z such consecutive intervals the resulting totals, the "synodic arcs", form again an arithmetic progression of constant difference d. If Z such basic intervals reach beyond one of the end points of the semicircles of the ecliptic one reflects the excess of the increment d on a fixed value M or m respectively. The final result is a linear zigzag fuction of period $P = \Pi/Z$ computable with the rules of "System B". Hence "System B" is based on the least sophisticated way of taking the empirical variations of the synodic arcs into consideration: one assumes simply a linear change in the length of all basic intervals between given limits.

4. To demonstrate these statements in the simplest direct fashion we begin with "normed" intervals of length 1, 2, 3, ... etc. until $\dfrac{\varPi}{2}$, which we denote for typographical reasons by π. We thus assume that one half of the circle which represents the ecliptic has the length $1 + 2 + \ldots + \pi$ and we define a sense of rotation such that we can distinguish between an "increasing" and a "decreasing" semicircle. All letters, a, b, k, etc. in the following represent integers.

We now add any Z consecutive intervals with the sole restriction that they all belong to the same semicircle and that the same holds also for the next Z intervals. Then we have a "synodic arc"

$$y_1 = (a + 1) + (a + 2) + \ldots + (a + Z)$$

followed by another synodic arc

$$y_2 = (a + 1 + Z) + (a + 2 + Z) + \ldots + (a + Z + Z).$$

Hence

$$y_2 = y_1 + Z^2$$

and

(1) $\qquad d = Z^2$

is the constant difference for any sequence of synodic arcs which lie inside one of the semicircles.

5. The smallest possible synodic arc within one semicircle begins with the interval 1 and is of length

$$1 + 2 + \ldots + Z = \tfrac{1}{2}Z(Z + 1) = \tfrac{1}{2}d + \tfrac{1}{2}Z.$$

If the arc belongs to the increasing semicircle it has an arc of exactly the same length as its neighbor on the decreasing semicircle. Thus we have two branches of linearly varying synodic arcs which come down with the slope $\pm d$ to the same value $\tfrac{1}{2}d + \tfrac{1}{2}Z$. Hence these two linear functions meet at a minimum which is $\tfrac{1}{2}d$ lower than the last value, hence

(2) $\qquad m = \tfrac{1}{2}Z.$

Similarly at the greatest possible synodic arc within each semicircle:

15*

(2a) $\quad \pi + (\pi - 1) + \ldots + (\pi - Z + 1) = Z\pi - \frac{1}{2}Z(Z-1)$
$$= \frac{1}{2}(\Pi + 1)Z - \frac{1}{2}d$$

and therefore for the intersection of the two branches

(3) $\quad M = \frac{1}{2}(\Pi + 1)Z.$

Hence we have a zigzag function with the amplitude

(4) $\quad \Delta = M - m = \frac{1}{2}\Pi Z$

and thus with a period

(5) $\quad P = \dfrac{2\Delta}{d} = \dfrac{\Pi Z}{Z^2} = \dfrac{\Pi}{Z}.$

6. If one begins the summation of Z basic intervals at an arbitrary point one need not end with a synodic arc with the last interval (π). Let us assume that the last synodic arc y_n on the increasing semicircle terminates with the interval $\pi - 1$, hence begins with the interval $\pi - Z$. Let A be the synodic arc which reaches from $\pi - Z + 1$ to π, hence with (2a)

(6) $\quad\quad\quad\quad A = \frac{1}{2}(\Pi - Z + 1)Z.$

Then
$$y_n = A - \pi + (\pi - Z) = A - Z.$$

On the decreasing branch the synodic arc next to the top begins now one interval higher than before in the symmetric arrangement, hence (cf. Fig. 1)
$$y_{n+2} = A - d + (\pi - Z + 1) - (\pi - 2Z + 1) = A - d + Z.$$
Since
$$y_{n+1} = y_{n+2} + d$$
we have for the two intervals which straddle the maximum

(7) $\quad y_n + y_{n+1} = 2A = (\Pi - Z + 1)Z.$

In the same fashion one can show that $y_n = A - kZ$ if k intervals $(0 < k < Z)$ are missing to π while $y_{n+2} = A - d + kZ$. If we define
$$y_{n+1} = y_{n+2} + d \quad \text{if} \quad 2k < Z$$
$$y_{n+1} = y_n + d \quad\quad \text{if} \quad 2k > Z$$
we find that the total of the two synodic arcs which precede and follow

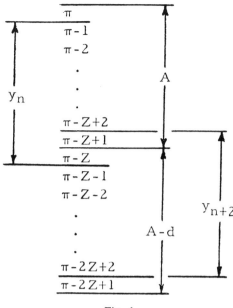

Fig. 1

the maximum is always $2A$. Similarly one finds near the minimum always

(8) $$y_n + y_{n+1} = Z(Z + 1)$$

But it follows from (1), (2), and (3) that (7) and (8) are the equivalent of

(9) $$y_n + y_{n+1} = \begin{cases} 2M - d \\ 2m + d \end{cases}$$

which is the standard reflection rule for linear zigzag functions.

It should be noted that this implies that one crosses an extremum simply by forming $y_n + d$ and reflecting the excess over M on M (and similar at m) but that one does not add after π the values π, $\pi - 1$, etc.

7. All that remains to be done is to free ourselves from the norm which we have so far adopted for the basic intervals by giving them the length $1, 2, \ldots, \pi$. The period is already correctly given by (5). If we call Δ_0 the normed amplitude (4) and Δ an empirically given amplitude, we have only to multiply the normed difference d_0, given by (1), with the factor

(10) $$c = \Delta/\Delta_0$$

and to add another appropriate constant b to cm_0 and to cM_0 in order to obtain the empirically required extrema m and $M = m + \Delta$.

Example: $\Pi = 4,8$ $Z = 9$ (i.e. lunar anomaly).

Normed parameters from (1), (2), (3), and (4):

$$d_0 = 1,21 \quad m_0 = 4;30 \quad M_0 = 18,40;30 \quad \Delta_0 = 18,36.$$

Empirically required (of course adjusted to arithmetical requirements):

$$m = 11;4 \quad \Delta = 4;49,20.$$

Consequently

$$c = \Delta/\Delta_0 = 0;0,15,33,20$$
$$d = cd_0 = 0;21$$

and

$cm_0 = 0;1,10$ hence $b = 11;4 - 0;1,10 = 11;2,50$
$cM_0 = 4;50,30$ hence $M = cM_0 + b = 15;53,20$
and $\Delta = 4;49,20$ as required.

NOTES

1. A. Aaboe, *On Period Relations in Babylonian Astronomy*. Centaurus vol. 10, 1965, p. 213–231 and A. Aaboe – A. Sachs, *Some Dateless Lists of Longitudes of Characteristic Planetary Phenomena from the Late-Babylonian Period*, Journal of Cuneiform Studies vol. 20, 1966, p. 1–33.
2. This concept was first introduced by van der Waerden in 1957 (under the name "Schritte") "in order to simplify" the computations for Mars (*Anfänge der Astronomie*, Groningen 1966, p. 187). But by saying (p. 188) that this concept "war also den babylonischen Rechnern nicht ganz fremd" he shows that he had not yet reached insight into the decisive role and historical significance of the method. He also misinterprets the historical situation when he says (p. 186) "Die Marstheorie ist sehr schwierig. Das liegt daran, dass Mars sich extrem ungleichmässig bewegt". On the contrary, it is the great eccentricity of Mars which makes it possible to arrange effectively the Π points in question in so simple a model as six arcs of two zodiacal signs each. It is by no means accidental that it was again Mars which provided Kepler with sufficiently great deviations from the Ptolemaic orbit to detect the correct one.

4. Greco-Roman

ON TWO ASTRONOMICAL PASSAGES IN PLUTARCH'S
DE ANIMAE PROCREATIONE IN TIMAEO.

1. The following investigations arose from a question of Prof. A. Bidez. He drew my attention to § 31 (1028 F) of the work mentioned in the title, in which Plutarch speaks about a " Chaldean " doctrine according to which the four seasons of the year can be arranged in certain harmonic proportions. The discovery of the unequal length of the four seasons is undoubtedly one of the most fundamental achievements of ancient astronomy because it is equivalent to the discovery of an inequality in the movement of the sun. The explanation of this inequality as apparent by assuming a certain eccentricity of the sun's orbit with respect to the earth is the basis for the ancient theory of the eccentric movements, a theory which finally led to Kepler's discovery of the elliptic orbits of the planets. On the other hand, the cuneiform astronomical tablets of the Seleucid period show that the inventors of these mathematical devices also were fully conscious of the fundamental rôle of an adequate description of the inequality of the movement of the sun in the prediction of the visibility of the moon's crescent and of eclipses. The careful investigation of every ancient statement about the unequal length of the seasons is therefore fully justified.

The first passage in question reads as follows: [1] Χαλδαῖοι δὲ λέγουσι τὸ ἔαρ ἐν τῷ διὰ τεττάρων γίγνεσθαι πρὸς τὸ μετόπωρον, ἐν δὲ τῷ διὰ πέντε πρὸς τὸν χειμῶνα, πρὸς δὲ τὸ θέρος ἐν τῷ διὰ πασῶν. " The Chaldeans say that spring makes a fourth with respect to autumn, a fifth to winter, an octave to summer." These musical harmonies can be represented by the proportions $\frac{4}{3}$, $\frac{3}{2}$, $\frac{2}{1}$, respectively; if we therefore denote the four seasons by s_1, s_2, s_3, s_4, respectively, beginning with spring as s_1, then one would offhand interpret the sentence in question as

$$(1) \qquad s_1 = \tfrac{4}{3} s_3 = \tfrac{3}{2} s_4 = \tfrac{2}{1} s_2$$

which is obvious nonsense because spring (s_1) is certainly not twice as long as summer (s_2), etc.

On the other hand, spring is actually the longest of the four seasons according to the following inequality:

[1] Plutarch, *Moralia*, VI, ed. Bernardakis, p. 202, 14 (1028 F).

455

(2) $s_1 > s_2 > s_4 > s_3$

while from (1) follows

(3) $s_1 > s_3 > s_4 > s_2.$

Plutarch's statement, although undoubtedly incorrect in its interpretation (1), has at least as a consequence the relation (3), which is correct except for the interchange of s_3 and s_2.

Having found at least a partial justification for Plutarch's statement, we might try to replace the obviously incorrect proportions (1) by some more reasonable expression. Now one need only remark that the four seasons actually differ very little in length in order to realize that all speculations about the relations between the seasons should sooner be directed towards the *deviations* from some constant time interval than towards the lengths of the seasons themselves. In other words, we may assume that the lengths of the seasons are considered as consisting of two parts

(4) $$s_1 = s + \sigma_1 \qquad s_3 = s + \sigma_3$$
$$s_2 = s + \sigma_2 \qquad s_4 = s + \sigma_4$$

namely a common part s and deviations $\sigma_1, \ldots, \sigma_4$. Then the meaning of the sentence in question would be [2]

(5) $\sigma_1 = \tfrac{4}{3}\,\sigma_2 = \tfrac{3}{2}\,\sigma_4 = \tfrac{2}{1}\,\sigma_3.$

From (5) and (4) follows

(6) $$s_1 = s + 12\sigma$$
$$s_2 = s + 9\sigma$$
$$s_4 = s + 8\sigma$$
$$s_3 = s + 6\sigma$$

where s and σ are still undetermined numbers, but common to all four numbers $s_1, \ldots, s_4$. We therefore must find six unknown quantities $s_1, \ldots, s_4, s,$ and σ satisfying the four conditions (6) and in addition to it the obvious relation

(7) $s_1 + s_2 + s_3 + s_4 = 1$ year.

Thus, with only five equations for six quantities *one* of them can be chosen arbitrarily.

---

[2] Here also, of course, we interchange the second and fourth place in Plutarch's statement.

The most natural assumption is evidently $\sigma = 1$, which means that the differences in question are not only multiples of the famous harmonic numbers 12, 9, 8, and 6 but *equal* to these numbers. Making this assumption and adopting $365\frac{1}{4}$ days as the length of one year, one can easily solve (6) and (7). The result, compared with the values accepted by Hipparchus,[3] are

$$
\begin{array}{lll}
(8) & s_1 = 82\frac{9}{16} + 12 = 94\frac{9}{16} & \text{Hipparchus } 94\frac{1}{2} \\
& s_2 = 82\frac{9}{16} + 9 = 91\frac{9}{16} & 92\frac{1}{2} \\
& s_3 = 82\frac{9}{16} + 6 = 88\frac{9}{16} & 88\frac{1}{8} \\
& s_4 = 82\frac{9}{16} + 8 = 90\frac{9}{16} & 90\frac{1}{8}
\end{array}
$$

The coincidence is not perfect, yet so close that there can be little doubt that we are on the right track. Thus Plutarch refers to the theory that the following relation holds:

$$
\begin{array}{llll}
(9) & s_1 = s + 12 & & s_4 = s + 8 \\
& s_2 = s + 9 & & s_3 = s + 6
\end{array}
$$

which makes the remainders harmonic numbers, respectively yielding fourth, fifth, and octave by their proportions.

That the numbers found do not exactly agree with Hipparchus' values is not surprising. Hipparchus determined the length of the seasons by careful observations and did not hesitate to introduce fractions of days. We have, however, a better chance to find the right numbers by going back to older and less elaborate systems which expressed the length of the seasons by an integral number of days.[4] And indeed, one of them, the system of Callippus (*ca.* 330 B. C.) [5] shows a very close relationship to the formulae (9) if we assume $s = 83$. Then we get

$$
\begin{array}{lll}
(10) & s_1 = 83 + 12 = 95 & \text{Callippus } 95 \\
& s_2 = 83 + 9 = 92 & 92 \\
& s_3 = 83 + 6 = 89 & 89 \\
& s_4 = 83 + 8 = 91 & 90
\end{array}
$$

I have no doubt that this contains the solution of our problem. Having realized that the inequality of the seasons according to Callippus can be brought into the form (10) with three har-

---

[3] Ptolemy, *Almagest*, III, 4.

[4] See e. g. W. B. Dinsmoor, *The Archons of Athens* (Cambridge, 1931), p. 318, note.

[5] Cf. A. Böckh, *Über die vierjährigen Sonnenkreise der Alten* (Berlin, 1863), p. 46.

monic increments 12, 9, and 6, the Pythagorean philosophers did not hesitate to improve the work of nature by replacing 90 by 91 in order to obtain complete " harmony."

In attributing these speculations to the so-called Pythagoreans, we consider Plutarch's reference to the " Chaldeans " as unhistorical. We know that Babylonian astronomy of the Seleucid period used values for the lengths of the seasons very close to Hipparchus' values [6] and therefore equally unsuitable to the set (10) of Plutarch's numbers. No traces of a Babylonian theory which could be brought into relation with the number-mysticism of Plutarch's source are preserved. There exists, on the other hand, a certain Greek tradition which assumes the " Chaldean " origin of the harmonic proportions; [7] here also, proofs of its reliability are lacking.

2. The second passage to be discussed precedes the report about the harmonic qualities of the seasons and deals with the unequal length of the days during the year. We read: [8] τοῦ δ' ἡλίου περὶ τὰς τροπὰς ἐλάχιστα καὶ μέγιστα περὶ τὴν ἰσημερίαν ἔχοντος κινήματα, δι' ὧν ἀφαιρεῖ τῆς ἡμέρας καὶ τῇ νυκτὶ προστίθησιν ἢ τοὐναντίον, οὗτος ὁ λόγος ἐστίν· ἐν γὰρ ταῖς πρώταις ἡμέραις λ' μετὰ τὰς χειμερινὰς τροπὰς τῇ ἡμέρᾳ προστίθησι τὸ ἕκτον τῆς ὑπεροχῆς, ἣν ἡ μεγίστη νὺξ πρὸς τὴν βραχυτάτην ἡμέραν ἐμποιεῖ, ταῖς δ' ἐφεξῆς τριάκοντα τὸ τρίτον, τὸ δ' ἥμισυ ταῖς λοιπαῖς ἄχρι τῆς ἰσημερίας "Because the movement of the sun has its minimum at the solstices and its maximum at the equinox, it substracts or adds to day and night according to the following proportion: during the first 30 days one sixth of the difference between the longest night and the shortest day is added, one third in the following thirty (days), one half during the rest until the equinox." These sentences obviously contain different astronomical errors. The sun's velocity cannot be a maximum in two opposite points of its orbit and a minimum 90° distance from the maxima. Furthermore, the extremes of the sun's velocity have nothing at all to do with the solstices and equinoxes, which depend only on the inclination of the ecliptic toward the equator; but it is

[6] Cf. F. X. Kugler, *Babylonische Mondrechnung* (Freiburg, 1900), pp. 83 ff. and O. Neugebauer, *Untersuchungen zur antiken Astronomie*, III, pp. 206 ff. (*Quellen u. Studien zur Gesch. d. Mathem.*, B, III [1938]).
[7] Cf. Iamblichus, *In Nicomachi arithm. introd.*, ed. Pistelli, p. 118, 23 f.
[8] 1028 F = p. 202, 4, ed. Bernardakis.

interesting to notice that here the problem of the inequality of the sun's movement is mentioned, which is the basis for the inequality of the seasons as we have mentioned above. The following remark about the intervals of 30 days is also not quite correct because instead of 30 days one should rather speak about 30 degrees travel of the sun. The final paragraph is the really interesting part, although it too contains an error. The increase of the length of the days from winter solstice to equinox is said to be $\frac{\Delta}{6}$, $\frac{\Delta}{3}$, $\frac{\Delta}{2}$, respectively, if $\Delta = M - m$ ($M =$ longest, $m =$ shortest day). The error here lies in taking Δ instead of $\frac{1}{2}\Delta$, i. e., instead of the difference between the longest day and equinox.[9] But correcting this carelessness, we obtain a theorem which is well known in Greek time-reckoning, explicitly formulated in Cleomedes, *De motu circulari orbium caelestium,* I, 6.[10] This associates Plutarch, or rather, his source, with a known geographical doctrine, represented not only by Cleomedes but also by Gerbert, Martianus Capella, and *Pap. Michigan,* III, 149, a doctrine which can finally be traced back to Babylonian astronomical tablets of the first, second, and third centuries B. C.[11] Plutarch is now the earliest representative of the appearance of this theory in Greek literature.

O. NEUGEBAUER.

BROWN UNIVERSITY.

[9] This is the inverse error to the first mentioned: the extremes of the sun's velocity were assumed to be only 90° distance instead of 180°; now the increase Δ is taken for diametrically opposite points of the sun's orbit instead of points 90° apart.

[10] Ziegler, p. 50. Cf. Neugebauer, " Cleomedes and the Meridian of Lysimachia," *A. J. P.,* LXII (1941), pp. 344 ff.

[11] The details of these relations are discussed in Neugebauer, " On some astronomical Michigan Papyri and related problems of ancient Geography and Astronomy," *Trans. Am. Philos. Soc.,* XXXII (1942). The Babylonian theory referred to is the so-called " system B " in the terminology of the quoted paper.

Reprinted from
American Journal of Philology
43 (4), 455−459 (1942)

The Early History of the Astrolabe

Studies in Ancient Astronomy IX

BY O. NEUGEBAUER *

1. Introduction and Summary. Few astronomical instruments have enjoyed as much popularity as the astrolabe. Even today the artistic qualities of many Persian and Arabic astrolabes have given these instruments a place in museums and private collections. Beautifully illustrated articles and books on the astrolabe have been published but this material hardly goes back beyond the 10th century. About the earlier history, very contradictory statements can be found in the literature. While classical scholars tend to trace the idea of the astrolabe back to Apollonius and Archimedes (about 200 B.C.) or even to Eudoxus (about 350 B.C.), one finds in a recent book [1] the statement, "En résumé, l'astrolabe est méditerranéen. Son histoire va du VIᵉ au XVIIᵉ siècle de notre ère."

The present paper is an effort to establish a more solid foundation for the history of the astrolabe in Greek astronomy. I shall show that the astrolabe as an instrument was known to Ptolemy (150 A.D.[2]) — "astrolabe" here and in the following always means the "plane astrolabe" and never the armillary sphere — and I shall also demonstrate that the contents of a work of Theon Alexandrinus (375) on the astrolabe is preserved through the treatise of Severus Sebokht on this subject (written before 660 [3]). Finally the place of Johannes Philoponus (530) in this tradition can be established.[4]

In this way the history of the astrolabe as an instrument can be followed with certainty back of Ptolemy. There arises the question of tracing the underlying theory of stereographic projection still farther back. I think good reasons can be given for assuming that it was known in the time of Hipparchus (150 B.C.).

The discussion of the above mentioned points does not require much space, assuming the underlying texts as known. Actually the available publications require com-

---

* Brown University, Providence, R. I.
[1] Michel, TA p. 13.
[2] Dates are meant only to indicate the main period.
[3] Nau [2] p. 227.
[4] For the date of Philoponus see the results of a study by Meyerhof, quoted *Isis 18* (1932) p. 447 f.

ments on many points in order to be fully understandable. I have therefore added appendices of which especially the discussion of the letter of Synesius (400) had to be expanded to greater length.

2. The Astrolabe. For the reader who is not familiar with the theory of the astrolabe the following remarks will suffice to allow him to understand the main features of its construction.[4a] The celestial sphere is projected from its South pole onto the plane of the equator. Thus the solstitial circles and the equator are mapped into concentric circles, preserving angular distances on these circles. The innermost circle is the image of the summer solstitial circle, the outer circle represents the winter solstitial circle and forms the rim of the instrument. This rim is divided into 360°, measuring equinoctial time (right ascension).

The ecliptic is a circle which touches the solstitial circles. In the instrument the ecliptic is represented by the "spider," a circle which carries the 12 divisions of the zodiac (of course no longer equidistant) and which turns around the center of the instrument, i.e. the North pole.

For a given geographical latitude the horizon is mapped onto a circle which intersects the equator at two diametrically opposite points. The parallels of equal altitude (called almucantarats by the Arabs) form a family of circles which, with increasing altitude, converge toward a point which is the image of the zenith. All these circles lie in the upper part of the instrument, bounded by the horizon. Below the horizon, 11 curves are drawn which represent the seasonal hours for the given latitude. The sixth curve is simply a straight line, passing through the center, and representing the meridian. These hour curves can be found as follows. Draw a circle with the North pole as center and consider only the arc below the horizon, limited by the intersections with the horizon. Divide this arc into 12 sections of equal length. This gives 11 dividing points, not counting the endpoints on the horizon. If the sun travels on the given circle, which is a parallel circle to the equator, then the dividing points represent the seasonal hours of night for the day in question. If we repeat this construction for all parallels between the solstitial circles (these included) we obtain the hour curves as loci of the dividing points.

If we know the longitude λ of the sun, e.g. from a table, and if an hour of the night is given, we can turn the point λ of the spider on the corresponding hour line and know the exact position of the zodiac for the given moment. If an hour of daylight is given, the point $\lambda + 180$ lies on the corresponding hour line.

The spider, at its point of contact with the outer rim,[5] has an "index" which measures the rotation of the ecliptic in right ascension. The spider also has a certain number of pointers, projecting into the interior or exterior field of the ecliptic. The endpoints of these pointers correspond to the projections of important fixed stars, such as Vega, Arcturus, Spica etc. These stars move with the ecliptic around the North pole. The name "spider" is obviously derived from this web which turns above the disc of the instrument.

The reverse of the instrument is used as a sighting apparatus for measuring altitudes. By suspending the disc of the astrolabe on a vertical diameter, its plane falls in the plane of a circle of altitude. Around the center moves a sighting device, the "diopter," which can be pointed toward the sun or toward a star. Then the altitude can be read on a circle on which one endpoint of the diopter moves. The result of such a measurement can again be used for determining the corresponding position of the ecliptic on the celestial sphere. One only has to place the point λ of the spider on the proper circle of altitude.

3. Ptolemy. The design of an astrolabe is a simple problem of descriptive geometry. All that one needs is to construct the projection of the sphere onto its equatorial

---

[4a] For figures see, e.g., Drecker [2] p. 19 f.   [5] This point is, of course, the winter solstice, ♑ 0°.

plane by using this plane as plane of construction. All details of this construction are given in Ptolemy's "Planisphaerium,"[6] including the numerical values for the climate of Rhodes ($\phi = 36$).

While it cannot be denied that Ptolemy possessed all theoretical elements for the construction of an astrolabe, it has often been questioned whether he knew of a mechanical execution. Yet we read in chapter 14 of the "Planisphaerium":[7]

...huius series habet equidistantes zodiaco, quousque assignent loca stellarum fixarum, qua ratione ea contineat id, quod in horoscopio instrumento aranea uocatur:	the parallels to the zodiac, as far as they determine the positions of the fixed stars; hence it will contain what is called in the horoscopic instrument the spider."
"The present part of our discussion concerns	

The use of the old name "horoscopic instrument" instead of "astrolabe" excludes the possibility of a mediaeval interpolation. Thus there cannot be any doubt that Ptolemy knew the instrument which is later called "the astrolabe."[8] This is fully confirmed by Synesius (cf. p. 248) and Philoponus (cf. p. 253).

4. Theon Alexandrinus, Philoponus and Severus Sebokht. Ya^{c}qūbī (875) assigns[9] to Ptolemy the following four works on

(1) the armillary sphere
(2) the plane astrolabe

(3) the "canon"
(4) the Almagest.

The Fihrist[10] (987), al-Qifti (1200),[11] and Bar Hebraeus[12] (1250) quote Theon as the author of works on

(1) the armillary sphere
(2) the astrolabe

(3) the "tables"
(4) the Almagest.

The two last works are actually preserved; the "canon" of Ya^{c}qūbī corresponds, of course, to the "Handy Tables." Suidas[13] mentions among other writings of Theon No. 3, the "Handy Tables," and No. 2, εἰς τὸν μικρὸν ἀστρόλαβον ὑπόμνημα "Memoir on the little astrolabe."[14] No Greek author seems to know about No. 1, neither for Theon nor for Ptolemy. Nevertheless we have no reason to doubt Ya^{c}qūbī's extensive table of contents. The only question which remains is whether Ya^{c}qūbī has substituted Ptolemy for Theon. Klamroth[15] and, following him, Honigmann,[16] were inclined to favor this assumption. Against this might be pointed out that Philoponus[17] speaks of Ptolemy's drawings of the hour curves, a subject which is not found in the "Planisphaerium" but which is necessary in the description of the astrolabe. In this case Suidas's term ὑπόμνημα could be taken in its later sense[18] of "commentary." In favor of a substitution of Ptolemy for Theon speaks the fact that Ya^{c}qūbī's No. 3, the "canon," is certainly Theon's work on the "Handy Tables" and not Ptolemy's.

---

[6] Opera II p. 225 ff. The Greek title was probably ἐξάπλωσις ἐπιφανείας σφαίρας "unfolding of a spherical surface" as suggested by Kauffmann in RE 2, 1801,11 and Honigmann, SK p. 186 note 3. Cf. also the Fihrist sub Pappus: "plane representation of the sphere" (Suter, Fihr. p. 22). Suidas (ed. A. Adler, IV, 254,7) has ἅπλωσις ἐπ. σφ. "simplification etc."

[7] Heiberg p. 249,20 ff. Transl.: Drecker [1] p. 271.

[8] Delambre, HAA II p. 446, says only vaguely: "la seconde partie a pour object les circles parallèles à l'écliptique, et la construction de cette piece du planisphère qui est connue sous le nom d'araignée."

[9] Klamroth [1] p. 18 ff.

[10] Klamroth [1] p. 20. Suter, Fihr. p. 21.

[11] Steinschneider [1] p. 341 f.; Wiedemann

[1] p. 254 f.; al-Qifti is dependent on the Fihrist.

[12] Klamroth [1] p. 20 note 1. Bar Hebraeus follows al-Qifti; cf. Lippert, Stud., p. 39 ff.; p. 43.

[13] ed. Adler II p. 702,15 f.

[14] Houzeau-Lancaster, Bibl. I, 1 p. 631 No. 3073 mentions two manuscripts in the Laurentiana, both under the name of Theon of Alexandria. One is a Persian translation of Theon's "De astrolabio," the other an Arabic translation of Theon's "Instrumentum astronomicum." Only an investigation of these manuscripts can decide the validity of their titles.

[15] Klamroth [1] p. 20.

[16] Honigmann, Sk, p. 187.

[17] Hase [1] p. 139,13. Drecker [2] p. 29.

[18] Cf. the discussion in Rome CA p. 4 note.

This follows from Ya'qūbī's table of contents in which he mentions a chapter on the theory of an oscillation of the vernal point between ♈ 0° and ♈ 8° with a speed of 1° in 80 years.[19] This chapter is actually preserved in Theon's little commentary to the handy tables [20] but is not found in Ptolemy's introduction to his handy tables.

We shall show presently that the table of contents given by Ya'qūbī for No. 2 agrees so closely with Severus Sebokht's treatise on the plane astrolabe that there can be no doubt that Sebokht's work is taken from the same work which Ya'qūbī describes. Thus Sebokht has preserved a work of either Ptolemy or of Theon. Now Sebokht always distinguishes between "the philosopher, who wrote on the astrolabe" and Ptolemy, whom he quotes for his "canon." [21] Thus "the philosopher" is Theon and the close agreement between Ya'qūbī and Sebokht shows either that Theon followed Ptolemy very closely or that Ya'qūbī substituted Ptolemy for Theon. For us the main fact remains that Sebokht has preserved the contents of Theon's work on the astrolabe.

5. Comparison of Ya'qūbī and Severus Sebokht (and Philoponus). We will now compare Ya'qūbī and Severus Sebokht. We add Philoponus to this comparison because we shall see that he too depends on the same tradition, based on Theon or Ptolemy.

For the table of contents from Ya'qūbī, I follow Klamroth's translation though I use modern technical terms.[22] For Sebokht and Philoponus we are fortunately not dependent upon titles of chapters alone but we know their contents. Because the titles are sometimes rather misleading I have always followed the real subject matter. The numbers of chapters are quoted for Sebokht as in Nau's text, for Philoponus I follow for the sake of convenience the numbering used by Tannery,[23] though no such numbers are given in the original.

Ya'qūbī	Sebokht	Philoponus
Description of the astrolabe	Description of the astrolabe	
Rim, discs, spider, diopter	Discs, rim, spider, diopter	2. Diopter
Dismantling and assembling	Assembling	
Circles of equal altitude, inclination (?)	Circles of equal altitude	3. Description of the discs: circles of equal altitude
Description of:		
the discs for each climate, length (of hours) and latitude of climate	the discs, the 5 zones	longest daylight, latitude of climate
the stars		
hour-lines	hour-lines	hour-lines
rising and setting on the horizon	rising and setting on the horizon	rising and setting on the horizon
the ecliptic		4. the ecliptic
the solstitial circles	the solstitial circles	equator and solstitial circles [24]
the points ♑ 0°, ♋ 0° ♈ 0° and ♎ 0°	the points ♑ 0°, ♈ 0°, ♎ 0° and ♋ 0° the centers (κέντρα) [25]	

---

[19] Klamroth [1] p. 26. This phenomenon was later called "trepidation" in contrast to the "precession" of constant speed.

[20] Delambre HAA II p. 625 f; Halma, Tab. man. I p. 53.

[21] It can be shown that he is referring to the "handy tables," not to the tables of the Almagest.

[22] E.g. zodiacal sign for "Burg" or rising time for "Aufsteigung" (ἀναφορά).

[23] Tannery, Mém. sci. IX p. 341 ff. The numbers in Gunther, AW I p. 61 ff are one lower up to No. 6, and are the same from No. 7 to No. 15. (The unpublished MS of the Serai, Deissmann 19, fol. 292 ff., distinguishes a "preface" and 14 sections. Consequently all numbers given by Tannery should be lowered by one. [Note added in proofs.])

[24] Belonging to chapter 3.

[25] These are the points of rising and setting and of upper and lower culmination.

This introductory part itself shows that Sebokht covers exactly the ground described by Yaᶜqūbī. Also the arrangement is practically the same, whereas Philoponus has a much longer description of the diopter while he is shorter in the remaining material. Furthermore Philoponus has subdivided his material into three chapters whereas both Yaᶜqūbī and Sebokht give a sectioning of the second part only.

In continuing this comparison we split the second part into two groups because Philoponus shows no parallel at all to the second group.

Ya'qūbī	Sebokht	Philoponus
Use of the astrolabe	Use of the astrolabe	
1. Check of the astrolabe	VII. Check of the astrolabe	
2. Check of the diopter	VIII. Check of the diopter	
3. Hour of daylight, rising point of ecliptic	I. Hour of daylight, the centers. Example: ♋ 0° $h = 30°$ then 3rd hour, ♌ 5 rising	5. Hour of daylight. Example: ♈ 20° $h = 30°$ 6. The hour lines 6a. Fractions of hours 7. The centers
4. Hour of night, rising point of ecliptic	II. Hour of night from star, III. from moon	8. Hour of night from star 9. if near culmination
5. Longitude of the sun	IV. Longitude of the sun	12. Longitude of the sun 13. if near solstices
6. Longitude of the moon and of the 7(!) stars	V. Longitude of the moon and the planets	14. Longitude of planets
7. Latitude of the moon	VI. Latitude of the moon	
8. Rising times of the 12 zodiacal signs and for 7 climates	IX a. Rising times and setting times for zodiacal signs for given climate. X,XXV. Rising time of the horoscopic sign. XVI. Identity for rising times	10. Rim of astrolabe. Example: rising time of ♏ in climate III
9. Rising times for sphaera recta (i.e.) right ascension (of zodiacal signs)	IX b. Rising times for sphaera recta	
10. Length of night and daylight for given season and given climate	XI. Length of daylight, seasonal hours for daylight and night	11. Length of daylight, seasonal hours for daylight and night
11. Length of the "day" for a star and time between rising and setting	[XII. missing]	
12. Longitude and latitude of stars	XIX. Longitude and latitude of stars XXI. Latitude of stars. First and last visibility	
13. Precession: 1° in 100 years [26]	[XX. missing]	
14. Obliquity of the ecliptic	XXII. Obliquity of the ecliptic XXIII. Declination of the sun	15. Declination of zodiacal signs and of sun, moon and planets

The parallelism between Yaᶜqūbī and Sebokht is obvious. The only major change consists in a transposition of a few chapters and some variants in dividing chapters.

---

[26] This is the value of Ptolemy (e.g. Almagest VII, 4). Yaᶜqūbī's "100 lunar (!)-years" is, of course, a mistake caused by being used to the Arabic calendar.

Chapter 8 of Yaᶜqūbī is multiplied by Sebokht. All that it is necessary to say about the determination of the rising times of zodiacal signs for a given climate is said in the first half of chapter IX. But Sebokht repeats the same thing in chapter X, using astrological terminology, and once more, out of place and in a rather obscure form, at the very end of the treatise in chapter XXV. Also out of order is chapter XVI where Sebokht pretends to deal with rising times of sphaera recta. Actually he discusses an identity for rising times which must hold for any climate, illustrating it by means of Ptolemy's (handy) tables. This whole discussion has nothing to do with the astrolabe and is obviously a later addition, either of Sebokht or of a copyist.

Also Philoponus has subdivided the original chapters and made additions of his own. Yet he omitted exactly one-half of Yaᶜqūbī's chapters. An addition of his own is obviously the very lengthy description of the use of the diopter in chapter 5. The main topic, however, the determination of the hour from a measured altitude, fell rather short. Philoponus adds at the end of 6 and in 6a a trivial discussion of how to determine the fractions of hours. Also perfectly trivial are his additions 9 and 13 of artificially chosen cases which never will occur in practice.

Thus Philoponus is a much poorer editor of Theon's work than Sebokht. For the following geographical section he remains silent.

Ya'qūbī	Sebokht
15. Geographical latitude of cities	XIII. Geographical latitude of cities
16. Geographical longitude of cities	XIV. Geographical longitudes of cities
	XV. Difference of local noons
17. Latitude of every climate	XVIII. Latitude of climate for a given disc
18. Latitude of ones own climate	XVII. Latitude of climate from observation of sun (or stars)
19. Latitude of any climate and any city	
20. The 5 zones	XXIV. The 5 zones of the celestial and terrestrial sphere

This completes our comparison and again we find full agreement between Yaᶜqūbī and Sebokht. One can even understand the apparent duplication of chapters 17, 18, and 19 of Yaᶜqūbī's list. Chapter 17 apparently dealt with the practical question of finding the climate for which a given disc was drawn (Sebokht XVIII). In 18 and 19, however, the determination of the geographical latitude was discussed by means of observations. Sebokht mentions in XVII both sun and stars, omitting, however, the stars in the text. In Theon these two cases were probably the subject of 18 and 19 respectively.

6. Features of the Common Source. Now that we have seen that Sebokht preserves the contents of Theon's work on the little astrolabe while Yaᶜqūbī gives the accurate arrangement, one can use Philoponus and Sebokht for finding smaller details which belong to the common Theonic source. Only two examples will be mentioned. In chapter I Sebokht mentions as an example that the altitude of the sun was found to be 30°. The same value is quoted as an example in Nos. 5, 5a and 7 of Philoponus. But while Sebokht carries out the examples in all details, Philoponus makes no use of the numbers he quotes. Obviously $h = 30°$ was given as an example in the original treatise but only Sebokht preserved it completely.

Another feature of the common source is the name "equinoctial tropic" for the equator. This rather illogical term caused Philoponus to make a far-fetched explanation [27] while Sebokht uses it without questioning.[28] I know of only one sun-dial, from Samos, where the equator is called ἰσημερινὴ τροπή.[29] We can now add Theon as a representative of the same terminology.*

[27] Hase [1] p. 150,19; Drecker [2] p. 39.
[28] Nau [1] p. 97.
[29] Theophaneides [1].
* Also Hippolytus, Refut. omn. Haeres. V, 13 Wendland p. 107, 4 ff. and many others.

Also for the practical construction of the instrument information can be obtained from the comparison of Philoponus and Sebokht. The discussion of Nos. 3 and 15 of Philoponus and of Severus XVIII presuppose that the equator and the summer solstitial circle were only drawn in the part below the horizon. This is a rare feature in preserved samples.[30]

7. Terminology. It is well known that "astrolabe" in the Almagest always means the "armillary sphere."[31] The same holds for the "Hypothyposis" of Proclus,[32] who follows the Almagest closely. That Theon's "little astrolabe" denotes the "plane astrolabe" has been seen already by Lippert.[33] Philoponus, following his teacher Ammonios,[34] already uses "astrolabe" alone, as is customary in the Middle Ages.

8. The Time before Ptolemy. According to the Fihrist, "Ptolemy was the first who made a plane astrolabe, though there are people who say that (plane astrolabes) were made before his time, though this cannot be stated with certainty."[35] This statement of the Fihrist still characterizes the situation. The strongest argument for assuming at least the knowledge of the stereographic projection for Hipparchus seems to me to lie in the structure of Ptolemy's "Planisphaerium" itself. If this work would have been written in order to explain the construction of an astrolabe, it would have sufficed to give the geometrical construction of the centers and radii of the circles involved.[36] Actually, however, Ptolemy everywhere proceeds to a numerical computation, not only of the position of the centers and the length of radii but he also computes rising times which would be given directly by the instrument. Also the accuracy of the computation exceeds by far the practical need for a mechanical execution. Thus it is evident that the main purpose of the "Planisphaerium" is to demonstrate how the problems of spherical astronomy could be solved by means of plane trigonometry alone. We know on the other hand that Hipparchus had no spherical trigonometry at his disposal, as is clear from the Spherics of Theodosius, but that he nevertheless was able to deal successfully with problems of spherical astronomy. Thus it is very plausible to assume that he knew the use of stereographic projection. This is confirmed by the often quoted statement of Synesius, a pupil of Theon's daughter Hypatia, that Hipparchus was the first who investigated the unfolding of a spherical surface (σφαιρικῆς ἐπιφανείας ἐξάπλωσις).[37] It seems to me mere speculation, however, to assume Archimedes or Apollonius to have been the inventors of this method because the main theorem, the preservation of circles, can be proved by means of theorems from the "Conic Sections."

9. Vitruvius's Description. The history of instruments, based on stereographic projection, in the time before Ptolemy, has as one secure foundation Vitruvius's description[38] of an astrolabic clock in which the ecliptic is moved around once a day by means of water. In this case the ecliptic is no longer the rim of a "spider" but of a complete disc engraved with the zodiacal signs and the constellations north of the ecliptic. A fragment of such a disc from the first or second century A.D. has actually been found in Salzburg, and shows clearly the difference in size of the zodiacal signs.[39]

---

[30] Cf., e.g., Gunther, AW I pl. XXV f.
[31] For details cf. Tannery, Mém. Sci. 4, p. 244 or RE 2, 1799, 27 ff. and Rome [1] p. 79.
[32] Chapter VI, ed. Manitius p. 198 ff.
[33] Lippert, Stud. p. 43. Quoted Rome CA p. 4, note.
[34] His work "On the construction of the horoscope and the astrolabe" is still preserved in the Serai (Deissmann, FFS p. 62) but unpublished. [Added in proofs: see, however, below Section 14]
[35] Suter, Fihr. p. 41. The introductory sentence of this chapter of the Fihrist is rendered by Gunther, Early Science in Oxford, II p. 189 as follows: "There were Astrolabes in early

times (even Planispheres)." This is a mistranslation of Suter's sentence: "Es waren die Astrolabien in der frühern Zeit eben (Planisphärien)."
[36] Ptolemy's "Analemma" is a good example of how he proceeded directly from the geometric construction to instrumental application when this was his goal.
[37] Cf. above note 4.
[38] Vitruvius, De arch. IX,8 ed. Krohn p. 220,13–221,10. It is to be regretted that the translation in the Loeb series is based on a book written in 1584 and disregards the results of Bilfinger, Rehm, Diels and others.
[39] Diels, AT (3) pl. XVII and p. 213 ff.

The circumference was perforated with 15 holes for each sign. The rest of the instrument was restored by Rehm [40] from Vitruvius, who describes the hour-lines and the circles of declination as wires mounted in front of the rotating zodiacal disc. In other words, the "spider" is now the fixed system of curves which we find engraved on the discs of the ordinary astrolabe. The position of the sun is indicated by a peg (bulla) which is moved by hand into the hole which corresponds to the longitude of the sun. Thus it is certain that the mechanical execution of an astrolabic clock was known at the beginning of our era.[41]

The term "spider" ($\dot{\alpha}\rho\dot{\alpha}\chi\nu\eta$) for a fixed or movable network is not sufficient evidence for the simultaneous use of stereographic projection. Thus nothing can be concluded from Vitruvius's remark [42] that "the astronomer Eudoxus or, as some say, Apollonius" had invented the spider. His reference [43] to a "conical spider" shows the more general use of the term.

APPENDIX

In the following I discuss special sections of the sources, insofar as they concern the history of the astrolabe.

10. Ptolemy, Planisphaerium (Latin trsl. of lost Arabic trsl. of lost Greek original, ed. Heiberg; German trsl. Decker [1]).

Like all works of Ptolemy, the "Planisphaerium" is written according to a very carefully planned scheme. We can distinguish four main sections. Section I, Nos. 1 to 3: introduction about the general method of stereographic projection. Section II, Nos. 4 to 13: the zodiac, the rising times, both in graphical and numerical treatment. Section III, Nos. 14 to 19: representation of the ecliptic coordinates within the greatest of the always invisible circles. Section IV, No. 20: practical execution of a planisphaerium.

Sections III and IV are not too well preserved and already caused difficulties for the Arabic translator.[44] It might be tempting to assume that the fourth section originally also contained rules for the practical construction of an astrolabe. Yet I think this hypothesis can be rejected by the observation that this would imply the addition of new theoretical sections describing the construction of the circles of equal altitude and of the hour-curves. Actually Ptolemy describes in Section IV the construction of a star map, which is bounded by the greatest of the always invisible circles and in which the curves of equal latitude are drawn. This ecliptic system seems to have been combined with an equatorial system, which has the advantage that all meridians are mapped into straight lines. A similar mixture of ecliptic and equatorial coordinates is known from Hipparchus.[45] It seems to me that it is possible that there exists a relationship between the method of stereographic projection and this particular choice of coordinates which are readily given by the planisphaerium. A reflex of a discussion about the merits of the two different coordinate systems is still to be found in Synesius.[46] He also uses the greatest always invisible circle as the boundary for his instrument.[47]

Nos. 10 ff. Rhodes ($\phi = 36°$) is the standard example, not only here but also

---

[40] Rehm [1]; see esp. p. 43 fig. 19.
[41] I disregard here the so-called "astrolabe of Antikythera" because I am not sure that this instrument is based on stereographic projection. According to Theophanidis (Praktika tes Akademias Athenon 9 (1934) p. 140–149) it is a fragment of a planetarium, using cogwheels and eccenters. Stereographic projection, however, does not preserve circles which lie in the plane of the ecliptic but do not have the same center.

Also the date of this instrument is rather uncertain. Cf. for the literature Hartner [1] p. 2531 note 3.
[42] IX,8 ed. Krohn p. 218,4 f.
[43] p. 218,10.
[44] Cf., e.g., *Isis 9*, p. 271 note (2).
[45] For a careful discussion of Hipparchus's stellar positions cf. Vogt [1].
[46] Cf. below p. 250.
[47] Cf. below p. 250.

in the Almagest, as Ptolemy himself remarks (p. 242,2).[48] The same example is chosen by Severus Sebokht, i.e. Theon.[49]

For the obliquities of the ecliptic are given: p. 229,10 "partibus 23 punctis fere 51"; p. 234,6 "partibus 23 punctis 51 secundis 20"; p. 259,13 "24 fere gradibus." The second value is the accurate value which is also used in the Almagest (I,15; cf. also I,12). The two other values are called "approximately" (fere = ἔγγιστα).[50] For the last value an explicit motivation is given by the fact that 24 has the same divisors 2, 3 and 6 as 30°. Thus the same number of divisions can be used in the practical execution for zodiacal signs and equatorial degrees. This is the parallel to the μονομοιριαίοις, διμοιριαίοις, τριμοιριαίοις ἀστρολάβοις in Philoponus [51] and the διπλοῦς and τριπλοῦς instruments of Sebokht.[52] This shows that it is wrong to consider the occurrence of the value ε = 24° as a criterion of high antiquity.[53]

11. Synesius (Migne PG 66,1577–1588; English transl.[54] Fitzgerald, Syn., p. 258–266).

In the following I shall collect the astronomical passages from the letter of Synesius to Paeonius. Though the translation by Fitzgerald is very useful I think that a better understanding can be reached in several passages. The references to the text are to the edition in Migne PG 66.

1581 D/1584 A. "Astronomy itself is a venerable science, and might become a stepping stone to something more august, a science which I think is a convenient passage to mystic theology, for the happy body of heaven has matter (ὕλη) underneath it, and its motion has seemed to the leaders in philosophy to be an imitation of mind. It proceeds to its demonstrations in no uncertain way, for it uses as its servants geometry and arithmetic, which it would not be improper to call a fixed standard of truth.[55] I am therefore offering you a gift most befitting for me to give, and for you [Paeonius] to receive. It is a work of my own devising, including all that she [Hypatia], my most revered teacher, helped to contribute, and it was executed by the best hand to be found in our country in the art of the silversmiths."

1584 B, C. "The unfolding of a spherical surface, preserving identity of relations (λόγων) in difference of figures (σχημάτων), Hipparchus of olden times darkly discussed (ἠνίξατο), and he was the first to direct his energies to this question. But we, if it is not more than it befits us to say, have finished the weaving of this tissue even to the fringes, and have perfected it. The problem had been neglected in the long intervening time. The great Ptolemy and the divine band of his successors were content to have it as their one useful possession, for the 16 stars made it sufficient as instrument to know the hours of the night (νυκτερινὸν ὡροσκοπεῖον). Hipparchus transposed only these stars and inserted them in the instrument (τῷ ὀργάνῳ)."

This is the locus classicus for ascribing to Hipparchus knowledge of the astrolabe.[56] Actually it is very difficult to interpret this report of Synesius intelligently. Only two essential steps are possible in the development of the astrolabe. The first is the discovery that stereographic projection preserves circles; the second is the use of the stereographic projection of the celestial sphere in an instrument which imitates the rotation of the celestial sphere with respect to the given horizon. The first step is the only difficult one. After it has been made, the construction of the centers and radii of the projections is a simple matter. Thus, if Hipparchus had stereographic projection at all, he had at his disposal also the remaining contents of the "Planisphaerium."

---

[48] Cf. also Theon's commentary to the Almagest II, 2 (ed. Rome p. 614,5 ff.) ; Delambre HAA II p. 569.

[49] A Byzantine astrolabe of 1062 shows this climate as the permanent disc. Cf. Dalton [1] p. 138 f.

[50] In the "Analemma" we find the approximation "partes proxime 23 dimidiam et tertiam" (ed. Heiberg p. 215, 14f.) i.e., 23;50°. Cf. also "Geography" 1,24. The "Handy Tables" use 23;51° (cf. Delambre HAA II p. 629).

[51] Hase [1] p. 131.

[52] Nau [1] p. 89.

[53] Nau [1] p. 298 (1).

[54] The German translation by B. Kolbe, "Der Bischof Synesius von Cyrene als Physiker und Astronom" (Berlin 1850) is without value. Also W. S. Crawford, "Synesius the Hellene" (London 1901) contributes nothing to the understanding of the section on the "astrolabe."

[55] This is taken from the introduction to the Almagest (Heiberg p. 6, 17 ff.).

[56] Cf., e.g., Delambre, HAA II p. 453.

Synesius's remark that Hipparchus "darkly discussed" the problem can therefore at best mean that Synesius had difficulties in understanding a writing on the subject by Hipparchus. What he says about Ptolemy is rhetorical nonsense, and his own contributions cannot have been significant because Ptolemy's "Planisphaerium" contains all the necessary theory and the clock described by Vitruvius and Theon's work on the "little astrolabe" contain everything needed for the use of the instrument. There is simply no room left for any essential contribution by Synesius either to the theory of stereographic celestial maps or to the astrolabe as an instrument.

It has often been stressed that Synesius describes not an astrolabe but only a celestial map. Though the name "astrolabe" does not occur in his letter he consistently speaks about an "instrument" (ὄργανον), a term which never means a stationary map. Whatever the case may be, it is clear from our passage that he knew of an instrument used for the determination of hours and that he thought that Hipparchus had already used it. His mention of 16 main stars is paralled by "17 or more stars" in Philoponus (No. 8). A Byzantine astrolabe of 1062 [57] shows 14 stars; [58] an Arabic instrument of about 950 gives 15 stars. [59] Nicephoros Gregoras (about 1330) says that no more than 12 stars should be shown on the spider "in order not to obscure the background of the parallels on the discs," [59a] and this is repeated by Isaac Argyros (1367/8). [59b] Preserved drawings show 13 stars in a MS of Nicephoros Gregoras and 9 stars in Isaac Argyros. [59c]

1584 D/ 1585 B. "Considering the problem of unfolding worthy of study for its own sake we worked it out and composed a treatise and studded it thickly with the necessary abundance and variety of theorems." [Obviously a theoretical treatise now lost, which accompanied the gift to Paeonius, is alluded to here.] "Then we hastened to translate our conclusions (λόγους) into material form (ὕλη) and produced a beautiful image of the cosmic width. This very approach gives us the means of cutting a plane surface and a uniform cavity (ὁμαλὴν κοίλην) in the same relation (εἰς τοὺς αὐτοὺς λόγους). And as we think that any sort of cavity is more nearly related to the completely spherical, we have hollowed the breadth by pressing it in (?) and we have taken care of the rest such that the appearance of the instrument may remind the intelligent observer of the reality."

This passage causes great difficulty because it indicates that Synesius represented the celestial sphere not on a plane but on a concave surface, thus spoiling the basic idea of the whole method. The following passages, however, fit rather well the ordinary stereographic projection. Thus Synesius might have modified stereographic projection by transferring it simply to a shallow concavity. "For those stars which are classified in six magnitudes we have entered, preserving their relative alignments." This passage is decisive for the interpretation of the "instrument." The entry of stars down to the sixth magnitude involves, according to the Almagest, more than 1000 stars, whereas an astrolabe will rarely go much beyond 40 or 50 stars. "And of the circles, some we drew (completely) round, some we drew across." The first type of circles probably consists of the concentric circles of the equator and its parallels, the second type belongs either to the ecliptic system or to the horizon system, both of which intersect the first group. "But all (circles) we divided by degrees, making the 5-degree lines larger than the (single) degree lines." Actually no division into degrees is needed except for the outer rim (equatorial degrees) and for the ecliptic if one does not assume also azimuths indicated on the horizon and its parallels. "We also

---

[57] Dalton [1]; also Gunther AW I p. 104 ff. It is incorrect to say that this astrolabe shows no Arabic influence. The little bird above λύρα is the "falling eagle" of the Arabic astrothesy (cf. Gundel in RE 13, 2491, 29 ff.). The constructor of this astrolabe calls himself "Sergius the Persian, holding a consul's rank." Also the stylistic argument is not valid. Two early Arabic astrolabes (about 950 and 990 A.D.) show a design almost identical with the Byzantine astrolabe (Gunther, AW I, Pl. LII).

[58] Dalton is wrong in saying that 9 stars belong to the northern, 5 to the southern hemisphere. The astrolabe shows clearly that both Betelgeuse and Procyon (represented by the two middle pointers on the bridge) lie correctly north of the equator. Gunther, AW I p. 105 has taken over this error.

[59] Gunther, AW I p. 232.
[59a] Delatte, AA II p. 204,19 ff.
[59b] Delatte, AA II p. 250,9 ff.
[59c] Delatte, AA II figs. 4 and 8 respectively.

enlarged the inscriptions indicating the numbers at these lines; and what is below is on silver, the black giving the appearance of a book. But they were not all cut equidistant (ὁμόστοιχος), either individually or with each other. Some are cut equal in size, others irregularly and unequally as to appearance but actually (τῷ λόγῳ) regular and equal." Indeed, equatorial degrees are mapped equidistant, whereas ecliptic degrees are azimuths appear not equidistant. "This was a matter of necessity, so that the different figures should agree. By this reason also the great-circles, drawn through the poles and the signs of the tropics, though remaining actually (τῷ λόγῳ) circles, have become straight lines by the shift of viewpoint." This shows that we are dealing with stereographic projection from the south pole. "The antarctic circle has been inscribed larger than the great-circles and the relative distances of the stars have been lengthened according to this projection (κατ᾽ ἐκεῖνο τῆς ἐξαπλώσεως)." We see here that Synesius's instrument was bounded by the "antarctic circle," i.e., the greatest of the always invisible circles, as discussed by Ptolemy in the "Planisphaerium" No. 14 and by Sebokht chapter XXIV.

Synesius now makes some remarks about epigrams written in the space free of stars outside of the image of the antarctic circle. One of these epigrams is usually ascribed to Ptolemy.[60] The other was written by Synesius and "it seeks to tell the astronomer alone what advantage he can get from the instrument."

Then he continues (1585 C/ 1588 A):

"It (the instrument) professes to show the positions (ἐποχάς) of the stars not with respect to the zodiac but with respect to the equator. For it has been shown by geometric construction (διὰ τῶν γραμμάτων) that this is impossible. And it says that the declinations (λοξώσεις) are given, I mean of the parts of the zodiac with respect to the equator. And for all of them the rising times (συναναφοράς), that means the number of zodiacal degrees which cross with a number of equatorial degrees the same horizon."[61]

Here are described problems which can be solved with the "instrument." One concerns the declination of points of the ecliptic. This problem is discussed by Theon (No. 14), by Philoponus (No. 15), and by Severus Sebokht (chapters XXII, XXIII). In the astrolabe it is solved by means of the intersections of the circles of equal altitude with the meridian. By turning also the corresponding point of the ecliptic into the meridian, the difference in altitude towards the equator can be read on the instrument; and for the meridian, differences of altitudes and of declinations coincide. The second problem concerns rising times, a problem again known from the treatises on the astrolabe (cf. above p. 244) as well as from Ptolemy's "Planisphaerium" (Nos. 8 to 13). It can be solved directly by the astrolabe by measuring on the outer rim the equatorial degrees which correspond to the motion of a zodiacal sign across the eastern horizon. Both problems speak very distinctly in favor of seeing in the "instrument" of Synesius an astrolabe, just as the reference to the stars of sixth magnitude points to a star map. The only way to reconcile this apparent contradiction seems to me the assumption that Synesius's instrument was of the type of the Vitruvian clock. If the movable "spider" represented the horizon system, then the background could carry a detailed star map. This map contains the whole visible part of the sky, indicating the ecliptic, the equator and its parallels. The circumference of the "spider" represents here the horizon which touches the outer rim of the instrument, the antarctic circle, and a small inner circle, the greatest always visible (or arctic) circle. This might also be the explanation of the statement that the stars are referred to the equator, not to the zodiac, though the wording of the text, taken literally, makes no sense because there is nothing impossible in the use of ecliptic coordinates.

The letter concludes with Synesius's poem, which contains nothing of interest,

[60] Anthol. Pal. IX, 577.

[61] The text has here, erroneously, "equator" instead of horizon.

except that it mentions again the uneven spacing on equal circles, the declinations of the zodiac, and the "centers" at noontime. This again points toward an instrument with movable parts and not to a permanent star map.

It should finally be remarked that Nicephoros Gregoras (about 1330) several times refers to Synesius as the author of a treatise on the "astrolabic instrument" while in his own letters he follows closely the letter of Synesius to Paeonius.[61a] Nicephoros states that his own treatise on the astrolabe (cf. below section 14) was guided by Synesius's work whereas Philoponus's writings were lost and no work in Greek or a "Barbaric" language was known to him.

12. Severus Sebokht (Syriac text and French translation: Nau [1]; English translation of the French translation: [62] Gunther AW I p. 82–103).

Introduction. Nau p. 92 it is stated that the disc represents the whole visible hemisphere above the horizon. This is correct for Ptolemy's planisphaerium, which is bounded by the greatest always visible circle and for the instrument of Synesius. The astrolabe of Sebokht, however, is bounded by the winter solstitial circle. His error is probably due to the condensing of a more complete discussion in Theon.

The interest in the astrological "centers" (the culminating points and the rising and setting points of the ecliptic) is very visible in Sebokht but is also seen in Philoponus and Synesius. It is therefore plausible to assume that already Theon mentioned the "centers."

Ch. I, Example (Nau p. 275 ff). For climate IV ($\phi = 36$) longitude ♋ 0° and altitude 30° of the sun one finds on the instrument: ♌ 5 rising, ♒ 5 setting, ♈ 24 in upper culmination, ♎ 24 in lower culmination; 3rd hour of daylight. Sebokht likes to check his results from tables. Here this is condensed in the one short sentence that if one transfers the rising and setting points to sphaera recta, then one obtains ♈ 24 as culminating point. What is meant by this sentence is the following. Put ♌ 5 and ♒ 5 on the eastern and western horizon respectively. It has been found that the index of the spider (the point ♑ 0°) has moved $112° = 90° + 22°$. Consider the meridian line as horizon of sphaera recta. Then the index has moved 22° across the meridian of sphaera recta, or 22° is the is rising time of ♈ 0° at sphaera recta. From the tables (e.g. Almagest II,8) one finds that ♈ 24 has this rising time;[63] thus this point is in the meridian, as is found by the instrument.

Ch. VI, Latitude of the Moon. (Nau p. 280 f.) The method is in practice very inaccurate. Also theoretically the meridian is not the correct separating line, which is a curve determined by the points of contacts between ecliptic and circles of altitude.

Ch. VII, Check of the Astrolabe (Nau p. 281 ff.). First method. Let λ be the given longitude of the sun. The "Tables of Ptolemy," obviously the "Handy Tables," give for each climate three columns. Column 1 gives λ. Column 2 gives the rising time $\rho(\lambda)$ counted from ♈ 0°. Column 3 gives the corresponding length $h(\lambda)$ in degrees of a seasonal hour. If $c(\lambda)$ denotes the corresponding length of daylight, then

$$c(\lambda) = \rho(\lambda + 180) - \rho(\lambda)$$

holds; consequently the following identity must be satisfied:

$$\rho(\lambda + 180) = 12 \cdot h(\lambda) + \rho(\lambda).$$

The right-hand side can be found in the tables, the left-hand side on the instrument. The two values should agree within one degree.

Second method. Find the rising times of single signs for sphaera recta by measuring the movement of the index while the sign passes the meridian. The result should agree with the table for sphaera recta.

[61a] Bezdeki [1] p. 308,32 to p. 309,9 and p. 252,16 to 32. French translation Guilland, Corresp. p. 249 f. and p. 95 f.

[62] Adding some errors.
[63] More accurately 22;11.

It is at the end of this chapter that Sebokht says that these methods control simultaneously the astrolabe and the tables because "le canon de Ptolémée est fait d'après l'astrolabe" as Nau p. 283 translates. From this Nau concluded (p. 62) that the astrolabe existed before Ptolemy. Though I do not doubt this fact as such, I think that the passage in question cannot be taken so seriously. The particle *mn* means "from" and is used of places, of origin (descended from), of time (from the beginning) etc.[64] but not strictly in the sense of "after." When Sebokht says that the tables were made "from the astrolabe," he obviously means that they were made by measuring the rising times directly on the instrument. Though this is, of course, possible, in principle, it is nevertheless historically incorrect because we know from the Almagest that Ptolemy computed at least the main entries by means of spherical trigonometry.

Ch. IX. Rising times. The special case of sphaera recta (Nau p. 285/286) uses again, as in the second method of Ch. VII (above p. 251), the meridian for the horizon of sphaera recta. This shows that originally the meridian was the only straight line engraved on the disc whereas all preserved astrolabes, at least to my knowledge, show also the perpendicular straight line.

Ch. XIV and XV. Geographical Longitude (Nau p. 292 f.). This chapter is somewhat in disorder. The reference to solar eclipses is wrong because only lunar eclipses can be used for the determination of the difference in local time.

The reference to a previous chapter must mean Ch. III. The moon can be used in both places to determine the hour of night. The text operates with culminating degrees of the ecliptic, probably in order to introduce equinoctial time.

In Ch. XV degrees measured by the diopter are taken instead of right ascensions measured by the index of the spider.

The example has nothing to do with the astrolabe but quotes the classical lunar eclipse of B.C. 331 Sept. 20 observed in Arbela at 5^h, in Carthage at 2^h (Ptolemy, Geogr. I,4,2).* Sebokht quotes, however, 6^h and 3^h respectively for reasons unknown to me.

Ch. XVI (Nau p. 292 f.). According to its title this chapter concerns the rising times for sphaera recta. Actually it deals with an identity which must hold for any climate. Using the notation of p. 251 one finds

$$6 \cdot h(\lambda) + \rho(\lambda) = \tfrac{1}{2}(\rho(\lambda + 180) + \rho(\lambda)) = 180.$$

Sebokht uses as example "the first degree of Cancer" which is, according to ancient terminology, the equivalent of $\lambda = \text{\reflectbox{S}}0°$. Sebokht quotes

$$h(\lambda) = 18;7 \qquad 6 \cdot h(\lambda) = 108;42 \qquad \rho(\lambda) = 72;22$$

which would give for $6 \cdot h(\lambda) + \rho(\lambda)$ the result 181;4 and not 180 as he states. The values quoted are found in Theon's "Handy Tables" for $\text{\reflectbox{S}}1°$.[65] The deviation from the expected identity is due to the inaccuracy of the interpolation in these tables.

Ch. XVIII. Geographical Latitude (Nau p. 294). The rule is as follows. Put $\Upsilon 0°$ on the meridian; then the line of altitude passing through this point gives $90 - \phi$. This shows that the astrolabe had originally not drawn the equator above the horizon. This is confirmed by chapters XXII and XXIII and Philoponus Nos. 3 and 15.

Ch. XIX. Latitude of a star (Nau p. 296). This method erroneously identifies the zenith with the pole of the ecliptic. Here follows a lacuna of the original from which our present manuscript was copied. Also chapter XXI is obscure.

Ch. XXII and XXIII. Declination (Nau p. 297 ff.). The method followed shows that neither equator nor winter solstitial circle was drawn above the horizon.

The text says erroneously that $\Upsilon 0°$ should be placed on the meridian below the

---

[64] Cf. J. Payne Smith, A compendious Syriac dictionary, Oxford, Clarendon, 1903 p. 280. Cf. also Nöldeke, Syrische Grammatik § 249.

[65] Halma, Tables manuelles astron. de Ptolémée et de Theon II p. 28/29.
* Pliny NH 72: 8^h and 6^h.

earth (instead of above) and that ♋0° moves on the winter solstitial circle south of the earth.

Ch. XXIV. The 5 zones (Nau p. 299 ff.). The zones for $\phi = 36$ are:

1. Arctic zone: $36°$ to both sides of the North pole
2. Summer tropic: $30°$
3. Equatorial zone: $48°$ ($= 2\epsilon$)
4. Winter tropic: $30°$
5. Antarctic zone: $36°$ ($= \phi$).

This division is mentioned once more as made "by the Philosopher" i.e., Theon. A "third" division is ascribed to Ptolemy. Obviously the only difference is that $\epsilon = 24$ is replaced by $\epsilon = 23;51$. Sebokht, however, modified the values in the wrong way.

Sebokht: 1   twice 36;9°   instead of: twice $\phi = 36$
                 2   30° $90 - (\phi + \epsilon) = 30;9$
                 3   twice 23;51
                 4   30°   30;9
                 5   36;9°   36

The value $\epsilon = 23;51$ occurs also in Ptolemy's "Planisphaerium" (cf. above p. 248).

Ch. XXV. Nau p. 302 f. This chapter again concerns rising times and is clearly out of place, perhaps a later addition. The rule given seems to state that

$$h(\lambda) = 90 - \rho(\lambda) = a(\lambda)$$

where $a(\lambda)$ is the right ascension of λ. I do not see how to interpret the text correctly.

13. Philoponus (Greek text badly edited by Hase [1]; additions: Tannery, Mém. sci. 4, p. 241–260; French trsl.: Tannery, Mém. sci. 9, p. 341–367; German trsl.: Drecker [2] [66]).

No. 3. (Tannery p. 346 f.) Here it is explicitly stated that "some instruments" show equator and summer solstitial circles only below the horizon. We have concluded this already from Sebokht (cf. above p. 252). Philoponus does not realize that the rule which he gives immediately afterwards for determining the obliquity of the ecliptic (assumed to be $24°$) presupposes that these two circles are not drawn in the upper part of the disc. Cf. also No. 15.

At the end of this chapter (Tannery p. 347) Philoponus remarks that "in some astrolabes" the outer rim (ἡ ἔξωθεν ἴτυς) is divided into $360°$. At first sight it seems paradoxical to assume that some astrolabes should not have had this division. The explanation is found in ch. 10 where it is explained in detail that the "one-degree instruments" do not have a holder (δοχεῖον) and that consequently the circumference of the disc itself is divided into $360°$.

No. 6 (Tannery p. 351). Here Philoponus refers to Ptolemy as having drawn the hour lines in the lower hemisphere only. This again proves that Ptolemy knew the astrolabe.

The lengthy discussion of the determination of fractions of hours is in all probability Philoponus's own contribution. One example is incorrect; cf. Drecker [2] p. 31 note 12.

No. 9 (Tannery p. 356 f.). This section concerns observation near the meridian and is probably Philoponus's invention. The example, ♈20° observed in an altitude of $70°$, supposes a geographical latitude of about $28°$.

No. 10 (Tannery p. 358). Here Philoponus speaks about the rising time of Scorpio in the third climate, i.e., Lower Egypt ($\phi = 30;22$). This is obviously an example of his own. No consequences, however, are drawn from his dates.

---

[66] The English translation published in Gunther AW I p. 61–81 is based on Hase's text only. The translator was not familiar with the astronomical terminology and thus rendered many passages wrongly.

No. 13 (Tannery p. 361 ff.). The problem concerns the determination of the longitude of the sun near the solstices. Again we are dealing with an addition of Philoponus himself. The example assumes that $\mathfrak{S}0°$ culminates at $90°$, but $\mathfrak{S}10$ and $\mathcal{X}20$ at $87°$. The first value would require $\phi = 24$, the second $\phi = 26;30$. Obviously Philoponus was not aware of the implications of the numbers he invented for his examples.

No. 15 (Tannery p. 366). The obliquity of the ecliptic is again assumed to be $24°$. The method employed for the determination of declinations assumes again that there is no equator drawn in the upper hemisphere (cf. p. 252).

14. Pseudo-Ammonios.[67] In note 34 I have referred to an unpublished treatise of Ammonios, preserved in the Serai. Through the kindness of Prof. Ritter, Istanbul, I have obtained photographs of this text.[68] Though it is called "Treatise of the astronomer Ammonios on the construction of the horoscope or the astrolabe," its author is beyond any doubt Nicephoros Gregoras. This is proved by comparison with Delatte's edition of the latter author's earlier work on the astrolabe (written before 1335). Indeed MS-*S* of Delatte is also ascribed to Ammonios whereas at least four other MSS give Nicephoros Gregoras as the author. There can be no doubt that he is the real author. This follows not only from the correspondence of Nicephoros, quoted by Delatte, but also from the longitudes of Altair and Regulus[69] whose values agree only with A.D. 1300 but not with the time of Ammonios, about A.D. 500. Also the constant use of Byzantium as geographical location excludes Ammonios.

I cannot offer any explanation for the strange substitution of the name Ammonios for Nicephoros. From Nicephoros's letters it follows that he utilized Synesius.[70] Ammonios is nowhere mentioned. Especially in his letter to Cavasilas,[71] Nicephoros follows almost verbatim the letter of Synesius to Paeonius. Thus it is not impossible that his treatises on the astrolabe reflect details from the lost treatise of Synesius. The small number of stars to be shown on the spider might be explained in this way.

The following concluding remarks are intended to illustrate the relation of the new MS (henceforth called *T*) with the MSS utilized by Delatte. I have counted 71 deviations of *T* from Delatte's text (the scholia excluded). Among those one is in common to each of the following MS: codd., *ABOC, ABOS, C, OC*. Four are in common with *O*, 12 with *S*, 14 with *CS*. Thus it is clear that *T* is closely related to *CS* and especially to *S*, with which it shares not only the reference to Ammonios but also the scholion 2, which forms the end of *T*. Among the deviations of *T* from all other MSS, I quote only the following ones which are of interest. Delatte p. 195, ⅓ ἀστονόμου] ἀστρονομικοῦ *T*; 196,20 περιφερείαν] σφαῖραν *T*; 204,7 τοῦ Ζυγοῦ] τῶν χηλῶν *T*. This terminology also points to a much older source. Otherwise the zodiacal signs are written by symbols, not in words, with the exception of the quoted case and 203,14 κριοῦ. As figures, there are given only Delatte figs. 2 and 3, the first one being more elaborate than the one reproduced by Delatte but closely related to Delatte's fig. 6 from Isaac Argyros.

Because scholion 2 is only known from *S*, I give all deviations of *T*. Delatte p. 210,12 post διαγράψαι add. διὰ τό; 210,12 κύκλων om.; 210,14 δύο]β′; 210,23/30 δωδεκατημορίων] ιβ′τημορίων; 210,32 αὐτοῦ — τοῦ om.

BIBLIOGRAPHY

Bezdeki [1] St. B.—, *Nicephori Gregorae epistulae XC.* Ephemeris Dacoromana, Annuario della scuola Romena di Roma 2 (1924) p. 239–377.

Dalton [1] O.M.D.—, *The Byzantine astrolabe at Brescia*, Proc. of the British Academy, 1926, p. 133–146.

---

[67] Added in proofs.
[68] Topkapi Sarayi Ahmed III, Deissmann 19, fol. 282–291.
[69] Delatte, AA II p. 208,5 and 10.

[70] Quoted by Delatte, AA II p. 190 ff. Cf. also Guilland, Corresp.
[71] Guilland, Corresp. No. 155 (p. 249). Greek text: Bezdeki [1] p. 308,32 to 309,9.

Deissmann, FSS Adolf D. — , *Forschungen und Funde im Serai*, Berlin-Leipzig, De Gruiter, 1933.

Delambre, HAA D. — , *Histoire de l'astronomie ancienne*, Paris 1817; 2 vols.

Delatte, AA II Armand D. — , *Anecdota Athenensia et alia. Texts grecs relatifs à l'histoire des sciences* (= Bibliothèque de la faculté de philosophie et lettres de l'université de Liége, Fasc. 38) Liége-Paris, 1939.

Diels, AT Hermann D. — , *Antike Technik* (³), Teubner, Leipzig-Berlin, 1924.

Drecker [1] J.D. — , *Das Planisphaerium des Claudius Ptolemaeus*, Isis 9 (1927) p. 255–278.

Drecker [2] J.D. — , *Des Johannes Philoponus Schrift über das Astrolab*, Isis 11 (1928) p. 15–44.

Fihrist see Suter, Fihr.

Fitzgerald, Syn. Augustine F. — , *The letters of Synesius of Cyrene*, Oxford Univ. Press, 1926.

Guilland, Corresp. R. G. — , *Correspondance de Nicéphore Grégoras*. Collection Byzantine, Association Guillaume Budé, Paris, Société d'édition "Les Belles Lettres", 1927.

Gunther, AW Robert T. G. — , *The astrolabes of the world*, Oxford Univ. Press, 1932; 2 vols.

Halma, Tab. man. N.H. — , *Commentaire de Théon d'Alexandrie, sur les tables manuelles astronomiques de Ptolémée* [and similar titles for parts II and III] Paris 1822, 1823, 1825.

Hartner [1] Willy H. — , *The principle and use of the astrolabe*, in: Arthur Upham Pope, *A survey of Persian art*, vol. III p. 2530-2554 (Pls. 1397-1402), Oxford Univ. Press, 1939.

Hase [1] H.H. — , *Joannis Alexandrini, cognomine Philoponi, de usu astrolabii ejusque constructione libellus*, Rheinisches Museum für Philologie (²) 6 (1839) p. 127-171; [also as separate publication, Bonn 1839. I have not seen this special publication].

Honigmann, SK Ernst H. — , *Die sieben Klimata*, Heidelberg, Winter, 1929.

Houzeau-Lancaster, Bibl. J.C.H. — et A.L. — , *Bibliographie générale de l'astronomie*, Bruxelles, Hayez, 1882, 1887, 1889; 2 vols.

Klamroth [1] M.K. — , *Ueber die Auszüge aus griechischen Schriftstellern bei al-Ja'qûbî. IV. Mathematiker und Astronomen*. ZDMG 42 (1888) p. 1-44.

Kolbe, Syn. Bernhard K. — , *Der Bischof Synesius von Cyrene als Physiker und Astronom*, Berlin, Stargardt, 1850.

Lippert, Stud. Julius L. — , *Studien auf dem Gebiete der griechisch-arabischen Uebersetzungsliteratur*, Braunschweig, Sattler, 1894.

Michel, TA Henri M. — , *Traité de l'astrolabe*, Paris, Gauthier-Villars, 1947.

Migne, PG J.-P.M. — , *Patrologiae cursus completus, series graeca.*

Nau [1] F.N. — , *Le traité sur l'astrolabe plan de Sévère Sabokht*. Journal Asiatique (⁹) 13 (1899) p. 56-101, 238-303; [also as separate publication, Paris, Leroux, 1899].

Nau [2] F.N. — , *Notes d'astronomie syrienne*, Journal Asiatique (¹⁰) 16 (1910) p. 209-228.

Philoponus see Drecker [2]
Hase [1]
Tannery [1] and [2]

Proclus, Hypotyp. *Procli Diadochi hypotyposis astronomicarum positionum*, ed. C. Manitius, Bibl. Teubneriana, Leipzig, 1909.

Ptolemy, Almag. *Claudii Ptolemaei opera quae exstant omnia*, vol. I, *Syntaxis mathematica*, ed. J. L. Heiberg, 2 vols., Bibl. Teubneriana, Leipzig, 1898, 1903.

Ptolemy, opera II *Claudii Ptolemaei opera quae exstant omnia*, vol. II, *Opera astronomica minora*, ed. J. L. Heiberg, Bibl. Teubneriana, Leipzig, 1907.

Ptolemy, Tetrab. *Claudii Ptolemaei opera quae exstant omnia*, vol. III,1, *Apotelesmatica*, ed. F. Boll and Ae. Boer, Bibl. Teubneriana, Leipzig, 1940.

Ptolemy, *Tetrabiblos*, edited and translated into English by F. E. Robbins. The Loeb Classical Library, 1940.

RE *Pauly's Real-Encyclopädie der classischen Altertumswissenschaft*. Neue Bearbeitung.

Rehm [1] O. Benndorf — E. Weiss — A. R. — , *Zur Salzburger Bronzescheibe mit Sternbildern*, Jahreshefte des österreichischen archäologischen Institutes in Wien 6 (1903) p. 32-49.

Rome, CA A.R. — , *Commentaires de Pappus et de Théon d'Alexandrie sur l'Almageste*. Rome 1931, 1936, 1943 [= Studi e Testi 54, 72, 106].

Rome [1] A.R. — , *L'astrolabe et le météoroscope d'après le commentaire de Pappus sur le 5ᵉ livre de l'Almageste*, Annales de la société scientifique de Bruxelles, sér. A, t. 47 (1927) p. 77-102.

Severus Sebokht see Nau [1]

Steinschneider [1] M.S. — , *Die arabischen Uebersetzungen aus dem Griechischen. II. Mathematik*. ZDMG 50 (1896) p. 161-219, 337-412.

Suidas *Suidae Lexicon* ed. Ada Adler, Leipzig, Teubner, 1928-1938; 5 vols. [= *Lexicographi graeci*, vol. I].

Suter, Fihr. Heinrich S. — , *Das Mathematiker Verzeichnis im Fihrist des Ibn Abî Ja'kûb an-Nadîm*, Abh. zur Geschichte d. Mathematik 6 (1892) p. 1-87 (= Z. f. Math. u. Phys. 37, Supplement).

Synesius see Fitzgerald, Syn.
Kolbe, Syn.

Tannery, Mém. sci. Paul T. —, *Mémoires scientifiques* Toulouse — Paris, 1912 ff.

Tannery [1] Paul T. —, *Notes critiques sur le traité de l'astrolabe de Philopon*, (1888). Mém. sci. 4, p. 241–260.

Tannery [2] Paul T. —, *Jean le grammarien d'Alexandrie*, (*Philopon*), *Sur l'usage de l'astrolabe et sur les tracés qu'il présente*, (1927). Mém. sci. 9, p. 341–367.

Theon Alexandrinus see Halma, Tab. man. Rome, CA.

Theophaneides [1] Bas. D. Th. —, *Enepi-graphon heliakon horologion ek Samou*, Archaiologikon Deltion 12 (1929) p. 236 f.

Vogt [1] H.V. —, *Versuch einer Wiederherstellung von Hipparchs Fixsternverzeichnis*, Astronomische Nachrichten 224 (1925), col. 17–54.

Wiedemann [1] Eilhard W. —, *Beiträge zur Geschichte der Naturwissenschaften. III.* Sitzungsberichte der Physikalisch–medizinischen Sozietät in Erlangen 37 (1905) p. 218–263.

ZDMG Zeitschr. d. Deutschen Morgenländischen Gesellschaft.

Reprinted from
ISIS
40 (3), 240–256 (1949)

THE ASTRONOMICAL ORIGIN OF THE THEORY OF CONIC SECTIONS

OTTO NEUGEBAUER

Professor of the History of Mathematics, Brown University

(Read November 20, 1947)

INTRODUCTION AND SUMMARY

Appolonius's theory of the conic sections (about 220 B.C.) is undoubtedly one of the masterpieces of ancient mathematics and will remain one of the great classics of mathematical literature. Very little, however, is known about the origin of the theory of conic sections as such. It is well known that the familiar names of these curves, ellipse, hyperbola, and parabola, originated from Apollonius's method of attack, which consists in applying the methods of "geometrical algebra" to the discussion of these curves. Apollonius obtains his curves by intersecting a fixed skew circular cone by a plane of variable angle. We also know that this approach is very different from the earliest known method to obtain conic sections. Menaechmus, a pupil of Eudoxus, is credited with the discovery of the conic sections (about 350 B.C.). These curves were obtained, however, by a very peculiar construction. The cone is a right circular cone; the intersecting plane is always perpendicular to one of the generating lines of the cone, and the three types of curves are obtained by varying the angle at the vertex of the cone.

The strange condition of perpendicularity of the intersecting plane always seemed to me to point to only one explanation, the theory of sundials. The generating line must be the "gnomon," the intersecting plane is the plane on which the shadow is cast. I shall demonstrate in the following section that a very simple type of sundial satisfies all requirements which lead to the above definition of conic sections.

THEORY OF THE SUNDIAL

In this section I shall develop the consequences of the following hypothesis. A gnomon of length 1 is adjusted in such a way that it always points at the sun when it culminates. The plane onto which the shadow is cast is perpendicular to the gnomon. We want to find the length of the shadow as function of time.

Let us first say that this arrangement has obvious advantages. In the case of the equi-

noxes the gnomon lies in the plane of the equator. The receiving plane always intersects the plane of the horizon exactly in the East-West line. The shadow at equinoxes travels a straight line from West to East. If α is the angle which the sun is distant from noon, then the length s of the shadow is simply $\tan \alpha$ and can be found graphically without any theory (*cf.* fig. 1). If

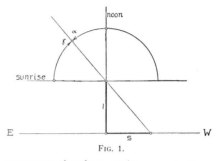

FIG. 1.

we measure time from sunrise, an angle $90 - \alpha = \gamma$ corresponds to the moment α before noon. Thus $s = \cot \gamma$. In the case of a solar declination different from zero the above description is no longer accurate. The deviation, however, remains obviously small because the adjustment of the gnomon towards the culminating point of the sun keeps the deviation of the shadow surface from the plane of the daily solar orbit within narrow limits. Hence we may expect that $s = \cot \gamma = \tan \alpha$ requires only small corrections. We furthermore see that the receiving plane needs only to be of small dimensions perpendicular to the main West-East path of the shadow. Finally the adjustment towards the culminating point is very easy to control; one must merely prevent the noon shadow from becoming visibly different from zero. Thus the whole construction is very simple in practical execution.

For our purposes, it is essential, however, to develop the accurate theory for arbitrary declination δ between $+ \epsilon$ and $- \epsilon$, ϵ being the obliquity

PROCEEDINGS OF THE AMERICAN PHILOSOPHICAL SOCIETY, VOL. 92, NO. 3, JULY, 1948
Reprint   *Printed in U. S. A.*

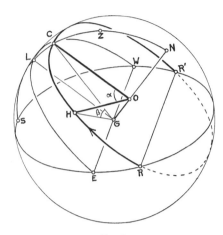

FIG. 2.

of the ecliptic. Figure 2 represents the celestial sphere with the shadow-casting point G of the gnomon as center. ELW is the equator, RESWR' the horizon, RHCR' the daily orbit of the sun, culminating at C. Hence CGL = δ the declination, and CG the direction of the gnomon. Assume the sun in H. The angle α, measuring the distance from noon is given by α = COH, where O is the center of the parallel circle RHCR'. The length s of the shadow is then given by s = tan β, where β = HGC is the angle between the ray HG and the direction of the gnomon GC. All we have to do is to find β as function of α.

In figure 3 the parallel circle RHCR' is once

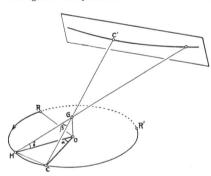

FIG. 3.

more represented—though, for the sake of simplicity, in horizontal position. A line perpendicular to its plane at the center O must meet the point of the gnomon G. Hence CG is the extended gnomon. From figure 2 it follows that OCG = CGL = δ.

It is now easy to express the angle β = HGC as function of α = COH. If r = OH is the radius of the parallel circle we have on the one hand (fig. 3)

$$ CH = 2r \sin \frac{\alpha}{2}. $$

Because $CG = HG = \dfrac{r}{\cos \delta}$, we have on the other hand

$$ CH = \frac{2r}{\cos \delta} \sin \frac{\beta}{2}. $$

Thus $\sin \dfrac{\beta}{2} = \cos \delta \cdot \sin \dfrac{\alpha}{2}$. Because $s = \tan \beta$ we compute also $\tan \dfrac{\beta}{2}$. By a simple computation one finds

$$ (1) \qquad \tan \frac{\beta}{2} = \frac{\cos \delta \sin \dfrac{\alpha}{2}}{\sqrt{1 - \cos^2 \delta \sin^2 \dfrac{\alpha}{2}}}. $$

This answers our question.

CONSEQUENCES

(A) Because cos δ = cos (− δ) we obtain the same shadow length for declinations symmetric to the equator, especially for δ = ε and δ = − ε. Our sundial shows equal shadow lengths for both solstices.

(B) Because the geographical latitude has no influence on δ, our sundial gives the same shadow lengths for all localities on the earth.

(C) Because shadows are only cast on the receiving plane when |β| < 90° we find for |δ| = ε the following limit for α. We substitute in formula (1) the value $\tan \dfrac{\beta}{2} = 1$ and find

$$ (2) \qquad \cos \alpha_0 = 1 - \frac{1}{\cos^2 \epsilon}. $$

Because 1/cos² ε > 1 we see that cos α₀ < 0 and therefore α₀ > 90°.

The angle c = ROC measures the half length of daylight for a solar declination δ. If c > α₀ the sundial does not operate for the interval from $\alpha = c$ to $\alpha = \alpha_0$. The farther we go north the longer is c for the summer solstice and the later after sunrise the sundial can be used. For latitudes in the eastern basin of the Mediterranean, however, this is of little interest.

(D) Because δ is small, thus $\cos \delta$ close to 1, we see from (1) that $\tan \frac{\beta}{2}$ is not very different from $\tan \frac{\alpha}{2}$. This confirms our expectation that the length of the shadow is in all cases close to $\tan \alpha$.

(E) Example. For $\epsilon = 23;50$ we obtain the following table for the shadows at summer solstice.

α	β	s
0°	0°	0
20	18;17	.33
40	36;28	.74
60	54;26	1.40
70	65; 3	2.15
80	72; 1	3.08
90	80;36	6.04
100	88;58	55.44
108	95;28	

The limiting angle is $\alpha_0 = 101°$ ($\beta = 90°$). The greatest possible value for α at Babylon is about 108°. Hence the sundial is in practice always usable. At the winter solstice, α reaches only the value of about 72°. At that time the shadow at sunrise begins with a length only slightly more than twice the length of the gnomon. At summer solstice the shadow at sunrise is 6. Thus we see that our sundial has the additional advantage of avoiding very long morning shadows for at least half of the year.

THE CONIC SECTIONS

It follows from figure 3 that the shadow-casting rays HG form the generating lines of a straight circular cone of slant δ. The shadow cone is the other sheet of this cone with the vertex G. The shadow of G falls onto a plane perpendicular to the generating line which continues CG; this continuation GC' is the gnomon.

Thus the shadow of G travels on a conic section one point of which is the foot C' of the gnomon. The form of the conic section depends on the slant δ or on the angle $180 - 2\delta$ at the vertex G, which is the only variable parameter for the discussion of the shadow curve.

This corresponds exactly to the definition of the conic sections given by Menaechmus. The only generalization beyond our description lies in the fact that δ is no longer restricted to the interval form $-\epsilon$ to $+\epsilon$ if we want to obtain more than hyperbolae. This generalization is trivial as soon as one investigates from a geometrical point of view the curves travelled by the shadow.

CONCLUSION

Though I feel confident that the above explanation gives the real motivation for the early Greek theory of conic sections, I must admit that I do not know of the existence of sundials of this type. The majority of Greek sundials with a plane as receiving surface assume either a horizontal or a vertical position of this plane. The only exception is one sundial in the British Museum whose receiving plane coincides with the equator plane,[1] while the gnomon points towards the North or South pole of the celestial sphere. The shadow travels on circles for all declinations different from zero, whose radius is the same for opposite signs of δ, again independent of the geographical latitude. This sundial does not work, however, for the equinoxes and, in practice, for a considerable interval before and after the equinoxes. One might say that this rather unpractical clock represents the other extreme of a development whose opposing end is indicated by the arrangement discussed in the preceding sections. The common basis of the theory of both clocks is an arrangement as given in figure 2 though interchanging the role of gnomon and receiving plane.

---

[1] *Cf.* Joseph Drecker, Theorie der Sonnenuhren, chap. V, Berlin, De Gruyter, 1925.

Estratto dalla **RIVISTA DEGLI STUDI ORIENTALI**

pubblicata a cura dei Professori della Scuola Orientale nella R. Università di Roma

VOLUME XXIV

ASTRONOMICAL FRAGMENTS IN GALEN'S TREATISE ON SEVEN-MONTH CHILDREN

R. Walzer published in vol. XV of this journal [1] an interesting article *Galens Schrift " Ueber die Siebenmonatskinder "* in which he made available, together with a German translation, the complete Arabic text of a short treatise of Galen which was until then only known in a fragmentary Greek version. This treatise contains several astronomical references, among which two concern the work of Hipparchus. Because a collection of fragments of the astronomical writings of Hipparchus is a much needed desideratum for the historian of ancient science it seems to me desirable to clarify as much as possible the details in these references.

From p. 347, and from p. 350, 214 ff. we know that according to Hipparchus half a year contains 182 days 15 hours " and a little fraction of about one 24 th of an hour ". This would correspond to a year of 365 1/4 + 1/288 days. The Almagest, on the other hand, has preserved for us a direct quotation from Hipparchus's book on the length of the year where it was stated that a tropical year contains 365 1/4 days *minus* " about 1/300 " of a day (Almagest III, 1, 207, 24 f., Heiberg). If Galen, as is plausible, was really quoting from Hipparchus, one must assume that he erroneously added instead of subtracting. On the other hand the assumption of a year of more than 365 1/4^d is not isolated in the Greek tradition. For instance Vettius Valens IX, 11 (353, 10 ff. Kroll = CCAG `5,2, p. 127, 17 ff.) quotes several values exceeding 365 1/4^d. Especially the Βαβυλώνιοι are mentioned with a year of 365 1/4 + 1/144^d. This is twice the excess mentioned by Galen and agrees exactly with the value for the sidereal year deducible from Hipparchus's constant of precession.

We now can turn to the question of the length of the synodic month. Both the Greek and the Arabic texts are corrupt where they state the excess of a (mean) synodic month over 29 1/2 days. The Greek text has (p. 354,19 f.)

$$\frac{1}{30} + \frac{1}{20} + \frac{1}{27.000} + \text{a small part}$$

---

[1] *Rivista degli Studi Orientali*, XV (1935), pp. 323–357.

O. Neugebauer

whereas the Arabic version gives (p. 342 lines 195 ff.)

$$\frac{1}{372} + \frac{1}{27.000} + \text{a very small part.}$$

Both texts agree that this accurate value is due to Hipparchus and that the small remainder is not worth mentioning.

Walzer followed Schoene in emending the τριαχοστόν (1/30) of the Greek text to τριαχοσιοστόν (1/300). He furthermore assumes a missing seven and thus restores

$$\frac{1}{327} + \frac{1}{27.000}.$$

The same restoration he obtains for the Arabic text by assuming a permutation of 27 into 72. He thus obtains for the mean synodic month the value $29^d\ 12^h\ 44.037^{min}\ 3\ 1/5^{sec}$, which he finds in sufficiently good agreement " mit dem von Rehm ... mitgeteilten Wert von $29^d\ 12^h\ 44^{min}\ 31/3^{sec}$ ". Unfortunately Walzer committed an error in determining the decimal point when he transformed 1/327 day into minutes. The result is not 44.037^{min} but 4.4037^{min}.

In order to restore the text correctly we must start from the correct value for the length of the mean synodic month as it was used by Hipparchus, his Babylonian predecessors, and by astronomers of the Middle Ages [1]. This value, expressed in days and sexagesimal fractions [2] of a day, is

$$29 ; 31 , 50 , 8 , 20 .$$

The excess beyond $29 ; 30$ days is consequently $0 ; 1 , 50 , 8 , 20$. Because $0 ; 0 , 0 , 8 = 1/27.000$ and $0 ; 0 , 50 = 1/72$ we see immediately that we should expect

$$\frac{1}{60} + \frac{1}{72} + \frac{1}{27.000} + \text{a small fraction}$$

where the " small fraction " stands for $0 ; 0 , 0 , 0 , 20 = 1/648.000$. Consequently the 1/72 of the Arabic text and the 1/27.000 of both versions are correct. The error of the Arabic text in writing 300 instead of 60 is explained simply enough by assuming that the original manuscript used

---

[1] KUGLER, *Babylonische Mondrechnung*, p. 24 and p. 111. GEMINUS, *El. astr.*, VIII, 43 (116,22 f. Manitius). PTOLEMY, *Almagest*, IV, 2 and IV, 3 (271,11 and 278,5 f. Heiberg). Al-Bīrūnī, *Chronology*, VII (143,24 tr. Sachau). MAIMONIDES, *Mishnah Torah, Kiddush Hachodesh*, VI, 3. Cf. also FELDMAN, *Rabbinical mathematics and astronomy*, London 1931, p. 123 ; S. GANDZ, *Complementary fractions in Bible and Talmud* (Am. Ac. Jewish Research, 1945, p. 148). This list makes no claim of completeness.

[2] Notation: numbers preceeding the semicolon are integers, numbers following the semicolon are sixtieths and sixtieths of sixtieths etc. Thus $0 ; 1,50 = 1/60 + 50/3600$.

number signs instead of spelled–out words. It would follow that we are dealing with the erroneous writing of $š = 300$ شـ for $s = 60$ سـ. Hence the Arabic texts undoubtedly agreed with the classical value for the length of the synodic month.

It is clear that also the Greek text must have had an expression for $1/60 + 1/72$ which was distorted into $1/30 + 1/20$. That these values are incorrect is obvious for two reasons. First, their total is $0 ; 2 + 0 ; 3 = 0 ; 5$ thus much larger than $0 ; 1 , 50$. Secondly, the smaller fraction should follow, not precede, the bigger fraction. Also Schöne's emendation to $1/320$ leads to the totally wrong result $0 ; 0 , 11 , 15$ which is not surprising in view of the fact that Schöne thought that he was dealing with the sidereal year (!) and not with the synodic month [1]. The real explanation of the preserved numbers 30 and 20 seems to me to lie in the ordinary use of sexagesimal fractions. The original value in Hipparchus's text was, as we have seen, $29 ; 31 , 50 , 8 , 20$. Galen, or a later copyist, wanted to translate this technical writing into popular language. The main part $29 ; 30 = 29 \ 1/2$ days was mentioned before. In the remainder $0 ; 1 , 50 , 8 , 20$ he dropped the last $0 ; 0 , 0 , 0 , 20$ as being " a very small fraction ". Then he replaced $0 ; 0 , 0 , 8$ correctly by $1/27.000$. Finally he expressed $0 ; 1 , 50$ as " a sixtieth of a day and $5/6$ (of a sixtieth) ". The ordinary writing for $5/6$ is either $1/2 + 1/3$ or, using sexagesimal fractions, ϰ′ λ′ (corresponding to our notation $0 ; 30 + 0 ; 20$) which is exactly the arrangement in our text. Some editor misinterpreted ϰ′ λ′ as $1/30 \ 1/20$ and wrote τριαϰοστὸν ϰαὶ εἰϰοστόν. Thus the only error in the archetype of the Greek version consists in the omission of a word or, more likely, of a symbol for a sixtieth. The only difference between Greek and Arabic version consists in replacing $5/6$ of $1/60$ by the equivalent $1/72$ but both represent the same well–known value for the length of the synodic month.

O. Neugebauer.

---

[1] H. Schöne, *Galens Schrift über Siebenmonatskinder. Quellen und Studien z. Gesch. d. Naturwiss. u. d. Medizin*, 3,4 (1933), p. 137 [345].

THE HOROSCOPE OF CEIONIUS RUFIUS ALBINUS.

In 1894 Theodor Mommsen suggested[1] that Ceionius Rufius Albinus was the person to whom the horoscope contained in the second book of the Mathesis of Firmicus Maternus referred. Circumstantial evidence led Mommsen to the conclusion that it was the *praefectus urbi* of the year 336/337 A. D. whose horoscope is discussed in detail by Firmicus. Mommsen did not, however, submit his thesis to the final test: whether or not the astronomical data agree with his hypothesis. It is the purpose of the present note to fill this gap. At the same time I wish to point out how easily problems of this type can be solved without going into a great many unnecessary details which are usually invoked in the dating of horoscopes by professional astronomers who are not familiar with the techniques of ancient astronomy and astrology, techniques which by their approximative character make quite meaningless the application of modern high precision tools.

Our horoscope contains as data only the zodiacal signs in which the seven planets and the Horoscopus, the rising point, are located. As starting-point we use the positions of Saturn and Jupiter in Virgo and Pisces respectively. I know of twelve Greek horoscopes with Saturn in Virgo, three of which show Jupiter in Virgo also. Their dates, A. D. 65, 124, and 125 respectively,[2] are too far away from the critical years around 300

---

[1] *Hermes*, XXIX, pp. 468-72 = *Gesammelte Schriften*, VII, pp. 446-50.
[2] These texts are Vettius Valens, II, 21; VII, 5; *P. Fouad* 6 respectively.

which alone are interesting for our problem. With *P. Harris* 53, however, we reach the year 245 A. D. but both planets are one sign farther ahead than required. The common period of Saturn and Jupiter is 59 years, thus the same situation will prevail in 304 A. D. Jupiter moves one sign per year; thus 303 will give the right sign for this planet and it can be hoped that Saturn is one sign back also. Thus 303 is our only chance before 336 A. D. Other possibilities are either 59 years earlier or later, and thus incompatible with Mommsen's hypothesis.

The first step consists in finding the approximate positions of Saturn and Jupiter in 303 A. D. We now have to consider the position of the sun in Pisces. This requires a date shortly before the vernal equinox. I choose 303 March 1 because the sun is then in the middle of Pisces (about Pisces 12). For this date one can find the mean longitudes of Saturn and Jupiter by two additions of triplets of numbers,[3] and one more addition gives the required positions as Virgo 24 and Pisces 26 respectively. Thus 303 is possible. We repeat the same process for Mars and find Aquarius 5, again in agreement with the data of the text. Though the longitudes computed so far may be wrong by several degrees, March 303 is certainly a possible date.

Narrower limits are obtainable when we consider the longitude of the moon, which was located in Cancer. We again use mean motion in longitude alone and find by adding twice three numbers each[4] that the moon was in the middle of Cancer either on February 15 or on March 14 of 303 A. D., i. e., either 13 days before or 13 days after our preliminary date, at which the sun was near Pisces 12. Because the sun moves about 1° per day the earlier date will barely lead to a position in Pisces whereas the second is still well inside this sign. Consequently we compute the longitudes of Venus and Mercury for the more plausible date, 303 March 14. We now use the tables quoted in note 3 to their full accuracy (requiring the addition of six numbers for each planet) and find for Mercury Aquarius 28, for Venus Taurus 11. The text gives Aquarius and Taurus respectively.

This result also tells us that we need not check the second

[3] Denoted by a_1, a_2, a_3 in the *Genäherte Tafeln für Sonne und Planeten* by P. V. Neugebauer, *Astronomische Nachrichten*, 248 (1932), cols. 161 ff.
[4] Denoted by L_1, L_2, L_3 in the *Tafeln zur astronomischen Chronologie*, II, by P. V. Neugebauer (Leipzig, 1914).

possibility, February 15, because one month back Venus cannot have reached Taurus. Finally, change from March 1 to March 14 can only improve on the position of Mars and will not change appreciably the longitudes of Jupiter and Saturn. Thus A. D. 303 March 14 satisfies all requirements. Because we have placed the moon only approximately in the middle of Cancer one must consider not only March 14 but also March 13 or March 15 as equivalent dates. Knowing from Firmicus that Scorpio was rising while the sun was located at the end of Pisces, we see that the hour of birth must have been about 9 p. m. Computing the longitude of the moon for this hour of March 13 or March 15 shows that the moon was entering Cancer in the first case, leaving it in the second; consequently only 303 March 14 remains as the date of the birth.

We have thus removed the only serious possible argument against Mommsen's conclusions, namely that the age of the person in question might not fit the other data. We may now from the opposite point of view as well see in the perfect agreement of all external data with the data of the horoscope an explicit confirmation of the fact that horoscopes in ancient astrological literature were not artificially made up examples but constitute a valuable source both of historical and astronomical information.

O. NEUGEBAUER.

BROWN UNIVERSITY.

303 March 14

	Tuckerman				Text
♄	177. 6	≈	♍	27; 40	♍
♃	355. 4	≈	♓	25, 30	♓
♂	310. 9	≈	♒	11, 0	♒
♀	39. 4	≈	♉	9; 30	♉
☿	326. 5	≈	♒	26, 30	♒
☉	353. 8	≈	♓	23; 50	♓
☾	101	≈	♋	11	♋

Reprinted from
American Journal of Philology
74 (4), 418–420 (1953)

ON THE "HIPPOPEDE" OF EUDOXUS

By O. Neugebauer

FEW astronomical theories have exercised so deep and lasting an influence on human thought as the discovery of Eudoxus that the motion of the planets can be explained, at least qualitatively, as the combination of uniform rotations of concentric spheres about inclined axes. The sphericity of the universe, the fundamental importance of uniform circular motion, must have appeared from then on as an established fact. Combined with Aristotle's idea of the "prime mover" the universe could be understood as one great system, truly geocentric. No wonder that this theory held its fascination for almost two thousand years over the minds of philosophers and even astronomers, in spite of the fact that serious difficulties were apparent almost from the start. The theories developed by specialized astronomers, like Hipparchus or Ptolemy, had to their credit a far superior agreement with the observational data. Nevertheless, the deeply rooted human conviction that simplicity and beauty are criteria of truth kept the hope alive that the homocentric spheres, albeit in some modification, may represent correctly the plan of the creator.

It is well known that Eudoxus' theory was completely forgotten in the interval between the destruction of all ancient and medieval theories of the universe by Galileo and Newton and the beginning of historical studies. It was not until Schaubach (1802) and Ideler (1828/1830) that the theory of Eudoxus found serious consideration. The details of its mechanism were finally revealed by Schiaparelli (1874) in a classical paper[1] which has been cited ever since and its figures reproduced. If I return here to the discussion of the Eudoxian theory it is not to make any addition to the factual results obtained by Schiaparelli. I feel, however, that one essential gap was left open in Schiaparelli's discussion. His derivation of the fundamental qualities of the curve on which a planet moves (notably that it is the intersection of the sphere with a cylinder), is so complicated that it seems impossible to identify it with Eudoxus' argument (around 360 B.C.), centuries before the invention of spherical or plane trigonometry. I have always felt that it must be possible to "see" the results of the combined spheri-

---

[1] G. Schiaparelli, *Scritti sulla storia della astronomia antica*, v. II, p. 3–112.

225

cal motions which form the essence of the model.   I shall demonstrate
that this is indeed the case.   A most elementary geometric consideration gives all details of the motion in question.   I have no doubt that
we have here the explanation of how the "Hippopede" was discovered.

The geometrical discussion which I shall set forth is entirely independent of astronomical concepts.   It is only for the benefit of the
modern reader that I shall add the basic astronomical considerations
which lead to the formulation of the problem.   A reader who is familiar
with this astronomical background can pass directly to the discussion
of Fig. 2.

The problem which Eudoxus was to solve consisted in the explanation of two phenomena, characteristic for the planetary motion as
seen in the sky: first, the fact that the planets from time to time reverse the direction of their movement, they become "retrograde"; and
second, this motion is not effected in a plane (or great circle) but deviates from the line of symmetry, the ecliptic; the planets have a "latitude."   From the manner in which these two problems were solved
by Eudoxus, it is perfectly clear how he must have argued: since the
motion in latitude never exceeds fixed limits, it suffices to have the
planet move on a plane of fixed inclination to the ecliptic.   If we suc-
ceed in keeping the planet in a limited area, not only as far as the latitude is concerned, but also with respect to the motion along the ecliptic
(the motion in "longitude"), then it must move on some closed curve and
consequently its longitude will become periodically retrograde.   But
this is all we need: since the planet shows a certain mean motion in
longitude covering in a given period the whole ecliptic, we only have
to put our latitude-retrogradation mechanism into a sphere which rotates with the speed of the mean motion in order to obtain retrograde
loops and deviations in latitude periodically[2] all along the ecliptic.

Thus we have to solve the following problem: the planet which
moves between fixed limits in vertical direction on its inclined plane
should not be able to escape a limited interval in horizontal direction.
Evidently this may be accomplished as follows: if the planet moves
on its inclined plane forward by an angle α, rotate the whole inclined
plane back through the same angle.   Then it is clear that the planet
is essentially captured; in the limiting case in which the "inclined"
plane coincides with the horizontal plane the planet would be moved
exactly as much backward as it would go forward on its own circle,
thus it would remain at the same place.   If, however, the inclination is
small, the planet will somehow wobble around a mean position and this

---

[2] The period of the retrogradations is the "synodic period," but this need not concern
us here.

is exactly what we want.   Thus the general plane of the model is now
clear (Fig. 1):   let the planet move on the inclined plane, which is the

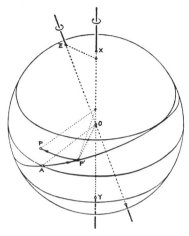

Fig. 1

equator of one sphere, and turn this sphere as a whole backward with
the same speed.   Problem: determine the character of the path of
the planet.

In order to investigate the motion of the planet P we consider it as the
result of the two rotations[3] described in Fig. 1.   We begin the motion
at A where the inclined orbital plane intersects the horizontal plane.
If then AP' corresponds to a rotation of the amount α around the axis
$O\Xi$, then we turn this axis back by the same amount α.   This brings
P' on a parallel circle to the final position P.   It obviously suffices to

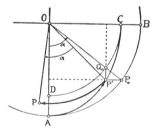

Fig. 2

investigate the positions of P for only one quadrant of the motion of
P' because after having reached maximum deviation, P will descend in
a symmetric fashion toward the horizontal plane and so forth.

[3] The method of resolving a motion in components was well known; cf. Schiaparelli,
l.c., p. 105, note (1).

We now look at this motion from the pole X of the horizontal plane, that is, we consider the projection of the motion on the horizontal plane (Fig. 2). The quadrant of the horizontal plane under consideration is the circular arc AB. The inclined plane projects itself onto the curve AC—that it is an ellipse need not be known—with C as point of maximum deviation. Let α be the angle of rotation on the inclined plane counted from OA. If the planet were moving in the horizontal plane, the angle α would bring it from A to P_0. Since it moves, however, in an inclined plane, we have only to tilt the arc AP_0 around the axis OA in order to see that the point P' must be located somewhere on a line P_0P' which is perpendicular to OA. Furthermore, all distances on the inclined plane from the line of intersection OA are foreshortened in the same ratio as the radius OC. In order to find P' we have therefore nothing to do but to make $P_0Q_0 = BC$ and draw $Q_0P' \parallel OA$.[4] Thus P' is known for any given α.

This suffices to determine the locus of P. All we have to do is to turn P' backward by the same angle α by which P_0 was moved forward from A (cf. Fig. 3). AP_0, $P'P$, Q_0D are circular arcs with com-

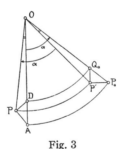

Fig. 3

mon center O. We consider the triangle with vertex P and base AD. This triangle is obviously congruent with the triangle $P'P_0Q_0$ since it is obtained by a rotation around O through the angle α. Hence the angle at P is always a right angle. Hence the projection P of the planet moves on a semicircle of diameter AD while P' moves from A to C.[5] This solves our problem: the planet itself moves on the sphere on a curve which is the intersection of the sphere with a straight circular cylinder perpendicular to the horizontal plane and of diameter AD.[6] This is indeed the "Hippopede" of Eudoxus (Fig. 4).

[4] All we did was, of course, to utilize a well-known construction of points of an ellipse with axes OA and OC.
[5] Because the angle at Q_0 and therefore also at D is again α, we see that the angular velocity of P on its circle is twice the angular velocity of P' around O.
[6] $AD = BC$ is obviously identical with the level difference of the poles Ξ and X.

In the final arrangement, Eudoxus let the axis XOY of the "horizontal plane" be a diameter of the ecliptic. Then the two loops extend along the ecliptic and their width corresponds to the motion in latitude. The motion in longitude is obtained by rotating the ecliptic around its

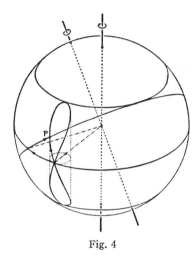

Fig. 4

poles with the proper mean motion. A fourth sphere produces the daily rotation. This completes the model.

Brown University

Reprinted from
Scripta Mathematica
19 (4), 225–229 (1953)

COMMUNICATIONS ON PURE AND APPLIED MATHEMATICS, VOL. VIII, 641–648 (1955)

Apollonius' Planetary Theory

By O. NEUGEBAUER

Brown University

There is no one center of all the celestial circles or spheres.

Copernicus, Commentariolus (trsl. E. Rosen, p. 58)

§1. Ptolemy's "Almagest", written about 140 A.D. and itself a work of great mathematical skill, has preserved interesting references to investigations of Apollonius, three centuries earlier, on the cinematics of eccentric and epicyclic motion. After Eudoxus, about 360 B.C., had succeeded in giving a qualitative explanation of the apparent planetary orbits by means of concentric spheres, rotating about inclined axes[1], Apollonius seems to have been the first geometer to utilize plain circular motion. What we know, through Ptolemy, about his theory seems to indicate that the theory was not far enough developed to account, e.g., for the dependence of the arcs of retrogradation on the longitude. Indeed, the complexity of the empirical data is so great that Hipparchus (about 150 B.C.) refrained from formulating a consistent planetary theory, a task finally accomplished by Ptolemy. To us the relative merit of Apollonius' or Ptolemy's theories for the representation of empirical data is no longer of great interest. Apollonius' discussion, however, of the appearance of stationary points in the planetary motion seems to me worthy of being remembered for its intrinsic mathematical elegance.

§2. Before turning to a sketch of Apollonius' arguments, let us recall some general facts concerning the ancient theory of planetary motions. It is well known that the longitudes of a planet which moves in a Kepler ellipse of eccentricity e can be represented with an error of the order of e^2 by a uniform circular motion on a circle whose center is the second focus of the Kepler orbit, and whose radius is $a + e$, a being the major half axis of the ellipse. For example, the Hipparchian-Ptolemaic description of the solar orbit by means of an eccentric model produces errors in longitude of at most $45''$, thus far below ancient observational accuracy.[2] The distances are much less well represented but no direct measurement of planetary distances was possible for the ancients, and therefore all emphasis was laid on the correct representation of longitudes only. It is easy to see that the effects of the comparatively small values of planetary

---

[1]For a description of the leading idea in Eudoxus' mechanism cf. Neugebauer, On the "Hippopede" of Eudoxus, Scripta Mathematica 19, 1953, pp. 225–229.

[2]This is not yet the best possible approximation since Ptolemy's value for the eccentricity is .042 instead of .034. In 900 A.D. al-Battānī reached .035 which reduces the maximum error in longitude to **43′**.

latitudes can be ignored in first approximation.[3] Thus within the accuracy of
ancient observations a correct planetary model would consist in an arrangement
as represented by Figure 1 in which S represents the sun, $\overline{E}$ and $\overline{P}$ the centers

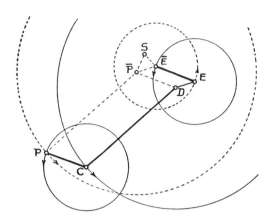

FIGURE 1.

of the eccenters of the earth E and the planet P. Since we finally wish to
describe the planetary positions as seen from E, we need only transform E to
rest by giving to $\overline{E}$ (now called "mean sun") the opposite angular velocity
which E had with respect to $\overline{E}$. Thus we have reached a model of eccentric
epicycles which in essence underlies the final Ptolemaic theory.

§3. We now turn to the theory of Apollonius, which we shall formulate
for an exterior planet. At its foundation lies an empirical fact which was already
fully utilized for the arithmetical theories of the Babylonian astronomers of
the Seleucid period: the number of sidereal rotations of a planet plus the number
of synodic periods (e.g., occurrences of conjunctions) equals the number of
years. For example, in 79 years Mars performs 42 rotations in longitude and

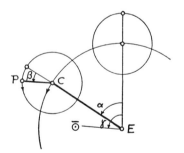

FIGURE 2.

---

[3]This point is discussed in detail by Ptolemy, Almagest XIII.

comes in conjunction with the sun 37 times.[4]   This fundamental relation is equivalent in an epicyclic model (Figure 2) to the condition

(1) $$\alpha + \beta = \gamma$$

where α is the angular distance of the center of the epicycle from the apogee (the "mean longitude"), β the angular distance of the planet from the apogee of the epicycle (the "anomaly") and γ the longitude of the "mean" sun, since we assume that all angles vary proportionally to time.   It is this simplified model which will be used henceforth.[5]

Apollonius does not restrict himself to an epicyclic model.   In the theory of the lunar motion the equivalence of eccenter and epicyclic motion is easy to demonstrate by means of a simple parallelogram construction (Figure 3).[6]

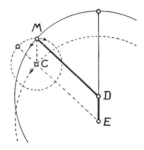

FIGURE 3.

The moon in M on an epicycle describes an eccentric circle whose eccentricity equals the radius of the epicycle.   Exactly the same construction in the case of an outer planet leads to an eccenter model (Figure 4).   The relation (1) shows that the center D of the eccenter moves with the velocity of the mean sun in positive direction around the observer in E, whereas the planet on the eccenter moves in the opposite sense away from the apogee of the eccenter.

The equivalence of these two models, epicycle and eccenter, is characteristic for Apollonius' discussion.   It is obvious that Apollonius was fully aware of the cinematic equivalence of the heliocentric and geocentric description since it

---

[4]Almagest XII, 1 or IX, 3.   The same in Babylonian cuneiform texts.

[5]It may be remarked that the relation (1) does not determine the sense of rotation of the planet on the epicycle.   To distinguish between these two possibilities Ptolemy uses the theorem that the mean motion of the planet appears in the points of its orbit which are 90° distant from the apogee.   The observed time intervals between extremal and mean motion then require a motion on the epicycle in the same sense as the motion of the epicycle on the deferent (Almagest IX, 5).   This implies, furthermore, that the radius of the epicycle CP is always parallel to the direction under which the mean sun appears from E. — That the wrong sense of rotations had also been assumed in antiquity is shown by a papyrus of the Michigan collection (No. 149).

[6]Almagest IV, 5.

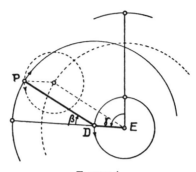

FIGURE 4.

only amounts to the question which of the vertices of the parallelogram is kept fixed.

§4. We now come to Apollonius' theorem about the stationary points:

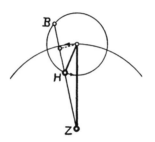

FIGURE 5.

In the epicyclic model (Figure 5) a planet is stationary in the point H of its epicycle if for an observer in Z

(2a)
$$\frac{\frac{1}{2}BH}{HZ} = \frac{v_1}{v}$$

where v_1 represents the angular velocity of the center of the epicycle, v the velocity of the planet on the epicycle. For the eccentric model (Figure 6) the condition for a station H as seen from T is

(2b)
$$\frac{\frac{1}{2}BH}{HT} = \frac{v_2}{v}$$

where v_2 is the velocity of the center of the eccenter, v of the planet.

The proof combines the two cases by using the same circle either as eccenter or as epicycle. This is made possible by showing that the observer T in the interior of the circle in the case of the eccenter is related to the point Z in the exterior (representing the observer in the case of the epicycle) by a transformation with reciprocal radii. It is then shown that the point H defined by (2a)

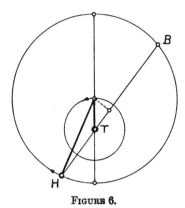

FIGURE 6.

or (2b) separates the points on the circumference in its neighborhood into two classes: points with direct motion and points with retrograde motion. The point H is therefore a station.

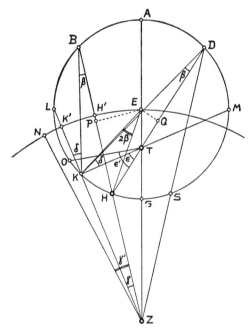

FIGURE 7.

The details of the proof are as follows. First the relation (cf. Figure 7)

(3)
$$\frac{AZ}{GZ} = \frac{AT}{TG}$$

315

is derived from the assumption that GH = GS. This construction establishes the fact that the two positions of the observer, Z and T, are related to one another by inversion on the given circle[7] of diameter AG. The points H and S will be the two stations.

Similarly it is shown that

(4)
$$\frac{DZ}{SZ} = \frac{DT}{TH}$$

and from it, by an easy transformation,

(5)
$$\frac{PH}{HZ} = \frac{QT}{TH}.$$

We now assume that H and S were chosen such that (2a) is satisfied,

$$\frac{PH}{HZ} = \frac{v_1}{v},$$

and therefore, using (5)

$$\frac{QT}{TH} = \frac{v_1}{v} \quad \text{or} \quad \frac{QT + TH}{TH} = \frac{QH}{TH} = \frac{v_1 + v}{v}.$$

But, according to the basic relation (1) for the motion of an outer planet, we have

$$v_1 + v = v_2$$

and therefore

$$\frac{QH}{TH} = \frac{v_2}{v}$$

which is the condition (2b) for the eccentric model. Thus we have shown that the points H and S satisfy the conditions of our theorem for epicyclic and eccentric model alike, Z and T playing the role of the observer respectively.

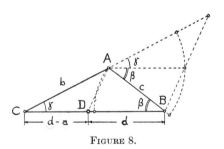

FIGURE 8.

---

[7]This is, of course, only a special case of the pole-polar relation which is discussed, in full generality, by Apollonius in his "Conic Sections" (III, 37).

In order to establish H and S as stations the following lemma is proved. If in a triangle (Figure 8)

(6a) $$c \leqq d < a$$

then

(6b) $$\frac{d}{a-d} > \frac{\gamma}{\beta}.$$

(The proof is based on the comparison of the circular sectors with vertex A and the corresponding triangles.)

We now turn to the case of the epicycle model and assume (Figure 7) that H is defined by (2a) whereas the planet may be situated in K somewhere between B and H. For any such point K,

$$BK < BH < BZ$$

holds. Thus, from lemma (6),

(7) $$\frac{BH}{HZ} > \frac{\gamma}{\beta}$$

and, with (2a),

(8) $$\frac{\frac{1}{2}BH}{HZ} = \frac{v_1}{v} > \frac{\gamma}{2\beta}.$$

Let N be a point on the deferent[8] such that its angular distance γ' from H satisfies

$$\frac{\gamma'}{2\beta} = \frac{v_1}{v}$$

which implies that the epicycle moves from H' to N in the same time in which the planet moves from K to H. From (8) we see that $\gamma' > \gamma$, that is to say, that the planet still gains in longitude while moving from K to H. Thus K belongs to an arc of direct motion.

For the case of the eccenter, (7) is again used in the form

$$\frac{BZ}{HZ} > \frac{\beta+\gamma}{\beta} = \frac{\delta}{\beta}$$

(Figure 7). Using (4) and $\epsilon = \beta + \delta$ one finds

$$\frac{\frac{1}{2}DH}{TH} = \frac{v_2}{v} > \frac{\epsilon}{2\beta}$$

---

[8]Obviously following an old tradition, the figures in Halma II p. 322 as well as in Heiberg II p. 458 (and all translations and commentaries known to me) place N on the epicycle. That this is wrong is evident, however, if one chooses for K the point of contact for the tangent from Z to the epicycle.

which again indicates that the direct motion of the eccenter exceeds the opposite motion of the planet.[9]

Similarly it is shown that points between H and G have a retrograde motion. Thus H separates direct from retrograde motion.

In a concluding remark, attention is drawn to the fact that the condition (2a) implies that the velocities v_1 of the epicycle and v of the planet must satisfy

$$\frac{v_1}{v} < \frac{EG}{GZ}$$

since otherwise no chord ZHB can be drawn in accordance with (2a). Thus the existence of stationary points depends on the ratio of the velocities and eccentricities which need not be fulfilled a priori. It is unknown whether the empirical data about the lengths of the retrograde arcs were used by Apollonius to determine the numerical values of the parameters of the models. The methods used by Ptolemy are certainly his own since they are adapted to his models with eccentric deferents and equants.

Copernicus, who greatly complicated Ptolemy's planetary theory by introducing secondary epicycles, demonstrated that the center of the earth's orbit, that is, the mean sun, may be used with the same results as the center of the earth for the computation of the observed phenomena. He also repeated[10] Apollonius' proof for the location of their stationary points for the case of the epicycle. To this extent Copernicus did not go beyond what was known to his predecessors. We have no evidence from antiquity, however, for his most important discovery, namely, the use of the radii of the planetary epicycles as measure of their parallax.

Received May 20, 1955.

---

[9] The point O in Figure 7 is defined by $(\epsilon'/2\beta) = v_2/v$.
[10] De revol. V, 35.

NOTES ON HIPPARCHUS

O. NEUGEBAUER

Even the most casual discussion of ancient astronomy will not fail to call Hipparchus of Nicaea in Bithynia "the greatest astronomer of antiquity." It is obvious enough that classifications of greatness are usually void of any precise meaning, though it is equally evident that they will remain a stock phrase in the histories of science. It is perhaps not useless, however, to underline how little we actually know about Hipparchus' astronomy and its relation to his predecessors and followers.

Rehm, in his article on Hipparchus in Pauly-Wissowa, wrote in 1913, "Eine Sammlung der Fragmente liegt bisher nicht vor, abgesehen von den geographischen ... Bei den astronomischen würde es wenigstens dankenswert sein, diejenigen beisammen zu haben, welche außerhalb der ganz auf Hipparch's Vorarbeiten aufgebauten Syntaxis des Ptolemaios überliefert sind." Here an expert in Greek scientific literature clearly felt that essentially we are reading Hipparchus when we read the *Almagest*, written about three centuries after Hipparchus. But it is plain that this cannot be the case.[1] We have the explicit statement of Ptolemy that his planetary theory is his own and that Hipparchus refrained from developing one because of the complexity of the phenomena which had to be explained. A careful study of the *Almagest* everywhere confirms Ptolemy's claim. Thus we know practically nothing about Hipparchus' attitude toward the analysis of planetary motion.

The story that the Catalogue of Stars in the *Almagest* is simply obtained from Hipparchus by adding $2\frac{2}{3}°$ for precession to all longitudes is also untenable,

---

[1] This has been clearly seen by Norbert Herz in his *Geschichte der Bahnbestimmung von Planeten und Kometen* I, Leipzig 1887, p. 86 ff.

although it has been frequently repeated since it was invented by Delambre[2] in 1817. Here again, Ptolemy's statements were fully confirmed by Boll's investigations, in which he showed that Hipparchus' list contained many fewer stars than Ptolemy's catalogue[3], and H. Vogt[4] demonstrated, on the basis of the Aratus Commentary by Hipparchus, that almost all Ptolemaic positions are independent of the data known to be Hipparchian.

In the theory of the moon, the famous "second inequality" was discovered by Ptolemy, and all that Hipparchus could possibly have known is the simple epicyclic theory.

In spherical astronomy, Ptolemy operates mostly with the methods developed by Menelaus, who lived two centuries after Hipparchus. What remains as possibly Hipparchian is the elementary plane trigonometry and the table of chords. Thus the determination of the solar eccentricity from the inequality of the seasons and of the distance of sun and moon are[5] the only integral pieces of Hipparchian tradition which were incorporated as a whole in the *Almagest*: roughly 10 pages out of 600 in Heiberg's edition.

This rapid summary of the contents of the *Almagest* suffices to show that only a quite minor part can be taken at face value as Hipparchian. The methodological approach must therefore be exactly the opposite: only a very minute investigation of all references to Hipparchus in comparison with the whole related discussion can lead to a real appreciation of Hipparchus' methods and knowledge.[6]

The situation is further complicated by the fact that we cannot ignore Hipparchus' predecessors, Greek as well as Oriental. Again, from the *Almagest*, we know that half a century before Hipparchus, Apollonius had proved important theorems concerning epicyclic and eccentric motion. The gap between the results obtained by Apollonius and the methods from which Ptolemy took his departure is so narrow that I see no possibility of answering in concrete terms the question: in what direction did Hipparchus advance beyond Apollonius?

One may hope to obtain some information about Hipparchus' astronomy from the astrological literature. Cumont went so far as to say[7] that Hipparchus' name "doit être placé en tête des astrologues comme des astronomes grecs." The main

[2] J. B. J. Delambre, *Histoire de l'astronomie ancienne* I, p. XXXII; p. 183; II, p. 264, etc.

[3] F. J. Boll, *Bibliotheca Mathematica*, 3. Folge. vol. 2 (1901) p. 185–195.

[4] H. Vogt, *Astronomische Nachrichten* 224 (1925), col. 17–54.

[5] *Almagest* III, 4 and V, 15.

[6] As an example of what I have in mind, I can quote a discussion by O. Schmidt of *Almagest* IV, 9 ("Determination of the Epoch of the Mean Motion of the Moon in Latitude" [Danish]) in *Matematisk Tidsskrift* B, 1937, p. 27–32. Here it becomes clear how much more sophisticated than Hipparchus' procedure is that of Ptolemy.

[7] *Klio* 9 (1909) p. 268.

source of this statement is a chapter in Hephaestion[8] (who lived more than 500 years after Hipparchus) on a form of astrological geography according to which the single countries are not associated, as usual, with the zodiacal signs directly but with the parts of the stellar pictures (right and left shoulder of the ram, etc.). This doctrine, of which no trace seems to be preserved elsewhere, is ascribed to "Hipparchus and the old Egyptians."[*] Equally little is to be gained from a list of zodiacal signs and motivation of their names[9] often found in astrological texts and ascribed to Hipparchus. The only astronomical information contained in one of these versions is a primitive (strictly linear) scheme for the variation of the length of daylight[10]. Finally, Vettius Valens[11] describes a Hipparchian "handy method" for the determination of the longitude of the moon and the planets. This method is extremely crude and is not much more than an estimate of the type that any Hellenistic astronomer would have been able to give at any time. In its preserved form, the method is based on the era of Augustus.

Thus we are left with the same result: we know extremely little about Hipparchus' theoretical astronomy. We are in a much better position so far as observations are concerned. Not only do we have at our disposal in the commentary to Aratus an authentic work of Hipparchus which contains a great number of observations, but also the *Almagest* which contains detailed references to Hipparchian observations[12], though their number is not very great—fewer than twenty.

This does not exclude, however, that Hipparchus himself relied on older data, particularly from Mesopotamia. At the very outset of his studies of Babylonian astronomy, Kugler realized[13] that the Hipparchian values for the length of the mean synodic month and for the mean anomalistic month were attested in cuneiform texts of the Seleucid period. This showed that fundamental empirical data of Hipparchus' luni-solar theory were simply taken from Babylonian astronomy.

It seems to have escaped the notice of historians that the same holds also for Hipparchus' planetary periods, which are quoted by Ptolemy in the *Almagest* IX, 3. To establish this fact we have to digress somewhat in the classification of cuneiform astronomical texts of the Seleucid period. It is convenient to divide

[8] A. Engelbrecht, *Hephaistion von Theben und sein astrologisches Compendium*, Wien, 1887, p. 47, 20; 60, 30 etc.

[9] Published in two versions by E. Maass, *Analecta Eratosthenica*, Philol. Untersuchungen 6 (1883) p. 141–149.

[10] It is also found in Porphyrius (d. 304); cf. *Cat. Cod. Astrol. Graec.* 5, 4, p. 209.

[11] I,19 and probably I,20.

[12] The quotations in the *Almagest* need not be complete, as A. Rome has shown from two eclipse observations preserved in greater detail in Theon's commentary to book III (ed. Rome, p. 828 ff.).

[13] F. X. Kugler, *Babylonische Mondrechnung*, Freiburg 1900, p. 111.

[*] Since the second half of this statement is notoriously wrong, our confidence in the first half cannot be too great.

this material into two major groups, the first of which contains the strictly mathematical astronomical texts, devoted to the prediction of lunar and planetary phenomena.[14] The second group, about five times as numerous as the first, in part contains observational data and can be divided into four major subgroups, following an analysis given by A. Sachs in recent years.[15] The planetary periods play a particular rôle in one of these sub-groups, the so-called "goal-year texts." These texts combine, for a given year (the "goal year"), planetary phenomena which happened a planetary period earlier. These fixed periods are

for Saturn	59 years
for Jupiter	71 and 83 years
for Mars	79 and 47 years
for Venus	8 years
for Mercury	46 years.

The duplicity in the case of Mars and Jupiter is explained by A. Sachs, who observed that the first mentioned year number is associated with the synodic period, the second with the sidereal period of these planets. It is exactly these five periods which Ptolemy quotes as used by Hipparchus for his determination of the planetary mean motions, values which then were slightly refined by the application of Ptolemy's own theory over much longer intervals of time.

The fact that Hipparchus chose the periods used in Babylonian goal-year texts as the basis for his investigations is in all probability not accidental. It was well known to the Babylonian astronomers that the above-quoted periods were not exact. The mathematical astronomical texts, e.g., consider as accurate the following periods[16]

for Saturn	265 years
for Jupiter	427 years
for Mars	284 years
for Venus	1151 years
for Mercury	480 years.

The occurrence of exactly these numbers in the astrological literature[17] leaves no doubt that also this part of the Babylonian theory was familiar to the Greeks. The reason for Hipparchus' not using these larger and supposedly exact periods undoubtedly lies in the fact that he was aware of their fictitious accuracy. It is

[14] Cf. now O. Neugebauer, *Astronomical Cuneiform Texts*, London, 1955, 3 vols.
[15] "A Classification of the Babylonian Astronomical Tablets of the Seleucid Period." *J. of Cuneiform Studies* 2 (1950) p. 271–290.
[16] Cf., e.g., my *Astronomical Cuneiform Texts*, p. 283.
[17] Cf. P. Tannery, *Mémoires Scientifiques* 4, p. 261 ff. [1892] and F. Boll, *Byzant. Z.* 7 (1898) p. 599.

clear that they were not the result of direct observations over so long intervals of time but resulted from the combination of smaller, slightly inaccurate periods, such that the corrections were eliminated as nearly as possible in the combined period. For the Babylonian theoretical texts these resulting larger periods were naturally chosen as the basis of further computation. Hipparchus remained as close as possible to the primary empirical material of which the goal-year texts are the reflection.

If the trend of our analysis is at all sound, then we must say that there is very little space left for really significant innovations in theoretical astronomy between Apollonius and Ptolemy. But wherever we are confronted with facts, we see Hipparchus at work to provide observations and to arrange them for proper analysis by later generations. It is our good luck to be able to see in the *Almagest* how Ptolemy utilized this material with supreme skill, adding to it an enormous empirical material of his own, not to speak of the incredible amount of numerical calculations which underly the tables of the *Almagest*. It is not because of philosophical prejudices that the Ptolemaic system dominated astronomy for about 1500 years but because of the solidity of its empirical foundations.

Reprinted from
The Aegean and the Near East. Studies Presented to Hetty Goldman, 292–296
J.J. Augustin, Locust Valley, New York, 1956

Reprinted from

ISIS

VOLUME 50, PART 1, NUMBER 159

MARCH 1959

Ptolemy's Geography, Book VII, Chapters 6 and 7

By O. Neugebauer *

PREFACE

AT the end of Book VII of Ptolemy's "Geography" one finds the discussion of a representation of the terrestrial globe in the plane as seen by an observer who is placed in the plane of the parallel of latitude which passes through Syene. The text is difficult to understand and — at least to my knowledge — has been discussed in detail only twice: by Johannes Werner in 1514 (Latin translation and commentary) and by Mollweide, who presented a German paraphrase and an extensive discussion in his article "Mappierung-skunst des Claudius Ptolemaeus" in Zach's *Monatliche Correspondenz* for 1805. Delambre, *Histoire de l'astronomie ancienne* II, (Paris 1817) p. 530, speaks about "d'inutiles efforts" of Werner to reconstruct the text though Werner had reached a much better understanding than Delambre, who was always ready to consider a work of Ptolemy as bungled. Mollweide's excellent publication seems to have remained unknown to Delambre. Halma's French translation (Paris, 1828) does not go beyond Delambre. The English translation given by E. L. Stevenson in his monumental work (New York Public Library, 1932) is still worse since its author did not understand, even remotely, what he was translating.[1]

The following is an attempt to give a consistent translation of the two chapters in question and to explain the underlying procedures. I am following essentially the text as published by Nobbe (1843) vol. II, p. 181 to 190. For the present chapters, with their coherent mathematical discussion, the absence of a critical edition is fortunately of less importance than for the purely geographical sections. I did compare Nobbe's text with the versions published by Erasmus (1533) and by Halma (1828) but since it was not my intention to produce a new edition of the text I did not go back to the manuscripts themselves. I had, however, the good fortune of an extended correspondence with Professor Aubrey Diller, who kindly communicated to me the results of his comparison of six manuscripts.[2] This shows that much remains to be done for the establishment of the text itself but it also confirmed my initial statement that an understanding of the procedures in VII, 6 and 7 can be reached on the basis of the printed text.

* Brown University.

[1] I can only subscribe to Professor Diller's review in *Isis*, 1935, 22: 553–539. Modern monographs on related subjects, e.g., by Schlachter, Gisinger, Brendel, completely ignore this chapter of the "Geography."

[2] Including Erasmus's text and Urbinas gr. 82, which is available in reproduction in Codices e Vaticanis selecti, vol. 19 (Leyden-Leipzig, 1932).

TRANSLATION

Chapter 6

Representation of the Ringed Sphere [3] together with the Oikoumene [4]

1.[5] The plan of the whole composition (of this work) has so far (been concerned with the) commensurate (type of the mapping). It will therefore not be out of place (now) to set forth how one can depict on a plane the hemisphere of the earth in which the oikoumene is located, surrounded by a ringed sphere. Whereas several persons have made an attempt to give such a demonstration, they seem to have made very unreasonable use of it.

2. We propose to represent the ringed sphere in a plane including a part of the earth. For the position of the eye we assume (a point of) the straight line which is common to the meridian through the solstitial points — which also divides into half the length of our oikoumene — and the parallel which is drawn on the earth through Syene — which in turn approximately halves the width of the oikoumene.

3. The ratios of the dimensions of the ringed sphere and of the earth and of the distance of the eye must be chosen such that the whole known part of the earth is visible in the space between the equatorial ring and the solstitial ring, the southern one of the (two) semicircles of the zodiac being located over the earth in such a way that it will not obscure (something of) the oikoumene which is located in the northern hemisphere.

4. Under these assumptions it is clear that the said meridians appear in one straight line, coinciding with the axis, since the eye falls into their plane; also, by the same reason, that the parallel of Syene appears perpendicular to it whereas the remaining of the given circles appear turned concavely toward these straight lines, the meridians toward the (line) through the poles, the parallels toward the (line) through Syene, and — as is obvious — the more (so) the farther they are distant from them.

5. How one can proceed with the graphical representation which is as much as possible similar to the visual impressions will be (made) handy for us in the following fashion.

6. Let ΑΒΓΔ be the meridian which, in the ringed sphere, goes through the equinoctial points; let its center be Ε, its diameter ΑΕΓ, where Α represents the north pole, Γ the south pole. Let the arcs ΒΖ and ΔΗ and also ΒΘ and ΔΚ be laid off corresponding to the distance of the solstices from the equator; similarly the (arcs) ΑΛ and ΑΜ and also ΓΝ and ΓΞ at the distance of the arctic and of the antarctic (circle) from the poles; and let Ο be the point at which the diameter of the summer solstitial (circle) intersects ΑΕ.

7. Since the parallel through Syene must have a position between Ε and Ο, and since the ratio of the arc (which is contained) between the (parallel) through Syene and the equator to a quadrant is about 4 to 15,[6] and the ratio of half of ΕΟ to ΕΑ is about 4 to 20, and therefore [7] ΕΑ will be 4/3 of the radius of the earth. Let now ΕΠ be taken 3 parts as ΕΑ contains 4; with Ε as center and ΕΠ as radius let the circle ΠΡ, which encloses the earth, be drawn in the same plane. A straight line, equal to ΕΠ, should be divided into 90 equal parts corresponding to one quadrant; {the (distance) ΕΣ should be made 23 ½ ⅓

---

[3] Scholion: "Remarks: A ringed sphere is a sphere which is made of rings and not solid, and is therefore also called hollow." In modern literature such a system of rings, representing the principal circles of the celestial sphere, is called an "armillary sphere."

[4] That is, the inhabited part of the earth.

[5] The division in sections follows Nobbe, who introduced it for convenient reference.

[6] Or 24 to 90; thus the obliquity of the ecliptic is assumed to be 24°.

[7] This makes no sense. The ratio 4 to 20, accepting Diller's text, is numerically correct (while Nobbe gives 4 to 3), and confirms an emendation by Mollweide (p. 508). Nevertheless the ratio 4 to 20 assumes the knowledge of ratio $R/r = 4/3$ which it is supposed to explain. Halma reads "The ratio of half of ΕΑ to ΕΟ is 4 to 3 and ΟΑ is ¾ of the radius of the earth." Thus he not only changes the letters in the figure (though in part following the ed. princ.) but replaces ἐπίτριτος = ⁴⁄₃ by ¾. All this does not help to explain the procedure. "Tout cela est à peu près inintelligible" says Delambre (*Histoire de l'astronomie ancienne*, II p. 532), who suggested this change in the text.

and ET}[8] 16 ⅓ ¹⁄₁₂ [9] and EY 63 parts. And let ΦΣX be drawn at right angles to EΠ (thus representing) the parallel through Syene.

8. Consequently the point T will be the point through which has to be drawn the parallel which determines the southern limit of the oikoumene, symmetrically located to Meroe,[10] and Y the point through which will be drawn the parallel which determines the northern limit and which passes through Thule.

9. Let a point a little to the south from T be taken, like the point Ψ and connect Ψ with Δ; the extended lines ΣX and ΨΔ will meet in Ω. If we then imagine the given circles in the plane which passes through the solstitial points and the poles and through the eye at Ω, the (straight lines) drawn from Ω through M and H and Δ and K and Ξ will produce intersections, according to our assumptions, on AΓ through which will be drawn the arcs of the five circles as seen (from Ω); for example the point Ψ through which has to be drawn the equator which passes through Δ. The lines drawn from Ω through Λ and Z and B and Θ and N [11] will produce the intersections on AΓ through which will be drawn the arcs of these parallels across the earth.[12]

10. Similarly, if we take for the parallels to be drawn on the earth the appropriate distances from the equator on ΠP, e.g., Y and T; and the corresponding intersections of the straight lines from Ω with the semi-circle ΠXP, and the points opposite to them located on the parallels — then we will have the points through which shall be drawn the sections of the said parallels like ATB [13] and ΓΔY.[13a]

11. On these we take the distances of the meridians to be drawn on both sides of TY and on the straight lines ΦX in the proper ratios of the three parallels we will draw through the corresponding three points the sections of the given meridians, e.g., those which limit the longitudes (of the oikoumene) namely, ϚZH and Θ and E.

12. The amount (of detail) to be depicted on the earth must be adapted to the size of the drawing.

13. At the design of the rings, one has to watch that each goes through the said four points, in an egg-shaped form and not ending sharply where they meet the outermost circle, in order not to give the impression of a break,[14] but it should be given a consistent direction even if the convexities which end the ellipse fall outside the circle which encompasses the figure; this also appears to happen with the real (rings of a globe).

14. One must also take care that the circles are not merely (represented by simple) lines but with an appropriate width and in different colors and also that the arcs across the earth (be given) in a fainter color than those near the eye; and that of (apparently) intersecting parts those which are more distant from the eye be interrupted by the nearer ones, corresponding to the true position on the rings and on the earth; and that the zodiac with its southern semicircle which passes through the winter solstice, lies above the earth, while the northern (semicircle) which passes through the summer solstice be interrupted by it (the earth).

15. We also shall inscribe at fitting places the appropriate names (of the rings) and also at the circles on the earth the numbers of the distances and hours as indicated in the representation of the oikoumene; (and) at the outermost circle the familiar names of the winds in conformity with those on the ringed spheres and the method of marking at the given five parallels and the poles.

[8] I follow here the text of the editio princips (by Erasmus, 1533). Nobbe gives εο instead of εσ, εγ instead of ετ, and 63 instead of 23 ½ ⅓.
[9] Or 16;25.
[10] With respect to the equator.
[11] Text (ed. pr. and Nobbe) has here H; omitted by Halma.
[12] Nobbe p. 184, 29 has πέντε where Urbinas gr. 82 fol. 59(58)ᵛ I,44 gives πέρατα, while Diller suggests πέραν.
[13] Text (ed. pr. and Nobbe) has here ATM; cf. Mollweide p. 510, note**). Halma: ΔTB,

ΘEK, ZΣH; Diller ATB.
[13a] Text (Nobbe) ΓΛΔ; Diller from Vat. Gr. 191: TTΔ. In Fig. 4 the points of contact with the circle ΠXPΦ would be Γ and Δ for the northern parallel, A and B for the southern.
[14] The printed editions have here the unintelligible word κατάματος. Professor Diller drew my attention to the fact that Urbinas gr. 82 fol. 59(58)ᵛ II,14 gives κατάγματος. The correct interpretation was given already by Werner in his Latin translation, Nürnberg, 1514.

Chapter 7 [15]
Outline of the Unfolding

1. The outline will be appropriate for this unfolding and (also serve as) a summary. This mapping onto a plane of the ringed sphere with the encompassed earth is supposed to have a position according to which the eye lies in the straight line which is the intersection of the meridian through the solstitial points — under which also lies the dividing longitude for our oikoumene — and the parallel drawn on the earth through Syene — which also approximately halves the width of the oikoumene.

2. The ratios of the size of the sphere and of the earth can be seen in the space left between the equator and the summer solstitial circle while the southern semicircle of the zodiacal ring is located below the earth in order not to obscure something of the oikoumene which is located in the northern hemisphere. Thus the said meridians appear as one straight line coinciding with the axis when the eye falls into their plane.

3. And by the same reason the parallel through Syene remains perpendicular to the given (meridian) whereas the remaining circles appear turning their concavities toward these straight lines, the meridians toward (the straight line) through the poles, the parallels toward the line through Syene, the more (so) the farther they are distant from each other, such that the arctic circle is more inclined toward north than the summer solstitial circle, but the winter solstitial circle is more inclined toward the south than the equator, and the antarctic circle more than the winter solstitial circle.

4. And the known part of the earth will be represented such that it is not surrounded by the Ocean from any other place but only at the boundaries drawn toward the Northwest — and the Northwind and of Libya and Europe, in agreement with older works.

COMMENTARY

Ch. VI, 1. Ptolemy's two conic projections, explained in Book I chapter 24, had as their goal to preserve as much as possible correct ratios of distances ($\sigma\nu\mu\mu\acute{\epsilon}\tau\rho\omega\varsigma$ in the sense of preserving measurements), e.g., along parallels of latitude and along meridians or radii. Now, however, it should be shown how to construct a picture of the appearance of the terrestrial globe, surrounded by rings which represent the principal circles.

2. The eye is placed onto the line which lies in the intersection of the two planes which halve the oikoumene in latitude and longitude, the parallel of Syene and the central meridian $L = 0$.

3 to 5. The globe should be placed in such a way below the rings that the central meridian passes through the solstitial points and that the southern semicircle of the ecliptic is turned toward the observer, in order not to obstruct the visibility of the quadrant which carries the oikoumene.

Under these conditions the appearance of the globe with its surrounding rings should be given.

6 and 7. A cross-section is constructed (Fig. 1) through the sphere of the rings (diameter $A\Gamma$) and the globe (diameter ΠP) in the plane of the central meridian. ΛM and $N\Xi$ are the planes of the rings which represent the arctic circles ($\pm \phi = 90 - \epsilon$), ZH and ΘK the solstitial rings ($\phi = \pm \epsilon$), $B\Delta$ the equator. The first problem consists apparently in the determination of the ratio R/r of the radius of the ringed sphere to the radius of the globe. The condition which must be satisfied is stated at the beginning of 3: the rings should not obscure the area of the oikoumene.

[15] Halma, p. 83, has here two misprints: in the Greek text H for Z, in the translation, XVII for VII.

At the beginning of **7** it is said that the latitude of Syene is about 24° and that consequently $R/r = 4/3$. No proof is given but we shall show in the discussion of the following sections that Ptolemy's conditions lead by a very simple geometrical reasoning with necessity to this ratio.

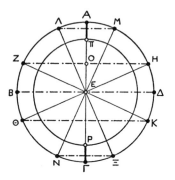

Fig. 1

7 to 9. The next step consists in the determination of the distance ΣΩ (Fig. 2) of the eye Ω on the parallel of latitude ΣX of Syene such that the rings do not obscure the map of the oikoumene. At this point a strange inconsistency is introduced into the problem. While the representation of the rings is constructed as a true perspective picture as seen from Ω, the oikoumene is not represented as the corresponding perspective representation of the circles on the globe but as a map which preserves latitudes on the central meridian. Thus Ptolemy divides the radius EΠ = r of the globe into 90 parts and makes EΥ = 63 parts for the northern boundary of the oikoumene (parallel of Thule) and EΣ = 23;50 parts for the parallel of Syene. Finally T with ET = 16;25 determines the southern limit of the oikoumene (parallel of Anti-Meroe).

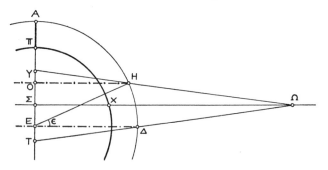

Fig. 2

This map of the oikoumene should now be within the projections of the solstitial circle (H) and the equator (Δ). The extremal position of the eye Ω is obtained by intersecting TΔ with ΣX at Ω. If we finally connect Ω with H, the nearest point to Ω of the solstitial ring, we find that it is projected just outside Υ, the northern limit of the oikoumene, as required. Thus Ω or any point nearer to X can be used as center of projection.

This fact can be used in inverse order to restore the proof that $R/r = 4/3$. Operating with degrees of geographical latitude we count r as 90; then we are given four points, T, E, Σ, and Υ with distances TE = 16;25, EΣ = 23;50, ΣΥ = 39;10. The angle HEΔ = ε is assumed to be 24° and ΩXΣ ⊥ TΥ. From this the distances EΔ = EH = R and ΣΩ = x should be determined in such a fashion that ΩΔ extended meets T, and ΩH similarly Υ (cf. Fig. 2). This leads to the following two conditions for R and x. From the similar triangles ΩΣΥ and ΔET it follows that

$$\frac{x}{\Sigma\Upsilon} = \frac{R}{ET} \quad \text{or} \quad \frac{x}{40;15} = \frac{R}{16;25}.$$

Secondly, from the triangles ΩΣΥ and HOΥ,

$$\frac{x}{\Sigma\Upsilon} = \frac{HO}{O\Upsilon} = \frac{R \cos \epsilon}{ET - R \sin \epsilon} = \frac{R \operatorname{crd} (180 - 2\epsilon)}{2 ET - R \operatorname{crd} 2\epsilon}.$$

or

$$\frac{x}{39;10} = \frac{R \operatorname{crd} 132}{126 - R \operatorname{crd} 48}.$$

Using the table of chords in the Almagest I,11 one finds

$$\operatorname{crd} 48 = 0;48,48,30 \qquad \operatorname{crd} 132 = 1;49,37,32$$

and thus

$$R = \frac{48;24,23}{0;24,24} = 119;0 \approx 120 = \frac{4}{3}r$$

which is indeed the ratio adopted by Ptolemy.[16]

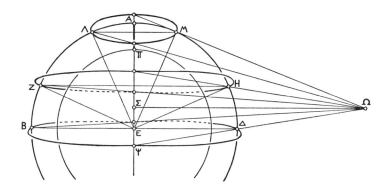

Fig. 3

9 and **10.** For the actual picture, Ω is not taken at its greatest possible distance at which Δ is projected into T but somewhat nearer to the globe, such that the equator is represented by an ellipse which has a vertex Ψ slightly to the south of T. From Ω chosen in this way are now drawn straight lines to all the endpoints of the rings of the sphere (Fig. 3), producing at their intersections with the line AEΓ the endpoints of the corresponding minor axes of the ellipses into which the rings are projected.

The picture of the parallel through Syene is the straight line ΦΣX (Fig. 4). There remain to be constructed the curves which represent the northern and southern limits of the oikoumene. To this end Υ and T are connected with Ω and the intersection with the semicircle ΠXP are considered to be points of these curves to which also belong the points which are symmetric to them with

[16] The value x = 292 is not needed since Ω is found graphically.

respect to ΠP. It is not said whether these points are considered the points of contact of ellipses with the circle of diameter ΠP, or vertices, or whatever else. Since the points T and Υ are not the result of a real perspective projection, one cannot make sure which interpretation would be correct.[17] It is plausible, however, to assume the circle ΠΣP as the contour for the representation of the globe [18] and consequently make the points in question the points of contact with ellipses.

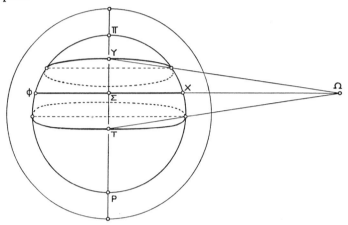

Fig. 4

11. The meridians are drawn — probably as circle arcs — through three points which are determined on the northern and southern limit of the oikoumene and on the parallel ΦΣX. Thus the representation is again a mapping preserving distances on three parallels [19] and not a perspective picture. Apparently the following is assumed. If the length of a quadrant on the equator is 90 units, then a quadrant at latitude φ has a length of 90 cos φ. Since ΠP = 180, one takes on the perpendicular through Υ 90 · cos 63 = 41 units, on the perpendicular through Σ 90 · cos 24 = 82 units, finally at T 90 · cos 16;30 = 81

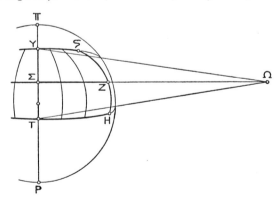

Fig. 5

---

[17] Werner and Mollweide (and all later authors who copied their figures more or less skilfully) assumed without question circular arcs.

[18] Although this is, strictly speaking, not correct.

[19] A similar but not identical mapping is discussed in Book I, Chapter 24 of the "Geography."

units. The circle arc ⊊ZH [20] represents the eastern boundary of the oikoumene (L = 90). The other meridians are drawn through corresponding equidistant points on the three parallels of Υ, Σ, and T. Fig. 5 illustrates the central and eastern part of the map obtained by this construction.

12 to 15. In the concluding paragraphs, Ptolemy gives instructions for the execution of the drawings. The methods of descriptive geometry which Ptolemy here and elsewhere [21] uses would allow him to construct accurately also the major axis of the ellipses into which the rings and the parallel circles are projected. In practice, however, this would be of very little value since almost all of these ellipses are very flat. Consequently, it suffices to know the smaller diameters and two points near the ends of the major axes. Ptolemy expressly warns of the mistake of considering the points on the surrounding ring as the vertices — a mistake still commonly made by modern books in illustrations of the celestial sphere.

Ch. VII, 1 to 4. Chapter VII is an almost verbatim repetition of the introductory paragraphs of chapter VI. It would not be surprising if it were the result of an early corruption in the manuscript tradition.[22]

BIBLIOGRAPHY

Text: editio princeps [by Erasmus]: Claudii Ptolemaei Alexandrini . . . *De Geographia libri octo* . . . Froben, Basileae Anno MDXXXIII.

Halma: *Traité de géographie de Claude Ptolémée d'Alexandrie*, traduit pour la première fois, du grec en français sur les manuscrits de la Bibliothèque du Roi, Paris, 1828.

Nobbe: *Claudii Ptolemaei Geographia.* Leipzig, 1843–1845 [editio stereotypa in 3 parts, Leipzig, 1898–1913].

Discussion:

Mollweide, Mappierungskunst des Claudius Ptolemaeus, *Monatliche Correspondenz zur Beförderung der Erd- und Himmels-Kunde 11* (1805), pp. 505–514; *12*, pp. 13–22.

See also W. H. Stahl, *Ptolemy's Geography, a select Bibliography*, New York Public Library, 1953.

[20] Following Nobbe's text. The northwestern endpoint would be E, the southwestern Θ.
[21] E.g., in the "Analemma."
[22] Cf. Berger, *Geschichte der wissenschaftlichen Erdkunde der Griechen*, 2te Aufl., Leipzig 1903, p. 638, n. 1. For the composite character of Ptolemy's "Geography" see Leo Bagrow, "The Origin of Ptolemy's Geographia," *Geografiska Annaler*, 1943, pp. 318 to 387, in particular pp. 368 ff.

Scripta Mathematica, v. 24, 1959

The Equivalence of Eccentric and Epicyclic Motion According to Apollonius

By O. NEUGEBAUER

1. INTRODUCTION

W E HAVE from antiquity several short references to astronomical investigations of Apollonius of Perga, and all of them specifically refer to the moon. Ptolemaeus Chennus (about 100 A.D.) says[1] "Apollonius, who lived in the time of Philopator,[2] became most famous as an astronomer; he was called ε since the figure of ε is related to the figure of the moon which he investigated most accurately." Then Vettius Valens, who wrote about 160 A.D.,[3] says[4] that he used the tables "of Hipparchus for the sun, of Sudines, Kidenas, and Apollonius for the moon, and also Apollonius for both types (of eclipses)." Finally the "Refutation of all Heresies" (written about 230 A.D.) quotes a figure for the distance from the surface of the earth to the moon proposed by Apollonius.[5]

The only substantial information about Apollonius' astronomical work comes from the Almagest. In Book XII,1 is found a long section concerning the determination of the stationary points in the apparent planetary motion following a method of Apollonius.[6] The most conspicuous feature of this discussion is the method of transforming eccentric and epicyclic models into each other by means of an inversion on a fixed circle which serves either as eccenter or as epicycle. It seems

---

[1] Quoted by Photius about 870 A.D. in his "Bibliotheca," 190, ed. Bekker (Berlin, 1824), p. 151b, 18–21. The passage is quoted in Heiberg's edition of the works of Apollonius, II, 1893, p. 139, Frgm. 61.

[2] Thus 221 to 205 B.C. Apollonius was born under Ptolemaeus Euergetes (246 to 221 B.C.); cf. opera II, p. 168, 6f. (Heib.). He died probably around 170 B.C. This date was suggested by W. Crönert, *S. B. Akad. d. Wiss.*, Berlin, 1900, 2, p. 958, on the basis of the biography of Philonides, the geometer, mentioned by Apollonius in the preface to Book II of his Conic Sections. Philonides lived in Laodicea in Syria; cf. U. Köhler, *S. B. Akad. d. Wiss.*, Berlin, 1900, 2, p. 999f.

[3] Cf. Neugebauer, "On the Chronology of Vettius Valens' Anthologiae," *Harvard Theol. Rev.*, 47, 1954, p. 65–67.

[4] Anthol. IX, 11 ed., Kroll, p. 354, 5f.; also *Catal. Cod. Astrol. Gr.*, V, 2, p. 128,14.

[5] *Hippolytus Werke III*, ed. Wendland, 1916, p. 41, 13 (= Apollonius opera II, p. 139, Frgm. 6')) and p. 42,19. For the problem of the authorship of the "Refutation," cf. Pierre Nautin, Hippolyte et Josipe, Paris, 1947 (Études et textes pour l'histoire du dogme da la trinité).

[6] Ptolemaeus, opera I, 2 ed., Heiberg, p. 450,10 and 456,9.

Received by the editor, July 11, 1958.

5

to have escaped attention, however, that Ptolemy refers to exactly the same method also in Book IV, Chapter 6, on the occasion of his discussion of the determination of the epicycle radius (or of the eccentricity) of the simple lunar theory.[7] I have no doubt that also this section belongs to Apollonius. Not only do both sections form a perfect unit, being a special case of Apollonius' theory of the pole-polar relation discussed in Book III of his Conic Sections,[8] but Ptolemy in his note in the lunar theory refers explicitly to a later use—obviously meaning the theory of planetary motion.[9] By combining both sections we can obtain a fairly clear picture of the methods used by Apollonius for the determination of the fundamental parameters in the lunar and planetary theory.

In the following I shall give a summary, in modern terms, of the relevant sections in the Almagest. It is, of course, not possible to say in every specific instance whether a conclusion has been made in exactly this form by Apollonius or was modified by Ptolemy. Nevertheless the inner coherence of the single steps is so great that only very minor changes are permissible without depriving the whole structure of its meaning.

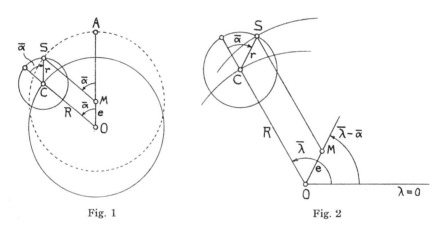

Fig. 1   Fig. 2

2. THE EQUIVALENCE OF THE SIMPLE MODELS

Neither Ptolemy's theory of the planetary motion nor his lunar theory is restricted to a simple epicyclic or eccentric motion. In the planetary theory the uniformity of motion is controlled from an

---

[7] Opera I, 2, p. 306,19 to 307,18. "Simple lunar theory" always means the model with only one inequality.

[8] In particular Theorem 37f. (Apollonius, opera I, p. 402ff., ed. Heiberg).

[9] This has been remarked by Manitius in his German translation of the *Almagest* (Teubner 1912), I, p. 223,2.

eccentric point (the so-called equant), symmetric to the observer, and for the moon and for Mercury a special mechanism is introduced in order to produce a periodic variation in the amplitude of the anomaly. None of these devices occurs in the discussion under consideration here. We are obviously dealing with the first, though most essential, level in the development of a theory which undertook to explain the phenomena by means of motions in a plane with constant angular velocities.   Thus we are dealing either with a simple epicyclic motion or with eccenters with fixed or with rotating apsidal lines.

The equivalence between models of this type is very easily recognizable.   If we consider (Fig. 1) a point S on an epicycle moving in such a fashion that the epicyclic radius r = CS remains parallel to a fixed direction OA, then it is clear that the motion of S can also be interpreted as a motion on an eccentric circle of radius R = MA and eccentricity e = OM.[10]

A model of this type describes the solar motion with sufficient accuracy.   In the lunar motion the mean longitude $\bar\lambda$ increases about $13;11°$ per day, the anomaly, $\bar\alpha$ however, only by about $13;4°$.[11] Consequently we obtain equivalence between an eccentric and an epicyclic motion (Fig. 2) only if the vector OM moves with the difference velocity.   Thus we need for the moon a movable eccenter.[12]

For the planets the motion on the epicycle proceeds in the same direction as the mean motion of the center C of the epicycle[13] (Fig. 3).

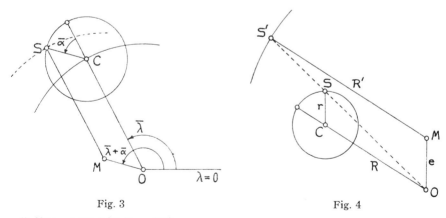

Fig. 3   Fig. 4

[10] *Almagest*, III, 3 (Heib., p. 225).
[11] If we introduce a consistent notation for the sense of rotation, namely positive = counterclockwise (i. e., increasing longitudes) we should say that the anomaly "decreases." We shall, however, follow generally the accepted terminology which makes no such distinctions but speaks of increase of direct and retrograde motions.
[12] *Almagest*, IV, 5 (Heib., p. 296).
[13] *Almagest*, X, 6.

For the outer planets CS must be parallel to the direction from O
to the mean sun, for Venus and Mercury OC itself is the direction to
the mean sun. Obviously the corresponding eccentric motion now
requires an eccenter that moves with the sum of mean motion and
anomaly.

In all these cases it is obvious that the radius r of the epicycle has the
identical length as the eccentricity $e = OM$ and that the radius R of the
deferent becomes the radius R' of the eccenter. In the Almagest,
however, we find the equivalence of the two models discussed not only
under the assumption that

$$r = e \qquad R = R' \tag{1}$$

but also if only the proper ratios are equal:[14]

$$\frac{r}{R} = \frac{e}{R'}. \tag{2}$$

Since it is evident that the object S' (Fig. 4) appears from O under the
same angle as S even if we enlarge the scale of one of these models such
that (2) is satisfied, one gets the impression that the generalization of
(1) to (2) is one of these trivial case-distinctions which make ancient
mathematical discussions so clumsy. I think, however, that is not
correct here. If we choose to make the radius R' of the eccenter equal
to the radius r of the epicycle, then we obtain from (2)

$$e \cdot R = r^2. \tag{3}$$

This allows of an important interpretation. Since we have made R' =
r we can interpret the same circle either as epicycle or as eccenter
(Fig. 5). If we consider it as epicycle, then the observer O is located
outside it at a distance R = OC. If, however, the circle serves as
eccenter, then the observer is to be placed inside it at Ō at a distance
ŌM = e. The relation (3) then implies that O and Ō are related to
each other by a tranformation with reciprocal radii on the fixed circle.
And this is Apollonius' statement.[15]

3. LUNAR MOTION

We have now seen that an apparently trivial generalization of the
direct equivalence proof is the basis for the possibility to transform
one model into the other by an inversion on a circle which serves
simultaneously as epicycle and as eccenter. The possibility of such a

---

[14] *Almagest*, III, 3 (Heib., p. 228); IV, 5 (Heib., p. 298ff).
[15] *Almagest*, XII, 1 (Heib., p. 452, 20f.) in the equivalent form $(R + r)/(R - r) = (r + e)/(r - e)$.

transformation is first mentioned in the Almagest Book IV,6 in connection with the problem of determining the radius of the lunar epicycle. The corresponding figure in the extant text is incomplete, and the numerical data are not given by Ptolemy. They are found, however, in Theon's Commentary to the Almagest, but again no mathematical justification is given. It is only for Ptolemy's discussion of Apollonius' theory of the planetary stations that the real background of the method becomes clear. In this way it is possible to restore all the arguments which lead to the formulation in the case of the lunar motion and then, conversely, fully to appreciate the importance of this method for the planetary theory in general.

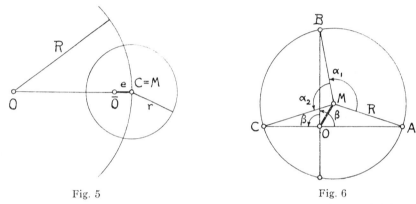

Fig. 5   Fig. 6

The discussion of the cinematic qualities of a certain model is of very little astronomical interest as long as one is not able to determine the numerical values of its characteristic parameters. These parameters fall into two groups of different character. The first class consists of the mean values which are associated with the progress of the celesterial bodies—these numbers are independent of any concept of geometrical model and can be determined, at least in principle, by simple counting of coincidences, e.g., of the type "m oppositions of a planet occur in n years." Actually m and n are usually numbers which are much too large to be determined directly in this form, but appropriate rational approximations lead to very accurate results which form the basis of Babylonian mathematical astronomy. These parameters reappear in Greek astronomy and can be taken for granted in the time of Apollonius since they were in use in Mesopotamia several decades[16] before Apollonius. The second group of parameters

[16] It should be underlined that the margin is rather narrow and it is at most by a century that the origin of Babylonian mathematical astronomy antedates Apollonius. The rapid development of Babylonian astronomy at the beginning of the third century B.C. is probably the main cause for the corresponding development of Greek astronomy.

are the parameters which determine the geometric configuration of models which were invented to explain the phenomena. One parameter in this group necessarily remains arbitrary in Greek astronomy, e. g., the radius R of the deferent, usually normed to be 60 units, since all computations are made sexagesimally. Thus the main problem for the simple models used by Apollonius consists in the determination of the epicyclic radii or their equivalents, the eccentricities. It is in this context[17] that Ptolemy mentions the procedure of inversion which connects the two types of models.

Only for the sun and for Venus and Mercury is it rather obvious what type of observation is required to determine the eccentricity or the epicycle radius. For the sun it suffices to measure the unequal amounts of time needed to travel each of three quadrants with the observer O as center (cf. Fig. 6). Since we always assume constant angular mean motion with respect to the center M of the circular orbit, we know not only the angles $\beta = 90°$ at O but also the angles α_1 and α_2 at M. The coordinates of M with respect to O are thus simply given by [18] R sin- $(\alpha_1 - \beta)$ and R sin$(\alpha_2 - \beta)$; thus OM $= e$ can be expressed in terms of R $= 60$. In the case of Venus and Mercury the radius of the epicycle is given by $r = R \sin \delta_0$, where δ_0 is the maximum elongation of the planet from the sun.

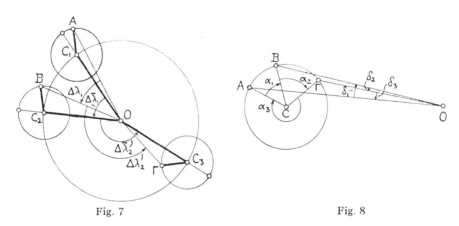

Fig. 7   Fig. 8

For the moon and for the outer planets, however, the problem is much more complicated. Fortunately Ptolemy gives a detailed description of its solution for the epicyclic lunar model. I shall briefly describe this procedure and then also give its parallel for the eccenter. This will finally lead us to a reconstruction of an analogous

[17] *Almagest*, IV, 6.
[18] In Greek terminology we have to replace R sin θ by $1/2$ crd 2θ where crd 2θ represents the length of a chord in a circle of radius R $= 60$ subtended by an angle of 2θ at the center.

method for the outer planets.

The procedure in the simple lunar theory[19] is based on the observation of three lunar eclipses.[20] These observations furnish us with the differences $\Delta\lambda_1$ and $\Delta\lambda_2$ between the observed longitudes of the consecutive eclipses and also with the corresponding time intervals Δt_1 and Δt_2 (cf. Fig. 7)[21] We are given furthermore the values for the daily mean motion in longitude $\bar{\lambda}$ and in anomaly $\bar{\alpha}$. Thus we can compute, using Δt_1 and Δt_2, the corresponding differences $\Delta\bar{\lambda}_1$, $\Delta\bar{\lambda}_2$ and $\Delta\bar{\alpha}_1$, $\Delta\bar{\alpha}_2$. These angular differences suffice to determine the radius r of the epicycle, assuming the radius of the deferent to be $R = 60$.

In order to verify this statement we turn the second and third position of the epicycle back to the first position (Fig. 8). Thus we obtain three points A, B, Γ on a circle of radius r with given angles $\alpha_1 = \Delta\bar{\alpha}_1$, $\alpha_2 = \Delta\bar{\alpha}_2$, $\alpha_3 = 360 - (\alpha_1 + \alpha_2)$ at C. At the point O these three points are seen under the angles $\delta_1 = \Delta\lambda_1 - \Delta\bar{\lambda}_1$, $\delta_2 = \Delta\lambda_2 - \Delta\bar{\lambda}_2$ and $\delta_3 = -(\delta_1 + \delta_2)$. Thus our problem has now taken the following form: determine the position of O such that three points of a circle appear from C under the angles α_1, α_2, α_3, from O under the angles δ_1, δ_2, δ_3.

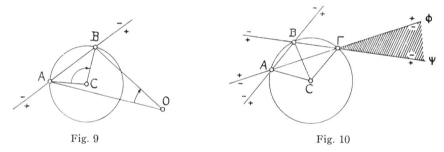

Fig. 9   Fig. 10

The first step to its solution consists in a purely topological consideration. In order that an arc AB appear from O under the same sense of rotation as from C, the point O must be located in the half-plane denoted by $+$ in Fig. 9. Since in our specific case $\delta_1 > 0$, $\delta_2 < 0$, $\delta_3 < 0$, we see that O must be located in the sector $\phi\Gamma\psi$ of Fig. 10.[22]

Having thus found the proper relative position of O with respect to the three points on the epicycle, one can proceed to the trigono-

[19] This term always means the pre-Ptolemaic form of the theory, ignoring the second inequality.

[20] *Almagest*, IV, 6.

[21] For the sake of clarity Fig. 7 does not correspond accurately to the data of the eclipses quoted by Ptolemy, though all essential features are preserved.

[22] This also guarantees the uniqueness of the solution, since the locus of all points from which AB and BΓ are seen under given angles are circles which can only have one intersection inside $\phi\Gamma\psi$.

metric solution.   One introduces the auxiliary line AEO (Fig. 11) which makes at E with EB and EΓ known angles $\beta_1 = \alpha_1/2$ and $\beta_2 = (\alpha_1 + \alpha_2)/2$, respectively.   Therefore also $\gamma_1 = \beta_1 - \delta_1$ and $\gamma_2 = \beta_2 - \delta_2$ are known, and BE as well as ΓE can be computed in terms of EO.   Thus we know in the triangle BEΓ two sides and the angle $\beta_2 - \beta_1 = \alpha_2/2$.   Hence one can find also BΓ in terms of EO.   But BΓ = crd α_2 for $r = 60$; thus EO can now be expressed in terms of $r = 60$.   Furthermore ΓE = crdα' and since ΓE is known we also know α' and thus AE = crd $(\alpha_1 + \alpha_2 + \alpha')$ in terms of $r = 60$.   Thus we know now not only EO but also AO and thus the product AO·EO. But[23]

$$\text{AO·EO} = (\text{R} + r) \cdot (\text{R} - r) = \text{R}^2 - r^2$$

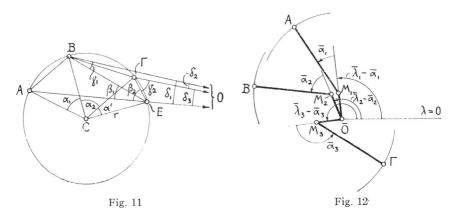

Fig. 11   Fig. 12

a relation which allows us to find R in terms of $r = 60$ and thus finally also r in terms of R = 60.[24]

It is in connection with this trigonometric procedure that Ptolemy remarks[25] that for the case of the eccenter the same circle (and we may add: the same three points A, B, Γ) can be used for finding the eccentricity, the only difference being that the observer must be located inside the circle.   Ptolemy also adds a figure to illustrate the trigonometric procedure for this case, but the figure in our text is incomplete,[26] nor does Ptolemy tell us which angles should be assigned to the observer. In Theon's Commentary the figure is complete (though in mirror image of Ptolemy's arrangement) and the numerical values for the angles at the observer are given by $\beta_1 = \alpha_1 + \delta_1$, $\beta_2 = \alpha_2 + \delta_2$ where

[23] *Euclid*, III, 36.
[24] The result is close to $r = 5^1/_4$.
[25] *Heiberg*, I, p. 306, 7ff.
[26] *Heiberg*, I, p. 306.

the α's and δ's have the same numerical values as in the epicyclic case.

In order to check these statements we start from the same configuration as given in Fig. 7, but applying to all three eclipses the
principle of Fig. 2 in replacing the sides OC, CS of the parallelogram
OCSM by the sides OM, MS (Fig. 12). We now again combine all
three cases in one eccenter by moving M_2 and M_3 to M_1. Then we
obtain three points, A, B, Γ on the circumference of the eccenter with
angular differences $\bar{\alpha}_2 - \bar{\alpha}_1 = \Delta\bar{\alpha}_1 = \alpha_1$, $\bar{\alpha}_3 - \bar{\alpha}_1 = \Delta\bar{\alpha}_2 = \alpha_2$ (Fig. 13).

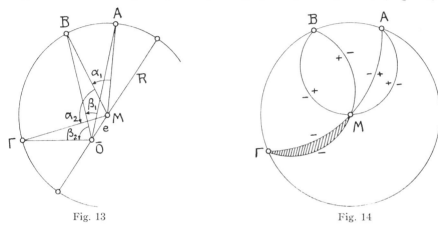

Fig. 13   Fig. 14

This is exactly the arrangement as in Theon's figure, but inverse to the
arrangement in Ptolemy's figure. This shows that Theon followed
our process, while Ptolemy operated directly with a transformation of
reciprocal radii from Fig. 8 which keeps the points on the circumference
fixed. The results are, of course, the same except for a change in
counting the sense of angles.

The identification of M_2 with M_1 requires a rotation of the amount
$\Delta\bar{\lambda}_1 - \Delta\bar{\alpha}_1$. Before this rotation, A and B were seen from O under the
angle $\Delta\lambda_1$ (cf. Fig. 7), thus the remaining angle at M_1 is $\Delta\lambda_1 - (\Delta\bar{\lambda}_1 - \Delta\bar{\alpha}_1) = \alpha_1 + \delta_1$. Similarly for B and Γ. This confirms Theon's
values. For the eccenter model our problem has now taken the
following form: given a circle with three points on its circumference
seen from the center M under angles α_1 and α_2, find the position of a
point $\bar{O}$ inside the circle such that these points appear under given
angles $\beta_1 = \alpha_1 + \delta_1$ and $\beta_2 = \alpha_2 + \delta_2$. The equivalence with the
epicyclic model then tells us that $\bar{O}$ is related to O by a transformation
with reciprocal radii on the given circle.

Exactly as before we have to determine the area in which O must be
located in order to satisfy $\delta_1 < 0$, $\delta_2 > 0$, $\delta_3 > 0$.[27] Now the locus of all

[27] The signs of the δ's are inverted because we have changed the direction of the α's.

points which see an arc AB (cf. Fig. 14) under the same angle as it is
seen from M is a circle which passes through A, B, and M. This
circle separates the domain for which $\delta_1 > 0$ from $\delta_1 < 0$. Thus Ō
must be located in the crescent-shaped area ΓM.[28] Finally we can
find Ō by a trigonometric process which is the exact analogue to the
previous one.[29]

4. PLANETARY MOTION

Apollonius' application of the transformation with reciprocal radii
to the problem of stationary points in the planetary motion provided
the starting point of our whole discussion. The occurrence of the
same method allows us to ascribe to Apollonius also the determination
of the radius of the lunar epicycle. We shall now discuss the analogous
problem for the outer planet.

As throughout in these investigations we are dealing with the epicyclic motion in its simplest form (Fig. 3), i. e., with the observer O
located at the center of the deferent and with a uniform motion of the
center C of the epicycle around O. In this model Apollonius determined the points at which the direct motion of C with the angular
velocity v_1 is exactly compensated by the motion of the planet S on
the epicycle, a motion whose angular velocity v with respect to C is
also known since CS must coincide with the direction from O to the
mean sun.[30]

The procedure by which Apollonius arrived at his result is mathematically of great interest.[31] For our present purposes, however, it
suffices to quote the main result (cf. Fig. 15) for the case of the epi-

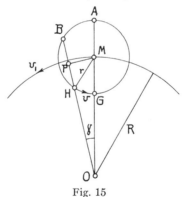

Fig. 15

[28] Since a transformation with reciprocal radii preserves circles, straight lines are mapped
into circles through M and Fig. 14 is the mirrored inversion of Fig. 10.
[29] Cf. below Section 5, note 36 (p. 18).
[30] *Almagest*, X, 6.
[31] I have described his method in a paper published in the *Communications on Pure and
Applied Mathematics*, 8, 1955, p. 641–648.

cyclic model: if H appears from O as a stationary point of the planetary motion then

$$\frac{PH}{HO} = \frac{v_1}{v} \qquad (1)$$

assuming that

$$\frac{v_1}{v} \lessgtr \frac{r}{R - r} \qquad (2)$$

since otherwise no stationary points exist in this motion.

The intrinsic interest of such a theorem concerning the most striking phenomenon of planetary motion is obvious. Ptolemy uses it in Book XII on the Almagest for the construction of tables for the stationary points, though the application of Apollonius' theorem to Ptolemy's refined planetary model is not strictly legitimate.

I think, however, that the usefulness of his theorem lay for Apollonius in a quite different direction. A table of stationary points can certainly not have been his goal since his simple epicyclic model was *a priori* unfit to account for the observable variation of the retrograde arcs. What must have been the first concern of any theory of epicyclic motion is the determination of the radius of the epicycle.

We have seen how much was required to solve this problem for the lunar motion. The answer was found as soon as one was able to determine in Fig. 11 the product $AO \cdot EO$ for the extended chord AE across the epicycle, since this product has the value of $(R + r) \cdot (R - r)$ and we assume $R = 60$. Now the theorem concerning the planetary stations concerns exactly the same configuration: the chord BHO is drawn across the epicycle (Fig. 15) and thus

$$BO \cdot HO = (R + r) \cdot (R - r). \qquad (3)$$

Unfortunately we do not know the value of this product but only the ratio resulting from (1):

$$\frac{BH}{HO} = \frac{2v_1}{v} \text{ or } \frac{BH + HO}{HO} = \frac{BO}{HO} = \frac{v + 2v_1}{v} \qquad (4)$$

Thus the determination of the radius r of the epicycle requires only finding a bridge which leads from (4) to (3).

Before showing that it is very easy indeed to use (4) in order to reach the form (3), I wish to make two remarks which should clarify the historical and the mathematical background of our problem. First, Ptolemy's method for determining r does not give us any clue to the solution of the same problem for the simple model of Apollonius. Ptolemy has to take into account all details of his model: eccentricity of the observer as well as equant. Thus he first has to find the eccentricity of the orbit by means of three oppositions which give him the positions of three points on the eccenter.[32] Thus this part is reduced to the eccenter model of the lunar theory (Fig. 13) except for an iteration process due to the introduction of the equant. Only after all parameters of the eccenter have been determined can a direct observation of the planet in relation to its position at opposition (which fixes the center of the epicycle) be utilized to determine the radius of the epicycle.

Secondly: though it is now clear that Apollonius did not have to face any of the complications of the Ptolemaic procedure, it is also easy to see that the knowledge of the ratio (4) or its equivalent

$$\frac{PH}{HO} = \frac{v_1}{v} \tag{1}$$

will not suffice to determine r. If in Fig. 16 OM = R = 60 and

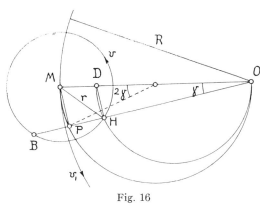

Fig. 16

$$\frac{MD}{DO} = \frac{v_1}{v}$$

then any chord OHP will satisfy (1) if H is the intersection of OP with the semicircle of diameter OD and P is a point of the semicircle of diameter OM.[33] Thus for every position of H there exists an epicycle of radius r = MH on which H appears as a stationary point if M moves

[32] *Almagest* X,7; XI, 1; XI,5.
[33] Because PM is parallel to HD. It may be remarked that exactly the configuration of Fig. 16 was studied by Apollonius in his work on "Contacts"; cf. Pappus, Collections VII, 100 (ed. Hultsch p. 824,17; trsl. Ver Eecke p. 636f.).

on the deferent with the angular velocity v_1 and H on its epicycle with the velocity v. Condition (2) indicates that D lies in the interior of the epicycle.

The preceding consideration shows that Apollonius' theorem about the stations alone cannot suffice to determine r, but it makes it evident that the problem is determined as soon as the angle γ between OH and OM is known. But this is a datum which it is most natural to observe in any investigation of planetary stations, namely, the half arc of retrogradation.[33a] Though this arc is in fact not of constant length, its mean value will produce at least a fair estimate for the size of the epicycle. We know that Hipparchus did not dare to propose a definite planetary model because of the complexity of the observational data. It is evident, however, that at least a preliminary answer to the question about the size of the epicycles had been attempted since Ptolemy mentions in the same context[34] different types of epicyclic models which are constructed to explain the planetary phenomena and to tabulate their positions. Thus I shall show how easily r can be found by using Apollonius' theorem concerning stationary points.

Indeed from (1) it follows that

$$\frac{PH + HO}{HO} = \frac{PO}{HO} = \frac{v + v_1}{v}$$

But

$$PO = \frac{1}{2} \operatorname{crd} (180 - 2\gamma)$$

and hence

$$HO = \frac{v}{2(v + v_1)} \operatorname{crd} (180 - 2\gamma).$$

Because of (4) this gives us also

$$BO = HO \frac{v + 2v_1}{v} = \frac{v + 2v_1}{2(v + v_1)} \operatorname{crd} (180 - 2\gamma)$$

and thus finally with (3)

$$R^2 - r^2 = BO \cdot HO = \frac{v(v + 2v_1)}{4(v + v_1)^2} \operatorname{crd}^2 (180 - 2\gamma).$$

From this relation, r can be computed, since the velocities and γ are given and $R = 60$. This proves our statement that Apollonius'

[33a] For the role which this element plays in Babylonian Mathematical astronomy cf. now A. Aaboe, On Babylonian Planetary Theories, *Centaurus* 5 (1958) p. 209–277, in particular p. 265 and p. 245.

[34] *Almagest* IX, 2.

theorem combined with the measurement of the arc 2γ of retrogradation leads directly to the value for the radius of the epicycle.

5. HIPPARCHUS

The goal of Book IV of the Almagest seems to have been reached with the table of corrections for the first lunar inequality in chapter 10. Instead of proceeding from here directly to Book V and the discussion of the second lunar inequality, Ptolemy gives a long appendix to the subject of Book IV, namely, data for two triples of eclipses by means of which Hipparchus had determined the eccentricity e of the lunar orbit or the radius r of the lunar epicycle.[35] Strangely enough, Hipparchus found widely different values for e and r, and Ptolemy reveals in some short paragraphs the causes for this flagrant contradiction to the mathematical equivalence of the two models, to wit, (a) incorrect determination of the time intervals Δt between the eclipses and (b) the use of different triples of eclipses for the two models.

The second objection obviously needs some qualification. If we assume that e and r are constant and since we know that the eccenter and the epicycle model are exactly equivalent, it should not make any difference which model we are using for the determination of these parameters. In practice, however, the situation is somewhat different. The observed data are not free of errors, and therefore different triples may lead to somewhat different values of e or of r, particularly since the results are very sensitive to small variations of the initial data. This is illustrated by Ptolemy's own computations. By using two different triples of eclipses he finds for r the values $5;13$ and $5;14$, respectively, Theon, in his commentary, computes for the first triple the value of e and again finds, by some fiddling, the value $5;13$[36]. In fact one should use for each triple the equivalence of the two models as a check for the accuracy of the computations.

It is extremely unlikely that Hipparchus intended to disprove the equivalence theorem of Apollonius through numerical examples taken from two different triples of eclipses. The only plausible motive for Hipparchus' investigation seems to be the question whether the eccentricity of the lunar orbit remains constant or not. Indeed, the variability of the planetary retrogradations shows that for the planets

[35] The eclipses in each triple (383/2 B.C. and 201/0 B.C., respectively) are 6 months apart and therefore are close to a diameter of the ecliptic. For the first triple this diameter is the diameter of the solstices, for the second triple the diameter of the equinoxes. This suggests that these eclipses were originally selected in connection with the problem of determining the equinoxes and solstices.

[36] Cf. A. Rome, *Commentaires de Pappus et de Théon d'Alexandrie sur l'Almageste*, III (Studi e Testi 106, Città del Vaticano, 1943), p. 1056,24 to p. 1060,10 and in particular p. 1060 note (2).

a model with only one inequality cannot suffice to explain the phenomena. It is natural to ask a similar question also for the lunar motion, and Hipparchus' results (obtained from either of the two equivalent models) must have led him to the conclusion that also the lunar epicycle shows a variable diameter.

In the light of this discussion it becomes evident that it was for Ptolemy a matter of great importance to disprove the validity of Hipparchus' results. Hipparchus as well as Ptolemy had investigated the problem of the apparent variation of the lunar epicycle during its revolution.[37] Both men must have been keenly aware of the fact that the simple theory had to be revised. Ptolemy finally succeded in bringing order into the bewildering empirical data which sometimes agreed, sometimes disagreed in various degrees with the simple theory, by discovering that the second inequality depended on the elongation of the mean moon from the sun. But then it was clear that this new inequality could *not* be detected by the investigation of eclipses. It is for this reason that Ptolemy added a final chapter to Book IV in order to protect his new theory against the argument that Hipparchus had found a variation of the eccentricity even for the syzygies. Only after having shown this as the result of an incorrect procedure could he present in Book V the data which had led him to his great discovery of the second lunar inequality, the so-called "evection."

Ptolemy's supplementary chapter to Book IV is, however, not only of interest for the understanding of the development of the lunar theory. It also furnishes us with a possibility to check Hipparchus' numerical computations. Ptolemy quotes the following data from Hipparchus' study: from three consecutive Babylonian eclipses (383/2 B.C.) he found for the ratio of the radius R' of the eccenter to the eccentricity e the value $R/e = 3144/327;40$ or $e \approx 6;15$ for $R' = 60$. Similarly he obtained from three eclipses observed in Alexandria in 201/0 B.C. the ratio $R/r = 3122;30/247;30$ or $r = 4;45$ for $R = 60$ in the case of the epicycle, as compared with Ptolemy's $e = r \approx 5;15$. Ptolemy explains these discrepancies from incorrect values for the time differences Δt and corresponding longitudinal differences $\Delta\lambda$. The values given for the time differences Δt allow us to compute the differences $\Delta\bar\lambda$ of mean longitudes and $\Delta\bar\alpha$ of anomaly. Combining these figures with $\Delta\lambda$ we can also compute the differences $\Delta\lambda - \Delta\bar\lambda$ and thus all the angles $\alpha_1, \alpha_2, \alpha_3, \delta_1, \delta_2, \delta_3$ and $\beta_1, \beta_2, \beta_3$ which are needed for the determination of e and r.

If one carries out these computations one finds the following results: in the first case $R'/e \approx 3374/362$ or $e \approx 6;26$ for $R' = 60$; in the

[37] Cf. the references to Hipparchian observations in Book V of the *Almagest*.

second case $R/r \approx 3810/322$ or $r \approx 5;4$ for $R = 60$.   Thus we have:

Hipparchus recorded:	$r = 4;45$	$\Delta = -0;30$
Hipparchus accurate:	$r = 5;4$	$-0;11$
Ptolemy:	$r = e = 5;15$	0
Hipparchus recorded:	$e = 6;15$	$+1;0$
Hipparchus accurate:	$e = 6;26$	$+1;11$

If, however, we substitute for both eclipse triples the values given by Ptolemy, we obtain in the first case $e = 5;16$, in the second $r = 5;13$, in excellent agreement with Ptolemy's values of $r = 5;13$ or $5;14$ obtained from Babylonian eclipses and from eclipses observed by himself.   For the Hipparchian results we learn that they are due not only to observational deviations but also to computational inaccuracy which may be ascribed to any place in the long way from his trigonometric tables to the square root and other approximations required by the problem at hand.

6. EPILOGUE

It is perhaps permissible to place the preceding discussion into a larger framework.   A better understanding of the character and extent of Apollonius' contributions to mathematical astronomy makes it increasingly clear that his work constitutes the real beginning of this branch of astronomy.   Shortly before his time we have the crude attempts of Aristarchus to determine the distances of sun and moon. Aristarchus based his procedure on the solution of one right triangle and on the measurement of one angle under conditions most favorable to gross errors.   In contrast, Apollonius' method of determining the eccentricity of the lunar orbit (or the equivalent radius of the epicycle) from three lunar eclipses is a masterpiece of mathematical analysis and is capable of making full use of the best available elements of lunar and solar motion.

That Apollonius must be very close to the beginning of mathematical astronomy in further demonstrated by the fact that his older contemporary, Archimedes, had still proposed a planetary system with relative distances based solely on numerology.[38]   On the other hand, Apollonius' planetary and lunar theory uses only one inequality, as is to be expected in a first attempt at a mathematical description of the celestial motions.[39]

[38] Known through Hippolytus in his "Refutation of All Heresies" IV,8–11, ed. Wendland p. 41,18–43,27; cf. also Archimedes, Opera II, p. 552ff. ed. Heiberg; P. Tannery, *Mém.*, *Sci.*I,p. 393.

[39] One could think that the Hindu theory of variable epicyclic diameter (*Surya Siddhanta*, II, 34–38) is an attempt to describe the variability of the retrograde arcs.   But the extrema lie in the quadrants instead of in the apsidal line only, and the amplitudes are much too small.

Eudoxus' brilliant demonstration of the possibility of a qualitative representation of planetary motion by means of concentric spheres, rotating with constant angular velocities, was certainly a fundamental step beyond the vague cosmogonies put forth by philosophers.   But the Eudoxian model contains too small a degree of freedom to be adaptable to the exigencies of actual observations.   The main importance of Apollonius' approach lies in the fact that it freed astronomy from the concept of geocentric spheres and allowed for variable geocentric distance and for centers of rotations outside the earth.   And it is quite obvious that the free use of the transformation which interchanges eccenters and epicycles implies the insight into the equivalence of all four vertices of the corresponding parallelogram to serve as centers of rotation, an insight which contains the cinematic equivalence of geocentric and heliocentric motion.   It is indeed exactly this transformation which is used over and over again by Copernicus to demonstrate the equivalence of his system with the system of Ptolemy.

Thus it seems to me that all the evidence points to Apollonius as the founder of Greek mathematical astronomy which provided the starting point for all further progress in the understanding of our planetary system.   If I am right in this estimate of the historical events, then we may say that a rapid development led from the first attempts of Aristarchus to Apollonius.   More than a century had passed since the first mathematical approach to a planetary theory had been made by Eudoxus, only to be calcified as part of philosophical doctrines concerning the structure of the world.   And far more remote from anything that may properly be called astronomy are the speculations during the centuries preceding Eudoxus, though ancient authors and modern scholars alike did their best to fill the vacuum.

BROWN UNIVERSITY

MELOTHESIA AND DODECATEMORIA

Otto NEUGEBAUER – Providence, R. I.

In carrying out an enterprise of such large dimensions as the " Catalogus Codicum Astrologorum Graecorum " it was, in practice, necessary to draw sharp lines somewhere for the detailed investigation of manuscripts. It is therefore not surprising to find occasionally within otherwise purely astronomical texts sections which actually belong to the material of the CCAG. This is the case with a table entitled περὶ μελῶν ζωδίων " On the limbs of the zodiacal signs " contained in Cod. Vat. gr. 208 fol. 129v, 130r and Cod. Palat. gr. 137 fol. 83 (¹). Since the doctrine which is presented by this table does not otherwise seem to have come to light, it is of some interest to make this text available.

The general scheme may be illustrated by the case of Cancer. Twelve places in this sign are associated with parts of the body:

2º	head	17º	left claw
5	face	20	back
7	mouth	22	abdomen
10	shoulder blades	25	hands
12	throat	27	feet
15	right claw	30	haunches.

The numbers show alternating differences 2 and 3. Obviously this is the result of dropping in every second case the fraction ½ in an original sequence 2 ½, 5, 7 ½, 10, etc. The latter numbers are

---

(¹) For Cod. Vat. gr. 208 cf. Io. MERCATI-P. FRANCHI DE' CAVALIERI, *Codices Vaticani Graeci*, Rome 1923, p. 256. The knowledge of the version in Pal. gr. 137 I owe to HEIBERG, *Ptolemaeus opera II*, Prolegomena p. CXCVI. No details are given in the catalogue *Codices manuscripti Palatini graeci*, Roma 1885. I have not investigated whether other versions of this text exist. For the planetary melothesia see A. OLIVIERI, *Memorie della R. accademia di archeologia, lettere e arti di Napoli*, 5 (1936) p. 19-58. See table one on pages 271 and 273.

another text :
*Marc. 325, f. 165*v

*Another vesion : Vindob. phil. gr. 87 fol. 46*v

the well-known " dodecatemoria ", i. e., the subdivision of each zodiacal sign into a " microzodiac " of twelve parts of 2 ½° each. As

περὶ μελῶν ζωδίων

μοῖ ραι	Κριοῦ	μοῖ ραι	Ταύρου	μοῖ ραι	διδύμων
ε ϛ	κεφαλή τράχηλος	ζ[1) ε	κεφαλή πρόσωπον	β ε	κεφαλὴ τοῦ πρὸς βορείον πρόσωπον
ι ιβ	μέτωπον καρδία	ζ ι	τράχηλος μέτωπον	ϛ ι	καρδία στῆθος
ιε ιη	ὦμοι ὀπίσϑια	ιβ ιε	καρδία ὦμοι	ιβ ιδ	χεῖρες πόδες
κ κγ	κοιλία γόνατα	ιη κ	ὀπίσϑια κοιλία	ιε ιϛ	ῥόπαλον ἀνὰ μέσον τοῦ ⚏[5)
κϛ κη	πόδες ὀπίσϑιοι αἰδοῖα	κδ[2) κϛ	γόνατα πόδες ὀπίσϑιοι	κ κβ	κεφαλὴ τοῦ νοτίου κοιλία
λ	οὐρά	κη λ	αἰδοῖα οὐρά	κε λ	πόδες γόνατα ἰσχία

μοῖ ραι	καρκίνου	μοῖ ραι	λέοντος	μοῖ ραι	παρθένου
β ε	κεφαλή πρόσωπον	δ[3) ε	κεφαλή πρόσωπον	β ε	κεφαλή πρόσωπον
ϛ ι	στόμα ὠμοπλάται	ϛ ι	στόμα καρδία	ϛ ι	χεῖρες στάχυς
ιβ ιε	τράχηλος χεῖλη δεξιά	ιβ ιε	τράχηλος ὦμοι	ιβ ιε	ὠμοπλάται στῆθος
ιϛ κ	χεῖλη εὐώνυμα νῶτος	ιϛ κ	ὠμοπλάται νῶτος	ιϛ κ	μάζος δεξιός μάζος εὐώνυμος
κβ κε	κοιλία χεῖρες	κβ κ[4)	κοιλία γόνατα	κβ κε	κοιλία πούς δεξιός
κϛ λ	πόδες ἰσχία	κϛ κθ	πόδες αἰδοῖα	κϛ κθ	πούς εὐώνυμος ἀστράγαλος
		λ	οὐρά	λ	ἰσχίον

1) P correctly β
2) P incorrectly κϛ
3) sic (also P) instead of β
4) P correctly κε
5) P correctly τῶν

Table one: upper half

Sachs and I have shown, this is one of the few astrological doctrines
which can be traced back to Mesopotamia (¹).

(¹) NEUGEBAUER-SACHS, *The Dodekatemoria in Babylonian Astrology*
Archiv f. Orientforschung 16 (1953) p. 65 f.

The above-mentioned numbers are correctly preserved only for Cancer, Capricorn, and Aquarius, whereas scribal errors have more or less obscured the underlying scheme in the other cases. Nevertheless the original pattern is clearly the pattern of the dodecatemoria.

The parts of the body associated with these twelfths form a curious mixture of the well-known zodiacal " melothesia " and specific features of the constellation in question. The ordinary arrangement — although subject to minor variations — associates the following parts of the body with the zodiacal signs (¹), beginning with Aries:

1	head	7	buttocks
2	throat	8	pudenda
3	shoulders, arms	9	knees
4	chest	10	haunches, loins
5	stomach	11	tibia, ankles
6	abdomen	12	feet.

Obviously, our lists follow the same principle for the twelfth-divisions of each individual sign, although modified by the underlying zodiacal configuration. The two " bicorporeal " signs, Gemini and Fishes, are divided into two sequences each. In Virgo the " Ear of Corn " (Spica) is mentioned, in Scorpio the Sting, in Sagittarius the Bow, etc., and a Tail occurs whenever possible. Yet, these parts of constellations are not related to their actual positions within the sign. For example, the Heart of Leo (which should be Regulus) and Spica are both associated with the 10th degree in their respective sign, although Regulus has a longitude of 2;30⁰ in Leo, Spica of 26; 40⁰ in Virgo for the time of Ptolemy (²). This proves our point since the longitudinal difference of stars is not influenced by precession. Thus we see that we are dealing here with a purely astrological speculation without any contact with astronomical reality. The time of origin (³) and the method of application remain unknown.

(¹) Cf., e. g. CCAG 5,4 p. 216, 24 ff; 8,1 p. 20 (figure); BOUCHÉ-LECLERCQ, L'astrologie grecque p. 319; J. DE VREESE, Petron 39, Amsterdam 1927, p. 198 ff.

(²) Almagest VII, 5.

(³) In favor of a relatively high age of the doctrine under consideration speaks perhaps the occurrence of the ' Club ' ῥόπαλον in Gemini, which is also mentioned in a text probably related to Balbillus (first century A. D.): CC AG 8,4 p. 244, 2.

The new table, however, permits us to recognize it as the ultimate source of sections in an excerpt of Rhetorius (about A. D. 500) from Teucros (probably first cent. B. C.) concerning the qualities and

περὶ μελῶν ζωδίων

μοῖ/ραι	ζυγοῦ	μοῖ/ραι	σκορπίου	μοῖ/ραι	τοξότου
β / ε	κεφαλή / πλάστιγξ εὐώνυμος	β / ε	κεφαλή / μέτωπον	β / ε	κεφαλή / πρόσωπον
ζ / ι	ἰσχία εὐώνυμα / πλάστιγξ δεξιά	ζ / ι	πρόσωπον / χεὶρ δεξιά	ζ / ι	στόμα / τράχηλος
ιβ / ιε	πόδες / μηρὸς εὐώνυμος	ιβ / ιε	χεὶρ εὐώνυμος / τράχηλος	ιβ / ιε	χεῖρες / τόξον
ιθ / κ	γνώμων ζυγῶν / καρδία	ιϛ / κ	νῶτος / κοιλία	ιϛ / κ	πόδες ἐμπρόσθοι / νῶτος
κβ / κγ	χεὶρ δεξιά / χεὶρ εὐώνυμος	κβ / κε	πούς δεξιός / πούς εὐώνυμος	κβ / κε	κοιλία / πόδες ὀπίσθια
κε / κϛ	πλευρὰ δεξιά / κοιλία	κη / λ	κέντρον / διάστημα κέντρου	κη / λ	ἰσχία / οὐρά
λ	ἐφηβαῖον				

μοῖ/ραι	αἰγοκέρωτος	μοῖ/ραι	ὑδροχόου	μοῖ/ραι	ἰχθύων
β / ε	κέρατα / κεφαλή	β / ε	κεφαλή / πρόσωπον	β / ε	στόμα / κεφαλή
ζ / ι	πρόσωπον / στόμα	ζ / ι	νῶτος / καρδία	ζ / ι	βράγχη / τράχηλος
ιβ / ιε	καρδία / τράχηλος	ιβ / ιε	χεὶρ δεξιά / ὕδωρ τὸ χεόμενον	ιβ / ιε	νῶτος / τὸ ἀνὰ μέσον
ιϛ / κ	ὠμοπλάται / χεὶρ δεξιά	ιϛ / κ	χεὶρ εὐώνυμος / κοιλία	ιϛ / κ	στόμα τοῦ βορείου / κεφαλή
κβ / κε	νῶτος / κοιλία	κβ / κε	μηρία / μηροί	κβ / κδ	στῆθος / τράχηλος
κϛ / λ	ὀπίσθια / οὐρά	κϛ / λ	πόδες / ἰσχία	κϛ / λ	κοιλία / τὸ μεταξὺ τῶν)('[1]

¹) P: μεταξὺ τῶν ἰχθύων

Table one: lower half

powers of the 12 zodiacal signs (¹). For each sign is given, among others, a paragraph for wich the case of Taurus may serve as an example (²): " There rises: from 1º to 3º the head, from 4º to 7º the horns, from 8º to 10º the throat, from 11º to 13º the chest, form 14º to 18º the loins, from 19º to 21º the haunches, from 22º to 24º the

(¹) CCAG 7 p. 194 to 213 (variants in CCAG 5,4 p. 123 ff).
(²) CCAG 7 p. 197, 24 to 27.

18

feet, from 25º to 27º the tail, from 28º to 30º the claws ". Obviously the compiler of this list (Rhetorius? or Teucros?) no longer understood that he was dealing with rounded dodecatemoria. Thus he changed the numbers, adopting the scheme " from a to b, $b + 1$ to c, from $c + 1$ to d " etc., thus reducing their number to nine (¹). Furthermore, he did not understand the purpose of the list and simply took it as a list of consecutively rising parts.

It is only through modern scholarship that we can see the growth of astrological doctrine at work, although still without accurate chronological basis. The original theory of the " melothesia ", that is, the rules of the zodiacal signs over parts of the body, was extended and duplicated along with the subdivision of the zodiac. Thus we obtain a decanal melothesia, with 10º as units, from it a melothesia for single degrees (²), and now a melothesia for the dodecatemoria. Although basically trivial in conception, these constructions soon reached bewildering dimensions, sufficient to obscure their origin to their own adepts.

In the following schematic translation (³), I have followed the arrangement of the parts of the body, as found in the text, but some minor adjustments were required in order to obtain twelve entries in each column. The Greek text reproduces Vat. gr. 208; variants from Palat. gr. 137 are quoted as P. I am obliged to the Vatican Library for permission to use these manuscripts.

(¹) This number varies between 7 and 10 in the other cases.
(²) Cf. W. GUNDEL, *Dekane und Dekansternbilder*, Hamburg 1936, particularly p. 286, 287.
(³) See table two. I have not translated ἰσχία (usually ' hip-joint ' or ' haunches ') since its consistent location near the end of the list is suspect.

emend to
ἴχνια

notes (as suggested by W. Koch, Astrologische Monatshefte 1961 p. 86)
All MSS, V, P, and Marc., have clearly ἰσχία

Reprinted from
Analecta Biblica
3, 270–275 (1959)

	♈	♉	♊	♋	♌	♍	♎	♏	♐	♑	♒	♓
2°	[horns]	head	head of n.	head	head	head	head	head	head	horns	head	mouth
5	head	face	face	face	face	face	left scale	forehead	face	head	face	head
7	throat	throat	heart	mouth	mouth	hands	left ἰσχία	face	mouth	face	back	gills
10	forehead	forehead	chest	shoulder bl.	heart	ear of corn	right scale	right hand	throat	mouth	heart	throat
12	heart	heart	hands	throat	throat	shoulder bl.	feet	left hand	hands	heart	right hand	back
15	shoulders	shoulders	feet	right claw	shoulders	chest	left thigh	throat	bow	throat	pouring water	between (♓)
17	hind [parts]	hind parts	club between ♊	left claw	shoulder bl.	right breast	pointer	back	fore feet	shoulder bl.	left hand	mouth of n.
20	abdomen	abdomen	head of s.	back	back	left breast	heart	abdomen	back	right hand	abdomen	head
22	knees	knees	abdomen	abdomen	abdomen	abdomen	right left	right foot	abdomen	back	[hind] parts	chest
25	hind feet	hind feet	knees	hands	knees	right foot	right side	left foot	hind feet	abdomen	thighs	throat
27	privy p.	privy p.	feet	feet	feet	left foot	abdomen	sting	ἰσχία	hind [parts]	feet	abdomen
30	tail	tail	ἰσχία	ἰσχία	privy p. tail	ankle ἰσχία	pubes	ext. of sting	tail	tail	ἰσχία	between ♓

Table two

357

In rereading E. Norden's excellent monograph *Die Geburt des Kindes* [1] I felt that it might be useful to amplify a little the note on p. 61 which concerns the above-quoted famous passage from Vergil's Fourth Eclogue, line 61. Norden rightly says that "viel Ungereimtes" has been said about it and that its meaning is simply the statement that "das Kind war voll ausgetragen."

What I wish to add in support of Norden's explanation is a remark which must be obvious to anyone familiar with ancient chronological terminology. Outside of specific local calendars there are three types of "months" in common use: schematic months of 30 days (e. g. with regnal years and months), synodic months of about $29\frac{1}{2}$ days length (approximating the interval, e. g., between consecutive new moons), and finally sidereal months of about $27\frac{1}{3}$ days which represent the average interval between consecutive returns of the moon to the same fixed star. This latter concept is the natural analogue to the original concept of "year" as the return of the sun to the same constellation, as well as to the "sidereal periods" of the planets. Everywhere in ancient astrology one distinguishes between positions with respect to stars, i. e. positions in the zodiac, and "aspects" of the movable celestial bodies with respect to each other, in particular their conjunctions. For the moon the first case defines the sidereal months, the latter the synodic months. All this is common astronomical knowledge in antiquity, not caused by but reflected in everyday astrological practice.

It is this triplicity of definition of the term "month" that allows one to express the normal duration of pregnancy either as nine (schematic or synodic) or as ten (sidereal) months. Indeed these intervals are all close to 270 days. *P. London* 130, 208-10 reckons 276 days as duration of pregnancy; Vettius Valens, *Anthol.*, I, 23, 24 (Kroll) assumes intervals between 258 and 288 days, thus a mean value of 273 days [2]—to quote

[1] *Studien der Bibliothek Warburg*, III (Teubner, Leipzig, 1924); reprinted 1931.

[2] Cf. Neugebauer-Van Hoesen, *Greek Horoscopes, Memoirs of the Am. Philos. Soc.*, XLVIII (1959), pp. 23 and 28.

64

only two less commonly used sources.[3] According to the Babylonian astronomical texts of the Seleucid period one mean sidereal month amounts to 27 ;19,10 days,[4] thus ten months cover 273 ;13 days.

On the basis of these facts the ancient schemes for the duration of pregnancy are easily understood. For example, Proclus in his commentary to Plato's *Republic* (II, p. 59, 16-20 Kroll [5]) gives values for " nine-month-children " and for " seven-month-children." If one notices that in the first case " nine months " are the equivalent of ten sidereal months and that " seven months " means seven sidereal months the whole scheme becomes at once apparent:

$$\text{ἐννεάμηνα:} \quad \text{max.} \quad 288\tfrac{1}{3} = 10\cdot27\tfrac{1}{3} + 15$$
$$\text{mean} \quad 273\tfrac{1}{3} = 10\cdot27\tfrac{1}{3}$$
$$\text{min.} \quad 258\tfrac{1}{3} = 10\cdot27\tfrac{1}{3} - 15$$

and

$$\text{ἑπτάμηνα:} \quad \text{max.} \quad 206\tfrac{1}{3} = 7\cdot27\tfrac{1}{3} + 15$$
$$\text{mean} \quad 191\tfrac{1}{3} = 7\cdot27\tfrac{1}{3}$$
$$\text{min.} \quad 176\tfrac{1}{3} = 7\cdot27\tfrac{1}{3} - 15$$

In the second group the text has $216\tfrac{1}{3}$ instead of $206\tfrac{1}{3}$; all other numbers are correct.[6]

To put it in simple terms: the nine-month-children are born after ten (sidereal) months, the seven-month-children after seven (sidereal) months, both cases within limits of plus or minus one-half of a (schematic) month. Vergil's terminology is equally plain.

O. NEUGEBAUER.

INSTITUTE FOR ADVANCED STUDY.

[3] Additional references from the astrological literature in Bidez-Cumont, *Les mages hellénisés*, II, pp. 162 f., notes to O 14 and O 15a.

[4] Neugebauer, *Astronomical Cuneiform Texts*, I, p. 272 (London 1955).

[5] Reprinted in Bidez-Cumont, *loc. cit.*, II, p. 162.

[6] All three of Usener's emendations, accepted by Kroll and by Bidez-Cumont, are wrong.

Reprinted from
American Journal of Philology
84 (1), 64–65 (1963)

ON SOME ASPECTS OF EARLY GREEK ASTRONOMY

O. NEUGEBAUER

Professor Emeritus of the History of Mathematics, Brown University

(*Read November 11, 1971*)

SOME EIGHT YEARS ago a "Symposium on Cunei-form Studies and the History of Civilization" was held at a meeting of this Society. As one of the speakers on that occasion I presented a paper on "The Survival of Babylonian Methods in the Exact Sciences of Antiquity and Middle Ages" [1] in which I tried to distinguish as far as possible between those areas of ancient astronomy in which Babylonian influence was decisive and those which represent an independent development. In the present paper some sections of this earlier study will be amplified. Its main purpose, however, is methodological. The progress of modern astron-omy since Brahe and Kepler is inextricably con-nected with the ever-increasing accuracy and range of observational techniques and it therefore has seemed plausible to assume that a similar trend existed also in the first phase of astronomical development in the Greek world, that is, in the period from the beginnings in the fifth century B.C. to the crowning achievement, Ptolemy's "Almagest" in the second century A.D.

I think that this retrojection of conditions pre-vailing during the last five centuries into a fun-damentally different milieu, two millennia earlier, has resulted in a severe distortion of the actual situation and has deprived us of a better insight into the origin of scientific methods that are difficult enough to reconstruct from our frag-mentary sources. Furthermore, since most of the sources in question were made accessible through the industry and philological competence of the classical scholars of the nineteenth century we also have inherited much of their basic atti-tudes. Classicists during this period were still undisturbed by fields concerned with "Ueberresten von gemischter Art" [2] (e.g., Archaeology or Papyrology—not to mention oriental material), and so they could act *sicut Deus, scientes bonum et malum*. Thus it was simply taken for granted that "progress" from Eudoxus and Aristotle to Aristarchus and Hipparchus

could be measured by the increasing agreement with modern data and methods. Wilamowitz (in 1897) did not hesitate to declare that around 240 B.C. (in the reign of Ptolemy III Euergetes) "man arbeitete auf der Sternwarte Alexandreias an einem Fixsternkataloge"—although there exists no trace of organized observational activity before the Abbasid period. There is no need to multiply such examples of baseless anachronisms; they would easily fill another paper.

Instead, I shall make an attempt to describe a drastically different aspect that emerges from fragments of early Greek astronomy, i.e., from the period from Eudoxus (early fourth century B.C.) to Archimedes and Apollonius (i.e., to about 200 B.C.). I think it is essential for our under-standing of this early period to realize, first that its approach to fundamental problems of astron-omy is in many respects totally different from what we customarily consider to be "Greek" astronomy, and, secondly, that Greek mathematics and Greek astronomy progressed in quite distinct levels, a distinction which left its effects until deep into the Renaissance.

Two more introductory remarks. I shall ab-stain from giving the bibliographical references and the discussions of details which would be necessary to support statements made in this paper; I hope to do this elsewhere within a wider framework. Secondly, it is not my intention to present a complete picture of what I think we do know about early Greek astronomy. I have only selected certain topics which seem to me partic-ularly revealing for the situation in the formative period of Greek astronomy. But I shall far transgress the traditional chronological and geo-graphical framework of early Greek science simply because I am convinced that many a medieval and non-Greek source gives us important information about hellenistic origins.

1. MEASUREMENT OF TIME

The division of the day into 24 hours—itself the outcome of a complicated mixture of Egyptian, Babylonian, and Greek components—presents

[1] *Proc. Amer. Philos. Soc.* **107** (1963) : p. 528-535.

[2] Fr. Aug. Wolf, *Museum der Alterthums-Wissenschaft* **1** (1807) : p. 77.

PROCEEDINGS OF THE AMERICAN PHILOSOPHICAL SOCIETY, VOL. 116, NO. 3, JUNE 1972
Reprint   *Printed in U.S.A.*

243

TABLE 1

Month		hora		
		3	6	9
I et XII		17	11	17
II	XI	15	9	15
III	X	13	7	13
IV	IX	11	5	11
V	VIII	9	3	9
VI	VII	7	2	7

itself in two different forms: one, of greater popular appeal in the Mediterranean world, is the division of the time of daylight and night separately into 12 ("seasonal") hours each; the other operates with 24 hours of constant length which agree with the seasonal hours at the equinoxes, therefore called "equinoctial" hours and commonly used in astronomical contexts, though the separation seasonal/equinoctial is by no means equivalent to popular versus scientific usage. Even the same text may have both types simultaneously. The "shadow tables" to be discussed presently concern 12 seasonal hours for each month but they also give the length of daylight for the same months in equinoctial hours—without any distinction in the terminology.

It is obviously equinoctial hours that are meant when the calendar page for the month of June in the "Tres belles heures de Notre Dame" says, "*Les heurs de la nuit 6 et duiour* 18." This "Book of Hours" of the Duke of Berry (about 1400) presents us with a simple pattern for the variation of length of daylight from month to month during the year: the maximum $M = 18^h$ in June, the minimum $m = 6^h$ in December and a fixed increase or decrease of 2^h per month, i.e., a strictly linear variation between m and M. However, not only does the abrupt change from increase to decrease stand in obvious contradiction to the most elementary experience, but a maximum of 18^h corresponding to a geographical latitude of almost 58° (as tabulated, e.g., in the Almagest), correct about half-way between Copenhagen and Stockholm, but surely not in Paris or Bourges.

How these data got into a Book of Hours I cannot say. Only the antiquity of the pattern is evident: "linear zig-zag functions" are a characteristic feature in cuneiform astronomical texts of which we have a great variety from the centuries between Artaxerxes and Caesar. But this is by no means the earliest evidence for linear calendaric schemes: a hieratic papyrus, known as

the "Cairo Calendar," written in the Ramesside period (twelfth century B.C.) shows exactly the same scheme as the Book of Hours of 1400 A.D., i.e. the same extrema of 18^h end 6^h and a linear variation with 2^h each month. I can only once more admit my inability to explain the origin of the basic parameters.

Fortunately arithmetical schemes constitute the leading principle also in a group of texts, the so-called "shadow tables," where we can reach a fairly clear historical understanding. To begin with the European medieval tradition, we have, e.g., a "*Horalogium Horarum*" of the ninth or tenth century which gives in six pairs information of the following type: "*Januarius et december, hora 3 et 9 pedes* 17, *hora 6 pedes* 11." Here "*hora*" must mean seasonal hour, the 6th always representing noon. The "*pedes*" measure the length of the shadow of a man standing upright and using his own feet as units of length. The resulting scheme is very simple (table 1). The last noon shadow of 2 feet is an obvious arithmetical error, both with respect to the sequence of the noon shadows and the shadows 3 hours before or after noon that are always 6 feet $(= 1 + 2 + 3)$ longer than the noon shadow.

Many more shadow tables are preserved in Greek from the Byzantine period (thirteenth and fourteenth centuries). They lead us to distinguish two major types and to recognize a systematic error in many tables, e.g., in the above-given Latin example (table 1). The arrangement in six pairs of months is wrong: there should be seven entries, one for each extremum alone (e.g., December and June) and five pairs of equidistant months (e.g., I and XI or V and VII). This holds for both types of tables: the one which uses names of months for the entries, the other which does not depend on calendaric conventions but uses the solar positions in the signs of the zodiac. For this second type one has one single entry for Capricorn and Cancer each and five pairs of signs symmetric to the solstices.

Another solid group of shadow tables is preserved in Ethiopic codices, all of very recent date (e.g., seventeenth and eighteenth centuries) but undoubtedly copied from much older (presumably Coptic) sources. Except for the adaptation to the Ethiopic calendar and many scribal errors, these tables are closely parallel to the Byzantine ones. This parallelism is further emphasized by two peculiarities. Several tables mention for each month the length of daylight, and these numbers again form a linear zigzag function, now always with $M = 15^h$, $m = 9^h$. Since according to ancient geography $M = 15^h$ is characteristic for the "clima" of the Hellespont, and since Byzantine relations with Ethiopia are well attested for the early Middle Ages (e.g., sixth century), a transmission of Byzantine astronomy to Ethiopia seems evident here. A second element of parallelism is found in the textual preambles to the tables, which address a "King Philip." In astronomical context one will think, of course, of Philip Arrhidaeus whose regnal years constitute the basis for the "Era Philip," used, e.g., in the famous "Handy Tables" of Ptolemy and Theon.

I have no doubt that both conclusions are wrong. The preserved material amply suffices to restore the original pattern of the shadow tables (cf. table 2). Both the lengths of the noon shadows and the lengths of daylight form arithmetical sequences with difference 1. Since a daylight of 12^h is necessary for the equinoxes, such a sequence leads automatically to $M = 15^h$ and $m = 9^h$. In other words, these extrema are the consequence of a primitive arithmetical pattern, not the result of observations which we could then utilize to determine the underlying geographical location, e.g., Hellespont or Byzantium. In fact we have good evidence from Hipparchus and from Geminus that the ratio 15:9 was considered by Eudoxus, Aratus, and Attalus as representative for Greece in general. Thus our pattern does not belong to the Byzantine period but originated in early Greek astronomy in Greece, presumably at Athens. This conclusion is supported by another consideration. Shadow tables have nothing to do with a chronological era but they naturally belong to the "parapegmata," i.e. texts which associate the risings and settings of fixed stars in the course of the year with weather conditions—much in the way that our Farmer's Almanacs still do. Ptolemy, among others, wrote a whole treaties on these "Phaseis," and he cites his authorities for all predictions, e.g., "unwholesome air and turbulence according to Callippus, Euctemon and Philip; rain and thunder according to Eudoxus. . . ." Finally, Ptolemy gives the list of his authorities from Eudoxus to Caesar and mentions the regions where they obtained their climatic experiences. Here "Philip" is associated with the Peloponnesus, Locris, and Phocis, and there is little doubt that we are dealing with Philip of Opus, who flourished in the first half of the fourth century B.C. (well known because of his connection with Plato's "Epinomis").

TABLE 2

LENGTH OF SHADOW IN FEET FOR THE HOURS OF DAYLIGHT BEGINNING AT SUNRISE (NOON = 6^h), DEPENDING ON THE SOLAR POSITIONS BETWEEN WINTER SOLSTICE (♑) AND SUMMER SOLSTICE (♋)

hour	♐	♒ ≈	η ♓	♎ ♈	♍ ♉	♌ ♑	♋
1	28	27	26	25	24	23	22
2	18	17	16	15	14	13	12
3	14	13	12	11	10	9	8
4	11	10	9	8	7	6	5
5	9	8	7	6	5	4	3
6	8	7	6	5	4	3	2
7	9	8	7	6	5	4	3
8	11	10	9	8	7	6	5
9	14	13	12	11	10	9	8
10	18	17	16	15	14	13	12
11	28	27	26	25	24	23	22

length of daylight:	15	14	13	12	11	10	9	hours
noon shadow:	2	3	4	5	6	7	8	feet

Thus it seems fairly certain that the arithmetical patterns of table 2 for the shadow lengths and lengths of daylight originated in early Greek astronomy, and this conclusion is supported by the remark that the zodiacal patterns (which, in view of the Greek lunar calendars, are the original form) presuppose the Eudoxan norm which places solstices and equinoxes in the middle of the zodiacal signs. Again, Athens seems to be the plausible center for this development. A lucky accident allows us to show that shadow tables of the type discussed here appear already in the Ptolemaic period in Egypt. A papyrus fragment of an astronomical treatise (now in Vienna) has preserved a little corner of such a table, just enough to demonstrate the identity with the pattern of our table 2.

Having once established the basic structure of these tables it is no longer difficult to recognize their survival in more or less significant variations all around the Mediterranean medieval world: in an Armenian treatise, in Syriac, in Coptic, in Nubia (a Greek inscription in a temple at Taphis), in North Africa and Spain (in "Anwac̆" tables), and in monastic manuscripts of France and England. Incidentally we can now say that the apparent arithmetical error 2 in the last line of our table 1 is the only correct residue of the original pattern, whereas all other numbers are adapted to the faulty six-pair pattern that replaced the original $1 + 2 \cdot 5 + 1$ scheme. It should be noted, however, that this modification of the ancient Greek scheme is only a misguided arithmetical re-

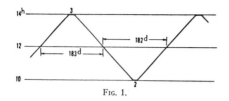

Obviously this is an attempt to describe by a simple numerical pattern the rapid increase of shadow lengths toward sunrise and sunset. The pattern is a difference sequence of the second order (or nearly so) and it is clear that it is only arithmetical expediency that determined these numbers, not any set of actual measurements, however crude.

arrangement and is by no means based on any empirical correction for different geographical situations.

It is easy to show that the assumption of a linear variation of the length of daylight was a common feature in astronomical and calendaric treatises of the early Hellenistic period. The calendar of P. Hibeh 27 (about 300 B.C.) gives the length of daylight day by day, increasing linearly during 180 days, decreasing in the same fashion for 180 days, and simply kept constant for 3 days at $M = 14^h$ and for 2 days at $m = 10^h$ (cf. fig. 1); the ratio 14:10 is the norm characteristic for "Lower Egypt" (Alexandria) in ancient geography. The multiples of the daily increment of $4^h/180 = 1^h/45$ are given for each day on the basis of the clumsy Egyptian rules for operations with unit fractions. Similar linear schemes are found throughout antiquity, e.g., with Porphyry (around A.D. 300) for $M = 15^h$, $m = 9^h$.

Strictly linear patterns are not the only arithmetical devices of early astronomy. The shadow tables, e.g., have for all months the same increase of shadow lengths per hour before or after noon (cf. table 2)

1, 2, 3, 4, 10(= 1 + 2 + 3 + 4).

Much more sophisticated arithmetical methods appear on the scene with closer contact with Babylonian astronomy during the Seleucid-Parthian period. About the details of this contact we know very little except for the evidence of fundamental Babylonian parameters in Hipparchus's lunar theory. This does not exclude earlier contacts, which are indeed suggested by the use of sexagesimal units about a century earlier (Eratosthenes). To the period of earlier borrowing probably also belong two types of arithmetical schemes (denoted as "System A" and "System B") which produce a quite satisfactory representation of the actual variation of the length of daylight during the year. Instead of assuming a linear variation of the length of daylight with its abrupt changes at the extrema, the underlying rising times of the subsequent zodiacal signs are brought into an arithmetical pattern, linear in "System A," with double difference at the equinoxes in System B. In this way the lengths of daylight become a kind of difference sequence of second order (as the sum of six consecutive rising times) with smooth approach to the extrema. Nevertheless all computations follow very simple patterns which are completely determined by the value of the longest daylight and the choice of the system, A or B.

Transplanted to Alexandria, these patterns had

TABLE 3

						Byz.		
Trigon.	I	II	III	IV	V	VI	VII	ΔM
	13^h	$13,30^h$	14^h	$14;30^h$	15^h	$15;30^h$	16^h	$= 0,30^h$
	Meroe	Syene	Lower Eg.	Rhodes	Hellespont	Mid-Pontus	Borysth.	$= 7,30°$

		I	II	III	IV	V	VI	VII	ΔM
A		$14^h = 3,30°$	$14;16^h = 3,34°$	$14;32^h = 3,38°$	$14;48^h = 3,42°$	$15;5^h = 3,46°$	$15;20^h = 3,50°$	$15;36^h = 3,54°$	$= 4°$
B		$3,32° = 14;8^h$	$3,36° = 14;24^h$	$3,40° = 14;40^h$	$3,44° = 14;56^h$	$3,48° = 15;12^h$	$3,52° = 15;28^h$	$3,56° = 15;44^h$	$= 0;16^h$
		I	II	III	IV	V	VI	VII	
	Syria								

a profound influence on Greek and medieval geography. Babylonian astronomy itself had never introduced (at least so far as we know) any element of geographical variation. The Babylonian schemes for the length of daylight are always based on the ratio 3:2, i.e., on the extrema $M = 216° = 14;24^h$ and $m = 144° = 9;36^h$. This parameter appears also in Greek geography as characteristic for "Syria" (cf. table 3) but the same pattern is expanded to seven (why seven?) exactly similar schemes such that the determining parameters M form a linear sequence of constant difference $4° = 0;16^h$. For each "clima," as these geographical steps were called, the computational method follows "System B" (incidentally, also the system to which the Hipparchian lunar parameters belong). A similar sequence of seven climata, but now centered in Alexandria and based on System A is first attested in a little treatise by Hypsicles (around 150 B.C.) with $M = 14^h$ as the point of departure (cf. table 3, middle section). The constant difference is the same as before. Both sequences are frequently found in Greek astrological literature and in the famous work of the poet Manilius who had pretentions to astronomical competence but mixed data from Babylon and System A with elements for Rhodes and System B.

A radical departure from these arithmetical patterns occurred after the invention of spherical trigonometry by Menelaos (around A.D. 100) which made it possible to compute from correct trigonometric relations the variation of the length of daylight for any geographical latitude. This astronomically correct theory is then again applied to a set of "seven climata," defined by a linear progression of longest daylight with constant difference $1/2^h$, beginning at $M = 13^h$ (Meroe in Nubia) and ending at $M = 16^h$ (Borysthenes = Dnjeper). The choice of M, instead of the geographical latitude, as characteristic parameter shows the strength of the arithmetical tradition.

This does not mean that the concept of geographical latitude (or altitude of the pole) did not play a role in Greek astronomy or geography. On the contrary, one can even get the impression that at an early time the latitude of Rhodes, $\varphi = 36°$, played a special role—often explained by the prominence of Hipparchus or Posidonius. But again numerology undermines such arguments. One of the earliest symptoms of Babylonian influence seems to be a sexagesimal division of the circumference of the circle—not into 360 degrees of the later standard, but simply a division into 60 parts such that a quadrant contains 15 parts.

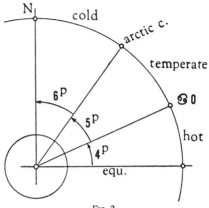

FIG. 2.

Luckily for such speculations, the obliquity of the ecliptic can be described fairly enough as representing the side of a regular polygon of 15 sides (i.e., $\epsilon = 24°$), hence covering exactly 4^p. The remaining 11^p of the quadrant can then be divided into 5^p plus 6^p with the "arctic circle" as boundary (cf. fig. 2), producing a neat numerical pattern with 4^p, 5^p, 6^p for each quadrant. This, however, assigns a fixed position to the arctic circle, that is to say, a distance of $6^p = 36°$ away from the pole and hence creating a situation demanding $\varphi = 36°$. Hence it seems at least conceivable that it is not a "school" or "observatory" which accounts for the prominence of this latitude but a cosmologic doctrine of "pythagorean" flavor.

2. COSMIC DIMENSIONS

From the period of early Greek astronomy we have three outstanding attempts at bringing cosmic motions and cosmic dimensions under the control of mathematical methods: the "homocentric spheres" of Eudoxus (and their modification by Callippus and Aristotle), Aristarchus's treatise on the sizes and distances of sun and moon, and some very strange schemes for planetary distances by Archimedes, as well as his "Sand-reckoner."

That the last-mentioned work is a work of mathematics and not of astronomy is obvious: in it the "universe" is of interest only in so far as its enormous but definite size furnishes the concrete space which nevertheless can be measured in accurately defined numerical terms. I consider it of vital importance for our understanding also of the other above-mentioned works to realize that their real goal is to demonstrate the power of the mathematical approach, not the solution of some specific astronomical problems.

The conceptual beauty and mathematical elegance of the Eudoxan homocentric spheres are undeniable, but equally evident is their inability to explain obvious details in the observable planetary motions. I think it is in vain that modern scholars have tried to reconstruct numerical parameters to be substituted in order to make the Eudoxan model represent the planetary motions with some semblance of truth. I think such numerical data never existed. And I confess that I consider it quite possible that exactly the same thing could still be said with respect to the sophisticated investigations of epicyclic and eccentric motions by Apollonius some 150 years later.

The same general attitude explains the much discussed disrespect of Aristarchus for observational data. We know from Archimedes that Aristarchus was aware of the fact that the apparent diameter of the sun and moon is about $1/2°$; nevertheless, in his treatise he computes with a value of $2°$, thus avoiding trigonometric difficulties with small angles. The basic mathematical idea to determine the ratio of lunar to solar distance is very neat (cf. fig. 3): if one knows the angle η of elongation between sun and moon when the latter appears exactly half illuminated, then one has only to solve one right triangle to obtain the desired ratio. Astronomically the method is totally impracticable: the moment of dichotomy cannot be determined with any accuracy, and the value of η is so near $90°$ that the difference falls below the limit of accuracy of visual observation—Aristarchus's value of $87°$ is purely fictitious. What really interests him in this treatise is, on the one hand, the trigonometry of the problem (at his time an undeveloped topic) and, secondly, an accurate mathematical investigation of a question which is again without any practical importance: how far is the terminator between light and darkness removed from the great circle which contains the radius moon-earth

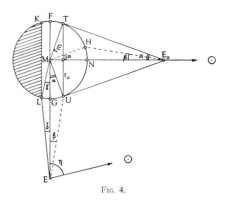

Fig. 4.

(ME in fig. 3). The answer that a minute eccentricity of the terminator cannot materially affect the ratio of the distance is, of course, obvious from the very beginning. But it is the rigorous mathematical aspects of the problem and not observational techniques which concern Aristarchus.[3] And his treatise ends without giving actual distances and sizes.

To determine the distance of sun and moon Hipparchus eventually introduced a method which is based on geometric conditions that prevail at a lunar eclipse and which are well suited for observational refinements. None of these methods were applicable to planetary distances, and ancient astronomers remained entangled in philosophical doctrines which eventually led Ptolemy to the model of nested spheres that dominated Islamic and western astronomy during the Middle Ages. In a completely different direction, however, lies a system of planetary distances, proposed by Archimedes, that escapes our understanding so far as the underlying principle is concerned.

It is a very peculiar accident which has preserved for us a record of this Archimedean theory. Hippolytus, in the first half of the third century,

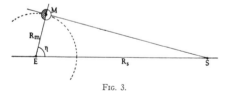

Fig. 3.

[3] The figure which belongs to the discussion of the terminator had been garbled already in antiquity, a fact not realized by Pappus (early fourth century A.D.) or by the modern editor, Sir Thomas Heath (1913). Actually two cases were condensed into one figure: one, the earth E at quadrature (cf. fig. 4), the other E_0 in the axis moon-sun in order to find the smallest terminator TU, assuming equal apparent diameter of sun and moon. The problem is to find how big $LG < GU$ appears from E. Pappus did not realize that EM must be perpendicular to E_0M and he therefore discusses the apparent size of GU as seen from E_0, which is absurd because this arc is invisible from E_0.

TABLE 4

	A		B	
From earth to	Stadia		From earth to	Stadia
Moon	5544130 =	1a + 2d		
Sun	55816195 =	11a + 3d	Venus	50815160
Venus	76088260 =	15a + 4d	Mercury	52688256
Mercury	126904455 =	25a + 7d	Sun	121604451
Mars	167448585 =	33a + 9d	Mars	132418581
Jupiter	187720650 =	37a + 10d	Jupiter	202770646
Saturn	227992715 =	45a + 11d	Saturn	222692711
Zodiac	248264780 =	49a + 12d		

$$a = 5 \cdot 10^6$$
$$d = 272065$$

Moon to Mercury
= Mercury to Zodiac
= 121360325 = 24a + 5d

Mercury to Saturn = Sun to Saturn
= 101088260

was perhaps the last Greek-educated bishop of Rome. He was of an uncompromising fighting spirit, not only directed against his more worldly competitors for the episcopal throne but also against heresies about which he composed a still extant treatise. Though not a heresy by itself, astronomy is full of speculations about the structure of the world and is thus potentially a preparation for heretical ideas. As an example of such useless speculations he gives the list of two sets of distances assumed by Archimedes (cf. table 4), unfortunately without any further details.

All that we can say is derived from the bare numbers. For no apparent reason we find two different arrangements: in A the sun is nearer to the earth than Venus and Mercury, in B it is located beyond these two planets. In the first scheme, Mercury is exactly at the midpoint between moon and zodiac (cf. fig. 5) and the distance Mercury-Saturn in A is exactly the same as the distance sun-Saturn in B. But the most surprising feature is the structure of the numbers in A. Each one of them is a linear combination of a multiple of $a = 5000000$ and of $d = 272065$ (cf. table 4). Why these numbers were chosen and why they were combined in this peculiar fashion is a complete mystery. All that the pious bishop has to say is that Archimedes should have taken numbers that satisfy harmonies postulated by

Plato. We must, alas, admit that we do not understand how Archimedes, the greatest scientist of antiquity, came to these numbers.

3. TRIGONOMETRY

Aristarchus's discussion of the solution of one right triangle demonstrates the absence of a systematic trigonometry in the time before Hipparchus. It is obviously nonsense that Theon ascribes to Hipparchus a work on chords in 12 books. The total of ancient trigonometry, plane and spherical, with proofs, tables, and applications, requires only two of the 13 books of the whole Almagest. How Pliny's error originated I do not know; at any rate it is not worth the effort to rescue this passage in Pliny by emendations. It is important, however, not to be misled by Pliny into the assumption of a very developed stage of trigonometric tables in the time of Hipparchus, comparable, say, to the table of chords in the Almagest computed in steps of 1/2°. On the contrary, I think we have good evidence that suggests a rather crude set of tables as the core of Hipparchus's trigonometry.

That Hipparchus had no spherical trigonometry in the proper sense is certain, simply because it did not become clear before Menelaus (in the first century A.D.) that spherical triangles must be made up exclusively of great circle arcs. This does not exclude, however, the possibility of solving certain problems of spherical astronomy by means of the so-called "Analemma" methods, to which also Ptolemy devoted an elegant treatise. The basic idea can best be described as using "descriptive geometry" for the transformation of three-dimensional configurations to two-dimensional problems which then only require plane trigonometry for their solution. These methods fortunately do not depend on the distinction between great circles and parallels which have such a natural place in spherical astronomy. The theory of sundials (as preserved, e.g., by Vitruvius) also makes use of the analemma methods. Though direct proof is lacking, it seems very likely that Hipparchus knew (if indeed he did not invent) these methods; it is certainly not accidental that analemma methods play a vital role also in Indian astronomy.

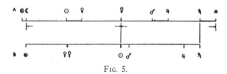

FIG. 5.

TABLE 5

M	Hipparchus		Ptolemy			
	a	Δa	a	φ	$a = \bar{\varphi} - \epsilon$	Δa
16^h	9 cubits		18°	48;32°	17;37°	
		3 cub. = 6°				5;30°
17	6		12	54;1	12;8	
		2 cub. = 4°				4
18	4		8	58	8;9	
		1 cub. = 2°				3
19	3		6	61	5;9	

In dealing with problems of spherical astronomy Hipparchus was not restricted to the Analemma; we know from data transmitted by Strabo that he also made use of arithmetical patterns that are characteristic for Babylonian astronomy. The problem in question concerns the altitude the sun can reach at the winter solstice at given geographical latitudes. The correct solution has been given by Ptolemy and the results are tabulated in the Almagest (cf. table 5, right half). Hipparchus also answered the question for latitudes where the longest daylight is 16^h to 19^h. The corresponding solar altitudes are given in "cubits," an angular measurement of 2° well attested as a Babylonian norm (cf. table 5, left). His answer agrees fairly well with the correct data, but the basis is obviously a simple sequence of second order with 1, 2, 3 as differences.

What Hipparchus's plane trigonometry looked like is not to be judged from Pliny but from Indian astronomy whose dependence on Hellenistic (and thus Babylonian) prototypes has long been recognized. Our insight into this process of transmission has moved one important step forward with the proper interpretation of two Greek papyri concerned with the motion of the moon: P. Lund Inv. 35a from the time of Nero-Domitian (A.D. 60 to 84) and P. Ryl. 27 written around A.D. 250. The methods displayed in these papyri are based on Babylonian parameters and procedures; their exact counterparts had been found two centuries ago by LeGentil in South India where he attempted to observe the Venus transits of 1761 and 1769 from the French colony at Pondicherry. He missed the transits (first because of war and then because of clouds), but he succeeded in getting native Tamil scholars to compute for him the circumstances of a lunar eclipse using rules which Cassini eventually was able to explain and which are, as we now know, identical with the methods in the above-mentioned papyri. It is certainly no accident that a site near Pondicherry, Arikamedu, was a center of Roman trade in the early imperial period. No doubt the interest in astrology, a discipline brought to its highest development in hellenistic Egypt, was the vehicle that transmitted also the strictly astronomical techniques to India.

It is not within the scope of this paper to describe how much we owe to Indian astronomy (e.g. through Varahamihira's Pancasiddhāntikā) for our knowledge of Babylonian and Greek astronomy of the hellenistic period. I shall only draw attention to some features of Indian trigonometry which, I think, reflect very accurately the type of Hipparchus's trigonometry.

In the only work of Hipparchus that has come down to us, his Commentary on Aratus (more a sharp critique than a "Commentary"), he uses peculiar spherical coordinates— arcs on the ecliptic in combination with arcs on circles of declination,[4] a system also used in Indian astronomy. As units he repeatedly uses "signs," i.e., twelfths of the circumference regardless of whether these arcs lie on the ecliptic or not. Again the same terminology is found in India.

Another peculiar terminology is revealed to us by Theon's Commentary to the Handy Tables (fourth century A.D.) according to which 15°-sections, i.e., 24ths of the circumference, are called "steps" ($\beta\alpha\theta\mu o\acute{\iota}$), in particular in relation to lunar latitudes and solar declinations. Once alerted to this concept, one finds many instances of its application from Roman Egypt to the late Byzantine period. It is particularly significant in the present context that the above-mentioned P. Ryl. 27 uses this concept in relation to the lunar argument of latitude. These "steps" also appear in the astrological literature, e.g., in Vettius Valens (second century A.D.) and in relation to planetary latitudes. An apparently very old connection exists with meteorology (i.e., to the same background from which the parapegmata and shadow tables come) since the steps of the four quadrants are related to wind directions.

Finally, we find in the geographical literature, e.g., in fragments from Eratosthenes, in Posidonius, and in Geminus, a 48-division of the circle, resulting in "parts" ($\mu\acute{\epsilon}\rho\eta$) of $7;30°$. All these units are more or less interwoven in our sources

---

[4] This peculiar system could well have been suggested by its convenience for stereographic projection which is fundamental for the "astrolabe"; but explicit evidence for Hipparchus is still missing.

and represent together a simple sequence of arcs:
30°, 15°, and 7;30°.

We can now turn to Indian trigonometry. In India the great step was made of changing the Greek "chords" to "half-chords" i.e., to sines in modern terminology:

$$\frac{1}{2} \, \mathrm{crd} \, \alpha = \sin \frac{\alpha}{2} \, .$$

Applying this transformation to the previously mentioned units of signs, steps, and parts, we obtain arcs that are multiples of 3;45°: this is exactly the norm in the Indian tables.[5] I have very little doubt that these tables are nothing but the transformation of the Hipparchian table of chords to a table of sines.

Through the intermediary of Islamic astronomy these Indian tables did reach medieval western Europe, in particular Spain and England, under the name of "kardaga," a crude rendering of a Sanskrit term meaning "half-chord." In this circuitous way, Hipparchian trigonometry returned to a part of the world from which it had originated well over a millennium before.

A concluding remark must be added concerning the practical use of tables for a trigonometric function, computed only for a few values of the argument, 3;45° apart for the sines, 7;30° for the chords (24 values in all). Here again the Tamil computers testify to the existence of a rational procedure. Their table of lunar latitudes progressing in steps of integer degrees of the argument of latitude is simply found by linear interpolation between the values computed for the multiples of 3;45° as argument. Once more we see at the very end of the "early" period of ancient astronomy the all-pervading convenience of the linear function. And it seems to me of methodological interest that Tamil astronomers of the eighteenth century can provide us indirectly with better information about early Greek astronomy than does Pliny's encyclopedia.

4. EPILOGUE

That we could use sources from the Roman imperial period until deep into the Middle Ages is due to more than a lucky accident of preservation of antiquated material. The extreme simplicity of the arithmetical methods, their entanglement with

a priori speculations of numerological character (politely called "pythagorean"), the lack of observational accuracy, all this makes early Greek astronomy ideally suited to the mental climate of the Middle Ages.

Our results imply a strong warning against periodization of cultural history. Contemporary with the haphazard steps of an elementary astronomy and mathematical geography are the most brilliant achievements of Greek abstract mathematical thought. The theory of irrational quantities by Theaetetus and Eudoxus, Aristotelian logic, Euclid, Archimedes's integrations, and Apollonius's conic sections are examples of mathematical structures whose significance was fully recovered only in modern times. Of the astronomy of the same period nothing remained that could be incorporated into the new kinematic astronomy that emerged from the work of Apollonius and Hipparchus in the second century B.C. While mathematics had clearly passed its peak, astronomy progressed to become an exact science of splendid methodology, including observational techniques, numerical and graphical methods, and theoretical optics. If one had only to rely on methodology, one would date Peurbach, Regiomontanus, Brahe, and Kepler in the century following Ptolemy. In fact it took two centuries to produce the competent mediocrity of Pappus and Theon who turned Ptolemy's work into a segment of higher scientific education, not much less sterile than the tradition followed by astrological practitioners who preserved for us so much of early Greek astronomy.

In our discussion, we have repeatedly referred to the background formed by Babylonian astronomy, whose influence is evident in the use of the sexagesimal system or in the basic procedures of arithmetical methods. Nevertheless, I think that one has to concede to early Greek astronomy a good measure of independence, in particular so far as geometrical considerations are concerned and also with respect to the trend to numerological speculation. How far the Greeks ever reached a detailed understanding of the refined Babylonian methods for the computation of lunar and planetary ephemerides is difficult to say. We only know of the use of basic Babylonian parameters by Hipparchus, and one may also conjecture that the idea of the tabulation of numerical material is an important borrowing from Mesopotamia. But the roads soon parted when geometric models became the basis of Greek astronomy which made possible a physical interpretation some 1,500 years later.

---

[5] I know of no proof for the commonly accepted explanation that 3;45° = 225′ was chosen because it is the largest arc for which sin $\alpha = \alpha$ numerically. Obviously this implies additional assumptions concerning the length of the radius and the accuracy of the tables.

ON THE ALLEGEDLY HELIOCENTRIC THEORY
OF VENUS BY HERACLIDES PONTICUS.

The *locus classicus* for ascribing to Heraclides Ponticus (and by implication to Plato) an epicyclic and heliocentric theory for the motion of Venus is a sentence in Chalcidius (ch. 110, ed. Wrobel, p. 176, 22-5) where it is said that Venus is located *interdum superior, interdum inferior sole*. This indeed seems to imply a variation of geocentric distance of the planet and hence to represent, in combination with the limited elongation from the sun, a heliocentric motion of Venus.

In fact, however, this interpretation ignores the existence of a corresponding terminology in early Greek spherical astronomy. The Greek original of *superior / inferior* is, of course, ἀνώτερον / κατώτερον which in the spherical astronomy of Theodosius [1] denotes positions in the ecliptic, equivalent to later προηγούμενος / ἑπόμενος. Any variation in depth is excluded by the very nature of spherical astronomy which exclusively concerns appearances on the celestial sphere. Hence the proper rendering of these terms is " ahead " and " behind " (in the ecliptic), not " nearer " or " farther."

Applying this terminology to the chapter in Chalcidius we may now translate the beginning as follows: " Heraclides Ponticus, when describing the circle (*circulum*) of Venus as well as that of the sun, and giving the two circles the same centre (*unum punctum*) and the same mean motion (*unam medietatem*), showed that Venus is sometimes ahead (*superior*), sometimes behind (*inferior*) the sun." What follows is a lengthy discussion of this variable elongation, reaching 50° on each side of the sun. The topic of the whole chapter thus becomes a rather trivial discussion of the shift between the morning-star and the evening-star phase of the planet and has nothing to do with any heliocentric cinematic model.[2] Whether Chalcidius himself used as illustration an epicyclic model or not (four centuries after Apollonius) is of no interest. The reference to Heraclides contains no such element.

O. NEUGEBAUER.

BROWN UNIVERSITY.

---

[1] Theodosius of Bithynia, first century B. C., *De diebus et noctibus*, I, 9; 10; II, 16 (ed. Fecht, *Gött. Abh.*, N.F. XIX, 4) ; cf. also the scholia Nos. 41 and 45: " he calls ἀνώτερον / κατώτερον what is nearer or farther with respect to the solstices." A motion in the direction of increasing longitudes always proceeds from ἄνω to κάτω.

[2] Prof. H. Cherniss drew my attention to a paper by G. Evans (*C. Q.*, LXIV = N. S. XX [1970], pp. 102-11) in which he came to essentially the same conclusion but without knowing of the simple confirmation by the terminology preserved in Theodosius.

Reprinted from
American Journal of Philology
93 (4), 600–601 (1972)

A GREEK WORLD MAP

(Pl. III 2)

It was only by accident that I came in contact with a group of illustrations in Byzantine manuscripts recently studied by A. Tihon. In an article *Les scolies des Tables Faciles de Ptolémée* [1] she discussed a collection of fifteen notes, more or less concerned with the use of the Handy Tables and related topics. In her introduction Dr. Tihon remarked that in several manuscripts the text of the scholia were followed by a group of diagrams among which she listed also *une carte géographique*. Since I had never seen a geographical map in a Greek manuscript I began to investigate this group of illustrations, using films I had at hand and several prints kindly put at my disposal by Dr. Tihon. I found this material of sufficient general interest to make it available for further study.

Before turning to a detailed description of the map in question I wish to submit what I consider as a likely date for the archetype (none of the nine manuscripts investigated is earlier than the 13th or 14th century). Two names which seem to me chronologically significant appear on the map of Egypt : Heptanomia and Hiera Sycamenos. The first represents an administrative district of Middle Egypt, established in the early Roman imperial period and well attested for centuries after. The second refers to a locality in Nubia, halfway between Aswān and Abū Simbel, [2] subject to Roman rule well into the third or fourth century. Finally a strictly pagan description of the Netherworld can also be taken as supporting a date in the first two or three centuries of our era. Thus our map may fairly secure be located in Egypt during the Roman imperial period.

The manuscripts used are :

La : Laur. gr. 28,1 (fol. 177ʳ)	M : Marc. gr. 314 (fol. 222ᵛ)
Lb : 28,7 (107ʳ)	Ox : Oxon. Canon. 32 (17ʳ)
Lc : 28,12 (296ᵛ)	P : Par. gr. 2390 (155ᵛ)
Ld : 28,47 (271ᵛ)	Sc : Scor. *Φ* I 5 = Esc. gr. 183 (31ᵛ)
	V : Vat. gr. 183 (21ʳ)

The stemma codicum established by A. Tihon is valid also for the map. It suffices here to say that Ld remains isolated (the map is left incomplete), that P, La, V form

---

[1] *Bulletin de l'Institut Historique Belge de Rome*, 43, 1973, p. 49-110.
[2] Modern el-Maḥarraqa ; cf. Porter-Moss, *Topographical Bibliography* VII, p. 51 and Map I.

one coherent group, and that the rest shows only insignificant variants. In Pl. III, 2 I reproduce M as a typical example ; Fig. 1 shows schematically the general location of the textual passages.

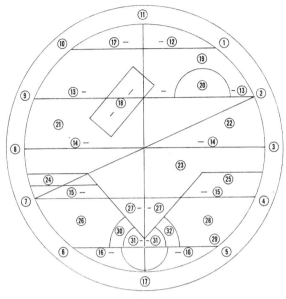

Fig. 1

Our « map » is not based on any rational method of projection ; we have to « read » the story the designer wished to convey. The outer ring gives the conventional scheme of directions for any horizon in the hellenistic world (text (1) to (10)). Then follows a vertical cross section of the terrestrial sphere, indicating the axis from pol to pol, the planes of the equator, the tropics, and the arctic and antarctic circles (text (11) to (17)). The space left inside the upper semicircle contains the real « map » : Egypt and Persia and to the south the Ethiopic and Indian Seas (text (18) to (22)). Below the equator a « Fiery Sea, not navigable » separates us from the « Ocean of the inhabited world of the antipodes »[1] to the south of the winter tropic (text (23), (26) to (28)). Finally we can look into the netherworld : in the surface of the « earth » ((24)-(25)) opens a deep chasm which leads down to the rivers of the netherworld and the Acherousian Marsh (text (30) to (32)).

[1] I translate by « antipodes » what would be more accurately the inhabitants of the Counter-Oicoumene.

Texts

Winds :

(1) βορέας

(2) καικίας ; in La, Ld, V placed one step too low

(3) ἀπηλιώτης ; omitted in La, Ld, V. On margin of Ld, P : ἰσημερία ; cf. (8)

(4) εὖρος

(5) φοινικίας ; in La, Ld, V incorrectly combined with (29) to γῆ φοινική

(6) νότος

(7) λίψ

(8) ζέφυρος ; on margin of La, Ld, P, V : ἰσημερία ; cf. (3)

(9) ἰάπυξ

(10) θρασκίας

Zones :

(11) βόρειος πόλος

(12) ἀρκτικὸς κύκλος

(13) θερινὸς τροπικός

(14) ἰσημερινὸς κύκλος ; cf. (3) and (8)

(17) νότιος πόλος

(16) ἀνταρκτικὸς κύκλος

(15) χειμερινὸς κύκλος ; cf. (28)

World Map :

(18) inside the rectangle : κάτω χώρας

ἑπτανομία

συήνη at intersection with axis and summer tropic

ἱερὰ συκάμενος

μερόη λίμνη

I do not know of a « Marsh » of Meroe, only of an « Island ». In La, P, V λίμνη
is replaced by λιβύη. Perhaps this is a scribal error since these manuscripts (as
well as Lc and Ld) begin (21) with λιβύη.

(19) περσίς

(20) περσικὸς κόλπος τῆς ἐρυθρᾶς. The semicircle which encloses these words
probably indicates that the « Persian Gulf of the Red Sea » extends to the north
of the summer tropic.

(21) αἰθιοπικὸς ὠκεανός, preceded by λιβύη in La, Lc, Ld, P, V

αἰθιοπικὸν πέλαγος

(22) ἰνδιακὸς ὠκεανός

(23) διάπυρον πέλαγος ἄπλωτον

ὠκεανὸς κατὰ τὴν ἀντοικουμένην. Both omitted in Ox

(26) ἀντοικουμένη ζώνη εὔκρατος, ἡ κατοικουμένη τῶν καταχθονίων.

Temperate zone for the inhabited world of the antipodes, (i.e.) the inhabited world
for the subterraneans.

(27) ἡ κατὰ τὴν ἀντοικουμένην θάλασσα. The Sea of the antipodes.
La, P, V add : ... σατηθῆς καλεῖται « it is called... » I do not know how to explain this name.
(28) ὁ μεταξὺ τόπος τοῦ χειμερινοῦ τροπικοῦ καὶ τοῦ ἀνταρκτικοῦ τῷ πλάτει σταδία μ̄ : The region between winter tropic and antarctic (circle) is in latitude 40 stades.
(29) γῆ. I see no reason for giving in (28) a terrestrial distance and its amount makes no sense. The manuscripts write either σταδία μ̄ γῆ or combine γῆ with (5). Perhaps μ̄ is a mistake for μύριοι (10000) and γῆ is a misreading for some numeral. At any rate γῆ makes no sense here in relation to (24) and (25).

Netherworld :

(24) (25) γῆ ; cf. (29). Sc omits (25).
(30) λήθης ποταμός ; omitted in Sc.
(31) πυριφλεγέθων ποταμός
(32) ἀχερουσία λίμνη.

A map of the here described kind must have become the prototype of a whole class of mediaeval world maps, preserved in Isidorus of Seville's *De natura rerum*. A sample from the 13th century is reproduced in *Mappemondes A.D. 1200-1500* (réd. Marcel Destombes) planche D. Four of the twelve winds in the outer ring are named : *Boreas, Eurus, Auster, Zephirus*. The equatoreal zone is crossed by a diagonal *Zodiacus*. The inhabited world is schematically divided in *Europa, Africa, Asia*. To the south of the equatoreal *mare magnum* is the *temperata, incognita* zone of the antipodes, but the Acherousian Marshes are gone.

As said before our Byzantine world map is associated with notes more or less related to the Handy Tables. Neither our map, however, nor several of the other illustrations around it, have anything to do with Ptolemaic astronomy. Nevertheless there exists some connection even with such high level works as the Handy Tables of Vat. gr. 1291, written in the 9th century. This beautiful manuscript shows already some adaptations to Christian usage, e.g. a circular table of « epacts », important for the Easter computus. This table covers the years from Diocletian 30 to 257 (i.e. A.D. 313/4 to 540/1) while marginal notes contain examples for the years Diocletian 239 (A.D. 522/3) and 541 (A.D. 824/5). Exactly this table appears also in several manuscripts which display our world map.

Another adaptation concerns Ptolemy's circular table of solar rising amplitudes for the seven climata. [1] An eighth ring for Byzantium has been inserted between clima V and VI in one of our manuscripts (Ld). The values for the azimuths are

---

[1] Opera I ed. Heiberg, folding table at the end (to Alm. VI, 11, p. 539).

simply the arithmetical means between the values in the rings above and below. All other manuscripts (including Vat. gr. 1291) give a separate diagram headed ὁρίζοντος καταγραφὴ τοῦ διὰ βυζαντίου with numbers which are independently computed, not simply found by interpolation.

Also two other groups of figures seem to me definitely related to pre-Ptolemaic astronomy, i.e. to a period which is still under the direct influence of Babylonian arithmetical methods. One set of such figures shows a V-shaped list of numbers which represent in a strictly linear fashion the variation of the lunar latitude in relation to the moon's motion in its orbit. As units of angular measurement are chosen βαθμοί, « steps », corresponding to arcs of 15° each. It can hardly be doubted that these steps were the units in Hipparchus' table of chords from which the Indian table of sines was derived [1]. Also these primitive concepts are reflected in the Handy Tables of Vat. gr. 1291, although not for the moon but for Mercury. This suggests the existence of tables of steps for all five planets.

The second group of figures consists of elaborate tables, in part again V-shaped, for magnitudes of lunar eclipses. The level of these tables is once more undoubtedly pre-Ptolemaic. The increase of the eclipse magnitudes is assumed to be strictly linear as function of the argument of latitude, measured in steps. Totality is represented by the magnitude of 12 digits, regardless of longer or shorter duration. As units of time serve thirtieths of the mean synodic month, a norm of Babylonian origin, also transmitted to India where it soon occupied a central position in all time reckoning.

These short remarks must suffice as proof that the small collection of diagrams and tables, accidentally attached to short astronomical notes, carries along remnants from a period some three centuries older than Ptolemy's Handy Tables. It is also clear that the world map is only the latest of these non-Ptolemaic accretions to our collection.

We have seen that our world map belongs to a type that survived to the end of the European Middle Ages. Another type reached the West through oriental (Islamic) sources. Vat. gr. 211 is a Byzantine manuscript (of the 13/14 century) which also harbors some folios of otherwise unrelated astronomical diagrams [2]. Among these we find two slightly different versions of a world map (fol. 120ʳ and 121ᵛ). This « map » consists of a circular area which contains in its lower half (that represents, following oriental custom, the northern hemisphere) seven parallel strips for the seven climata. A second, slightly eccentric circle, provides the space for the

---

[1] Cf. G. J. TOOMER, *The Chord Table of Hipparchus and the Early History of Greek Trigonometry.* Centaurus 18, 1973, p. 6-28.

[2] One of these figures displays the mechanism introduced by al-Ṭūsī (13th century) for obtaining straight oscillations by means of uniformly rotating circles, a device of central importance to Copernicus.

ὠκεανός. On the left we find ἀνατολή, ὁ παράδαισος, on the right δύσις, ἡ γῆ τῶν μακάρων, and the 7th climate is named ἡ τούλη. The lowest segment, between arctic circle and the pole, is βόρειον ἀοίκητος. Other details (e.g. the diameter of 3233 μίλια) do not concern us here, being oriental modifications. But the drawing on fol. 121ᵛ is accompanied by a long scholion which belongs to a very different milieu. After some discussion on geographical longitudes and latitudes (in degrees) the scholion concludes with a list of the boundaries for the seven climata, reckoned in degrees of latitude [1]:

I	II	III	IV	V	VI	VII
15°	23	30	36	41	45	48

These latitudes differ (excepting IV) from the trigonometrically determined values in the *Almagest* (II, 6 and 8). But their structure is nevertheless plain ; one need only look at the intervals

$$8 \quad 7 \quad 6 \quad 5 \quad 4 \quad 3$$

to see that we have here again one of the arithmetical patterns which are characteristic for many topics in early hellenistic astronomy. Below the recognizable chronological sequence of the still extant documents lies a succession of significant, but totally irrational, events, a real πέλαγος ἄπλωτον.

Providence, R. I.   O. NEUGEBAUER

---

[1] The Vatican catalogue misreads (p. 264, fol. 121) τὸ ἕβδομον ἕ(ως) τῶν μ̄η̄ to ... ἐτῶν μη΄.

Pl. III

Gasr - el - Libia. Mosaïque du Phare.

Marcianus gr. 314 (fol. 222ᵛ). A Greek World Map.

Reprinted from
Hommages a' Claire Préaux, 312–317
L'Université de Bruxelles, 1975

5. Medieval-Renaissance

THE ASTRONOMY OF MAIMONIDES
AND ITS SOURCES

O. NEUGEBAUER, Brown University, Providence, R. I.

TABLE OF CONTENTS

THE ASTRONOMY OF MAIMONIDES
AND ITS SOURCES

INTRODUCTION

MATHEMATICAL astronomy owes an enormous debt to the institution of lunar calendars. The apparently simple question whether a month will be full or hollow, i. e., whether the new crescent will be visible on the evening of the 29th or of the 30th day, led Babylonian astronomers of the fourth century B.C. to ingeniously constructed arithmetical devices which enabled them to compute ephemerides of great accuracy for the movement of the sun and the moon. We know practically nothing about the underlying concepts concerning the physical nature of the treated phenomena. In contrast, we are well informed about the geometrical interpretation which formed the basis of the corresponding theory of Greek astronomers, at least so far as reflected in the Almagest. Finally, Ptolemy himself brought the theory of the planetary movement to the same level which the lunar theory had reached centuries before.

In the course of this development, the problem of the visibility of the new crescent disappears almost completely from Greek astronomy. The originally lunar character of the local Greek calendars was apparently overshadowed in comparatively early times by institutions which depended on the civil administration. Thus Metons's attempt in 432 B.C. to introduce order and regularity into the Athenian calendar met with no success. Its principle, however, a simple cyclic arrangement for the intercalation of lunar months in order to obtain periodical agreement between solar and lunar phenomena, had the greatest influence on the future development. This is not the place to discuss the question whether the "Metonic cycle" was Meton's own invention or was of Babylonian origin. All that we know with certainty

322

is that this same cycle became the regulating principle of the Mesopotamian luni-solar calendar during the very same period, thus preceding by a century or a century and a half the successful attack of the visibility problem.

Usefulness and tradition kept this cyclic calendar alive for many centuries, spreading with small variations all over the ancient Near East. Its principle is reflected in the computation of Easter. Though the discussion of this problem was not free of the violence which is so conspicuous in the development of the various Christian doctrines, the focusing of a dogmatic interest on a calendaric problem contributed to the preservation at least of some astronomical tradition in mediaeval Europe.

We know very little about what happened to the theoretical part of Babylonian lunar theory. Lunar ephemerides are preserved almost to the very latest years of cuneiform writing on clay tablets, i. e. to the last decades B.C. Astronomers like Hipparchus who utilized Babylonian experiences and methods might have obtained their information from Greeks living in Mesopotamia, perhaps more so than from men of the type of Berosus, whose (admittedly very fragmentary) writings do not reflect any knowledge of technical details. For the Near East it might be agreed that Babylonian astronomy may have been preserved in Aramaic and Hebrew books, now totally lost. A hypothesis of this type needs specification if it is to be considered seriously. The extant Babylonian texts are "ephemerides," computed for specific years, or "procedure texts" containing rules in extremely condensed and often enigmatic form for the practical computation of these ephemerides. To understand them, a simultaneous oral tradition must be assumed. With the disappearance of this tradition and with the halt of the year-by-year computation of ephemerides, there was nothing left to be handed down in Aramaic or Hebrew. Only if we assume that the problem of lunar visibility itself remained of active interest, is the continuity of Babylonian tradition worth discussing as a hypothesis.

From this viewpoint it is only natural to investigate the Jewish calendar for its relation with Babylonian astronomy. Its strictly lunar character is well known; the conditions for

direct contact are most favorable, and, at the same time, the
need for accurate prediction instead of witnessed observations
must have been felt most urgently. In a paper published in 1919,
D. Sidersky tried to show that a Babylonian ephemeris for the
year 133/132 B.C. used criteria for the visibility of the new
crescent which were essentially the same as those still used by
Maimonides 1300 years later.[1] It was this relationship which
gave the original impetus to the present investigation. Nevertheless, we will not have to discuss Babylonian methods in the
following paragraphs. Sidersky had only one ephemeris at his
disposal and even this text was not complete. Neither the now
completed text nor additional ephemerides confirm Sidersky's
results. Though there undoubtedly exists a certain parallelism
between the Babylonian approach to the visibility problem and
the methods found in Maimonides's work on the Sanctification of
the Moon, there is no hope of establishing a direct connection
with Babylonian astronomy. The present paper is intended to
reach a much more modest goal, namely to establish a very close
relationship between Maimonides and Al-Battānī (ca. 900 A.D.),
at least so far as mathematical astronomy is concerned. As to
the purely calendaric part of Maimonides's work, the relation
with Jewish tradition is evident to such a degree that it can be
justly doubted whether Maimonides added anything of his own
to this part of his subject. At any rate, I do not feel qualified
to pass judgment on this question.[2] All parts, however, of
Maimonides's astronomical work offer so much of interest that
I shall not exclude those sections where I must leave the question
of authorship unanswered. Whatever the origin of any part of
Maimonides's treatise might be, the presentation of the material
shows everywhere the great personality of the author and
supreme mastery of a subject, worthy of our greatest admiration.

---

[1] Cf. the Bibliography under Sidersky [1].

[2] It is evident, however, that Friedländer's statement (Guide p. XXII)
"The section on the Jewish Calendar ... may be considered as his original
work" cannot be literally true. It suffices to quote Al-Bīrūnī, Chronology
of Ancient Nations, chapter VII (written 1000 A.D.) where all the essential
elements of the cyclic Jewish calendar are described.

MOLADOTH AND TEKUFOTH

1. When I wrote in the Introduction that Sidersky's assumption of a direct connection between Maimonides and Babylonian astronomy cannot be upheld, I did not mean to deny the existence of an indirect relation of a more general kind. It is evident, e. g., that the use of the division of the circle into 360 degrees and their sexagesimal parts is of Babylonian origin, as is also the division of the zodiac.

Less obvious, however, is the fact that the division of the hour in 1080 "parts" (chelakim) also points to the use of Babylonian units. Maimonides, in VI,2,[1] motivates the choice of this fraction by the fact that 1080 contains all integers from 1 to 10 as its divisors with the sole exception of 7. Yet this quality would be shared by 360, not to mention the fact that metrological units are not constructed in such an artificial way but are the outgrowth of practical needs and many compromises between disparate systems. In our specific case, we must go back to the Old-Babylonian unit of the "Barleycorn" (called še in Sumerian) which is 1/180 of the shekel. This relation was still kept alive in the Neo-Babylonian period[2] and is again attested in use in the ephemerides of the Seleucid period as 1/180 of the cubit.[3] Because 1 cubit corresponds to $2\frac{1}{2}°$ we find that 15° contain 1080 barleycorn. But 15° of the equator correspond to one hour, which gives the relation we have under discussion. It might be added that the "barleycorn" as a metrological unit for terrestrial distances is well attested in oriental sources.[4]

Another case of continued Babylonian tradition was already mentioned, namely, the Metonic cycle of intercalation. During

---

[1] References of this kind always mean chapter and section of the "Sanctification of the Moon."

[2] Cf. Sachs [1].

[3] Cf. Neugebauer [1].

[4] Without attempting completeness I quote: Old Armenian (7th or 8th century) Mžik [1] p. 43, p. 87. Bar-Hebraeus (ca 1250), Nau p. 178. Abū-l- Fidā (ca 1300), Reinaud I p. 264 ff. Cf. also Sauvaire [1] p. 482 and Neugebauer [2] p. 280.

19 years a 13th month must be added 7 times. In VI,11 Maimonides gives the numbers of these intercalary years within the
cycle as 3, 6, 8, 11, 14, 17 and 19. Exactly the same arrangement
is found in the Babylonian calendar of the Seleucid period[5] if we
identify the year which intercalates a second Ulul with the
year 17.[6]

Finally we must mention an important astronomical constant:
the length of the mean synodic month. If we transform the value
of 29½ days 793 parts, given by Maimonides in VI,3, into days
and sexagesimal fractions we obtain*

$$m = 29;31,50,8,20^{d}$$

This value is well known, e. g., from the Almagest (IV,2) and
is the basis for System B of the Babylonian lunar theory.[7]
It is equally fundamental for the Jewish calendar system.

Similar relations can be established for other constants,
though without adding any new insight into the transmission
of astronomical knowledge from Babylonian sources into Greek
and Arabic science. For Maimonides all these sources might be
called "indirect sources" whose actual origin was certainly unknown to him.

2. The above-mentioned value m for the "mean synodic
month" is a mean value in the proper sense of the word. Though
the length of the real synodic months shows considerable variation, a mean value can be obtained by dividing the total
length of a great number of synodic months by their number.
Without dicussing the details of this well known method, it
is clear that a value m of great accuracy can be reached if one
has a sufficiently long interval of observations at one's disposal.

The schematic calendar, discussed by Maimonides in the

[5] Kugler, SSB I, p. 212.

[6] Al-Bīrūnī, Chronol. p. 64/65 (transl. Sachau) mentions three different
countings of this same cycle as being used by the Jews. The years corresponding to a second Ulul are 16, 15 and 14 respectively, thus different from
Maimonides. The last counting is said to be of Babylonian origin and thus
most popular among the Jews.

* Here and in the following we use a semicolon to separate integers from
sexagesimal fractions. Thus $29;31^{d}$ means $29+31/60$ days. Similarly $5;20,6°$
$= 5°20'6''$ etc.

[7] Kugler, BMR p. 7, p. 24, p. 111.

chapters VI to X of the Sanctification of the Moon (and similarly in an earlier independent little treatise[8]), is based on this value m for the mean synodic month. The years of the world are counted from the first of Tišre of the first year, and this day is supposed to have fallen on the second day of the week, which we denote by (2), counting Sabbath as (7). The moment

year 1   Tišre 1 5^h 204^p

is denoted (VI,8) as "*molad Tišre*" of the first year (1^p being the above-mentioned "part"$=1/1080$ of one hour). Adding to this moment the constant amount

$$m = 29\tfrac{1}{2}^d \; 793^p$$

we obtain the "molad" of the next month (VI,15). Adding six times the value of m to the molad Tišre we obtain the "*molad Nisan*" of the first year

year 1   Nisan 1 9^h 642^p

which is the 4th day of a week. Proceeding in this fashion, we obtain a sequence of moladoth with constant interval m from molad to molad. Each day into which a molad falls is in principle the first day of a month; "in principle" means that there are certain rules (chapter VII) according to which the beginning of a month must be postponed for one or two days, e. g. if the molad falls on one of the days (1), (4) or (6) of the week (VII,1). These exceptions, however, influence only the beginning of individual months and do not disturb the regular arrangement of the moladoth.

It is customary to consider the moladoth as "mean conjunctions." We shall show presently that this is not accurate. To this end, we must first define the concept "mean conjunction." As before, we can find mean values for the movement of the sun and the moon, and we can introduce ideal celestial bodies, called "mean sun" and "mean moon," which move exactly with these mean velocities. The real sun and the real moon will alternatingly move faster or slower than the mean bodies. It is evident that the conjunctions of the mean sun and the mean moon will always be equidistant with the value m as interval. It is also evident that we know all these "*mean conjunctions*"

---

[8] Published by Dünner; written 1158 A.D.

if we decide about the moment of any one of them. Because the mean conjunctions show the same distance m from each other which was used for the definition of the moladoth, it suffices to investigate the relation between any one molad and the nearest mean conjunction. If *one* such pair coincides, *all* moladoth will fall on mean conjunctions. On the other hand, a single discrepancy proves separation between all moladoth and all mean conjunctions.

The choice which was made for the position of the mean conjunctions can readily be determined from the following remarks. In XII,4 we are told that in the year 4938 of the world Nisan 3 0^h, the mean sun was in ♈ $7;3,32°$; and from XIV,4 we learn that the mean moon was at the same moment in ♉ $1;14,43°$. From XIV,2 we know that the mean velocity of the moon is $13;10,35°/^d$, of the sun $0;59,8°/^d$. Thus the mean elongation is $12;11,27°/^d$. Because the elongation on Nisan 3 0^h was $24;11,11°$, it is evident that a mean conjunction fell very little less than 2 days earlier, i. e. in the first hour of Nisan 1. Because we are also told (in XI,16 or in XII,4 etc.) that the Nisan 3 of the epoch 4938 fell on the fifth day of a week, we know that the mean conjunction in question fell on Nisan 1 0^h 415^p and that this day was the third day of a week.

It is now easy to prove that the moladoth never coincides with mean conjunctions. The 4937 completed years from the beginning of the world up to the above epoch contain 259 cycles of 19 years (of 235 months each) and 16 additional years, 6 of which are intercalary years.[9] If we multiply the corresponding total number of months by the value of m and reduce the result modulo 7 days* we obtain 5^d 16^h 79^p. We now use the fact that according to definition

$$\text{molad Nisan of the year } 1 = (4)^d \; 9^h \; 642^p.$$

Thus we obtain by adding 5^d 16^h 79^p for the molad Nisan 4938

$$\text{molad Nisan of the year } 4938 = (3)^d \; 1^h \; 721^p.$$

But we have already seen that the corresponding mean conjunction fell on $(3)^d$ 0^h 415^p i. e. 1^h earlier. Hence all moladoth

---

[9] Cf. the rule in VI, 11 quoted above p. 326.

* This is the standard mathematical terminology for saying that we disregard multiples of 7. Cf. note * on p. 333.

follow all corresponding mean conjunctions by the same amount.[10]

The conventional statement that the moladoth are the mean conjunctions is thus proved to be false. This would not be too interesting in itself if Maimonides did not seem to·have overlooked the consequences of the definitions which he himself gave. In VI,1 he seems indeed to indicate that molad means mean conjunction.[11] The question arises how such a statement can be explained. It seems to me that the way to the answer can be found from the fact that Maimonides makes an equally wrong statement about the tekufoth. In IX,3 he says that the four "*tekufoth*" coincide with the entry of the sun into the signs of Aries, Cancer, Libra and Sagittarius respectively. On the other hand, in IX,2 the distance from tekufah to tekufah is defined to be exactly one quarter of the "solar year" (VI,4) of 365¼ days. But no ancient or mediaeval astronomer was unaware of the fact that the solar year is not exactly 365¼ days long and, most important, that the four seasons are of unequal length. This latter fact is the foundation for the determination of the eccentricity of the solar orbit, and it would be absurd to assume that Maimonides was ignorant of so fundamental an element. But we have even the explicit assurance that he had full insight into these facts. In X,1 he refers to the much more accurate year of R. Adda, and in XIII,10 he remarks that the anomaly of the solar movement can be used in turn to determine the length of the seasons.

These observations fit exactly the situation which characterizes the whole section on the "Sanctification of the Moon" in the Mishnah Torah. On the one hand, chapters VI to X teach a purely schematic determination of the beginning of the months. Chapters XI to XVII, however, contain the most accurate methods for the determination of the moment of real visibility which will only accidentally lead to the same result as provided by the schematic calendar. Thus it is not surprising at all to find in the calendaric section "molad" called "mean

[10] The exact amount of the difference can be easily computed but is hardly significant.

[11] Similarly in his earlier treatise on the calendar (cf. Dünner. p. 28).

conjunction" and the tekufoth identified with the characteristic ponts of the year while the astronomically exact definitions are reserved for the second part, leading to different results. In other words "molad" in the first part is used in its original historical sense, while the added explanation as "mean" conjunction is strictly speaking correct only in so far as the distances between these moladoth is exactly the distance between mean conjunctions. But Maimonides did not see any need to underline the fact that the astronomically accurate definitions would lead to slightly different moments — a fact which must have been evident to every astronomer of his time.

It is interesting to remark that the same situation prevailed in the Seleucid period in Babylonia. Kugler found already 50 years ago[11a] that the lunar ephemerides recognized the anomaly of the sun and consequently the inequality of the seasons. But it was only recently that it became evident that all solstices and equinoxes which we find in the so-called observation texts were not observed but computed according to a simple scheme.[11b] In this scheme the "tekufoth" are spaced evenly over the year and follow each other with constant time interval such that the same dates are restored after the completion of a 19-year cycle. Thus we find here exactly the same apparent contradiction as in Maimonides between a simplified calendar scheme and the accurate theory which is only used, however, when delicate problems like eclipses and first visibility are involved.

3. As a matter of fact it would be of no advantage to identify exactly molad and mean conjunction. In a real lunar calendar the conjunction precedes the evening of first visibility, which marks the beginning of a month, by a variable interval of, roughly, 20 to 50 hours. The second part of Maimonides's discussion shows the complicated methods which are necessary to pass from mean conjunction to actual visibility. A cyclic calendar was compelled to be satisfied with a spacing of the months which would be correct in the average. This is to say that the first day of a month should always contain a moment

[11a] Kugler, BMR p. 83 ff.
[11b] Neugebauer [3].

which is exactly one mean synodic month later than the corresponding moment in the first day of the preceding month. These points have, of course, constant distance from the mean conjunction. Using the mean conjunctions would only have meant adding a constant amount to each of them to reach the corresponding moment which falls in the first day of a month. Thus it was the simplest solution to disregard the mean conjunctions completely and to operate with moladoth falling within the first days of the months (providing, of course, for additional rules for "postponements" if ritual reasons so require).

The question might be asked what arrangement was made when the cyclic calendar was inaugurated. One may assume that the series of moladoth started with a certain real lunar month. This was apparently the opinion of Ideler who assumed that the moladoth were moved towards the evening of first visibility.[12] Schwarz, however, insisted[13] that all festivals were moved forward to fall on the (mean) conjunction. Actually, our computation shows that neither assumption is correct for the time of Maimonides because the mean conjunctions precede by only about one hour the moladoth. Maimonides most likely refers to this fact when he says (VII,8) that the real purpose of the "postponements" is of astronomical character. Indeed the postponements move the beginning of the month back, closer to the new crescent.

A complete explanation of the arrangement described by Maimonides would be possible only if we know the time of its origin. It is certainly connected with the assumptions made concerning the era of the world and the corresponding epochs. The choice of the zero hour for the tekufah Nisan of the first year agrees with our expectation. No such simple explanation can be given, however, for the epoch of the moladoth. The real starting point is evidently some later zero point, unknown to us. The possibility of later modification must also be kept in mind, especially because the epochs given by Al-Bīrūnī[14] do not agree with Maimonides.

[12] Ideler, Chronol. I p. 543 f.
[13] Schwarz, JC p. 58 f.
[14] Chronology p. 144 f. transl. Sachau.

Whatever the historical reason for the discrepancy between moladoth and mean conjunctions may be, it should be underlined that our result does not depend on any historical hypothesis. The difference between moladoth and mean conjunctions was found by using exclusively numerical values given by Maimonides in two different sections of the same work. All we did was to follow Maimonides's rules for computing the moladoth and comparing the result with the date of the mean conjunction directly derived from elements quoted by Maimonides. This comparison rests on simple arithmetic only and is totally independent of any historical considerations whatsoever. From the purely astronomical viewpoint our result is not surprising at all. The basic value m of the length of the mean synodic month and of the interval between consecutive moladoth is a fraction of a second too long. Thus a discrepancy between moladoth and mean conjunctions must accumulate, even if exact coincidence did exist at a certain moment. Arabic astronomers overcompensated the error, as is known, e. g., from al-Bīrūnī (Chronol. p. 143) and the divergence found in Maimonides obviously reflects the result of such adjustments as compared with a schematic computation with fixed moladoth. Thus our observation only confirms the statement of al-Bīrūnī that there are divergencies between "the theory of the Jews themselves" and "that of the astronomers" when computing the length of the lunar nonths.

4. There is no need for a detailed description of the arithmetical rules which determine the cyclic calendar.[15] It is evident that both molad and tekufah can be determined exactly by purely arithmetical operations because the initial values and the differences are given. The same holds for the days of the week because all remainder modulo 7 of the characteristic parameters are also known. This latter fact is utilized for the solution of our problem. It is easy to determine the weekday of the molad and of the tekufah Nisan. Then an estimate is

[15] To my knowledge it was C. F. Gauss, who first gave a consistent arithmetical rule for the computation of the 15th of Nisan and the 1st of Tišre (Gauss, Werke 6, p. 80 f;10,1 p. 560;11,1 p. 215–218 with commentary and references to subsequent discussions by A. Loewy).

obtained for the time between molad and tekufah, accurate enough to limit us to less than 7 days. Combined with the information about the days of the week, the date of the tekufah is determined uniquely.

We can illustrate this method with the example given by Maimonides in IX,7: find the molad Nisan for the year 4930. We denote the consecutive days of the week by $(1)^d$, $(2)^d$, ..., $(7)^d$ and prove: (a) the first of Nisan will fall on $(5)^d$; (b) the tekufah Nisan will also be $(5)^d$; (c) the tekufah Nisan falls at least 5 days after the molad Nisan, but less than 12 days. Assuming for the moment (a), (b) and (c) as granted, we know that Nisan $1 = (5)^d$ thus Nisan $6 = (3)^d$ which is the earliest possible date for the tekufah because of (c). But from (b) we know that the day of the tekufah is not $(3)^d$ but $(5)^d$. This $(5)^d$ can only be Nisan 8 because the next $(5)^d$ is Nisan 15, which is more than 12 days after Nisan 1. Thus the exact date of the tekufah Nisan of the year 4930 is Nisan 8.

The statements (a), (b) and (c) are easily proved.

(a). 4929 complete years contain 259 cycles of 19 years and 8 additional years, 3 of which are intercalary (VI,11). If m denotes the length of the mean synodic month as given in VI,3 (cf. p. 326) we have*

$$19 \text{ years} = 235 \; m \equiv 2^d \; 16^h \; 595^p \quad \text{mod. } 7^d \quad (VI,12)$$
$$12 \; m \equiv 4^d \; 8^h \; 876^p \quad \text{mod. } 7^d \quad (VI,5)$$
$$13 \; m \equiv 5^d \; 21^h \; 589^p \quad \text{mod. } 7^d \quad (VI,5)$$

Hence

$$(259 \cdot 235 + 5 \cdot 12 + 3 \cdot 13) \; m \equiv 1^d \; 3^h \; 412^p \quad \text{mod. } 7^d.$$

From VI,8 we obtain for the epoch

$$\text{molad Nisan year } 1 = (4)^d \; 9^h \; 642^p$$

and therefore we obtain by adding $1^d \; 3^h \; 412^p$

$$\text{molad Nisan year } 4930 = (5)^d \; 12^h \; 1054^p$$

which proves our statement (a).

---

* One uses the notation $a \equiv b$ mod. c (read "a congruent b modulo c") if the difference of a and b is divisible by c. Example: $36 \equiv 26$ mod. 5. Hence $a \equiv 0$ mod. c means that a is a multiple of c. This notation is useful in all cases where one is not interested in multiples of a given period, e. g., in multiples of 7 days in dealing with days of the week, or in multiples of 360° for points on the circle. Thus $370° \equiv 10°$ mod. 360°.

(b) Because

$$1 \text{ "solar year"} = 365^d \ 6^h \equiv 1^d \ 6^h \quad \text{mod. } 7^d$$

and

$$4929 \equiv 1 \quad \text{mod. } 28$$

we find that the tekufah Nisan of 4930 falls $1^d \ 6^h$ later than

$$\text{tekufah Nisan year } 1 = (4)^d \ 0^h \ 0^p \quad (IX,3)$$

thus in day (5) as stated above.

(c) According to IX,3 the tekufah Nisan of the first year of the world preceded the molad Nisan by $7^d \ 9^h \ 642^p$. The spacing of the tekufoth is based on the "solar" year of $365^d \ 6^h$. The spacing of the moladoth, however, is based on the mean synodic month, 235 of which correspond to 19 lunar years whose total is $1^h \ 485^p$ shorter than 19 solar years (VI,10). Consequently the initial relation between tekufah and molad is gradually changed, resulting after 259 cycles in a delay of the tekufah by $8^d \ 5^h \ 773^p$. Thus we know that, for the beginning of our cycle, the tekufah Nisan was about 8 days later than the molad. Each additional year increases this distance by about 11 days, but each intercalary year decreases it by about 30 days. In our specific case we have for the delay of the tekufah $8+8\cdot11—3\cdot30 = 6$ which proves our estimate (c).

SOLAR AND LUNAR MOVEMENT

1. The "Guide for the Perplexed" (completed 1190) gives us some information about Maimonides's astronomical education. We know from II,9 of the "Guide" that Maimonides had studied astronomy under the guidance of a pupil of Abu Bekr ibn al Zaig[1] and also that he had contact with the son of Jābir ibn Aflaḥ.[2] Both these astronomers are quoted by Maimonides as criticizing the Ptolemaic theory. Abu Bekr especially is said

[1] Called ibn Bādja; cf. Sarton, Introd. II,1 p. 183. Ibn Bādja died 1138/9.

[2] Sarton, Introd. II,1 p. 206. Duhem might be right when he considers Jābir's work as an Arabic translation of a Greek original, made by an unknown author who lived before Al-Battānī, i. e., before 900 (Duhem, Système du monde, II, p. 172 ff.). Duhem's argument is based on the consistent use of the Greek arrangement of the letters in proofs and the total absence of references to Arabic astronomy. Only the second argument is valid; cf. Gandz [1].

to have given plausible arguments for placing the spheres of Mercury and Venus beyond the sphere of the sun,[3] and he is also quoted for a system of his own which operated with eccenters only, thus avoiding epicycles.[4] Furthermore Thabit ibn Qurra[5] and Al-Qabīṣī[6] are quoted, and Ptolemy[7] is referred to several times.

Maimonides was, of course, also familiar with Aristotle, if only through the medium of Arabic philosophy. In many respects he was a follower of Aristotelean philosophy though he had to reject the idea of an eternal existence of the world in the past because this would have eliminated the creation ex nihilo, postulated by his religion. Consequently Maimonides emphasizes (especially in II,24 of the "Guide") the incompatibility of Aristotle's cosmic model of concentric spheres[8] and the Ptolemaic system of eccenters and epicycles. "The difficulty [of Aristotle's theory] is still more apparent when we find that admitting what Ptolemy said as regards the epicycle of the moon, and its inclination towards a point different both from the center of the universe and from its own center,[9] the calculations according to these hypotheses are perfectly correct, within one minute; that their correctness is confirmed by the most accurate calculation of the moment, duration, and magnitude of the eclipses, which is always based on these hypotheses."[10] The next sentence, however, shows that Maimonides was not really familiar with the system of homocentric spheres when he says: "Furthermore, how can we reconcile, without assuming the existence of epicycles, the apparent retrogradation of a

[3] Guide, II,9.

[4] Cf. Maimonides's cautious report in the Guide, II,24.

[5] He died 901. Cf. Sarton, Introd. I p. 599.

[6] Sarton, Introd. I p. 699. Guide II, 24, refers to a treatise of Al-Qabīṣī "On the distances" (not mentioned by Sarton nor by Suter and Renaud). The same work is probably meant in Guide III,14.

[7] Guide II,9 (Almagest IX,1); II,11 (Almagest III,3); II,24 (Almagest in general, V,5 and XIII,2, Heiberg p. 532).

[8] Developed previously by Eudoxos and Kallippos.

[9] This is Ptolemy's theory of the "Prosneusis" of the lunar epicycle (Almagest V,5).

[10] Guide II,24, following Friedländer's translation p. 198.

star with its other motions?" Indeed the homocentric spheres
were invented in order to explain this very phenomenon.

Maimonides was aware, however, that the differences of
the astronomical models must not be taken too seriously for
the explanation of the physical world "for he [the astronomer]
does not profess to tell us the existing properties of the spheres"
because "his object is simply to find a hypothesis . . . which
in its effects is in accordance with observation."[11] He also
realized the existence of objections against Ptolemy's theory
among Arabic astronomers though "even if it be correct that
[Abu Bekr] discovered such a system [without epicycles], he
has not gained much by it; for eccentricity is likewise as contrary
as possible to the principles laid down by Aristotle."[12] And
he ends: "It is on account of my great love of truth that I
have shown my embarrassment in these matters, and I have
not heard, nor do I know that any of these theories have been
established by proof."

No such doubts are voiced in the Mishnah Torah (completed
1177). The Ptolemaic arrangement of the planets is accepted
without restriction.[13] The computation of the accurate positions
of sun and moon, explained in the "Sanctification of the Moon,"
everywhere follows Ptolemaic principles, making extensive use
of eccenters and epicycles.[14] Hence we can distinguish three
layers in Maimonides's astronomy. First, the philosophical
problems related to cosmogony, where he is very cautious and
accepts no system as definitely proved. Second, the practical
methods of cyclic calendaric computation, undoubtedly following
Jewish tradition. Third, mathematical astronomy. It is the
investigation of this last part that will be the object of the
subsequent discussion.

Off hand, it is to be expected that Maimonides followed
Arabic astronomers, even if he had Arabic translations of the

[11] Guide II,24 (Friedländer p. 198) and II,11 (Friedländer p. 167).
[12] Guide II,24 (Friedländer p. 196).
[13] Book I, Chapter III, 1 (Hyamson p. 36b).
[14] Sarton's statement (Introd. II,1 p. 373) "Maimonides rejected epicycles
and eccentric movements as contrary to Aristotelian physics" is wrong both
in fact and in explanation.

Almagest at his disposal. Indeed, we shall show that there exist very close relations between Maimonides and Al-Battānī. This was discovered by Nallino; quoting numerical agreement for values of the mean solar movement, he says in the preface of his edition of Al-Battānī[15] "tacite nostrum astronomum sequitur celeberrimus Maimonides." We shall confirm this result for the whole theory of the solar and lunar movement. The theory of visibility, however, deviates from Al-Battānī, as Nallino has also seen.[16] I do not know the source of this special section.

2. For a comparison of numerical material contained in chapters XI to XVI of the Sanctification of the Moon with the tables of Al-Battāni we must give a short account of their arrangement. Vol. II p. 75 ff of Nallino's edition of the "opus astronomicum" contains, e. g., tables for the mean movement of the sun and the moon for the following entries: single days from 1 to 30, single aequinoctial hours from 1 to 24, and "Roman" (i. e. Julian) years in steps of 20 up to 100 and in steps of 100 up to 600. All values are given in degrees, minutes and seconds only. These values are rounded off from more accurate values which are mentioned in the heading of the tables for years (p. 77). The mean solar movement, e. g., is said to be $0;11,10,14,35,31,30°$ (mod. 360°) for 20 Julian years whereas the table itself only gives

20	0;11,10	
40	0;22,20	
60	0;33,31	etc.

If one checks with the largest given number (600 years) where the influence of the smaller units must be most visible we find that the original number must be corrected to $0;11,10,14,35,39,30$ instead of . . . ,31,30. Similar scribal errors are unfortunately very frequent in our text, according to Nallino's preface (vol. II

[15] Nallino I, p. XXXIV. I found this remark after having practically completed the present study. Nallino did not utilize Baneth (published 1898, 1899, 1902, 1903) as he states explicitly (Nallino I [1902], p. XXXIV note 4). Baneth on the other hand, did not have Al-Battānī's tables at his disposal (Nallino II, 1907) and made only little use of Nallino I.

[16] Nallino I p. XXXIV: "Tamen in supputando arcu apparitionis Lunae novae, rationem sequitur faciliorem, sed minus exactam, Albateniana."

p. V). Most trivial errors were tacitly corrected by Nallino and Schiaparelli, including cases where the correct numbers could be found by comparison with Ptolemy, Theon, or even by means of modern computation (preface, vol. II, p. VI). This procedure was apparently adopted in order to avoid a cumbersome and often useless apparatus. It has the disadvantage, however, that we are not sure whether the rounding off in tables was not actually less consistent than it appear in the printed edition.

The tables in the Sanctification[17] are of a similar nature, only with slightly different entries. From XII,1 we obtain, e. g., the following table for the mean movement of the sun, where the left column means days:

1^d	$0;59,8°$	100^d	$98;33,53°$
10	9;51,23	1,000	265;38,50
29	28;35,1	10,000	136;28,20
		354	348;55,15 .

These values are not consistent with one another in the sense that they are not exact multiples of 0;59,8. The obvious explanation is, of course, that we are dealing here with rounded off multiples of a more exact initial value. Indeed, we can show even more. The values for 1, 10 and 29 days and for 354 days[18] are identical with the corresponding values in Al-Battānī's tables,[19] which, in turn are rounded off from multiples of 0;59,8,20,46,56,14.[20] The values for 100, 1,000 and 10,000 are not included in Al-Battānī's tables but can be derived directly from them as follows. We multiply the value for 30 days, given by Al-Battānī as 29;34,10, by 3 and add the value for 10 days, which is 9;51,23. Then we obtain exactly Maimonides's value

[17] We shall always speak of "tables" in order to avoid useless clumsiness of expression. There are no "tables" in the strict sense of this word in the Mishnah Torah, only lists of numbers.

[18] Erroneously assumed to be an error by Baneth [1] p. 43 note 5.

[19] Nallino II p. 75 and p. 20. The table of p. 22 should be identical with the table of p. 75 but contains two errors, overlooked by Nallino: 8;52,16 instead of 8;52,15 for 9 days and 9;51,25 instead of 9;51,23 for 10 days.

[20] This value is obtained from the value given for 20 Julian years (Nallino II p. 77) where 0;11,10,14,35,31,30 must be corrected to 0;11,10,14,35,39,30, as explained above p. 337.

98;33,53. From this the values for 1,000 and for 10,000 days are obtained by multiplication by 10 and 100.

The insight into this procedure of Maimonides is historically not without interest. In order to obtain the value for 100 days he simply took the rounded-off values for 30 and for 10 days, thus comitting an error $3\epsilon_1 + \epsilon_2$ if ϵ_1 and ϵ_2 are the errors of rounding off committed by Al-Battānī. This total error appears multiplied by 100 in the value for 10,000 days. It is an amusing accident that the result agrees much better with modern values than any other value from Ptolemy to Copernicus. Baneth,[21] who did not realize how Maimonides's tables were constructed, praised this result "als ein glänzendes Ergebnis" and conjectured that Maimonides compared observations of Al-Battānī with results of Hipparchus. We see now that Maimonides not only did not have the slightest intention to deviate from Al-Battānī but that he showed the same disregard for the cumulative effect of errors which can be recognized in almost all ancient and mediaeval astronomers.

3. In XII,3 the apogee of the solar orbit is said to move 1° in about 70 years. A similar statement is made in the first book of the Mishnah Torah, chapter III,8[22] which shows clearly that not an independent motion of the sun's apogee is meant but that we are dealing here with the general precession of the equinoxes. This is the traditional attitude, also held by Al-Battānī.[23] The numerical values given by Maimonides in XII,3 are as follows

in	10 days	0;0,1,30°
	100	0;0,15
	1,000	0;2,30
	10,000	0;25
in	29 days	about 0;0,4
	354	0;0,53

These numbers are consistent among themselves except for the two last ones where 0;0,0,21 and 0;0,0,6 respectively are dis-

[21] Baneth [1] p. 42 and [2] p. 259.
[22] Hyamson p. 37b.
[23] Cf. Nallino I p. 216.

regarded if one considers the value for 10 days as exact. If one asks oneself, however, in what time the movement of 1° will be accomplished one will not find 70 (Julian) years but 6,40,0 days or almost 66 Julian years. Thus Maimonides's "about 70 years" is an approximation of 66 years. This again leads to a close agreement with Al-Battānī, who assumes[24] a movement of 1° in 66 "Roman" (i. e. Julian) years, or 1° in 6,41,46;30^d. We will find this fully confirmed in the discussion of the position of the apogee for Maimonides's epoch (cf. below p. 344).

Following strictly Ptolemaic tradition the true anomaly of the sun is found (in XIII,3) from the mean anomaly by adding or subtracting the equation of center. The values, however, are not Ptolemaic but rounded off from Al-Battānī,[25] as is evident from the following comparison:

	Al-B.	M.		Al-B.	M.		Al-B.	M.
10	0;19,59	0;20	70	1;50,35	1;51	130	1;33,15	1;33
20	0;39,27	0;40	80	1;56,35	1;57	140	1;18,34	1;19
30	0;57,49	0;58	90	1;59,3	1;59	150	1;1,24	1;1
40	1;14,35	1;15	100	1;57,56	1;58	160	0;41,52	0;42
50	1;29,15	1;29	110	1;53,14	1;53	170	0;21,24	0;21
60	1;41,14	1;41	120	1;44,56	1;44	180	0	0

One sees that Maimonides everywhere follows his principle of rounding off, according to which 30 or more units of a lower order count as 1 of the next higher order.[26] The only exception is the second value, where one should expect 39 instead of 40.

From these elements the solar position can be computed for any time if one initial position is known. Because the question of the epoch will come up also in connection with the lunar movement, we postpone it in favor of a common treatment after the other elements of the lunar movement are introduced.

[24] Nallino I, p. 72.

[25] Nallino II p. 78.

[26] Cf., e. g., XIII,4. This principle, so familiar to us, is by no means common practice with ancient astronomers. Experience shows that one frequently either disregarded lower units, however close they might come to 1, or that only 45 or more lower units were taken for 1 of the next order.

4. In XIV,3 we find the following values for the mean anomaly of the moon, again with reference to days:

1^d	13;3,54°	100^d	226;29,53°
10	130;39,0	1,000	104;58,50
29	18;53,4	10,000	329;48,20
		354	305;0,13

The first three values, as well as the last value, are found in Al-Battānī.[27] The value for 10 days is inaccurate and should be 130;38,59. This better value is also attested in Al-Battānī[28] and implicitly used by Maimonides in his value for 100 days. If we add 130;38,59 to three times the value 31;56,58 for 30 days,[29] we obtain 226;29,53 as given by Maimonides. The value 305;0,13 for 354 days is also inaccurate and should be 305;0,14. Both values are attested side by side in Al-Battānī.[30] The values for 1,000 and 10,000 days are obtained from the value for 100 days by successive multiplication by 10.

For the mean motion of the moon we find in XIV,2

1^d	13;10,35°	100^d	237;38,23°
10	131;45,50	1,000	216;23,50
29	22;6,56	10,000	3;58,20
		354	344;26,43 .

All these values are either directly taken from Al-Battānī[31] or can be derived by the same process as in the previous case.

The last table of this kind concerns the retrograde movement of the ascending node of the lunar orbit (XVI,3):

1^d	0;3,11°	100^d	5;17,43°
10	0;31,47	1,000	52;57,10
29	1;32,9	10,000	169;31,40
		354	18;44,42 .

The first three values agree with Al-Battānī.[32] For the last value

[27] Nallino II p. 22 and p. 21.
[28] Nallino II p. 75.
[29] Nallino II p. 75 and p. 21.
[30] Nallino II p. 21 and p. 20 respectively.
[31] Nallino, II p. 75 and p. 20.
[32] Nallino, II p. 75.

Al Battānī gives correctly 18;44,41.[33] The remaining values are consistent among themselves but cannot be obtained by using Al-Battānī's table only, which ends with the value for 30 days.[34] If, however, one uses for the daily movement the more accurate value 0;3,10,37,24, ... which forms the basis of Al-Battānī's tables,[35] then one obtains again the values given by Maimonides for 100 days. The values for 10,000 and 10,000 days are obtained by simple multiplication by 10.

5. We can now return to the question of the initial values for the movement of sun and moon at the epoch which Maimonides had chosen (XI,16) to be 4938 Nisan 3 0ʰ (i. e., as will be shown presently, A.D. 1178 March 23), obviously close to the date of his writing, because the introduction to the Mishnah Torah calls the preceding year (1177) the "present date."[36] For this date the following initial positions are assumed

sun's apogee	♊	26;45,8°	(XII,4)
mean sun	♈	7;3,32	(XII,4)
mean moon	♉	1;14,43	(XIV,4)
moon' anomaly		84;28,42	(XIV,4)
ascending node		−180;57,28	(XVI,3).

The last four values allow for a direct comparison with tables given by Al-Battānī.[37] There we find these elements listed for given years of the Seleucid Era (called in Al-Battānī "Era of the Bicornute"[38] with additions for single days and hours. In introducing his epoch Maimonides equates[39] the year 4938 of the world with the year 1489 of the Seleucid Era (called here "Era of Contracts"). In Al-Battānī's tables we can obtain this year from two entries, Seleucid Era 1471, and 18 single years. In this way we find for the longitude of the mean sun:

[33] Nallino II p. 20 and 21.
[34] For 100 days one would obtain 5;17,44.
[35] This value is derived from the value given for 600 Julian years (Nallino II, p. 77).
[36] Hyamson p. 4.
[37] Nallino II, p. 72.
[38] Cf. Nallino I, p. 242.
[39] XI,16.

345;25,13+359;40,29 = 345;5,42. This position equals ♓15 and refers to noon[40] of Ādhār 1 (which corresponds, incidentally, to Julian March 14[41]). The mean sun at Maimonides's epoch Nisan 3, was in ♈7, hence 22° farther ahead than at Ādhār 1. Thus it is clear that we must add a movement between 22 and 23 days to the positions obtained from Al-Battānī in order to reach the epoch of Maimonides.[42] Repeating the same argument for the parameters of the moon and computing accurately, we obtain 22 days 6;50 hours as the best value which leads from Seleucid Era 1489 Ādhār 1 to Maimonides's epoch. The results are

	Al-Battānī		Maimonides		M.–Al.-B.
mean sun	♈	7;3,35	♈	7;3,32	— 0;0,3
mean moon	♉	1;14,41	♉	1;14,43	+ 0;0,2
moon's anom.		84;28,37		84;28,42	+ 0;0,5
asc. node		180;57,16		180;57,28	+ 0;0,12

The small deviations are in all probability due to the process of interpolation which is needed to obtain the movement during 50 minutes. I was not able, however, to reduce the differences simultaneously to zero by strictly consistent computations. Nevertheless it is clear that Maimonides's values are obtainable from Al-Battānī by a time difference of 22 days and 6;50 hours. The 22 days are explained by the difference between Ādhār 1 (= March 1) and Nisan 3 (= March 23). The 6 hours correspond to the time from Al-Battānī's noon epoch to Maimonides's evening epoch. There remain 50 minutes which in part are due to the difference in geographical longitude. Al-Battānī's elements are based on the meridian of ar-Raqqah in Mesopotamia. Jerusalem is assumed to have the longitude 66;30 while ar-Raqqah is placed at 73;15[43] i. e. 6;45° east of Jerusalem. The corresponding time difference is 27 minutes, leaving us with 23

[40] Nallino I, p. 72 (6).

[41] Nallino II p. 8 and p. 74.

[42] We could have obtained this result directly by the use of modern tables which give March 23 as the date of Nisan 3. It seems to me, however, preferable to show that our results can be obtained directly from the investigation of Al-Battānī and Maimonides alone.

[43] Nallino II p. 54 No. 273 and p. 41 No. 150.

minutes not yet accounted for. This amount corresponds most likely to the 1/3 of an hour which Maimonides assumes (XIV,6) to be required after sunset in order to make the new crescent visible. Thus it would have been more accurate to say that the epoch was chosen to be 0;20^h of Nisan 3 of the year 4938.

We have still left aside the initial position for the solar apogee. Al-Battānī states[44] that the apogee of the sun had at Sel. Era 1191 Ādhār 1 the longitude of about ♊ 22;15 and that 66 years correspond to 1° of its movement. We know that Maimonides's epoch is Sel. Era 1489 Ādhār 23. The interval of 298 years and 22 days between these two epochs contains $4\frac{1}{2}$ periods of 66 years and 1 year 23 days. The $4\frac{1}{2}$ periods correspond to a movement of the apogee of 4;30°. The remaining 388 days can be written as 4·100 — 12 days. According to the table in XII,3 the corresponding movement will be 0;1° — 0;0,1,48° = 0;0,58,12. Thus we obtain a total of 4;30,58° for the movement of the solar apogee between the two epochs. Unfortunately the position assumed by Al-Battānī is not accurately known. Considering Maimonides's value ♊ 26;45,8 as certain, we would obtain for Al-Battānī ♊ 22;14,10 whereas the tradition varies between ♊ 22;15 (cf. the above reference) and ♊ 22;14 (ibn Yūnus[45]); the latter value (or more accurately ♊ 22;14,16) can also be derived from Al-Battānī.[46] The present passage from the Mishnah Torah speaks in favor of the smaller value.[47]

6. Knowing the mean movement of the moon and knowing its position at the epoch we can compute the position of the mean moon for any moment whose time difference from the epoch is given. We have seen that the epoch was chosen in such a way that the delay of 20 minutes is already included. All that remains, therefore, is a correction for the variable length of daylight. If we measure time by means of equatorial degrees we can

[44] Chapter XXXIII, Nallino I p. 72.
[45] Caussin p. 154.
[46] The discussion Nallino I p. 214.
[47] This result has also been reached by Baneth [1] p. 50. By computing accurately he even obtains exactly 4;31,8 for the movement of the apogee. I do not think, however, that this procedure is historically correct.

say that the length of daylight at the equinoxes is 3,0°. If C denotes the length of daylight for any day of the year, we are interested in the difference $\delta = C-3,0$ measured in degrees. For a day when this difference has the value δ, sunset will be delayed by $\frac{1}{2}\delta° = 2\delta^h$. Because the mean elongation of the moon amounts to $12°/^d$, the elongation will increase during 1^h by $\frac{1}{2}°$ and therefore during $2\delta^h$ by $\delta°$. In other words, the value

$$\delta = C - 3,0$$

is identical with the correction of the mean elongation of the moon due to the variability of the length of daylight C. It is therefore our next goal to find the values of C for the latitude of Jerusalem.

Following ancient practice we can find the length of daylight from the table of ascensions in the proper climate. Al-Battānī's table of ascensions[48] contains for the third climate the latitudes $\varphi = 30;40$ and $\varphi = 33;37$. Linear interpolation for the latitude $\varphi = 31;40$ yields the ascensions for Jerusalem. This leads to a table of the following type:

	$\varphi = 30;40$	$\varphi = 33;37$	$\varphi = 31;40$
10	6;49	6;32	6;43
♈ 20	13;45	13;11	13;34
30	20;56	20;5	20;39

etc.

	$\varphi = 30;40$	$\varphi = 33;37$	$\varphi = 31;40$
30	3,0;0	3,0;0	3,0;0
10	3,11;33	3,11;50	3,11;39
♎ 20	3,23;9	3,23;43	3,23;20
30	3,34;50	3,35;41	3,35;7

etc.

These numbers indicate, e. g., for $\varphi = 31;40$ that $6;43°$ of the

---

[48] Nallino II p. 65. The entry for Saggittarius 250 and lat. 30° 40' must be 262° 16' instead 250° 16' as printed by Nallino.

equator rise with the first 10° of the ecliptic, 13;34° with 20° etc. Because 180° of the ecliptic rise during each day, the length of daylight is always given by the arc of the equator which corresponds to 180° of the ecliptic. If the sun stands in ♈ 0° we find, of course, 3,0 for the corresponding arc of the equator. For ♈ 10°, however, we lose 6;43° for the rising of the first 10° and gain 11;39 for the 10° in ♎. Thus the length of daylight will be 3,0 — 6;43+11;39=3,4;56°. For ♈ 20° we find 3,0 — 13;34+23;20=3,9;46°. Continuing in this way we are able to derive all values of C from 10 to 10 degrees from Al-Battānī's table for Jerusalem.[49] Subtracting 3,0° gives the values of the correction δ which we wanted to obtain. The result is presented in the graph of Fig. 1; the dots correspond to the values derived

in the above-described way from Al-Battānī's tables. The horizontal strokes represent the rounded-off values given by Maimonides in XIV,5. The agreement is so good that there can be no doubt that also here Al-Battānī's tables were the basis of Maimonides's values.

7. The procedure described in the preceding section is, of course, common practice among Greek and Arabic astronomers. Similarly the method followed by Maimonides to find the position of the true moon is shaped after classical examples. A detailed explanation is contained in Al-Battānī chapter XXX. Consequently Maimonides introduces in XV,1 the "double elongation." Because his final goal consists only in finding the ripeness of the new crescent he can restrict himself to limited elongations. Thus he assumes that the double elongation will not be less than 5° and not larger than 62°. The reason for these specific values will

---

[49] The longest daylight is found to be 14;5^h.

become clear when we discuss the limits for visibility.[50] For the present moment we accept these limits for the double elongation as granted. In V,3 Maimonides gives a table for the corresponding correction which leads from the true apogee of the lunar epicycle to the mean apogee. Al-Battānī gives a similar table[51] which in turn is based on Almagest V,8. This provides us with the following comparison:

Double elongation			Maimonides	Al-Battānī[52]		
6°	to	11°	1°	0;45°	to	1;38°
12	to	18	2	1;46	to	2;39
19	to	24	3	2;48	to	3;31
25	to	31	4	3;40	to	4;32
32	to	38	5	4;41	to	5;33
39	to	45	6	5;41	to	6;33
46	to	51	7	6;42	to	7;23
52	to	59	8	7;32	to	8;28
60	to	63[53]	9	8;36	to	8;44[53]

This shows that Maimonides divided the double elongation at such points where Al-Battānī's values change from 0;30 to 0;40. This principle is only slightly violated at the end where one should expect 53 and 61 instead of 52 and 60.

The next step consists in finding the quota of anomaly for all anomalies. Strictly speaking this problem should be solved for all elongations, and this is indeed the way followed by Ptolemy and Al-Battānī. Maimonides introduces here a convenient simplification. Because he deals only with small elongations (up to 31°), he can refrain from taking into account the influence of variable elongations. He therefore computes the quota of anomaly by assuming a fixed mean elongations of about 15°.

The procedure for finding the quota of anomaly including

[50] Cf. below p. 350.

[51] Nallino II p. 78 ff ("aequatio anomaliae").

[52] The values given here belong to the first and to the last value of the double elongation given in the first column. Al-Battānī's table proceeds in steps of single degrees.

[53] The value corresponding to 67° would be 9;30.

prosneusis was first described by Ptolemy in V,9 of the Almagest following the table in V,8. The same rules are applied by Al-Battānī in his chapter XXX[54] and in the corresponding tables.[55] If we assume a mean elongation of 15° we have to enter these tables with the value of 30° for the double elongation. In the 5th column of Al-Battānī's tables[56] we find the corresponding value of $c = 0;3°$ and the same is true for all entries from 26 to 31. It is therefore irrelevant whether 15° is exactly the value used for the mean anomaly by Maimonides. Ptolemy's values are more accurate in this column. For a double anomaly of 30° he gives $c = 0;3,24$ and each degree more or less would change this value by about $0;0,11$. We shall first follow Al-Battānī's tables and compute the quota of anomaly for an elongation of 30°. The corresponding value in the 6th column[57] is $M = 1;10$. Let q be the quota when the epicycle is in the apogee. Then $M = 1;10$ means that the quota of anomaly would be $q + 1;10°$ if the lunar epicycle would be at its perigee (which is the case for the quadratures) assuming a true anomaly of 30°. Actually the lunar epicycle is not in the perigee (elongation 90°), because we assumed an elongation of only 15°. Consequently only a fraction c of M will be the corresponding quota and we have already stated that $c = 0;3$ can be found in the 5th column. Thus the corresponding quota is not $q + 1;10$ but only

$$Q = q + 0;3 \cdot 1;10 = q + 0;3,30.$$

The value of q is given in the 3rd column[58] and has for the anomaly of 30° the value $2;19,45$. Thus we obtain

$$Q = 2;19,45 + 0;3,30 = 2;23,15.$$

The corresponding value given by Maimonides in XV,6 is $2;24$.

If we repeat this calculation with Ptolemy's tables we obtain for the elongation of 15° and the anomaly of 30° the quota

$$Q = 2;19 + 0;3,24 \cdot 1;10$$
$$= 2;19 + 0;3,58 = 2;22,58.$$

A similar situation prevails for the rest of the table. Though the values of Al-Battānī and Ptolemy are slightly different the final result is practically the same. The majority of Maimonides's

---

[54] Nallino I p. 50 ff.   [55] Nallino II p. 78.
[56] "Minuta addenda."   [57] "In longinquitate minima."
[58] "Aequatio simplex lunae."

values, however, are 0;1° greater than expected, especially in the first quadrant. I do not know how to explain this peculiar deviation. The following table shows the complete list:[59]

	Al-B.	M.		Al-B.	M.
10	0;49	0;50	100	5;8	5;8
20	1;37	1;38	110	4;59	4;59
30	2;23	2;24	120	4;39	4;40
40	3;5	3;6	130	4;11	4;11
50	3;43	3;44	140	3;33	3;33
60	4;15	4;16	150	2;47	2;48
70	4;41	4;41	160	1;55	1;56
80	4;58	5;0	170	0;59	0;59
90	5;7	5;5 (?)	180	0	0

8. The last element needed for the exact position of the moon is its latitude. Maimonides is following Arabic custom when he uses the ascending node as zero point of the argument of latitude[60] whereas the Greeks started from the point of greatest northern latitude.[61] We have already discussed the table in XVI,3 for the movement of the ascending node (above p. 341) and its derivation from Al-Battānī.

The values for the latitude itself, given in XVI,11, agree also with Al-Battānī,[62] though rounded off from three to two places. The only deviation is found for 20° where Maimonides gives 1;43 whereas Al-Battānī has 1;42,27.

VISIBILITY

1. The elongation of the moon from the sun is, of course, the essential element for the visibility of the new crescent the evening following conjunction. Maimonides therefore introduces

[59] Cf. for Maimonides's values Baneth p. 102 and Feldman p. 146.
[60] Counted in retrograde direction. Cf., e. g., Al-Khwārizmī, chapter 12 (Suter p. 12 and p. 55) and Al-Battānī, chapter 37 (Nallino I p. 75 f, p. 250 f., II p. 204).
[61] Cf., e. g., Almagest V,8 last column.
[62] Nallino II p. 78 ff last column.

(XVII,1) as "first longitude" the difference between the longitude of the true moon and that of the true sun. We shall denote this magnitude by λ_1. Similarly the "first latitude" β_1 is the latitude of the true moon, counted positive for northern latitudes, negative for southern latitudes.

Latitude and variable inclination of the ecliptic must be taken into consideration provided the elongation is not too small to render visibility impossible under all circumstances or too large to secure visibility unconditionally. In XVII,3 and 4 these limits are given as 9° and 24° respectively. Consequently only for first longitudes of the interval

(1) $$9° \leqq \lambda_1 \leqq 24°$$

is a closer investigation of the relative position of sun and moon necessary.

The true elongation λ_1 can differ considerably from the corresponding mean elongation. According to Al-Battānī's tables[1] the maximal anomaly of the sun is 1;59,10°, of the moon 5;1°. Hence it is possible that true and mean elongation deviate 7° from each other. The interval (1) for the true elongation therefore corresponds to an interval from 2° to 31°; thus the double elongation for which special considerations are needed varies from 4° to 62°. This is obviously the explanation of the limits 5° and 62° adopted by Maimonides in XV,2 and 3.[2] The value 5° instead of 4° suggests a lower limit 9;30 instead of 9 in (1). We shall see in the last section that all visibility limits are apparently affected by rounding-off, which is only in line with Maimonides's general tendency to present his material in the simplest possible form.

The limits (1) are not given in exactly this form by Maimonides. He distinguishes between the two halves of the ecliptic separated by the solstices. For the spring semicircle the limits are 9° and 15° respectively, for the autumn semicircle 10° and 24°. No boundaries of this kind are found in Al-Battānī.

[1] Nallino II, p. 81. Similar considerations are found in Almagest V,10 (Heiberg 396).
[2] Cf. above p. 346.

2. From now on we assume that the value of λ_1 is such that we must consider the additional influences. The first correction of λ_1 and β_1 is due to the parallax. The results

(2)
$$\lambda_2 = \lambda_1 - c_1$$
$$\beta_2 = \beta_1 - c_2$$

are called "second longitude" and "second latitude" respectively. The values of c_1 and c_2 depend, of course, on the inclination of the ecliptic. Though it is clear that we are dealing here again with rounded-off values only, we can estimate the values for $c = \sqrt{c_1{}^2 + c_2{}^2}$ which should represent the total parallax. Thus we obtain the following table:

	c_1	c_2	c
♈	0;59	0;9	1
♉	1	0;10	1
♊	0;58	0;16	1
♋	0;52	0;27	0;59
♌	0;43	0;38	0;57
♍	0;37	0;44	0;57
♎	0;34	0;46	0;57
♏	0;34	0;45	0;57
♐	0;36	0;44	0;57
♑	0;44	0;36	0;57
♒	0;53	0;27	0;59
♓	0;59	0;12	1

The lower values of c in the summer semicircle seem to indicate that it is assumed that the moon must be higher above the horizon during summer than winter. I do not see, however, how the asymmetry of our scheme with respect to the solstices can be explained. Baneth[3] has discussed the numerical values in detail without reaching a satisfactory solution.

[3] Baneth [I] p. 146 ff.

3. If M' denotes the apparent position of the moon (i. e. the true position of the moon influenced by parallax) Maimonides obtains in XVII 10,11 a "third longitude" λ_3 from the longitude λ_2 of M' by taking the arc of declination which passes through M' and intersecting it with the ecliptic in O (cf. Fig 2). Then

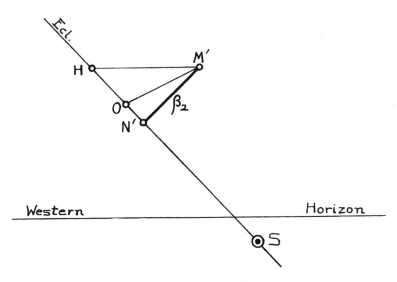

λ_3 is the distance from O to the sun S. If $M'N' = \beta_2$ is the apparent latitude of the moon and if $c_3 \beta_2 = N'O$, then we can write

(3) $\lambda_3 = \lambda_2 + c_3$

where c_3 is a coefficient depending upon the longitude of the moon.

The values and the signs given for c_3 leave no doubt that the above interpretation, given by Baneth,[4] is correct. It suffices to remark that $c_3 = 0$ for the solstices because the circle of latitude coincides in this case with the circle of declination and

---

[4] Baneth [I] p. 154 ff. Baneth's name, however, "Rektaszensionsunterschied" is misleading and not based on the text, which speaks loosely about a variation in the orbit of the moon. Cf. the translation by S. Gandz in a forthcoming volume of the Yale Judaica Series.

that $c_3 > 0$ for the spring semicircle, but negative for the autumn semicircle.[5]

It is difficult to understand the purpose of this step. One would expect that the time difference between sunset and disappearance of the moon would be needed. This problem could be solved by finding the intersection with the equator of a circle passing through M' and forming the angle $90 - \varphi$ with the equator.[6] Breaking this procedure into two steps, one could first find the intersection H of the above-mentioned circle with the equator and then ask for the arc of the equator which sets in the same time as the arc HS of the ecliptic. Indeed, the next step consists in finding the equator arc which corresponds to the ecliptic arc OS. Thus we can say that Maimonides determines the delay of setting of the point O instead of the moon. We shall return to this question in Section 5.

4. The problem of finding the arc of the equator which sets simultaneously with the arc $\lambda_3 = SO$ of the ecliptic is a variant of the classical problem of ancient astronomy: to determine the rising time of a given arc of the ecliptic. Because points which rise and set are diametrically opposite, our problem is easily reduced to a problem of rising times for which tables are available. In section 6 of the preceding chapter (p. 345) we obtained a table of ascensions for the latitude of Jerusalem by means of interpolation from Al-Battānī's tables. Thus we found, e. g., that the sign ♈ rises with $20;39°$ of the equator. At the same time the sign ♎ is setting. Consequently we know that $20;39°$ of the equator are setting simultaneously with the arc SO of Fig. 2 if SO coincides with the sign of ♎. Al-Battānī's tables would furnish similar information for arcs of $10°$ of length; linear interpolation suffices for still smaller parts.

Maimonides, in XVII,12, is satisfied with values for whole zodiacal signs, rounded-off to degrees. Thus he says, e. g., that for λ_3 in ♎ a "fourth longitude" should be found by subtracting

[5] For the single values and their comparison with computation, see Baneth [1] p. 156 or Feldman RMA p. 168.

[6] Feldman RMA p. 167 speaks of a circle "parallel to the horizon" which is, of course, impossible for a great circle.

$1/3$ of λ_3 from it. For $\simeq$ we would thus obtain $\lambda_4 = 20$. In general we can bring Maimonides's procedure to the form

(4) $\lambda_4 = \lambda_3 + c_4\lambda_3$

with coefficients c_4 depending on the zodiac. The following table shows the result of a comparison with the values obtained by interpolation from Al-Battānī

setting	Al-Battānī		Maimonides	
♈ or ♓	35;7		35	$c_4 = +1/6$
♉ or ♒	35;50		36	$+1/5$
♊ or ♑	34;41		35	$+1/6$
♋ or ♐	29;45		30	0
♌ or ♏	23;58		24	$-1/5$
♍ or ♎	20;39		20	$-1/3$
total:	180	total:	180	

The good agreement with Al-Battānī's values is not surprising because we are only using once again elements whose agreement we have already established in the correction for variable length of daylight.

5. The value of λ_4 indicates how much later than S the point O sets. In Fig. 3 this quantity is expressed by the equator

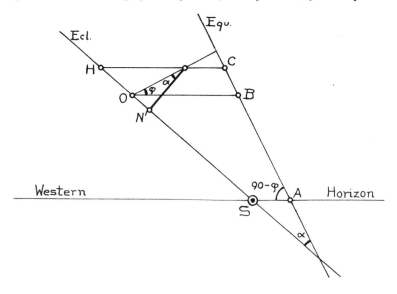

arc AB. If we now want to find the delay of the moon M' we must find the length BC. We may assume that all arcs are straight lines and BO and $CM'H$ parallel to the horizon. Then we find

$$M'O = \beta_2/\cos\alpha \qquad \text{and} \qquad CB = M'O \tan \varphi$$

thus

$$CB = \beta_2 \tan \varphi / \cos \alpha.$$

For Jerusalem we have $\varphi = 31;41°$, thus $\tan \varphi = .62$. The angle α between ecliptic and equator varies between $0°$ and $24°$, thus $\cos \alpha$ between 1 and $.91$. Hence $\tan \varphi /\cos \alpha$ varies between $.62$ and $.70$, with $.66 = \frac{2}{3}$ as mean value. This shows that

$$CB = \frac{2}{3}\beta_2$$

is a fair approximation for the additional delay of setting caused by the latitude of the moon.

The coefficient $\frac{2}{3}$ is called by Maimonides in XVII 13 the "quota of the geographical latitude" though he does not indicate how the geographical latitude has been used in order to arrive at this value. He furthermore replaces β_2 by the first latitude β_1, thus disregarding the parallax in latitude. Then he calls

(5) $b = \lambda_4 + \frac{2}{3}\beta_1$

"arcus apparitionis"[7] apparently because it is the fundamental quantity needed for deciding about the visibility of the moon.

Looking back at the computation of b we can say that b represents the equatorial arc which corresponds to the arc HS of the ecliptic in Fig. 2. The distance HS could have been expressed by $\lambda_2 + \beta_2 \cot \gamma$ where γ is the variable angle between horizon and ecliptic. Having found HS the same procedure which is used to determine λ_4 from λ_3 would lead from HS to b. It is difficult to see why this way was not followed. The determination of the variable angle γ cannot have been the reason because Greek spherical trigonometry had already solved this problem.[8] Maimonides's method is not only unnecessarily compli-

---

[7] Feldman translates "arc of vision" (p. 170) whereas Baneth [1] p. 167 uses "Sehungsbogen." This terminology is misleading because the "arcus visionis" is the negative altitude of the sun needed for visibility whereas b is an arc of the equator. Because the term used by Maimonides is formed from the same root as the Arabic term used by Al-Battānī (Nallino II p. 332), I use here Nallino's translation "arcus apparitionis."

[8] Almagest II,11.

cated but numerically very crude, especially in adding the constant amount $\frac{2}{3}\beta_1$ to λ_4. It is my impression that Maimonides depends in the whole section on visibility on much more primitive sources than in the computation of the position of the moon. This would be easily intelligible if Maimonides were following Greek methods because the Almagest does not contain rules for the solution of the visibility problem. For a follower of Arabic tradition, however, it is difficult to understand why lunar movement and lunar visibility should be treated differently. Yet there can be no doubt that Maimonides depends on the Arabic and not on the Greek tradition.

From the purely mathematical viewpoint the computation of b could have been condensed into a single rule

(6) $b = c_5 \lambda_1 + c_6 \beta_1 + c_7$

with coefficients c_5, c_6 and c_7 depending on the position of the moon in the zodiac.[9] Computing their values from the given values of $c_1, \ldots, c_4$, one realizes again the inconsistency in accuracy of the single steps. The coefficient c_3, e. g., is given in very detailed dependence on parts of zodiacal signs whereas most of the other coefficients are rounded-off values, constant for a whole zodiacal sign. It is very unlikely that inconsistencies of this type should be found in an astronomical work with extensive numerical tables like Al-Battānī's treatise.

6. In XVII,15–21 Maimonides states the final criteria of visibility. These are:

(7a) $b \geqq 14$ visibility certain
              $b < 9$ visibility excluded;

[9] The following rounded-off values will suffice to give the main trend of these relations:

	c_5	c_6	c_7		c_5	c_6	c_7
♈	1;10	0;12	1;5	♎	0;40	0;56	0;34
♉	1;12	0;22	1;9	♏	0;48	0;52	0;36
♊	1;10	0;34	1;6	♐	1	0;45	0;39
♋	1	0;45	0;54	♑	1;10	0;34	0;48
♌	0;48	0;52	0;42	♒	1;12	0;22	0;55
♍	0;40	0;56	0;36	♓	1;10	0;12	1;3

if, however, b lies between 9 and 14, then

(7b) $b+\lambda_1 \geqq 22$

is required for visibility.

For the further interpretation of these discussions, a graphical representation is most convenient. Because the visibility of the new crescent is made dependent upon the two quantities b and λ_1 only, we can represent all possibilities in a λ_1, b-plane. The condition (7b) says that the region of visibility lies above the line $b+\lambda_1=22$ and from (7a) we know that also the lines $b=9$ and $b=14$ belong to the boundary of the region of visibility. Finally we must remember that $\lambda_1=9$ and $\lambda_1=15$ are limits if the moon is on the spring semicircle, and $\lambda_1=10$, $\lambda_1=24$ on the autumn semicircle. Figs. 4a) and b) show the corresponding boundaries. It follows from these diagrams that the statement that $b > 14$ means visibility is trivial in both cases. The statement that $b < 9$ means invisibility, however, is trivial only in the case of the spring semicircle whereas it restricts visibility for the autumn semicircle. Fig. 4a) suggests furthermore the as-

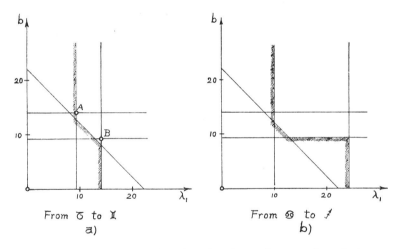

From ♉ to ♊   From ♋ to ♐
a)   b)

sumption that the small triangles formed near A and B are caused by nothing more than rounding-off of the limiting numbers. The triangels would, e. g., disappear if the boundary values for b and λ_1 would be changed to 8;30 and 13;30.

Maimonides's principle of rounding-off[10] would then lead to 9 and 14.

We can show that also the upper limits for λ_1, assuming visibility, are of no interest. It follows from (6) p. 356 that for a given latitude β_1 the "arc of vision" is a linear function of λ_1 alone. Because the latitude β_1 varies between $+5°$ and $-5°$ only, we can say that all possible values of b belong to a strip bounded by the two straight lines

$$b = c_5 \lambda_1 + (c_7 \pm 5c_6)$$

whose position depends on the zodiacal signs. Using the values given in note 9) of p. 356 leads to a graphical representation of all possible values of b and λ_1 for the two halves of the year (Fig. 5a) and b)). From these graphs it is evident that the

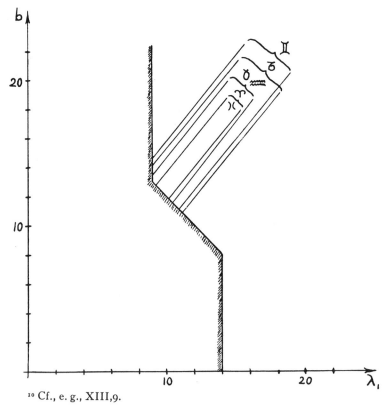

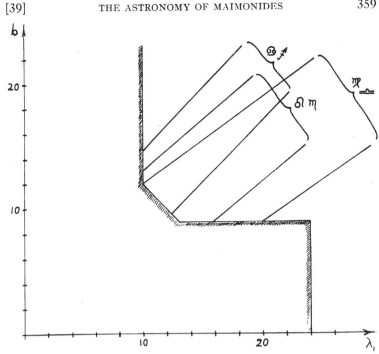

limits $\lambda_1 = 14$ and $\lambda_1 = 24$ respectively do not influence the region into which values of b and λ_1, actually can fall. On the other hand we see now that the condition $b > 9$ is very essential for the autumnal semicircle.

7. The most essential condition for visibility if, of course, the condition $b + \lambda_1 \geqq 22$. It expresses the fact that the time difference b between sunset and moonset can be smaller if the elongation increases. The reason for this is the fact that the width of the visible crescent is proportional to the elongation.

The same idea is reflected in Al-Battānī, though in a much more refined form.[11] As "arcus apparitionis fundamentalis" Al Battānī assumes the time difference $b_0 = 11;45°$. This corresponds to the delay of the setting of the moon under ideal conditions, i. e., both sun and moon being in the equator.

---

[11] For details cf. Nallino I p. 86 ff. and the commentary of Schiaparelli p. 266 ff.

Assuming mean motion for both bodies, the moon will have an elongation of $12;11°$ when setting. This is the elongation, obtained one day after conjunction, considered just sufficient for the visibility of the crescent. If the actual angular distance d between sun and moon, parallax included, is different from $12;11$, Al-Battani considers the ratio $12;11/d$ and multiplies b_0 by this quantity. If we call the result b'_0 and if b has the same meaning as with Maimonides the condition of visibility of Al-Battani is $b \geqq b'_0$. This shows that the details of expressing the relation between b and elongation are very different in Al-Battani and Maimonides. Al Battani furthermore considers the influence of the position of the moon on the epicycle, a factor which is completely disregarded by Maimonides. Also the methods for computing b are totally different in both authors.

Nallino gives a survey of different criteria used by Arabic astronomers prior to Al-Battani.[12] None of them corresponds to Maimonides's procedure. Only Kūshyār ibn Labban (about 1000 A.D.) uses a criterion of the form $d + h \geqq 18$ where d means the distance between sun and moon and h the depression of the sun at moon set. It therefore seems plausible to assume that Maimonides follows in the problem of visibility a Jewish tradition, uninfluenced by Arabic methods. It is very difficult to say whether such a Jewish tradition rests eventually on Babylonian methods. The ephemerides of the Seleucid period make it probable that a criterion of the type $b + \lambda_1 \geqq c$ was used, or, to be more cautious, the ephemerides certainly do not follow a criterion based on the elongation alone. Scantiness and bad preservation of texts combined with the omission of intermediate steps and the extensive use of rounded-off numbers make it extremely difficult to abstract accurate criteria from these ephemerides. Secure ground can be reached only with the discovery of procedure texts which state the rules of computation.

---

[12] Nallino I p. 268 ff.

BIBLIOGRAPHY

Abū-l-Fidā — see Reinaud.

Al-Battānī — Al-Battānī sive Albatenii opus astronomicum. Edit. C. A. Nallino. Pubblicazioni del Reale Osservatorio di Brera in Milano. N. 40. Parte I, 1903. Parte II, 1907. Parte III, 1899.

Al-Bīrūnī — The chronology of ancient nations . . of . . Albîrûnî. Transl. C. E. Sachau, London, 1879.

Al-Khwārizmī — Die astronomischen Tafeln des Muḥammed ibn Mūsā al-Khwārizmī . . . ed. H. Suter. Kgl. Danske Vidensk. Selsk. Skrifter, 7 R., hist. og filos. Afd. III,1, 1914.

Baneth [1] — E. Baneth, Maimuni's Neumondsberechnung Bericht über die Lehranstalt für die Wissenschaft des Judenthums 16 (1898), 17 (1899), 20 (1902), 21 (1903).

Baneth [2] — E. Baneth, Maimonides als Chronologe und Astronom. Moses ben Maimon. Sein Leben . . . Zur Erinnerung an den 700. Todestag . . . . herausgegeben von der Gesellschaft zur Förderung des Judentums. Vol. 2, Leipzig 1914, p. 243–279.

Caussin — Le livre de la grande table hakémite, observée par . . . ebn Iounis, ed. Caussin. Notices et extraits des manuscrits de la bibliothèque nationale 7, an XII [1803] p. 16–240.

Dünner — Lasar Dünner, Die aelteste astronomische Schrift des Maimonides. Thesis, Erlangen, 1902 (54 pp.).

Feldman, RMA — W. M. Feldman, Rabbinical mathematics and astronomy. London, 1931.

Friedländer, Guide — M. Friedländer, The guide for the perplexed by Moses Maimonides, 2nd ed., London 1928.

Gandz [1] — Solomon Gandz, Der Hultsch-Cantorsche Beweis von der Reihenfolge der Buchstaben in den mathematischen Figuren der Griechen und Araber, Quellen u. Studien z. Gesch. d. Math., B 2 (1931) p. 81–97.

HYAMSON — Moses Hyamson, The Mishnah Torah by Maimonides. Book I. New York 1937.

IBN YŪNUS — see Caussin.

IDELER, CHRON. — Ludwig Ideler, Handbuch der mathematischen und technischen Chronologie, 2 vols., Berlin 1825, 1826.

KUGLER, BMR — F. X. Kugler, Babylonischen Mondrechnung, Freiburg, 1900.

KUGLER, SSB — F. X. Kugler, Sternkunde und Sterndienst in Babel, Münster, 1907–1924.

MŽIK [1] — Hans v. Mžik, Erdmessung, Grad, Meile und Stadion. Studien zur armenischen Geschichte VI. [reprinted from Handes Amsorya 47 (1933)], Wien, 1933.

NALLINO — see Al-Battānī.

NAU — F. Nau, Le livre de l'ascension de l'esprit . . . par . . . Bar Hebraeus. [Bibliothèque de l'école des hautes études 121]. Paris, 1899, 1900.

NEUGEBAUER [1] — O. Neugebauer, Studies in ancient astronomy VII. Magnitudes of lunar eclipses in Babylonian mathematical astronomy. Isis 36 (1945) p. 10–15.

NEUGEBAUER [2] — O. Neugebauer, Untersuchungen zur antiken Astronomie III. Die babylonische Theorie der Breitenbewegung des Mondes. Quellen u. Studien z. Gesch. d. Math. B 4 (1938) p. 193–346.

NEUGEBAUER [3] — O. Neugebauer, Solstices and equinoxes in Babylonian astronomy during the Seleucid period. J. Cuneiform Studies 2 (1949) p. 209–222.

REINAUD — J. T. Reinaud, Géographie d'Aboulféda, traduite de l'arabe en français, Paris, 1848.

RENAUD [1] — H. P. J. Renaud, Additions et corrections à Suter "Die Mathematiker und Astronomen der Araber." Isis 18 (1932), p. 166–183.

SACHS [1] — A. J. Sachs, Two Neo-Babylonian Metrological Tables from Nippur. J. Cuneiform Studies 1 (1947) p. 67–71.

SARTON, INTROD. — G. Sarton, Introduction to the history of science. Carnegie Institution of Washington, Publication No. 376. Vol. 1, 1927; vol. 2, 1931.

SAUVAIRE [1] — H. Sauvaire, Matériaux pour servir à l'histoire de la numismatique et de la metrologie Musulmanes, IV, Journal Asiatique (8) 8 (1886) p. 479–536.

SCHWARZ JC — Adolf Schwarz, Der jüdische Calender, historisch und astronomisch untersucht, Breslau, 1872.

SIDERSKY [1] — D. Sidersky, Le calcul chaldéen des néoménies. Revue d'assyriologie 16 (1919), p. 21–36.

SUTER — see Al-Khwārizmī.

Reprinted from
Hebrew Union College Annual
22, 322–363 (1949)

[*From* Annals of Science, Vol. 8, No. 3, *September* 1952.]

HINDU ASTRONOMY AT NEWMINSTER IN 1428.

By Professor O. Neugebauer, Ph.D., LL.D., and Olaf Schmidt, Ph.D.
Brown University and Institute for Advanced Study.

Professor Thorndike has recently published and translated in this journal [1] sections from the Latin MS. Ashmole 191.II containing chronological and astronomical computations for the year 1428 and for the geographical latitude of Newminster. This short treatise is of interest in several respects, but most of all in its use of methods closely related to the Sūrya Siddhānta, the classical text-book of Hindu astronomy. It is this point which we wish to bring to the attention of the reader.

Obviously, one has to assume Islamic intermediaries for a contact of this kind between England in the fifteenth century and Hindu astronomy. Indeed, several Arabic authorities (especially ' Aomar ' in his book *De nativitatibus*) are quoted by our anonymous author and one may assume that it is ' Aomar's ' work [2] that provided some of the methods described in the text of 1428. This ' Omar ' is indeed well known in astrological writings, both Greek [3] and Latin [4], his real name being ' Umar ibn al-Farrukhān al-Ṭabarī [5]. He belonged to the early period of development of Islamic science (he died about 815) and this makes his relation to Hindu sources easily understandable.

The most obvious relationship between the anonymous of 1428 and Hindu astronomy lies in his remark that Alfonso began the year of the Flood on February 16, 3102 years before the Incarnation [6], a date which is obviously identical with the beginning of the era of the Kali yuga

---

[1] *Annals of Science*, 1951, **7**, 275–283.

[2] It was printed first in Venice, 1503, under the title *Omar Tiberiadis astronomi preclarissimi liber de nativitatibus et interrogationibus* (ed. L. Gauricus). No copy seems to be available in the United States. Even Steinschneider in Berlin in 1864 could not find a copy (*ZDMG* 1864, **18**, 180).

[3] Cf., e.g., *Catal. Cod. Astrol. Græc.* V, 1 p. 150, 31=XI, 1, p. 76, F. 265 (Οὔμαρ ὁ Τιβεριώτης) ; V, 3, p. 7, No. 20, F. 1 (Οὔμαρ ὁ τοῦ Φαρουχάν).

[4] Quoted by Albertus Magnus (d. 1280) in his *Speculum astronmicum*, the relevant chapters of which are edited in *CCAG* V, 1 p. 85–105. The complete text can be found in *Alberti magni opera omnia X* (ed. A. Borgnet), p. 629–650. For Omar cf. chapter VIII (*CCAG V*, 1, p. 95 ; *Opera*, p. 638).

[5] Suter, *Die Mathematiker und Astronomen der Araber*, p. 7, No. 13. Also *CCAG V*, 1, p. 143, note 1.

[6] Thorndike, *loc. cit.*, pp. 277 and 281.

commonly used in Hindu astronomy [7]. The interpretation of this Hindu era as the era of the Flood is common in Islamic sources. Exactly the same date, 3102 B.C. Feb. 17, appears as the equivalent of the date of the Deluge in the tables of al-Khwārizmī [8], which are preserved in the Latin translation by Adelard of Bath of the version composed by Maslama [9]. Abū'l Ḥasan Kūshyār (1000 A.D.) states that the authors of the old tables, of the Sindhind and of the Shah, used the era of the Flood to which all other eras were referred [10], and al-Bīrūnī discusses it in general [11] and mentions also the pretended connexion between the Deluge and the conjunctions of Jupiter and Saturn, a subject which is discussed at length by the English anonymous [12].

Far more significant than the use of the Hindu astronomical era is the method according to which our anonymous teaches how to find the times of ascension of arcs of the ecliptic. This fundamental problem of ancient spherical trigonometry is carefully discussed by Ptolemy in Books I and II of the *Almagest*. Finally tables are given for ten geographical latitudes beyond ' *sphaera recta* ' ($\phi=0$) which permit the finding of the required oblique ascension either directly or by simple interpolation. These latitudes are selected in such a fashion that the longest daylight progresses from zone to zone in steps of one-half hour. Similar tables are found in practically every astronomical work of the Middle Ages, Islamic or Western [13].

[7] Cf., e.g., *Sūrya Siddhānta* (henceforth quoted as S.S.), trsl. E. Burgess, reprinted edition of 1935, Univ. of Calcutta, p. 19. Burgess finds Feb. 17/18 as the day of epoch. One day's deviation is always explicable by the mode of counting. The difference between Feb. 16 and Feb. 18, however, is caused by using the entry into Aries of the true sun or mean sun respectively for the definition of the epoch.

[8] He died after 846.

[9] *Die astronomischen Tafeln des . . . al-Khwārizmī . . .*, ed. by Bjørnbo, Besthorn and Suter, *Danske Vidensk. Selsk. Schrifter* (7), *Hist.-fil. Afd.* 3, 1 (1914). Tab. 1 (p. 100) gives as its first entry the interval between diluvium and Yezdegerd : 3735 years 10 months 23 days. These data must be interpreted as referring to the Egyptian (and Persian) wandering years of 365 days. Then one obtains a difference of 1,363,598 days. The epoch of the era Yezdegerd being 632 June 16, one obtains for the era of the flood the epoch −3101 Feb. 17.

[10] L. Ideler, *Handbuch d. math. techn. Chronologie*, ii, 627 ff. Kūshyār's interval between flood and Yezdegerd is one day longer than in al-Khwārizmī, quoted in the preceding note.

[11] *Chronology of Ancient Nations* (written 1000), trsl. Sachau p. 27 ff. and the table on p. 133.

[12] The common period of 19 years 314 days 14;58,39,29 hours mentioned in the Latin text (Thorndike, *loc. cit.*, pp. 277 and 281) does not seem to occur with this accuracy in the above-quoted literature. Al-Bīrūnī in his *Book of instruction in the elements of the art of astrology* (written 1029) says : " The conjunction of Saturn and Jupiter which occurs once every 20 years is the conjunction par excellence " (No. 250 ; trsl. R. Ramsay Wright p. 150). Cf. also *Chronology of Ancient Nations*, trsl. Sachau, pp. 28 and 91.

[13] Cf., e.g., E. Honigmann, *Die sieben Klimata und die* πόλεις ἐπίσημοι, Heidelberg, Winter, 1929.

The procedure in the text of 1428 follows distinctly different lines. This is evident from the very beginning. The geographical latitude is not characterized by the length of daylight but by the length of the equinoctial noon shadow, the standard procedure of Hindu astronomy. Furthermore the computations make use of the Hindu sine-function which replaced the Greek method of using chords. And finally, most characteristic of all, the procedure is based on the "ascensional differences" which again constantly appear in Hindu spherical trigonometry. Thus, for the given latitude ϕ, the oblique ascensions are not found directly from ready-made tables but by means of the ascensional differences which relate the general case to the special case $\phi=0$ of *sphaera recta*. Thus the whole procedure is based on Hindu methods and not on the Western tradition.

In order to give an accurate account of the method in question [14], we shall use modern notation. Let R be the radius of the circle in which the trigonometric functions are defined—we shall see presently that R was chosen to be 150. If angles are measured in degrees, we shall denote the Hindu sine-function of an angle α by $\mathrm{Sin}\,\alpha$, thus

$$\mathrm{Sin}\,\alpha = R \cdot \sin\alpha.$$

Let A be the vernal point and B any point of the ecliptic with longitude λ less than 90°. By reasons of symmetry, all rising times can be found if the rising times of the first quadrant are known [15]. Let α be the rising time of the arc AB for sphæra recta, β the rising time for arbitrary latitude ϕ. Then β is found in the form

$$\beta = \alpha - \gamma \qquad \ldots \ldots (1)$$

where the correction γ, the so-called 'ascensional difference', is determined from

$$\mathrm{Sin}\,\gamma = c \cdot s \qquad \ldots \ldots (2)$$

s being the equinoctial noon shadow of a gnomon of length 12, and

$$c = \frac{R\,\mathrm{Sin}\,\delta}{12(R - \mathrm{Sin}\,\mathrm{vers}\,\delta)} \qquad \ldots \ldots (3)$$

where δ is the declination of B [16]. In the Sūrya Siddhānta [17] the two

---

[14] It is also described, e.g., by Suter in his commentary to al-Khwārizmī's tables, p. 73 ff.

[15] Cf., e.g., O. Neugebauer, "On some astronomical papyri and related problems of ancient geography", *Trans. Am. Philos. Soc.*, 1942, **32**, 251–263.

[16] The correctness of these formulæ can easily be seen as follows. Because

we have
$$R - \mathrm{Sin}\,\mathrm{vers}\,\delta = R\cos\delta,$$
$$c = \frac{R}{12}\tan\delta.$$

But $s = 12\tan\phi$, thus our formulæ (2) and (3) are equivalent with
$$\sin\gamma = \tan\phi\,\tan\delta.$$

last formulæ are combined into one. Our text uses them separately because (3) shows that the value of *c* does not depend on the geographical latitude. Consequently *c* can be tabulated for all points B between A and A+90° and the result is the *Tabula differentie ascensionum universe terre* " [18], quoted in our text. Once this table is computed and if the length *s* of the equinoctial noon shadow is known for the given locality, the rising time of an arc AB is simply found by multiplying first *s* by the value of *c* which is given in the table beside the longitude of B. With the result, one enters a table of sines and thus finds γ, which is the correction to be applied to the right ascension α in order to reach the oblique ascension β. Thus the whole procedure is based essentially on three tables only : a table of right ascensions, a table for the correction *c*, and a table of sines.

Fortunately, at least two of these tables are still preserved on the two pages immediately preceding our treatise. The only missing table is the table of right ascensions ; from it we know only three values which were quoted in the text (in addition to the trivial one $\alpha = 90$ for $\lambda = 90$) :

$$\lambda = 1° \qquad \alpha = 0;54,53,40$$
$$2 \qquad 1;49,33,20$$
$$30 \qquad 27;53$$

In al-Khwārizmī's tables we find 0;54,52 and 1;49,45 and 27;50,9, respectively [19]. The value 27;53 is certainly not explicable as a scribal error because it is used as 27;53,0,0,0,0 in the computation of β.

The first of the preserved tables which belong to our treatise is called " *Tabula altitudinis et umbre* " and need not be reproduced here because it is known both from the tables of Johannes de Lineriis [20] and of al-Khwārizmī [21]. The table is arranged in three triple columns, the first of which lists the degrees (" *gradus altitudinis* ") of $\bar{\phi} = 90 - \phi$. The two following columns give the corresponding values of *s* as ' *puncta umbre* '

Now let C on the equator be the point which rises simultaneously with B, and D the orthogonal projection of B onto the equator. Then DC is the ascensional difference γ and the angle at C is the geographical latitude $\bar{\phi}$ whereas BD is the declination δ of B. In the right triangle BDC we then have $\tan\bar{\phi} = \tan\delta/\sin\gamma$. Q.E.D.

[17] II, 61 to 63 and III, 44, 45.

[18] Thorndike, *loc. cit.*, pp. 278 f. and 282 f. Prof. Thorndike informs us that one should read p. 278, lines 2 and 5, *differentie ascensionum* instead of *directe ascensionum*.

[19] A third example, $\lambda = 2°$, is only mentioned in a very sketchy fashion at the end.

[20] Published by Maximilian Curtze in his " Urkunden zur Geschichte der Trigonometrie im christlichen Mittelalter " Bibliotheca Mathematica [3] 1 (1900) [pp. 321 to 416] p. 412. The MS. (henceforth quoted as L) is Cod. Basileensis F. II. 7, fol. 38 ff. (written 1432) ; for an older MS. in Erfurt, copied in 1323 from the original which was completed in 1322. cf. Curtze, p. 390.

[21] Tab. 60, *l. c.*, p. 174 ; henceforth quoted as K.

and '*minuta*' [22]. For $\phi=55$ we have $\bar{\phi}=35$ and with this entry we obtain $s=17;8$ as the length of the equinoctial noon shadow at Newminster. This is the only use made of this table in our texts.

The next table [23] (cf. p. 226) is called " *Tabula differencie ascensionum universe terre* " and is the table of our corrections c. It is arranged in three quadruple columns called " *Sinus differencie prime | secunde | tertie* " respectively. A second hand wrote above these columns in large letters $1 \cdot 7 \cdot | 2 \cdot 8 \cdot | 3 \cdot 9 \cdot$ thus indicating that the table not only holds for the first three zodiacal signs but also for the diametrically opposite signs. Below the columns we read, in the same second hand, $6 \cdot 12 \cdot | 5 \cdot 11 \cdot | 4 \cdot 10 \cdot$ which refer to the use of the table for the second and fourth quadrant. The first of the four subcolumns is headed by '*Gradus altitudinis*' which must be a mistake because the argument of this table is degrees of longitude, not of latitude. The three subsequent columns are headed '*minuta | 2 - a | 3 - a*' respectively and give the values of c [24]. A table of this kind existed in the original work of al-Khwārizmī, as is clear from the description which has survived in Adelard of Bath's translation [25]. Thus our anonymous author provided us with an interesting element for the understanding of the earliest Islamic astronomical tables.

Finally we find on fol. 138r a combined table of sines and declinations, called " *Tabula Sinus declinationis* " (cf. p. 226). This title explains the references in the text which appear either in the from of ' *Tabula sinus* ' or ' *Tabula sinus* [26] *declinationis solis* '. Again we have three main columns of 30 lines each. The first subcolumn is called " *linea numeri* [27] and contains the common arguments for both the following

---

[22] The following deviations between our table (fol. 137r, left half) and L and K should be noted : (*a*) for $\bar{\phi}=47$ our text has erroneously 11,21 where K and L give correctly 11,11 (*b*) at the beginning we have

here :	1	687;26	L :	687;26	K :	687;29 or 23	
	2	343;38		343;39		343;38 or 388,	383
	3	293;28		203;28		228;58 or 243;47	273;47
	4	171;42		171;42		171;36 or 181;36	

(*c*) otherwise the minutes in our table are greater than in L in 2 cases, greater than in K in 22 cases ; the opposite holds in 1 case for L, in 7 cases for K ; (*d*) finally :

$\bar{\phi}=$	8	here :	85;23	L :	85;28	K :	85;22	
	10		68;3		68;30		68;3 or	68;34
	21		31;16		31;19		31;15	
	79		2;20		2;30		2;19	

[23] Fol. 137r, right half.

[24] The entry for 84 should be 5,29,25 instead of 5,29,35.

[25] Bjørnbo-Suter, pp. 21 and 76 (cap. 26ᵃ).

[26] This reading was suggested to us by Prof. Thorndike instead of *summi* (p. 278, 8 and 12).

[27] The same terminology is found in Bjørnbo-Suter, Tab. 58 and 58ᵃ.

same table in Toledan Tables
Cod. Vindob. 2385 fol. 4v (=T)

Table of Sines (R=150)
same in fol. 3r, 3v
but Table of declinations based on ε = 23;51,0.

Tabula differencie ascensionis universe terre

1	0, 5, 18	31	2, 39, 41	61	4, 43, 33
2	0, 10, 35	32	2, 44, 31	62	4, 46, 39
3	0, 15, 52	33	2, 49, 19	63	4, 49, 40
4	0, 21, 10	34	2, 54, 6	64	4, 52, 35
5	0, 26, 27	35	2, 58, 49	65	4, 55, 24
6	0, 31, 44	36	3, 3, 31 1)	66	4, 58, 8
7	0, 37, 0	37	3, 8, 10	67	5, 0, 46
8	0, 42, 16	38	3, 12, 46	68	5, 3, 18
9	0, 47, 32	39	3, 17, 20	69	5, 5, 44
10	0, 52, 47	40	3, 21, 52	70	5, 8, 5
11	0, 58, 2	41	3, 26, 21	71	5, 10, 19
12	1, 3, 16	42	3, 30, 46	72	5, 12, 27
13	1, 8, 30	43	3, 35, 9	73	5, 14, 28
14	1, 13, 43	44	3, 39, 30	74	5, 16, 23
15	1, 18, 55	45	3, 43, 47	75	5, 18, 12
16	1, 24, 6	46	3, 48, 0	76	5, 19, 54
17	1, 29, 17	47	3, 52, 10	77	5, 21, 29
18	1, 34, 27	48	3, 56, 17	78	5, 22, 58
19	1, 39, 37	49	4, 0, 20	79	5, 24, 20
20	1, 44, 44	50	4, 4, 19	80	5, 25, 35
21	1, 49, 51	51	4, 8, 15	81	5, 26, 43
22	1, 54, 56	52	4, 12, 7	82	5, 27, 44
23	2, 0, 0	53	4, 15, 54	83	5, 28, 38
24	2, 5, 3	54	4, 19, 37	84	5, 29, 35 2)
25	2, 10, 4	55	4, 23, 16	85	5, 30, 4
26	2, 15, 4	56	4, 26, 51	86	5, 30, 30
27	2, 20, 3	57	4, 30, 21	87	5, 31, 1
28	2, 25, 1	58	4, 33, 47	88	5, 31, 19
29	2, 29, 56	59	4, 37, 8	89	5, 31, 30
30	2, 34, 49	60	4, 40, 23	90	5, 31, 34

1) T: 3, 2, 31 (incorrectly)

2) T: 5, 29, 25 (correctly)

16

Tabula Sinus declinationis

	Sinus	declinatio		Sinus	declinatio		Sinus	declinatio
1	2, 37, 5	0, 24, 0	31	77, 16, 22	11, 52, 45	61	131, 11, 35	20, 27, 36
2	5, 14, 5	0, 48, 0	32	79, 29, 17	12, 13, 40	62	132, 26, 32	20, 30, 48
3	7, 51, 2	1, 11, 50	33	81, 41, 45	12, 34, 23	63	133, 39, 5	20, 51, 39
4	10, 27, 50	1, 35, 51	34	83, 52, 42	12, 54, 53	64	134, 49, 10	21, 3, 9
5	13, 4, 25	1, 59, 47	35	86, 2, 52	13, 15, 12	65	135, 56, 45	21, 14, 12
6	15, 40, 45	2, 23, 40	36	88, 10, 5	13, 35, 16	66	137, 1, 15	21, 24, 56
7	18, 16, 50	2, 47, 30	37	90, 16, 20	13, 55, 7	67	138, 4, 32	21, 35, 11
8	20, 52, 35	3, 11, 19	38	92, 20, 17	14, 14, 41	68	139, 4, 40	21, 45, 2
9	23, 27, 55	3, 35, 5	39	94, 23, 15	14, 33, 6	69	140, 2, 15	21, 54, 25
10	26, 2, 50	3, 58, 46	40	96, 25, 50	14, 53, 11	70	140, 46, 15 8)	22, 3, 35
11	28, 37, 17 4)	4, 22, 28	41	98, 24, 32	15, 12, 3	71	141, 49, 42	22, 12, 16
12	31, 11, 52	4, 46, 0	42	100, 22, 10	15, 30, 41	72	142, 39, 32	22, 20, 29
13	33, 44, 32	5, 9, 30	43	102, 18, 0	15, 49, 2	73	143, 26, 45	22, 28, 17
14	36, 16, 52 5)	5, 32, 55	44	104, 11, 57	16, 7, 8	74	144, 11, 22	22, 35, 38
15	38, 49, 22	5, 59, 15	45	106, 3, 2	16, 25, 17	75	144, 53, 24 9)	22, 42, 36
16	41, 20, 44	6, 19, 29	46	107, 54, 12	16, 42, 29	76	145, 32, 40	22, 49, 6
17	43, 51, 21	6, 42, 38	47	109, 42, 20	16, 59, 49	77	146, 9, 20	22, 55, 44
18	46, 21, 9	7, 5, 40	48	111, 28, 25	17, 16, 25	78	146, 43, 20	23, 1, 59
19	48, 50, 6	7, 28, 24	49	113, 12, 25	17, 33, 46	79	147, 14, 40	23, 6, 11
20	51, 18, 10	7, 51, 25	50	114, 54, 20	17, 49, 48	80	147, 43, 57	23, 10, 46
21	53, 45, 49 6)	8, 14, 5	51	116, 34, 5	18, 6, 29	81	148, 9, 12	23, 15, 5
22	56, 11, 27	8, 36, 49	52	118, 12, 57	18, 21, 33	82	148, 32, 26	23, 18, 56
23	58, 36, 37	8, 59, 3	53	119, 47, 10	18, 36, 33	83	148, 52, 55	23, 22, 21
24	61, 0, 37	9, 21, 20	54	121, 21, 22	18, 51, 40	84	149, 10, 42	23, 25, 19
25	63, 23, 32	9, 43, 28	55	122, 52, 22	19, 6, 41	85	149, 25, 45	23, 27, 49
26	65, 45, 20	10, 5, 26	56	124, 21, 20	19, 21, 6	86	149, 39, 5	23, 29, 50
27	68, 5, 15	10, 27, 14	57	125, 48, 2	19, 35, 7	87	149, 47, 49 10)	23, 31, 44
28	70, 25, 15	10, 48, 52	58	127, 12, 27	19, 48, 48	88	149, 54, 32	23, 32, 2
29	72, 43, 17	11, 10, 19	59	128, 34, 30	20, 2, 7	89	149, 58, 40	23, 33, 13
30	73, 0, 0 7)	11, 31, 39	60	129, 54, 54	20, 15, 0	90	159, 0, 0	23, 33, 30

4) T: 28, 37, 57
5) T: 36, 16, 57
6) T: 53, 45, 29
7) T: 75, 0, 0

8) T: 140, 47, 15
9) T: 144, 53, 20
10) T: 149, 47, 40

tables, each of which has three columns called '*minuta* / 2 – *a* / 3 – *a*' for the table of '*sinus*' and '*gradus* / *minuta* / 2 – *a*' for the table '*declinatio*'[28]. For the table of Sines no direct comparison with al-Khwārizmī is possible because our table is based on the value $R = 150$ whereas al-Khwārizmī used $R = 60$, following the Hellenistic custom. The norm $R = 150$ reveals a direct relationship or our tables with the 'Toledan Tables' of al-Zarkālī, who seems to have been the first to have introduced it to the West[29]. This again brings us back to Hindu sources because $R = 150$ is used in the Khaṇḍakhadyaka of Brahmagupta[30], written about 665 A.D. and well known to Islamic astronomers[31]. The connexion with the Toledan Tables is confirmed by the use in our tables of the value 23;33,30 for the obliquity of the ecliptic. The "Canones Arzachelis sive regule super tabulas astronomie" quote even the author of this value[32] : "Jahiben fiilium Albumazaris" i.e. a son of the famous Abū Ma'shar (died 886) whose relation to Hindu astrology have been often demonstrated[33]. Thus we have again reached the period of early Islamic contact with Hindu astronomy.

[28] There are errors. Sin 30 should be 75,0,9 and not 73,0,0. The declination of 15° should be 5,56,15 instead of 5,59,15 ; similarly, for 39°, one should read 14,34,6 instead of 14,33,6 and for 62° the correct value would be 20,29,48 instead ot 20,30,48. At several occasions the differences reveal unevenness which cannot be explained by simple scribal errors.

[29] Delambre, Histoire de l'astronomie du moyen age (Paris 1819), p. 176. Braunmühl, Vorlesungen über Geschichte der Trigonometrie, I (Leipzig 1900), p. 79. Also Curtze, *l. c.*, note 20, p. 354.

[30] English translation by Prabodh Chandra Sengupta, Univ. of Calcutta 1934. The following Sines are found in I,30 and III,6 (*l. c.*, pp. 32 and 70) :

$\alpha = 15$	Sin $\alpha =$ 39	our text :	38,49,22
30	75		75,0,0
45	106		106,3,2
60	130		129,54,54
75	145		144,53,23
90	150		150,0,0

cf. also the '*Tabula Kardagarum sinus*' Curtze *l. c.* note 20 pp. 339 and 411.

[31] The index to al-Bīrūnī's 'India' gives 57 references to Brahmagupta's works, 16 to the Khaṇḍakhadyaka.

[32] Curtze, *l. c.*, p. 348. Also Delambre, Histoire de l'astronomie de moyen age, p. 176. For the time between al-Zarkālī (about 1080) and our author several cases of the use of 23;33,30 for ε are known ; cf. e.g., an anonymous treatise from Marseille of about 1140 (Duhem, Système du monde III, p. 214) or the "Compilatio" of Leopold of Austria, composed about 1270 (cf. Francis J. Carmody, Leopold of Austria " Li Compilacions de le Science des Estoilles ", Books I–III. Univ. of California Publications in Modern Philology, 33, pp. 35–102 [1947]).

[33] Cf., e.g., Dyroff in Boll, Sphaera (Leipzig, 1903), p. 482 ff.

Supplementary Remarks.

The ' cycles ' (*orbis*) of which the author speaks [34] are 360 years long, as can easily be deduced from the computations in the text. A unit of 360 years also occurs in Hindu time reckoning as " the year of the gods " [35].

Thorndike, p. 277, line 1, restore : 3376 annos et 31 dies [et 6 horas] as is required by the immediately following computations and confirmed by line 11. The other computations with the common period of Saturn and Jupiter seem to be in disorder.

Thorndike, p. 278, last line : 59 tertia etc. is wrong. One obtains actually 17;8,28,53,30,40. Prof. Thorndike informs us that the 8 minutes are a restored figure.

The author states at the end that the ascensional difference γ for $\lambda=2°$ is 1;9,16° after he had previously found for $\lambda=1°$ the value $\gamma=0;34,41°$. He then states that the oblique ascension for $\lambda=2°$ is 0;40,11,30 (as compared with 0;20,11,43 for $\lambda=1$). Hence the right ascension $\alpha=\beta+\gamma$ for $\lambda=2$ would be 1;49,33,20 (0;54,53,40 for $\lambda=1$). From al-Khwārizmī's tables one finds 1;49,45. Obviously the values in the text deviate too much from an essentially linear increase. One error is visible in the computation of β for $\lambda=1$ (one would find 0;20,12,40) and similar mistakes must have affected the remaining numbers [36].

[34] Thorndike, pp. 276 and 280.

[35] S.S. I,14 (trsl. Burgess, p. 9). Cf. also al-Bīrūnī, India XXXV (trsl. Sachau I, p. 350).

[36] Misprint on p. 283, lines 6 and 8 from the end : read 11″ for 2″.

Tamil Astronomy

A Study in the History of Astronomy in India

1. In 1825 Lieutenant Colonel JOHN WARREN published a book of over 500 quarto pages entitled *Kala Sankalita* with the subtitle *A Collection of Memoirs on the various modes according to which the nations of the Southern Parts of India divide time* (1). This truly remarkable work was compiled over a period of eleven years from information from natives, " originally intended for the sole use of the Honorable Company's College of Fort. St. George," but ultimately extended to a far-reaching study of Indian calendariography and astronomy. Though it was repeatedly consulted by BURGESS, THIBAUT, and others, its contents have by far not been adequately exploited for our knowledge of the development of Hindu astronomy. The present discussion of a single short chapter, pp. 334 to 348, will demonstrate this fact.

2. The path which brought me to the investigation of WARREN's work can be described simply enough. Ever since KUGLER's *Babylonische Mondrechnung*, it has been known that the ratio 2 : 3 of the length of shortest to longest daylight found in the Seleucid astronomical cuneiform texts was also used in India. THIBAUT, whose classical article on " Astronomie, Astrologie und Mathematik " was published the year before in BÜHLER's *Grundriss der Indo-Arischen Philologie und Altertumskunde* found these values correctly referring to a latitude of 34°, thus barely possible for India, but he warned against a hypothesis of Babylonian origin unless they were actually found in Babylonian texts—as actually happened in the following year.

A whole group of Babylonian-Indian parallels was recognized by SCHNABEL in 1924 (2) and in 1927 in two short notes, the first

---

(1) Professor F. EDGERTON kindly informed me that a copy of this work was in the library of the American Oriental Society in New Haven. I am grateful for the liberality with which this copy was put at my disposal for many months of study.

(2) *Z. für Assyriologie*, 35 (1924), p. 112 and 37 (1927), p. 60.

quoting planetary periods, the second the anomalistic period of 248 days of the moon.

This latter period appeared again in a Greek papyrus of the Roman imperial period, published in 1947 by ERIK J. KNUDTZON and myself (3), though combined with a second anomalistic period of 3031 days. Two years later I discovered that the rules for these methods were contained in another Greek papyrus, P. RYLAND 27 (4), written about 300 A.D. After the basic idea had become clear, I found the same procedure in the *Pañca Siddhāntikā*, published by THIBAUT and DVIVEDI in 1889. THIBAUT originally had great difficulties in understanding the passages in question and eventually found the solution in the *Kala Sankalita* of WARREN, who " after a long search for one of these mechanical computers " had found a " Kalendar maker residing in Pondicherry " who showed him how to compute a lunar eclipse by means of shells, placed on the ground, and from tables memorized " by means of certain artificial words and syllables." WARREN adds " and I found the Sashia thus introduced to me, competent to my object, for (as I wished) he did not understand a word of the theories of Hindu astronomy, but was endowed with a retentive memory, which enabled him to arrange very distinctly his operations in his mind, and on the ground." Thus his Tamil informer computed for him the circumstances of the lunar eclipse of 1825 May 31-June 1 with an error of + 4 minutes for the beginning, -23 minutes for the middle and -52 minutes for the end (cf. also p. 271). But it is not the degree of accuracy of this result which interests us here; it is the fact that a continuous tradition still survived in 1825, the earlier phases of which are found in the 6th century A.D. with VARAHAMIHIRA, in the 3rd century in the Roman empire, and in Seleucid cuneiform tablets of the second or third century B.C.

3. What we have said here about the theory of the motion of the moon is not an isolated phenomenon. Whole sections of the planetary theory of the *Pañca Siddhāntikā* have exact numerical parallels in Babylonian texts of the Hellenistic period. Though the corresponding Greek papyri are still missing, there can be

(3) *Bull. Soc. Royale des Lettres de Lund*, 1946-1947, II, p. 77-84.
(4) *Kgl. Danske Vidensk. Selsk., Hist.-filol. Medd.*, 32, 2 (1949).

little doubt that the way of transmission is the same in all cases, namely, through the astrological literature (5). Authors like VETTIUS VALENS and PAULUS ALEXANDRINUS (and probably also HIPPARCHUS) are much more closely related to the Babylonian arithmetical methods than the highly scientific work of PTOLEMY, whose *Almagest* is a much more independent work than one is usually willing to admit. Similarly, the *Sūrya Siddhānta* is a well rounded systematic treatise which leaves little room for the discovery of its historical predecessors except for some obviously far older and very primitive native doctrines, like the conception of the strings of air which push and pull the planets in their irregular motion. VARAHAMIHIRA's unique historical report, called *Pañca-Siddhāntikā* (6), allows us the possibility of discovering direct witnesses for the early contact with Hellenistic astronomy of the simpler type, in which arithmetical schemes are used directly for the prediction of astronomical phenomena, without going through the slow process of mathematical deduction from complicated geometrical models.

The discussion of the Hellenistic (and thus ultimately Babylonian) component of the *Pañca-Siddhāntikā* will be the subject of a subsequent communication. At present I shall restrict myself to the description of the lunar theory in so far as it appears in WARREN's report on the Tamil astronomers. While WARREN presented the method by means of a detailed numerical example, I shall follow the more convenient method of using the corresponding algebraic formulation which allows us easily to recognize the underlying theory.

4. As mentioned before, WARREN's informant performed all operations by means of shells laid out on the ground. His numerical system consists of the usual mixture of decimal and sexagesimal notation, a pernicious habit developed by the Hellenistic astronomers and still in use today. In order to simplify matters I shall be more consistent and shall ordinarily write numbers sexagesimally, integers and fractions alike. Whoever wants to follow Tamil practice exactly may either read

(5) This was seen as early as 1860 by J. B. BIOT in his review of BURGESS' *Sūrya-Siddhānta* (*Journal des Savants*, August to December 1860).

(6) Edited and translated by G. THIBAUT and MAHĀMAHOPĀDHYĀYA SUDHĀKARA DVIVEDI, Benares, 1889. Quoted in the following as " Thibaut."

WARREN's text or use shells or pebbles in the way described by
WARREN and represented by him in the following examples :

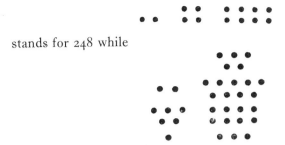

stands for 248 while

denotes 5 zodiacal signs (each 30°) 29° 58' 13". The first number
I shall write 4,8 (= 4 · 60 + 8), the second 2,59;58,13°, separating
integers from fractions by a semicolon.

Following WARREN, I shall speak of numerical " Tables,"
though they were known to his informant only by means of
mnemonic devices. Obviously there once existed written tables
upon which the oral tradition was built.

5. We now turn to the rules which lead to the determination
of the circumstances of a lunar eclipse. How an initial date
should be chosen is not explained by WARREN. In the taste of his
time he was for the most part interested in a comparison between
the results of the native procedure and the European answer.
Consequently he had the computation carried out for a date known
to him as the date of an eclipse. He mentions, however, that most
of the natives find the conjunctions and oppositions by means of
a 19-year cycle whereas the 223-month eclipse cycle seems to
have been introduced by Europeans. Indeed, no such eclipse
cycle is inherent in the subsequently described methods. The
syzygies can be found by any Hindu method, at least approximately.
Rules for finding the position of the nodes will be mentioned
presently (p. 269) and this suffices to rule out all those syzygies for
which eclipses are excluded. The remaining cases require
detailed computation as described by the text. Consequently it
is not at all surprising that WARREN's informant had no specific
rules for finding the initial date. We may therefore assume that
a date has been given for which the following computations should
establish the detailed circumstances of an expected lunar eclipse.

The main outline of the procedure to be followed is simple. First one finds for the day in question the true longitude (a) of the sun and (b) of the moon. Ordinarily the values obtained will not be exactly identical, but it is now easy to find by extrapolation the precise moment when this will be the case. For the same moment the longitude of the lunar node can be computed and thus the deviation of the line of nodes from the line of syzygies will be known. A simple table gives the corresponding latitude of the moon. Using this latitude and the lunar velocity for the day in question the eclipse magnitude can be found. This in turn gives the duration of the moon's travel through the shadow cone and thus the moments of first and last contact. This completes the description of the eclipse.

We shall now describe the details of this procedure step by step.

6. The first step consists in determining the number of days elapsed since epoch (ahargaṅa, henceforth called a). Owing to purely formal reasons, a is counted from the first day of the first week of the Kali yuga (3102 B.C.) (We may note in passing that the use of the planetary week is of Hellenistic origin whereas the consistent use of a purely sexagesimal division of the days is of ultimately Babylonian origin). It is furthermore assumed that one sidereal year contains $365;15,31,15^d$. If an exact number n of years were completed in the Kali yuga the number of days elapsed since the first day of the first week would be given by

$$n \cdot 365;15,31,15 - 2;8,51,15$$

where 2;8,51,15 days are subtracted in order to start with the entrance of the mean sun into Aries (cf. GINZEL, *Chronologie*, I, p. 343).

In general, however, the day for which an eclipse is to be expected will fall within the $n + 1$st year. This excess over a complete year is expressed in a peculiar fashion. We are supposed to know how many zodiacal signs the sun has completed, beginning with Aries, and how many days have elapsed since then. (This shows, by the way, that the beginning of the ahargana is supposed to coincide with the sun's being in the vernal point). From these two elements (the number of completed signs and the number of additional days), the number of days within the $n + 1$st year is computed as follows. A table is consulted which gives for each of the twelve zodiacal signs the number of days needed for the sun to traverse any sign. Similar tables existed also in Greek astrological texts (7). I reproduce here only the first two values (8)

$$\Upsilon \quad 30;55,32,1$$
$$\upsilon \quad 31;24,12,1$$

but the graph (1) on Plate I will clearly show its main trend. The time needed is greatest in Gemini, the solar apogee, and least in Sagittarius, its perigee. In this way the number z of days

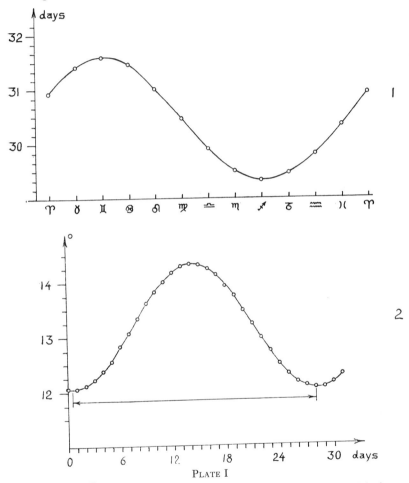

PLATE I

(7) *Catal. Cod. Astrol. Graec.*, vol. 11, 2, p. 133, 5 to 26. The published text contains several scribal errors, which can be emended, however, in a very satisfactory fashion. Cf. *Centaurus*, I (1951), p. 266-270.

(8) The complete table is given by WARREN. as Table III.

17

required to traverse the given number of signs can be found by simple addition. Finally, d days for the current sign have to be added. Thus the ahargana is given by

$$(1) \qquad a = n \cdot 365;15,31,15 - 2;8,51,15 + z + d$$

This number can now be considered as given. In WARREN's example it has the value $a = 8,19,48,33;2,18,15$.

A few remarks must be made about the structure of the above mentioned table for the solar motion [graph (1)]. By dividing its numbers by 30, one obtains the daily velocity in each sign. These values vary again in a sinusoidal curve, between about $0;56,56°$ and $1;1,20°$ per day. These extremal values agree well with the values given in an extensive table for the solar velocity (WARREN, Table XXVIII) where we find the extrema $0;56,50$ and $1;1,26$ respectively. The origin of tables of this kind cannot be exactly determined until we will have found a numerically identical replica in other sources. Tables of this character existed in many variations both in Babylonian and in Greek astronomy, and only detailed textual information can decide whether they were based on interpolation schemes or on geometrical considerations, i.e., on trigonometric tables. We shall meet additional tables of this kind in the further progress of our investigation. In all cases, we shall take them as granted, without investigating the method of their computation.

7. The method upon which the determination of the ahargana was based leads to an easy answer for the longitude of the sun at the given date. First, the number of years can be ignored because they represent complete rotations, beginning with Aries $0°$. Second, the number of completed solar months is identical with the number of complete zodiacal signs. Finally, there remains a number d of days elapsed since the completion of the last solar month. Let us assume that the solar velocity at a given date is w. A day-by-day table for w through a whole year was computed (WARREN, Table XXVIII) from which the values for every 8th day were excerpted. For example, for the solar month Taurus, one has

$$
\begin{aligned}
\text{day} \quad 1 \quad & 0;57,38 = 1 - 0;2,22 \\
9 \quad & 0;57,25 = 1 - 0;2,35 \\
17 \quad & 0;57,15 = 1 - 0;2,45 \\
25 \quad & 0;57,4 = 1 - 0;2,56.
\end{aligned}
$$

These values are taken as mean values for the solar motion during eight-day intervals. Thus the sun travelled during the first eight days 8° — 0;18,56, during the second eight days 8° — 0;20,40, then 8° — 0;22, and finally 8° — 0;23,28. In this way, we obtain the rule that the sun travels in Taurus during 8 days an arc of 8 degrees minus a small correction which is 0;18,56 or 0;19 during the first octade, then 0;20,40 or 0;21 in the second octade, etc. These corrections again form a table, memorized by the native computer, and reproduced by WARREN as Table XXVII. Its values deviate occasionally by one unit from the values which can be derived in the just-described fashion from Table XXVIII. These small discrepancies are probably due to rounding-off operations which are handled rather carelessly in computations of this kind.

In WARREN's example, we have $d = 20$ in month Taurus. Because $20 = 8 + 8 + 4$, we have to take the added corrections for the first octades, i.e., — 0;19 and — 0;21, thus — 0;40. The correction for the third octade is — 0;22 and therefore the correction for 4 days amounts to — 0;11, which must be added to the previously found correction of — 0;40. Thus the solar motion during these 20 days was 20° — 0;51 = 19;9 and the solar longitude is ♉ 19;9.

WARREN's informant computed with an apparently higher accuracy. For the ahargana we found a huge number of days, plus a fraction 0;2,18,15^{d} which is the result of fractional parts in the epoch, in the length of the solar year, and in the travel of the sun through complete zodiacal signs. Consequently we have used a small fraction of day 20 and need only 20 — 0;2,18,15 = 19;57,41,45 to come to the beginning of day 21. It is this value which he used in the above interpolation for the third octade (of course, only 3;57,41,45 as excess over 16 days) and he thus obtained for the solar longitude ♉ 19;6,48,5. The difference from our previous result is, of course, of no real significance in view of the crudeness of the 8-day scheme.

8. We now turn to the longitude of the moon. It is here that we meet the procedure which we mentioned in the introduction as establishing a direct contact with Hellenistic texts.

Before describing the details of this process, called " Vakyam " process, I shall outline its simple basic idea. Suppose we know that the moon was at its apogee (i.e., at its state of slowest possible motion) on a certain day and at a certain longitude. It is also known how much time elapses between two consecutive apogees—this time is called the " anomalistic month "—and how much the longitude of the apogee has progressed meanwhile. Thus we are able to compute all future moments of the apogee and all corresponding longitudes, thus especially the data for the last apogee which precedes the moment for which we expect an eclipse. If we now compute once and for all a table of lunar motion from one apogee to the next, we can directly determine from such a table the progress of the moon subsequent to the last apogee. Adding this amount to the longitude of the last apogee, we have found the longitude of the moon for the given moment.

This is the basic idea of the Vakyam process. For practical purposes, however, some modifications were introduced. First, one must realize that the length of one anomalistic month contains fractions of a day. Its length being approximately $27 \frac{5}{9} = 27;33,20^d$, one can get rid of these fractions by taking as the smallest unit nine anomalistic months or 248^d. During this time the lunar apogee moves forward $27;44,6°$ and thus we are capable of determining the longitude of the moon for steps of 248^d. For the intermediate days a table of 248 entries was computed (WARREN, Table XXVI), beginning as follows :

1^d	0^s	$12;3°$	$12;3°$
2	0	24;9	12;6
3	1	6;22	12;13
		etc.	

The first column gives the consecutive days since apogee. The third column gives the moon's daily motion and from this one obtains by consecutive addition the second column, in zodiacal signs and degrees, indicating the total progress of the moon. The first of the nine waves is plotted in graph (2) Pl. I.

In principle, this should solve our problem. If the date in question were n days after a given apogee, used as " epoch," we merely have to divide n by $D = 248^d$. The quotient δ would indicate that the apogee has moved $\delta \cdot 27;44,6°$ and the remainder would give the number of days for which the motion of the moon could be found in Table XXVI. The value of D, or better, the value A = 27;33,20^d for the anomalistic month is, however, not accurate enough to guarantee, for a large number of repetitions, a return to the apogee. Already the Babylonian astronomers of the Seleucid period considered A = 27;33,20 only as a convenient approximation of a more accurate value 27;33,16,26,54,... Consequently it is not surprising to find in the Greek papyri which use these methods another, larger period

$$C = 3031^d$$

during which the apogee moves $\Delta\lambda$ = 5,37;31,19,7°. The rules of the Tamil computer contain the same value C but assume a movement of 5,37;31,1°. For D the Greek texts assume a gain in longitude of 27;43,24,56° as compared with 27;44,6° of the Tamil rules.

The period C also appears in the *Romaka Siddhānta*, as is known from VARAHAMIHIRA's *Pañca Siddhāntika*, VIII, 5. Both C and D are also used in the *Pañca Siddhāntika* II, 1 to 4, a section which probably was contained both in the *Paulisa* and in the *Vāsishṭha Siddhānta*(9). The corresponding values for the increase in longitude are 5,37;31,12,42,... for C and 27;44,6 for D. The first value is close to the Greek value, the second is identical with the Tamil parameter.

By the same method the accuracy of the anomalistic period can be further improved. The Vakyam process gives two more groups

$$R = 4C + D = 12372^d \qquad \Delta\lambda = 4,57;48,10°$$
and
$$V = 1600984^d \qquad \Delta\lambda = 3,32;0,7°.$$

We have no direct witness for the use of the last two steps in the papyri at our disposal but at least for the existence of steps like R evidence can be found in the Greek papyri (10).

(9) THIBAUT, p. XI and p. XXXIII.

(10) The details are not clear. P. RYL. 27 gives two sets of rules, one operating with an epoch Augustus — 2 (= Cleopatra 20) XI 5, the second with Augustus 97 (= Nero 14) X 23. P. LUND, 35a, agrees with the second norm which shows one step D less than expected according to the first norm. Thus it seems as if one D was omitted sometime during this century in question. Because C = 11D + 303^d and R = 4C + D the same effect would appear by 10 repetitions of R, that is in about 340 years. Without new texts one can scarcely say more than that a shift in the position of D is recognizable and that its purpose is in all probability the same as in the formation of higher groups of the type R and V.

9. It follows from our general discussion that it is the purpose of the successive steps D, C, R, and V to obtain better approximations for the length of the anomalistic month. This can be checked explicitly by dividing the number of days contained in each step by the corresponding number of anomalistic months. In this way, one obtains

$$D = 4,8^d = 9 \text{ anom. months of } 27;33,20^d \text{ each}$$
$$C = 50,31^d = 1,50 \text{ anom. m. of } 27;33,16,21,...^d \text{ each}$$
$$R = 3,26,12^d = 7,29 \text{ anom. m. of } 27;33,16,26,11,...^d \text{ each}$$
$$V = 7,24,43,4^d = 16,8,22 \text{ anom. m. of } 27;33,16,58,...^d \text{ each.}$$

The value obtained from R for the length of the anomalistic month is very accurate and is identical with the Babylonian value for the first four places. The value obtained from V, however, is very suspicious. The sudden increase of the fourth place from 26 to 58 is very unlikely. Indeed, there are additional arguments against the correctness of V. From the values of $\Delta\lambda$ one can determine the mean motion of the moon assumed in each case. In this way, one obtains the following list :

D :	9 rotations	$+ 27;44,6°$	thus mean motion	$13;10,34,50,...°/^d$
C :	1,50 rot.	$+ 5,87;31,1°$	thus m. m.	$13;10,34,51,36,...$
R :	7,32 rot.	$+ 4,57;48,10°$	thus m. m.	$13;10,34,51,36,...$
V :	16,16,37 rot.	$+ 3,32;0,7°$	thus m. m.	$13;10,35,34,...$

Again V falls out of the order, even to such a degree that it exceeds the ordinary first approximation of $13;10,35$ by a considerable amount. Thus it seems practically certain that V is incorrectly handed down (11). Unfortunately, we have no check for this value in older sources.

10. We now can continue the numerical example of the lunar eclipse whose date fell $a = 8,19,48,33^d$ (ignoring fractions) after epoch. This number a is first divided by the period V, then the remainder by R, the remainder by C, and its remainder by D. Thus one obtains the following decomposition of a :

$$a = 1 \cdot V + 16 \cdot R + 0 \cdot C + 1 \cdot D + 129.$$

The application of the Vakyam process to the number a shows that the epoch was chosen in such a way that it not only represents

---

(11) As an additional argument, it may be mentioned that V seems not to be expressible in the form $aD + bC + cR$ with small integer coefficients a, b, and c. I realize, of course, that this is not a strictly mathematical statement.

the vernal point but also the moon in its apogee. This is confirmed by the subsequent computation. We multiply each $\Delta\lambda$
associated with V, R, C, D by its proper coefficient 1, 16, 0, 1,
respectively and obtain for the total motion of the apogee (disregarding multiples of 360°):

1 · V :	7^s	2;0,7°
16 · R :	2^s	24;50,40
1 · D :		27;44,6.

To this we add the lunar motion for the remaining 129^d, taken
from Table XXVI, where we find 8^s 24;18°. The total is 7^s
18;52,53 or Scorpio 18;52,53. This is the lunar longitude as
obtained by the Vakyam process. For the sun we have previously
found a longitude just beyond Taurus 19, thus very close to opposition as it should be the case for an eclipse. At the moment, of
course, we have not yet established more than closeness to full
moon.

11. According to our preceding discussion, one would expect
that we have now found the longitude of the moon. WARREN's
informant, however, added two small corrections to the preceding
result, one depending on the sun, the other on the moon. He was
not able to give a satisfactory explanation for these steps except
for indicating that the first correction was a correction for geographical longitude. We shall see to what extent this is correct.

The first correction is based on a small table (WARREN,
Table XLVII) which I give here in a slightly modified form while
WARREN's text contains some errors:

♈	+ 0;15	— 0;0,12	♎	+ 0;23	+ 0;0,12
♉	0;10	— 0;0,10	♏	0;28	+ 0;0,10
♊	0;7	— 0;0,6	♐	0;30	+ 0;0,4
♋	0;8	+ 0;0,2	♑	0;29	— 0;0,2
♌	0;11	+ 0;0,6	♒	0;26	— 0;0,6
♍	0;17	+ 0;0,12	♓	0;21	— 0;0,10

The argument of this table is the solar longitude; the first column
gives the correction for complete zodiacal signs, while the next
column concerns the interpolation for single degrees. WARREN's
informant gave in the first column the value 0;21 for Libra. But
the graph (3) Plate II shows clearly that this must be an error.
Yet, the error is old because the interpolatory coefficients transmitted to WARREN are adapted to this value. Thus one finds

in WARREN's text, in the second column, from Virgo to Scorpio the following irregular sequence : + 0;0,12 + 0;0,8 + 0;0,14. This again shows the necessity of correcting 0;21 to 0;23 in the first column. Another error in the column is — 0;0,4 for

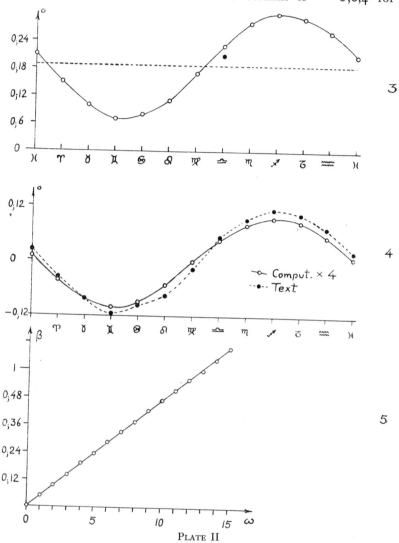

PLATE II

Sagittarius instead of + ' 0;0,4. These disorders prevented WARREN from realizing that both columns represent one single table.

We now can turn to the astronomical interpretation of this table. Graph (3) shows a minimum in Gemini, a maximum in Sagittarius, thus a variation corresponding to the anomaly of the sun which has its apogee in Gemini, its perigee in Sagittarius. It seems difficult to understand how the solar anomaly should have an influence on the longitude of the moon but there is one effect known to ancient astronomers which shows such a dependence : the " equation of time." All tables for lunar and planetary motion are necessarily based on mean solar time whereas actual observations are made in true solar time. If the true sun moves more slowly than the mean sun, it will be farther to the west and therefore in the local meridian sooner than computation with mean solar times indicates. Consequently the moon has less time to proceed westwards in the ecliptic, that is to say, the lunar longitude will be smaller than expected when the solar velocity is less than the mean. And vice versa in the opposite case.

Our table, to a certain extent, corresponds to this phenomenon. There is a smaller amount added to the lunar longitude when the sun is in apogee than in the case of perigee. But the mean value which we would expect is zero whereas our table shows a mean value of $+$ $0;18,45°$. This suggests splitting the correction under discussion into two parts : (a) the constant amount $+$ $0;18,45$ and (b) a correction depending on the solar anomaly as follows :

♈ — $0;3,45$	♎ $+$ $0;5,15$
♉ — $0;8,45$	♏ $+$ $0;9,15$
♊ — $0;11,45$	♐ $+$ $0;11,15$
♋ — $0;10,45$	♑ $+$ $0;10,15$
♌ — $0;7,45$	♒ $+$ $0;7,15$
♍ — $0;1,45$	♓ $+$ $0;2,15$.

The last figure in these numbers is, of course, not really significant because only one sexagesimal fraction is given in the original table.

We begin with the discussion of the constant correction $k = +$ $0;18,45°$. We can now make use of the hint that we are dealing with a correction for geographical longitude (WARREN, p. 130). This correction is expressed in degrees to be added to the lunar longitude. Thus k must be the product of the (mean) lunar velocity $\mu \approx 13;11^{o/d}$ and the time difference t due to the difference in longitude. Hence we obtain for t

$$t = \frac{k}{\mu} = \frac{0;18,45}{13;11} \approx 0;1,25^{d}.$$

In order to convert this fraction of a day into degrees of geographical longitude we must multiply by 6,0 and thus obtain

$$L \approx \frac{1,52}{13;11} \approx 8\frac{1}{2}^{o}$$

for the difference in longitude. Assuming the meridian of Ujjain (76° East) as the prime meridian for which the theory was originally developed, we obtain a longitude of about 84° East. This falls about four degrees to the east of Madras, but we do not know for what region the correction was computed nor do we know the ancient determination of relative geographical longitudes (12). The result seems to be good enough to leave no doubt that k represents the correction for a geographical displacement toward the east. Thus we have confirmed so far the information given to WARREN.

12.   There remains, however, the periodic correction which must take into account the equation of time. At first sight, a serious objection might be raised. Our correction depends on the solar anomaly alone; the equation of time, however, is not only caused by the difference of motion of true and of mean sun but also by the fact that the true sun moves in the ecliptic and that equal arcs of the ecliptic do not correspond to equal arcs of the equator in crossing the meridian. It is known, however, that this second component of the equation of time was ignored by Hindu astronomers until BHASKARA (13) (12th century). This strange fact holds, e.g., for the SŪRYA SIDDHĀNTA (14) but is in line with the oft-made observation that Hindu astronomy shows no influence from the ALMAGEST (15).

---

(12) According to the *Pañca-Siddhāntika* (III, 13), Benares is assumed to be 10° east of Ujjain (actually, it is 7°). If we reduce $8\frac{1}{2}^{o}$ in the same ratio we obtain for L slightly less than 6°, or a longitude less than 2° east of Madras.

(13) Cf. S. R. DAS, *Am. Math. Monthly*, 35, 1928, p. 541-543.

(14) Cf. BURGESS' commentary to S.-S., II, 46.

(15) This is confirmed by AL-BĪRŪNĪ (1030 A.D.), who intended to translate the *Almagest* into Slokas for the Hindus (India, XIII, vol. I, p. 137, trsl. Sachau).

It is not difficult to compute the numerical values one should expect for the velocity component of the equation of time. In WARREN's Table III (above p. 257) we are given the time t which the sun travels through a given zodiacal sign. Hence $v = \dfrac{30}{t}$ is the average velocity of the true sun in the sign. We need the deviation from the mean velocity $v_0 = 0;59,8^{o/d}$; thus we form $\dfrac{30}{t} - v_0$. This value gives the deviation of the true sun from the mean sun in degrees per day for a zodiacal sign. Hence the total deviation during t days for this sign is $\left(\dfrac{30}{t} - v_0\right)t = 30 - v_0 t$ degrees. If the true sun arrives at the meridian one degree after the mean sun, the moon gains in longitude the 360th part of its daily motion, or $\dfrac{13;10,35}{6,0} = 0;2,11,46^0$. This gives for the total correction for one zodiacal sign the amount of $(30 - 0;59,8 \cdot t)\,0;2,12$ degrees, where t is to be taken from Table III.

By this process one obtains the following table :

♈ — 0;1,3		♎ + 0;1,10	
♉ — 0;2,5		♏ + 0;2,1	
♊ — 0;2,32		♐ + 0;2,22	
♋ — 0;2,14		♑ + 0;2,8	
♌ — 0;1,18		♒ + 0;1,22	
♍ — 0;0,2		♓ + 0;0,13	

Comparing these numbers with the numbers on p. 265, one sees that the Tamil computer used corrections which are about 4 times greater than expected. The closeness of agreement between 4 times the computed values and the recorded values is shown in graph (4) Pl. II. I cannot explain the factor 4 which amplifies the correction caused by the equation of time even in its complete form to more than twice its possible amplitude.

13. Returning to our numerical example, we have found for the longitude of the sun the value ♉ 19;6,48. Thus we find in Table XLVII the correction + 0;15 for the completed sign of Aries. For the 19;6,48° in Taurus we have a correction of — 0;0,10 per degree, thus a total of $0;15 - 0;0,10 \cdot 19;6,48 = 0;15 - 0;3,11 = 0;11,49$.

449

14. The second correction depends on the lunar anomaly. We have to go back to the decomposition of the ahargana in complete anomalistic periods V, R, C, D plus a remainder d; in our example $d = 129$ (cf. p. 262). In this decomposition, D appears with an integer coefficient, say r (in our example $r = 1$). We know, furthermore, from Table XXVI (p. 261), the actual lunar velocity v at the day d (in our case, one finds for $d = 129$ the velocity $v = 13;46$). The mean lunar velocity is $\mu = 13;11$. One now forms the product

$$0;0,32 \cdot r\,(v - \mu) = U$$

and adds it to the previously obtained lunar longitude. For our example, one finds

$$U = 0;0,32 \cdot 1 \cdot (13;46 - 13;11) = 0;0,19.$$

In § 13 we have obtained the value $0;11,49$ for the first correction. Thus a total of $0;12,8$ has to be added to the lunar longitude of ♏ $18;52,53$ (cf. p. 263). This gives the final result

$$\lambda = ♏\ 19;5,1$$

for the longitude of the moon.

I cannot explain the reason for this last and very minute correction. Its maximum is $\pm\ 0;7,28^o$ ($v - \mu = \pm\ 1;10\ \ r = 12$) but ordinarily it will be much smaller, the moon being not too far from its mean motion and r being somewhere between 12 and 0. WARREN's informant, nevertheless, declared this correction to be "indispensable" (WARREN, p. 131) a statement which makes no sense in view of the minuteness of this correction. WARREN assumed a relation to the equation of time, but I cannot see how this would lead to a dependence on the coefficient r of D.

The factor $v - \mu$ could be motivated as follows. We took into consideration a correction for geographical longitude only insofar as the mean motion of the moon is concerned. Indeed, we found the difference t of local and mean time by dividing the constant $k = 0;18,45^o$ by the mean motion $\mu = 13;11^{o/d}$ (cf. p. 266). If the moon moves actually with a velocity v, then we should multiply $t = 0;1,25^d$ by the excess $v - \mu$. Thus one would expect a final correction of $0;1,25(v - \mu)$ whereas we are given the rule to compute $r \cdot 0;0,32(v - \mu)$. Thus the dependence on the number r of D's as well as the value of the numerical coefficient remain unexplained.

15. We now come to the second part of our main problem, the determination of the circumstances of a lunar eclipse. We have so far ascertained the longitudes of sun and moon and have found them nearly 180° different, as expected for a lunar eclipse. The next step needs no detailed comments. By simple extrapolation one can find the moment of exact opposition.

Similarly, no comment is required how to decide for a given locality into what part of night or daylight the opposition falls. The simplest method consists in the use of tables which give the length of daylight for every day of the year (WARREN, p. 340). In the case of our example, the opposition falls sufficiently long before sunrise to make the eclipse visible from beginning to end.

16. The next problem consists in finding the position of the ascending node for the day in question (its ahargana a, the distance from the epoch, having been found in the previous part). Then the rules followed by the Tamil computer can be represented by the following formula :

$$\lambda = 6,0 - \left\{ (a - c_1) - \frac{c_2 (a - c_1)}{c_3} \right\} \frac{30}{c_4} + c_5$$

where

$$c_1 = 7,24,27,46^{11}$$
$$c_2 = 9$$
$$c_3 = 47,10,9$$
$$c_4 = 9,26$$
$$c_5 = 0;40°.$$

It is easy to explain this formula. The last constant, c_5, is without theoretical interest. It is a so-called " bīja " or empirical correction, added to an older formula in order to compensate for an accumulated error (16). We shall ignore c_5 henceforth.

We furthermore replace $a - c_1$ by a single letter n because subtracting c_1 from a means simply to introduce a new epoch, c_1 days later than the epoch for the ahargana. Such a step is needed in order to start the computation from a known moment in which the ascending node had the longitude zero. Whereas the earlier epoch was — 3101 Feb. 18, we now operate with the epoch c_1 days later, i.e., + 1279 Nov. 16. It is plausible to assume that this comparatively late date coincides with the time when the

---

(16) Cf., e.g., *Sūrya-Siddhānta*, I, 9.

final correction c_5 was added. This corresponds roughly to the date of the latest similar empirical corrections in the tradition of the *Sūrya Siddhānta*.

The factor 30 outside the parentheses is introduced only in order to express the result in zodiacal signs instead of degrees. The 6,0 and the minus-sign express the fact that the nodal line has a retrograde motion. Its amount can be found by computing the single coefficients and then writing

$$\lambda = 6,0 - \left\{ n - \frac{1,0}{5,14,27,40} \, n \right\} \; 0;3,10,48,45,\cdots$$

This shows that the daily motion of the nodal line .was in first approximation assumed to be $-0;3,10,48,45^\circ$, to which was added a very small correction term $\dfrac{1,0}{5,14,27,40}$. The final result could have been written much more simply as

$$\lambda = 6,0 - n \cdot 0;3,10,48,9,\ldots$$

We can be glad that the native computers did not simplify their formulae because we would be unable to recognize in the final form the traces of earlier steps of development (17).

In the case of our example, the position of the ascending node is found to be ♏ 0;25,53

17. We now form the " argument of latitude," i.e., we find the difference ω between the longitude of the node and of the moon for the moment of opposition. With ω as argument we enter a little table (WARREN, p. 342) which gives for ω from 1° to 15° the corresponding latitude β. This table appears as

$\omega =$		$\beta =$
1°		$0;4,43^\circ$
2		$0;9,26$
3		$0;14,8$
	etc.	
15		$1;9,54$

The increase is practically linear [cf. graph (5), Plate II] except for small irregularities probably due to the process of interpolation

---

(17) The value in the *Sūrya-Siddhānta* is $-0;3,10,44,43,\ldots$ (BURGESS, ad I, 34) while PTOLEMY's value is $-0;3,10,41,15,\ldots$ (from *Almagest*, IV, 3). From the *Pañca-Siddhāntika*, III, 28 one can deduce the value $-0;3,10,44,14,\ldots$ and this book is in general based on the *Paulisa-Siddhānta*.

from a trigonometric table; the mean slope is $0;4,39,36$. The value $\beta = 1;9,54°$ for $\omega = 15°$ can be explained as $4;30 \sin 15°$ for which one finds $1;9,52,52$. Thus we see that this table is based on an extremal lunar latitude of $4;30°$. The same value is found in the old and in the modern *Sūrya-Siddhānta* as well as in the *Paulisa-Siddhānta* (18).

Using the above table one obtains, for our numerical example, $\omega = 11;20,52$ and hence the latitude $\beta = 0;53,9$.

18. The latitude β being known, the eclipse magnitude m is found in a process which can be described by the formula

$$(1) \qquad m = \frac{\left(\dfrac{v}{25} \cdot 5 \cdot \frac{1}{2} + \dfrac{v}{25}\right) \frac{1}{2} - \beta}{\dfrac{v}{25}}$$

where v is the lunar velocity which was found previously from Table XXVI. Totality is represented by $m \geq 1$.

Substituting $v = 13;46$ (cf. p. 268) and $\beta = 0;53,9$, one obtains for our eclipse $m = 0;8$ or, in modern terms, $m = 1;36$ digits or $m = 0.13$ decimally. Needless to say, such small partial eclipses will be very sensitive to the inaccuracies inherent in relatively primitive methods. OPPOLZER's " Canon " gives $m = 0.3$ digits and a duration of 36 minutes (N° 4690).

Before discussing the significance of the procedure which is represented by the above-given clumsy formula, we again condense the whole rule in an equivalent but much simpler form

$$(2) \qquad m = 1;45 - 25 \, \frac{\beta}{v}.$$

From this it follows that the greatest possible lunar eclipse is assumed to be of magnitude $1;45$ ($\beta = 0$). On the other hand, the ecliptic limits are given by $m = 0$ or by

$$\beta_0 = 0;4,12 \cdot v.$$

In order to find the corresponding arguments of latitude, we can use the table which relates β and ω, or the graph (5). Similarly, totality is reached for $m = 1$ or for all $|\beta| < |\beta_1|$ with

$$\beta_1 = 0;1,48 \, v.$$

---

(18) *Pañca-Siddhāntika*, IX, 5 (THIBAUT, p. 55 and p. XVIII) and III, 31 (THIBAUT, p. 21 and p. XXXV). *Sūrya-Siddhānta*, I, 68.

The following table gives the values for the lunar apogee
($v = 12;2^{o/d}$), for the mean distance ($v = 13;11^{o/d}$), and for the
perigee ($v = 14;19^{o/d}$), where the extremal values for v are taken
from Table XXVI.

	β_0	ω_0	β_1	ω_1
Apogee	0;50,32°	10;47°	0;21,40°	4;36°
Mean	0;55,20	11;50	0;23,43	5;2
Perigee	1; 0, 8	12;53	0;25,46	5;28

This table will allow the reader to make a comparison with the
known ancient and modern ecliptic limits.

19. The significance of formula (1) p. 271 will become clear
from Figure 1. Let A be the center of the shadow, E the center
of the moon at the moment of greatest phase. Then we have
approximately $AE = \beta$. Call

$r = \frac{1}{2}d$ the radius of the moon
$R = \frac{1}{2}D$ the radius of the shadow
v the lunar velocity in degrees per day.

Then the text assumes

(3)
$$d = \frac{v}{25}$$
$$D = \frac{v}{25} \cdot \frac{5}{2} = \frac{5}{2} \cdot d .$$

The proportionality of D and d is not strictly correct but may be
accepted as a working hypothesis with respect to the variation of

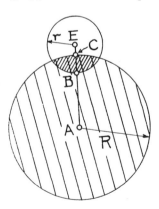

FIG. 1

d and D with the lunar distance (19). The relation $D = 5\ d/2$ has been used also by HIPPARCHUS (*Almagest*, IV, 9).

The magnitude of the eclipse is then measured by

$$CB = md \qquad\qquad\qquad 0 \leqq m \leqq 1.$$

From Fig. 1 we see that

$$\beta = EA = EC + CA = (r - md) + R$$

and therefore

$$md = R + r - \beta = \tfrac{1}{2}\,(D + d) - \beta$$

or

$$m = \frac{\tfrac{1}{2}\,(D + d) - \beta}{d}$$

and this is indeed formula (1) p. 271.

The relations (3) allow us to compute the values d and D which were assumed by those who constructed the theory. One finds

	d	D	$2;30\ d$
Apogee	0;28,53°	1;12,12°	1;12,12,30
Mean	0;31,37	1;19, 3	1;19,2,30
Perigee	0;34,21	1;25,54	1;25,52,30

20. We now can find the duration of the eclipse (cf. Fig. 2). We denote

τ	half duration
v	lunar velocity as before
w	solar velocity
δ	travel from middle of eclipse to outer contact.

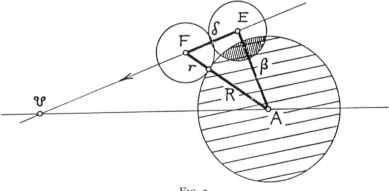

FIG. 2

---

(19) PTOLEMY finds for the perigee of the moon that $D = 2;36\ d$ (*Almagest*, VI, 5 ed. HEIBERG, p. 480).

Obviously

$$(4) \qquad \tau = \frac{\delta}{v - w} = \frac{1}{v - w} \sqrt{(r + R)^2 - \beta^2}$$

if we adopt a plane triangle as sufficient approximation for a small spherical triangle in the neighborhood of the nodes.

It is formula (4) which is used, step by step, by the Sashia Sami Naden, WARREN's informant. He finds

$$R + r = c = 0;57,49 \quad \text{thus} \quad c^2 = 0;55,42,46,1$$
$$\beta = 0;53,9 \quad \text{thus} \quad \beta^2 = 0;47,4,55,21$$

and from this, by endless rearrangements of his shells,

$$\delta = \sqrt{c^2 - \beta^2} = \sqrt{0;8,37,50,40} = 0;22,46$$

(indeed $0;22,46^2 = 0;8,37,19,16$). Furthermore

$$v = 13;46 \qquad w = 0;57,15 \ (20) \qquad v - w = 12;48,45$$

thus, from (4),

$$\tau = 0;1,46,36^{d}.$$

Adding and subtracting τ to and from the moment of opposition gives the moments of beginning and end of this lunar eclipse (cf. p. 253 for the result).

The *Sūrya-Siddhānta* follows essentially the same procedure but considers the results reached this far only as a first approximation, because the latitude of the moon does not remain constant during the eclipse (cf. S.-S. IV 14 and 15). It seems possible to me that this is indicative of an earlier period of origin of the Tamil tradition.

21. We conclude this paper with the following quotation from the next to the last paragraph of WARREN's book : " It was originally my intention to have added an example of a Solar Eclipse to the foregoing one; but family afflictions, and want of health, have prevented me from further gratifying the reader's curiosity with disclosures of Indian mysteries." At that point the matter will rest, I am afraid, for all time.

ADDENDUM. My last remark is fortunately incorrect. I have recently found that we possess an additional source for Tamil astronomy : LE GENTIL's *Mémoire sur l'Inde* published in the

---

(20) WARREN, Table XXVIII.

Mémoires for 1772, Part II, of the *Histoire de l'Académie Royale des Sciences* (Paris, 1776) p. 169-214 and 221-266. Exactly as WARREN more than 60 years later, LE GENTIL was shown by natives their method for the computation of eclipses, and fortunately the procedures for both lunar and solar eclipses were carefully recorded by LE GENTIL. Agreement with WARREN is perfect for lunar eclipses. Consequently it suffices to make the following additions.

N⁰ 11. The complete table for the first correction of the longitude of the moon is found on p. 232 with the correct value $+$ 0;0,4 against WARREN's $-$ 0;0,4 but with the other two errors, as expected (cf. p. 263).

N⁰ 16. LE GENTIL gives for the longitude of the lunar node the rule (p. 237 f.)

$$\lambda = 6,0 - \frac{a \cdot b_1 + b_2}{2c_3} \cdot 30 + c_5$$

where c_3 and c_5 have the same values as in WARREN's report whereas $b_1 = 10,0$ and $b_2 = 8,8,29,6$. It can easily be shown that b_1 has exactly the value which one would obtain from WARREN's formula if one condenses it to the above-given simpler form. The value of b_2 is a very close approximation of the expected value, disregarding complete rotations. I suspect that this deviation is due to LE GENTIL, who explains on p. 221 that he decreased the number of steps " parce que j'ai vu qu'on le pouvoit faire, sans nuire a la clarté que j'ai voulu répandre dans cette méthode."

N⁰ 17. LE GENTIL's informant gave three more values of the table for latitude (p. 265) :

16	1;14,24
17	1;18,54
18	1;23,24

The last pair would yield a mean difference of exactly 0;4,38.

N⁰ 20. The question whether a lunar eclipse can be seen at a given place requires the knowledge of the length of daylight. LE GENTIL p. 205 f. has preserved a Tamil table of the right ascensions of the zodiacal signs and of the corrections which lead to the corresponding oblique ascensions for Tirvalour (West of Negapatam, slightly less than 11⁰ latitude). The values are measured in sexagesimal fractions of one day. The right ascensions agree with the values which one obtains by using the values for the middle decans in *Almagest* II, 8. Similarly the values of oblique ascension result from interpolation for $\varphi = 11$ between the tables for $\varphi = 8;25$ and $\varphi = 16;27$ of the *Almagest*.

I shall return to the problem of solar eclipses at a later occasion.

(Brown University &
Institute for Advanced Study).   O. Neugebauer.

Reprinted from
OSIRIS
10, 252−276 (1952)

[Reprinted from *Isis*, Vol. 50, Part 4, No. 162 (December, 1959)]

Regula Philippi Arrhidaei

By O. Neugebauer *

FRANZ Cumont (in *Isis*, 1936, *26:* 8-12) discussed under this title a passage which the monk Hilduin had inserted in his biography of Saint Dionysius the Areopagite. Dionysius (who probably lived in the early 6th century[1]) is supposed to have consulted, in connection with a solar eclipse, "regulam Philippi Aridaei." Cumont is, of course, right in taking *regula* as the Latin equivalent for κάνών in the technical sense of "astronomical table." I also follow Cumont in rejecting suggestions of Fabricius and Böckh to emend the text and to ascribe the tables to Plato's pupil Philip of Medma, or Opus, instead of to Philip Arrhidaeus. Cumont himself thought of tables dedicated to Philip Arrhidaeus, the half-witted puppet king chosen to be Alexander's successor as a compromise in the struggle for power between Alexander's generals.

It seems to me that nothing speaks in favor of this conjecture. Otherwise completely unknown Greek tables for solar eclipses would have had to exist in the fourth century B.C. and still be in the hands of Saint Dionysius almost nine centuries later, at least according to a monk of the ninth century, "faussaire émérite," as Cumont calls him. A much simpler solution presents itself when one remembers that since Ptolemy's "Handy Tables" the "Era Philip" had become the standard era for tables of this type. All that we have to assume is that Hilduin, or his source, had vague knowledge that the astronomical tables, current at his time, took their departure from the reign of Philip Arrhidaeus. The confusion of the era of astronomical tables with their author is certainly not surprising.

In view of the amply attested use of the Handy Tables in Byzantine and Islamic astronomy, no special proof is required to demonstrate their existence in the sixth or seventh century. Since we have, however, an explicit reference to the use of these tables, I shall mention it here, especially since it was not fully understood in the original publication. I refer to a passage published in 1910 by F. Nau,[2] from a letter written about 662 by Severus Sebokht, Bishop of the monastery of Qenneshrē (on the East bank of the Euphrates opposite Jerablus) to a priest Basil in Cyprus, in answer to a question concerning the possibility of conjunctions of all seven planets. It is not far-fetched to assume that this problem originated in the discussion about the eternity of the world, in particular in the polemics of John Philoponus against Proclus. In this context Philoponus mentions a conjunction of the seven planets in Taurus, "in our

* Brown University, Providence, R.I.
[1] Jülicher, *Pauly Wissowa*, 5,1 col. 996-998 (No. 154). For the general background cf. Paul Peeters, La vision de Denys l'Aréopagite à Héliopolis, *Analecta Bollandiana*, 1910, *29:* 302-322.
[2] F. Nau, Notes d'astronomie syrienne, *Journal Asiatique*, 1910, sér. 10, *16:* 209-228.

477

time, in the year 245 of Diocletian," [3] i.e., in A.D. 529. It is exactly the same conjunction to which Sebokht refers in his letter, adding details not mentioned by Philoponus.[4] Severus Sebokht writes:

> In the year 245 of Diocletian . . . on the 27th of Pachon according to the Alexandrians, at the 12th hour of the day the seven planets were in conjunction in the sign of Taurus. And those who wrestle with such problems know how to compute with the Handy Tables by taking the given year, the month, the day, and the hour. Hence, one takes according to the Handy Tables, the year 851 in the table for 25-years, year 2 (in the table) of single years, the month Phaophi, the 10th day, and the 6th hour after noon. Then one. will certainly find that at this moment occurred a conjunction of the seven planets, namely, in Taurus, although not at the same degree.... (One finds in this way) : the sun in 29;28, Saturn in 25;14, Jupiter in 0;11,18, Mars in 12;13, Venus in 8;2, Mercury in 16;46 (of Taurus).[5]

It is clear that Sebokht is here describing correctly the use of the Handy Tables for finding the positions of the planets at the given date, Diocletian 245, Alexandrian Pachon 27 at sunset. As we mentioned at the beginning, the Handy Tables are based on the Era Philip and, of course, on Egyptian, not Alexandrian years. This explains the figures quoted by Sebokht, because

Diocletian 245 Pachon (Alex.) 27 sunset = Philip 853 Phaophi (Eg.) 10, 6 p.m.[6]

We have to enter the tables with the year 851 + 2 because 851 is the nearest number $\equiv$ 1 mod. 25, and the first set of tables is based on 25-year steps. Since the Handy Tables are extant,[7] we are able to check the results quoted by Sebokht and find for the longitudes in Taurus

	Sebokht	Handy Tables	error
sun	29;28°	29;28°	0;0°
Saturn	25;14	25;14	+0;0
Jupiter	0;11,18	0;19	[−0;1]
Mars	12;13	12;13	+0;0
Venus	8;2	8;5	−0;3
Mercury	16;46	15;49	1−0;3

In the case of Jupiter the value given by Sebokht must be wrong since the tables do not give seconds;[8] probably the 18 was intended as a correction of the 11. Thus the only serious discrepancy, amounting to one degree, occurs for Mercury and can be explained as a simple error in addition. The position of the moon is omitted.[9]

---

[3] Ioannes Philoponus, *De aeternitate mundi contra Proclum,* ed. Rabe (Leipzig: Teubner, 1899), p. 579, 14-18.

[4] This is reminiscent of the treatises on the astrolabe by Philoponus and Severus Sebokht who both copied Theon, but Severus more extensively than Philoponus. Cf. my article on The Early History of the Astrolabe, *Isis,* 1949, *40:* 240-256.

[5] I am following Nau's translation. Where

he misunderstood the procedure concerning the use of the table I had the help of A. Sachs for the Syriac text.

[6] Julian day 1914417 = A.D. 529 May 22.

[7] Using Halma's edition (Paris, 1823).

[8] Except for Regulus, for which one finds a longitude of 126;25,16°.

[9] One can expect, from modern computation, about Taurus 23.

COMMUNICATIONS ON PURE AND APPLIED MATHEMATICS, VOL. XIV, 593–597 (1961)

K. O. Friedrichs anniversary issue

Notes on Kepler

O. NEUGEBAUER

Brown University

One of the most frequently made statements about outstanding personalities in the development of scientific ideas is the assertion that their contemporaries were not able to appreciate their ideas. To subsequent generations, however, is somehow granted a deeper insight after the initial shock has been absorbed during the intervening years.

This may be so in some, or even in many cases. In two famous instances however, namely, with Copernicus and Kepler, a careful reading of their fundamental works without assuming Kepler's results in the study of Copernicus, or Newton's for the reading of Kepler, will do much more justice to the contemporary scholars who cannot be blamed too much for not introducing immediately upon reading a freshly published work, the radical modification to which Brahe and Kepler subjected the cumbersome eccenter and epicyclic machinery of a pseudo-heliocentric model based on utterly insufficient empirical data or for not replacing Kepler's theological astrophysics by a rational dynamics.

With this background in mind, I wish to give in this note some examples of those rather trivial obstacles which every careful reader of Kepler's publications had to meet on practically every page. If Kepler ever did make any attempt to give a final polish to his writings, one can only say that he was not very successful. The number of trivial computing errors is enormous[1], parameters are changed without explanation (usually belonging to different stages of investigation, e.g., concerning the motion of the apsidal line of Mars), references to observations accessible to no one else [2] are quoted sometimes in an incomplete form, sometimes for no evident reason, and so forth. To the historian of astronomy, this way of presentation is a blessing as great as a real disaster that leaves a city in total shambles is to the archaeologist. But contemporaries cannot be blamed if their reaction was different from ours.

The examples I have chosen come from the next to the last chapter (69)

---

[1] Only a small percentage is noted in the edition.

[2] It was not until Dreyer's edition of Brahe's works (Vols. 10–13, 1923 to 1926) that these observations became accessible, and even now there is often a great deal of work to be done before Brahe's records can be compared with the ecliptic coordinates given by Kepler.

of Kepler's "Astronomia nova", the famous work which contains the demonstration of the fact that Brahe's observations of Mars are incompatible with a circular orbit (of course eccentric and endowed with equant) but require an elliptic orbit and an equal-area theorem. My examples do not, however, concern Kepler's new theory because any discussion of this part would involve so many related topics that it would by far exceed the space of a short paper. Chapter 69, however, is only an appendix in which Kepler analyzes the effect of small observational errors in Ptolemy's theory of the solar motion which in turn influence the basic parameters for the mean motion of Mars. As a piece of critical analysis of the methodology of an ancient astronomer, this chapter is far superior to many a modern discussion of exactly the same topic. But since the problem is basically very simple we can also easily isolate the traces of hasty publication which must have left a reader with the impression that he hardly knew where he stood at the end of the discussion of a subject which was fully familiar to every contemporary astronomer. I could well imagine that the long periods of silence in the correspondence of Kepler's teacher Mästlin need not always be explained by the fact that the pupil had gone far beyond the teacher's capacity.

Kepler sets out to compare Ptolemy's elements for the solar motion[3] with his own in order to establish secular changes of the fundamental parameters for the sun and thus for Mars. Three elements are investigated: eccentricity, position of the apsidal line, and position of the equinoxes. Three alternatives (*bivia*) exist in these investigations, says Kepler, thus eight cases must be investigated. What is meant becomes clear only from the study of the subsequent pages: case 1 assumes all three parameters correct as given by Ptolemy; cases 2, 3, and 5 (*sic*) assume that only one parameter requires revision; cases 4, 6, and 7 that two parameters should be changed; case 8 modifies all three parameters. The corrections applied are always of the same amount, namely, a decrease of 20' in the eccentricity, an increase of $11\frac{1}{2}°$ in the position of the apsidal line, and an increase of $11\frac{1}{2}°$ in the longitude of Regulus reckoned from the vernal equinox. It is shown that corrections of this amount can be reconciled with the observational methodology of Hipparchus and Ptolemy.

Six ancient observations of Mars are known to Kepler: one from Aristotle, five from the Almagest. The Aristotelian observation (an occultation dated by Kepler to B.C. 357 Apr. 4) and a "Chaldean"[4] observation from the Almagest are supposedly of great interest for the latitude theory of Mars, but the first observation is never mentioned again, and the next four

---

[3]Solar and terrestrial motions are, of course, interchangeable; both Copernicus and Kepler freely use the classical terminology.

[4]There is no basis for this attribution by Kepler.

Ptolemaic observations contain no reference to latitudes[5], although one is put into Chapter 70 on latitudes. Only one observation remains for the discussion of the latitude and this case ends with a *non sequitur*. Thus the actual topic of Chapter 69 and half of Chapter 70 is the investigation of the influence of the above-mentioned changes in the parameters of the solar motion on the mean motion and initial position of Mars.

At the beginning of the Astronomia nova (in Chapter 7) Kepler told how he came to be entrusted by Brahe with the work on Mars because Longomontanus had failed to account for the observed latitudes of that planet. Kepler radically changed the whole program by setting out to determine with the greatest possible accuracy the actual form of the orbits of the earth and Mars. But he added as a relic of the original plan a last part (Chapters 61 to 70) which must have greatly helped to obliterate the impression that a definitive solution for the law of planetary motion had been discovered in the main part of the work.

I do not propose to discuss here the astronomical contents of Chapter 69. I will place myself in the position of a reader who, after a first reading of it, has realized what Kepler was driving at and what his basic data were. Now we shall see what we can make out of the numerical results of this investigation.

1. Three of the four Ptolemaic observations concern mean oppositions of Mars[6]. Kepler who, in contrast to Copernicus, operates really heliocentrically has to replace mean oppositions by true oppositions. In order to evaluate the influence of this change we need the time difference Δt for Mars to travel the distance between mean and true sun. At the moment in question (A.D. 130 Dec. 15) the longitudinal difference $\Delta \lambda$ between mean and true sun is 40', using Ptolemy's parameters. At opposition the planet is retrograde, thus the relative velocity of Mars with respect to the sun is the sum of solar and planetary velocity. Considering the positions in their respective orbits at the given date, Kepler finds a relative velocity of 84' per day. Thus $\Delta t = 24 \cdot 40/84 = 11\frac{1}{2}^{\text{h}}$. Kepler somewhere made a computing error and finds 8^{h}. Unfortunately, we are dealing with case 1, wich serves as a basis of comparison for all subsequent cases. Thus a large proportion of all subsequent numerical data are affected in the most pernicious way by this trivial first mistake. Since we are dealing with corrections of a few hours or a few minutes of longitude, the error is of the same order of magnitude as the results under investigation.

---

[5]To this list has to be added one Ptolemaic observation of precession: Kepler, Werke **3**, p. 418 from Almagest VII,2 (p. 14,8 Heib.).

[6]This is the classical method to eliminate one unknown parameter, namely the radius of the epicycle (or of the earth's orbit).

2. The time interval $\Delta_1 t$ between the first and the second true opposition is found to be 4 years and about $68\frac{1}{2}$ days, and similarly between second and third opposition $\Delta_2 t \approx 4$ years $97\frac{3}{4}$ days, thus in both cases approximately $4\frac{1}{4}$ years. The corresponding longitudinal differences are found to be $\Delta_1 \lambda = 68°23'$ and $\Delta_2 \lambda = 93°5'$, respectively. These values should now be corrected for precession, assuming the rate of $1°$ per century according to Ptolemy's norm. Thus for $4\frac{1}{4}$ years, precession would amount to $2'33''$, rounded by Kepler to $2'40''$. Thus he replaces $\Delta_2 \lambda = 93°5'$ by $93°2'20''$. But subtracting $2'40''$ from $\Delta_1 \lambda = 68°23'$ Kepler writes $68°21'20''$ instead of $68°20'20''$. In the final comparison the differences between Ptolemy's and Kepler's results remain below $2'$ and hence are of the same order as the computing error. This, of course, on top of the effects mentioned in the previous section.

3. In case 5, Kepler investigates the effects of a reduction of the solar eccentricity. Under the given circumstances this results in a reduction of the longitude of the true sun at the second opposition by $20'$ whereas the first and third remain unchanged since they are located very close to the apsidal line. Relying on his incorrect result in No. 1, that $40'$ of solar motion correspond to 8^h of relative motion, he concludes that a displacement of $20'$ requires 4^h motion of Mars or $4'$ in longitude. Accepting this, one sees that $\Delta_1 \lambda$ decreases by $4'$, $\Delta_2 \lambda$ increases by this amount. Kepler now compares these new positions first with the positions of Mars computed for case 1, correctly changing the longitude of the second opposition by $4'$. Then he recomputes the positions of Mars under the assumption of increased eccentricity in order better to account for the modified position. But coming to the end, he happened to pick the not-corrected longitude, and thus found a deviation of only $2\frac{1}{2}'$ where he should have found $6\frac{1}{2}'$.

4. In case 7 we change both solar eccentricity and apsides. Here it turns out that $\Delta_2 t$ from case 1 to case 7 decreases by $8^h 15^{min}$ (accurate computation would give $8^h 17^{min}$). Kepler says that during this time the epicyclic anomaly of Mars changes by $10\frac{1}{2}'$. But the tables in the Almagest IX,4 give for 8^h $0;10,28,52°$, for $8\frac{1}{4}^h$ $0;10,48,31°$. Since the final result is given to seconds, Kepler's estimate is too crude.

The effect of errors and inaccuracies on the final figures is such that not one of them can be considered significant. To this has to be added what Kepler himself calls the *"obscuritas"* of his writing. Only one example from Chapter 69 may suffice. In the discussion of the third observation in case 1, Kepler, in determining the velocity of Mars, gives to the planet the longitude of the sun instead of the point opposite to it. The result is nevertheless correct because the sun happens to be close to quadrature with respect to the aphelium of Mars and therefore the same holds for the point opposite the sun.

Thus the velocity of the planet is in both positions the same. Furthermore, the first and third observations are accidentally nearly opposite each other such that Mars will have for both observations practically the same velocity. Instead of simply stating this fact, Kepler prefers to assign the planet a position which it does not occupy and leaves it to the reader to detect the reason for what at first sight must appear a simple error.

These examples will suffice to illustrate the fact that a serious study of the above-mentioned discussion means a recomputation of all details from beginning to end. Kepler's radically new approach rightly made a deep impression on his contemporaries. But it is not surprising that he met little direct following until a new foundation was given to celestial mechanics.

Received September, 1960.

Notes on Ethiopic Astronomy

O. Neugebauer – Providence, R. I.

Gli studî sull'Etiopia son nati, si può dire, ieri ...
Conti Rossini, Rend. Accad. Lincei 8 (1899) p. 198

1. Introduction

In preparing a short survey of Greek astronomical papyri [1], to my great surprise I found in a Ptolemaic papyrus on weather prognostics, published in 1900 by C. Wessely, a short table for the determination of the hours from shadow lengths (not recognized as such in the original publication) which is the antecedent of a whole class of similar texts from the Byzantine period. These tables are of extreme simplicity: the (seasonal) hours of the day are associated with the length of the human shadow, measured in feet, under the assumption that the increment of the length of the shadow during the first hour before or after noon is always 1 foot, 2 feet for the next hour, then 3, then 4, and finally 10 feet. The noon shadow itself is supposed to vary linearly with a difference of 1 foot per month between a minimum value of 2 feet at the summer solstice and a maximum of 8 feet at the winter solstice [2]. There are good reasons to suppose that this primitive scheme originated in Athens in the fifth century B.C. and was transmitted, without essential modifications, to Alexandria whence it spread not only to mediaeval Byzantine and Latin treatises but also up the Nile, being attested in an inscription on a Nubian Temple on the tropic (in Taphis), perhaps written around A.D. 600.

After completion of this study I turned to an often postponed plan to inform myself, at least superficially, about Ethiopic astrono-

---

[1] Proceedings of the American Philosophical Society 106 (1962) p. 383-391.
[2] Cf. Neugebauer [1962] p. 32.

Orientalia — 4

mical texts, the existence of which was known to me from papers by
Rhodokanakis ([1]) and Grébaut ([2]). Being alerted to shadow tables I
first investigated texts of this kind in manuscripts which were kindly
placed at my disposal by the Vatican Library and by the Österreichi-
sche Nationalbibliothek. From these manuscripts it immediately
became evident that we have here another descendant of the Athenian shadow tables. Finally, three Coptic texts, published by U.
Bouriant ([3]), provide the link with mediaeval Egypt. All these texts
will be discussed in the following.

One of the reasons for my interest in the Ethiopic material was
the fact that the customary interpretation of the " gates " (ኅዋኅወ•)
of heaven as zodiacal signs is obviously untenable since, e.g. the moon
can never remain 7 or 8 days in the same sign as would be required by
our texts if the " gates " were identical with the signs of the zodiac.
I shall demonstrate in the following that the " gates " are fixed arcs
of the horizon, related in a very simple way with the rising- and
setting-amplitude of the sun during the course of one year.

Again there can be no doubt that the origin of this concept is
not Abyssinian but lies in the hellenistic world. This follows from
the fact that the " gates " appear prominently in the Book of Enoch
which plays such an important role in Abyssinian literature but which
is, as is well known, of hellenistic-palestinian origin ([4]). Thus also
here we are led back to a period for which our knowledge of astronomical concepts is utterly fragmentary.

If I have clarified in the following the tables of moon rise in relation to the " gates " and the tables of hours and shadow lengths I do
not pretend to have exhausted the astronomical material in Ethiopic
texts. On the contrary, even a cursory reading of the catalogues of
the major Ethiopic collections shows that there exist many problems
which need critical investigations. Only as an example I mention the
term kêkrôs which Dillmann in his Lexicon (col. 859) explains as
" sine dubio κίρκος circus, i.e. sexagesima diei pars ". The " sine
dubio " shows that Dillmann felt some doubt; rightly, because neither
κίρκος nor circus ever occurs in Greek or Latin astronomical litera-

---

([1]) Rhodokanakis [1906].
([2]) Grébaut [1918].
([3]) Bouriant [1898].
([4]) Cf., e.g., R. H. Charles, The Book of Enoch..., translated from
the editor's Ethiopic Text, Oxford, 1912.

ture, at least to my knowledge, while the meaning " sexagesima diei pars " for Ethiopic can be disproved from statements like the one found in cod. Vat. Aeth. 106 fol. 92^v ([1]): " (one year contains) 300 and [65] days, (i.e.) 6 and 5 kêkrôs " which can only mean that the sexagesimal equivalent of 365 is 6,5 (that is 6 times 60 plus 5), no fractions of days being involved ([2]). But there are many other cases where the meaning of kêkrôs remains obscure.

A much more serious problem to pursue would be the investigation of the astronomical contents of Abû Shâkir's large chronological work ([3]) but this requires a scholar with a less rudimentary knowledge of Ethiopic than mine.

At the risk of summarizing my impressions much too prematurely I would say that there are distinguishable three essentially different sources of Ethiopic astronomical concepts. First: of Greek origin are only the shadow tables, going back to the Ptolemaic period but still preserved in Coptic. Of Alexandrian-Christian origin is, of course, the Easter computus — an astronomically rather uninteresting subject. Second: Jewish-Hellenistic is the concept of the " gates " of rising and setting of the celestial bodies — transmitted to Ethiopia with the Book of Enoch. Third: Islamic influence is clear in all cases where zodiacal signs, lunar mansions, or planets are mentioned — their names are always transcribed from the Arabic. — In spite of the enormous mass of magical texts there seems to be practically no Greek or Islamic astrology extant. But details must await more serious study.

2. The " Gates "

In 1918 S. Grébaut published, in text and translation, from the Ethiopic MS 64 (Catal. Zotenberg) of the Bibliothèque Nationale in Paris a section which gives for each of the twelve Ethiopic months

([1]) Vat. Cat. p. 392, with my restorations.

([2]) The subsequent numbers are not fractions of days but hours and their fractions. The reading " 85 seconds " is, of course, wrong under all circumstances since no sexagesimal digit can exceed 59. The text has to be emended to [5];49,2,16,20^h. This gives for the tropical year the length of 365;14,32,35,40,50^d. This is confirmed by the rounded value found in cod. Borg. 1 fol. 3^v (Vat. Cat. p. 763); 365^d 5;49^h = 365; 14,32,30^d — a value found also in the Ḥakemî zîj of Ibn Yûriis.

([3]) Preserved, e.g., in Berlin (Dillmann's Verzeichniss No. 83), in the British Museum (Catal. Dillmann No. 36), and in the Vatican (No. 106).

the "gates" of moon rise (¹). The text for each month follows the same pattern which I illustrate for the first case, the 8th month of the Abyssinian calendar, Mîyâzyâ, roughly corresponding to julian April:

" Rise (of the moon) in (the month) Měyâzyâ: in the fourth gate 2 (days), in the fifth gate 2, in the sixth gate 8, in the fifth gate 1, in the fourth gate 1, in the third gate 2, in the second gate 2, in the first gate 8, in the second gate 6 (²), in the third gate 1, in the fourth gate 1. Total: 30 days, month of Měyâzyâ ".

Our Table I gives the complete information contained in this text, the lines representing the gates, the columns the months. The grand total of alternating full and hollow months is a lunar year of 354 days. The last column is obviously in bad disorder.

It is clear at first glance that we are dealing with an incongruous superposition of a lunar calendar on the julian (i. e. Alexandrian-Coptic) Abyssinian calendar. It is furthermore clear that such a simple arithmetical scheme for 354 days cannot take into account the lunar motion in latitude or anomaly. Thus we can only expect to come to an understanding of the whole pattern if we restrict ourselves to the simplest features of the mean motion of the moon. It is also evident that the great variation in the number of days cannot refer to the orbital motion of the moon. On the other hand the fact that gates 1 and 6 are associated with very long intervals (7 or 8 days) in contrast to 1 or 2 days for the intermediate gates suggests that we are dealing with a delay reminiscent of the slow change of daylight at the solstices. For the moon this can only mean a position near the solstitial points of the ecliptic, i.e. extremal distance of the points of rising with respect to the east point of the horizon. Thus we are led to an investigation of the relation of the dates of moonrise to the arcs of the eastern horizon which the moon crosses when rising. This reduces the problem to the question of finding a division of the horizon that leads to the numbers of the text. We shall now show that the answer is given by the simplest possible arrangement: six arcs of equal length covering the part of the horizon which is touched by the sun during the course of one year.

A few preliminary remarks are necessary. Since the text is obviously no more than a crude qualitative scheme we need not ope-

(¹) A duplicate is found in MS Berlin 84 fol. 30ʳ II, 19 to 31ʳ III, 2.
(²) Scribal error for 2, as follows from the total and from the general pattern.

rate with great numerical accuracy. We therefore may use $\varepsilon = 24°$ for the obliquity of the ecliptic; we also restrict the geographical latitudes to the interval from $\varphi = 31°$ (Alexandria) to $\varphi = 38°$ (Athens).

Table I

Gate	VIII	IX	X	XI	XII	I	II	III	IV	V	VI	VII	Gate
4	2												4
5	2	2											5
6	8	8	4	4									6
5	1	1	2	2	1								5
4	1	2	2	2	1	2							4
3	2	2	2	2	1	2	2					2	3
2	2	2	2	2	1	2	2	2				2	2
1	8	6	8	7	8	6	8	8	4	4		8	1
2	[2]	1	1	1	2	1	1	1	2	2	1	2	2
3	1	1	1	1	2	1	1	1	2	2	1	2	3
4	1	2	2	2	2	2	2	1	1	1	1	1	4
5		2	2	2	2	2	2	2	1	1	1	2	5
6			4	4	8	8	8	7	8	7	8	7	6
5				2	2	2	2	2	2	2			5
4					1	1	2	2	2	2			4
3						1	i	2	2	2			3
2							1	2	2	2			2
1									4	4	8	1	1
2											2	1	2
3												1	3
4													4
Total	30	29	30	29	30	29	30	29	30	29	30	29	Total

The following computations show that lower and higher values of φ would destroy the numerical agreement with the data found in the text.

A point of the ecliptic of longitude λ rises at an azimuth $EH = r$ from the east point E of the horizon (cf. fig. 1). The value of r is given by

$$\sin r = \frac{\sin \varepsilon}{\cos \varphi} \sin \lambda. \qquad (1)$$

The maximum of the rising amplitude is therefore

$$\sin r_{\max} = \frac{\sin \varepsilon}{\cos \varphi} .$$
(2)

From (2) is follows that $r_{\max}$ varies between 28° and 31° when φ varies between 30° and 38° (cf. fig. 2). Consequently we can simplify

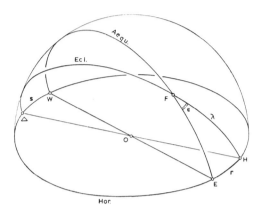

Fig. 1

our discussion by assuming for $r_{\max}$ the round value of 30°; that is to say, we represent each " gate " by an arc of 10° of the horizon ([1]). Thus the first gate reaches from 30° south of east to 20°, the second from 20° to 10°, the third from 10° to E, the fourth from E to 10° north, the fifth from 10° to 20°, and the sixth from 20° to 30° north

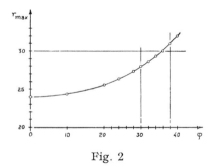

Fig. 2

[1] Accidentally one obtains exactly $r_{\max} = 30°$ for $\varphi = 36°$ and $\varepsilon = 23;51,37$ which is practically identical with the value used by Ptolemy (23;51,20). The data for fig. 3 are computed with these parameters.

of E.   Similarly we count the western gates from 1 to 6 from 30°
south of W to 30° north of W.

Under these assumptions it is easy to compute the ecliptic arcs
between the vernal point F and the rising point H (cf. fig. 1) if H is
10°, 20°, or 30° distant from E.   From (1) and (2) we obtain

$$\sin \lambda = c \cdot \sin r \qquad c = \frac{\cos \varphi}{\sin \varepsilon} = 1/\sin r_{max} \qquad (3)$$

and find, e.g., for gates 4, 5, and 6 (cf. fig. 3)

$$\begin{aligned}
F_4H_4 &= 20;20° \approx 20° \\
F_5H_5 &= 43;10° \approx 43° \qquad (4)\\
F_6H_6 &= 90°.
\end{aligned}$$

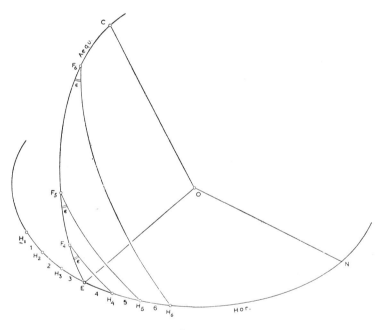

Fig. 3

The same arcs appear also for the first three gates:

$$F_1H_1 = 90° \qquad F_2H_2 \approx 43° \qquad F_3H_3 \approx 20°$$

and once more for the arcs between the horizon and the autumnal
equinox.

Let us now assume that sun and moon rise simultaneously at E. In other words we consider the case of a lunar month beginning at the vernal equinox. Since the moon travels in the mean about 13° per day in the ecliptic we know that after $1\frac{1}{2}$ days the moon will be at a distance of about 20° $= F_4H_4$ from the vernal point. Thus, if the moon rose first at E it will cross the horizon 1 day later before H_4, 2 days later beyond H_4, the boundary between gates 4 and 5. We will express such a situation by saying that the moon stays $1\frac{1}{2}$ days in gate 4. The moon will rise in H_5 when its distance from the vernal point has reached the amount $F_5H_5 \approx 43°$ or about 43° — 20° = 23° farther away than it was when it entered the fifth gate. These 23° require about 2 days travel. Now the moon rises in the sixth gate, increasing by 47° (from 43° to 90°) its distance from F until it rises at H_6 which lies at maximum distance $r_{max} = 30°$ from E. While the moon is now adding further to its distance from F it nevertheless remains in the sixth gate until it has approached the autumnal equinox to about 43°. Thus the moon has to travel a total of twice 47°, or 94°, until it crosses at H_5 back into the fifth gate. This requires about 7 days.

Now everything repeats itself in inverse order: 2 days in gate 5, $1\frac{1}{2}$ days in gate 4 and 3, 2 days in gate 2, 7 days back and forth through gate 1, then 2 days in gate 2 and $1\frac{1}{2}$ days in gate 3, coming back to E. This, however, does not represent the end of the lunar month because the sun is no longer in the equator at E but has gained in the meantime a distance of about 30° from F, hence rising beyond H_4 in the 5th gate. Thus the moon still has to cross the 4th gate and enter the fifth to a distance of about 30° from F, which requires about 2 days of lunar travel.

We have found for this lunar month the following time intervals for the single gates:

gate 4	$1\frac{1}{2}$ days	gate 2	2 days
5	2	1	7
6	7	2	2
5	2	3	$1\frac{1}{2}$
4	$1\frac{1}{2}$	4 and 5	2
3	$1\frac{1}{2}$	total	30 days.

This explains the pattern for the first month listed in our text:

gate 4	2 days	gate 2	2 days
5	2	1	8
6	8	2	2
5	1	3	1
4	1	4	1
3	2	total	30 days

The only difference consists in the elimination of fractions of days and a corresponding increase of the duration of rising in the extremal gates. Also the total is given as a full month of 30 days but for every second month it is reduced to only 29 in order to agree with the simple scheme of alternating full and hollow months in a schematic lunar year of 354 days.

The numbers for the subsequent months do not require lengthy explanations. The months near the solstices produce only extrema of 4 days in gate 6 (respectively gate 1) because these gates are traversed only in one direction if the month begins with maximum distance from E for sun and moon. The point of sunrise remains near the extremal rising amplitude because one month of solar motion does not bring the sun into the range of the next gate.

Small variations in the distribution of the numbers 1 and 2 in the middle gates are, of course, nothing more than inconsistencies in rounding or scribal errors accumulated during a long tradition. The last column is in total disorder, probably because the Ethiopic scribe attempted in vain to reconcile his twelve 30-day months plus epagomenal days with a lunar year of 354 days ([1]).

On the basis of the agreement between the text and the consequences derived from our assumptions we now can formulate our results in a general fashion without any reference to the trigonometric considerations which we needed for the reconstruction of the numerical data. The " gates " are simply sixths of the arc of the horizon which contains the points of sunrise in the course of one year. The numerical data agree with geographical latitudes between Lower Egypt and Greece. The " gates " being once established for a given locality by

---

([1]) Or perhaps he tried to produce a periodic pattern.

observing the total ortive amplitude of the sun during one year, the numbers of days of moonrise through each gate can be found empirically as crude mean values from the observed variation in the points of rising. These average values are then adjusted so as to agree with a schematic lunar calendar of alternating full and hollow months ([1]).

Nowhere in this scheme is explicit use made of the zodiacal motion of sun or moon; not even a measurement of arcs in specific units (degrees) is necessary. Thus we are dealing with an extremely primitive level of astronomy which shows no relation to the sophisticated Babylonian astronomy of the Seleucid period nor to its hellenistic Greek sequel. Of course no chronological conclusion should be based on such negative evidence for procedures which might well be of local Palestinian origin, uninfluenced by contemporary scientific achievements elsewhere.

In texts obviously influenced by Islamic astronomy, one finds formulations which probably gave rise to the incorrect identification of the " gates " with the zodiacal signs. For example we read in Vat. Aeth. 54,7 ([2]); " The sun rises through (the gate of) Aries (ḥâmâl) on the 17th of Magâbît (VII), i.e. the first of spring (rabîʿ) ([3]). (The sun stays in Aries) 30;43 days " ([4]). And: " On the 17th of Yakâtît (VI) the sun rises through the gate of Pisces (eḥût) ". But this is no more than to say that the sun rises through the same gate as the zodiacal sign of the month in question. We have to do with an adaptation of the standard concepts of Arabic astronomy to the traditional Ethiopic terminology.

3. The Astronomy of the Book of Enoch

It is now easy to understand the description of the lunar motion in the Book of Enoch (ch. 73). We see that the " gates " in the west are also counted from south to north and from 1 to 6 — of course identical in size with the eastern ones. The description in ch. 73,5 and 6 implies the case of the autumnal equinox. At the beginning

([1]) For refined schemes of this type in Egypt cf., e.g. Parker, Calendars, p. 14 ff.

([2]) Vat. Cat. p. 88 f.

([3]) A vernal equinox of Magâbît 17 = March 13 (jul.) is correct for the 14th century.

([4]) The corresponding daily motion is 0;56,36° which is quite reasonable for a position in Aries.

of the lunar month sun and moon set together exactly in the west, i.e. at the boundary of the gates 3 and 4. After 7 days the moon has reached maximum southern declination and therefore sets in gate 1. At full moon rising and setting of the moon take place again in the middle gates, while the last quarter corresponds to maximum northern declination, thus to a rising in gate 6. Similarly chapter 73,7 concerns the case of the vernal equinox when the moon at first quarter sets in gate 6. All this is an immediate consequence of the arrangement we have explained in the preceding section.

The solar motion is discussed in chapter 72, together with the variation of the length of daylight and night in the course of one

Table II

| Gate | Month | Days | Length of | |
			Day	Night
4	I	30	10	8
5	II	30	11	7
6	III	31	12	6
6	IV	30	11	7
5	V	30	10	8
4	VI	31	9	9
3	VII	30	8	10
2	VIII	30	7	11
1	IX	31	6	12
1	X	30	7	11
2	XI	30	8	10
3	XII	31	9	9

year ([1]). The purely schematic character of the latter is evident from the tabulation shown in our Table II. The author has no idea of the inequality of the seasons, a phenomenon which plays so prominent a role in the early phase of Greek astronomical discussion as well as in the consideration of the effects of the solar anomaly in Babylonian astronomy of the hellenistic period.

All this, as well as the concept of the zodiac as reference system for the solar motion, is unknown to the author of the astronomical

([1]) No specific calendar is mentioned; the months are consecutively numbered, beginning with I after the vernal equinox.

section which are embedded (in three slightly different versions —
chs. 73 and 74, 78 and 79, and 82) in the Book of Enoch. The learned
speculations of modern scholars about an early form of the Hebrew
calendar, based on a year of 364 days (with or without intercalations)
completely misunderstand the purely schematic character of our
text, equally apparent in the early phases of Babylonian and Greek
astronomy. Mathematical tools not being available, all that the
author can do is to describe to nearest integers the outlines of the
phenomena. The strict linearity of the variation of the length of
daylight — in obvious contradiction to the most elementary " observation " of the factual change — is a striking example of the primitivity of the methods at the disposal of this level. The results obtained can have no more than a qualitative resemblance to the actual
facts, calendaric as well as astronomical. It is amusing to see modern
authors looking for geographical regions where a ratio 2:1 for the
extremal daylight would be correct, ignoring the fact that nowhere
on earth can the scheme as a whole be based on reality (¹).

Table II also shows that the sun is assumed to rise each month
in another " gate ", except, of course, for the duplication near the
solstices. Using the same procedure as in the case of the lunar gates (p. 55) one finds that the boundaries for the solar gates should
have azimuths of about 15⁰ and 25⁰ to the north and south of E and
W instead of the 10⁰ and 20⁰ (and the common 30⁰) of the lunar gates.
It is possible that one made a distinction between solar and lunar
gates, or that our Ethiopic text represents a refinement over the
scheme given in the Book of Enoch. But the fact that in reality the
lunar latitude induces under all circumstances deviations of about
± 5⁰ from the mean values of our scheme makes such a refinement
quite useless. In other words it seems to me more plausible to assume
that one never worried about the slightly contradictory definitions
for the boundaries H_4 and H_5 (fig. 3) with respect to the moon or the
sun. One may add that 12 additional " gates " were distinguished for
the winds (ch. 76), extending over the whole circumference of the

---

(¹) The fact that linear schemes are common in ancient astronomy
and that the ratio 2 : 1 is also attested in Babylonia (Neugebauer [1945]
p. 13) seems to me not to constitute a sufficient basis for the assumption
of mutual contacts. Very primitive methods offer only little freedom of
choice. Operating with the simplest integer difference, 1, determines all
remaining numbers for a given ratio of the extrema or a given mean
value.

horizon. The four central gates (E, N, W, S) admit favorable winds, the remaining eight cause destruction. Finally there are many " windows ", which are not specified more closely, for the rising and setting of the stars (ch. 75,7-9). It is evident that the horizon is still the only reference system known to our author. This again has its parallels in Egypt and in early Greek astronomy without compelling us to assume mutual influences.

The Slavonic Enoch, translated from a Christian version in Greek (¹), shows that the meaning of the " gates " was no longer understood. The sun is now said to rise in each of the extremal gates (1 and 6) during 6 weeks, in all intermediary gates during 5 weeks. This device preserves at least the old schematic year of 52 weeks or 364 days together with some recognition of a longer interval near the solstices. But the moon completely upset the author's astronomy. He invented 12 gates each in the east and in the west and assigned to them, rather arbitrarily, numbers of days, e.g. 30, 31, 35, 22, etc., but such that the total again came to 364 days.

The revision made at the end of the 15th century modernized also Enoch's revelations (²). The sun retained 6 gates but conforms now to a julian year of 365 1/4 days. Also the moon behaves properly according to a lunar year of 354 days and a 532-year cycle. It was left to the scholarship of the 19th century to administer the *coup de grâce* to the gates by identifying them with the signs of the zodiac.

The length of daylight varies in the Slavonic Enoch within the limits of 9 and 15 hours, the classical ratio for the climate of the Hellespont (³). Exactly the same ratio is also attested in our Ethiopic sources, e.g. in *W* fol. 32ᵛ II,25 in a section showing clear islamic influences. Thus Kiev and Magdala both adopted values valid only for the latitude of Byzantium.

Another development is found in the Manichean tradition. There the " gates " appear to be replaced by "thresholds" of uneven width, related to the yearly solar motion (⁴). The details are not sufficiently clear, however, to justify a discussion in the present context.

(¹) Vaillant, Hénoch, p. VIII ff. The Christian origin of the Slavonic Enoch is denied by G. Scholem; cf. his " Ursprung und Anfänge der Kabbala ", Berlin 1962, p. 64 note 37.

(²) Vaillant, Hénoch, p. 91 ff.

(³) Almagest II, 8.

(⁴) Cf. Andreas-Henning, Mitteliranische Manichaica aus Cinesisch-Turkestan. S.B. Preuss. Akad. d. Wiss., Philos.-hist. Kl. 1932, p. 175-222. Henning, Ein manichäisches Henochbuch, ibid., 1934, p. 27-35.

4. Shadow Tables

I have used seven texts which contain eight shadow tables and usually a corresponding explanatory section: one from the Bibliothèque Nationale in Paris (Ab = Ethiop. 37 = Abbadie Catal. No. 37 = Chaine Catal. No. 37 = Conti Rossini [1913] No. 121), one from the British Museum (B = Br. Mus. Catal., Wright No. 397,15 = or. 816 fol. 43ᵛ to 44ᵛ; the tables in two versions, B_1 and B_2, on fol. 43ʳ), four from the Vatican Library (V_1 = Vat. Aeth. 119 fol. 56; V_2 = Vat. Aeth. 128 fol. 140f.; V_3 = Vat. Aeth. 123 fol. 66ᵛ; V_4 = Vat. Aeth. 171 fol. 91ᵛ), and one from Vienna (W = Rhodokanakis [1906] No. 25 fol. 22f.). V_1 and V_2 represent almost identical versions, characterized by the relative shortness of the explanatory text and by the contamination of the shadow tables with a list of the planets. W and B have a much longer, slightly different, textual section which is absent altogether in Ab, V_3, and V_4. The tables in Ab, B_1, V_1 and V_2, and of W have a single column for each month. B_2, V_3, and V_4, however, have only seven columns.

The seven-column arrangement combines in the same column one pair of months equidistant from the solstices. Thus one has five pairs of months with identical shadow lengths, and one column each for the longest and shortest, i.e. solstitial shadows. All Greek shadow tables from the Byzantine period share the error of compressing such a table into six columns for six pairs of months. Consequently one column of shadow lengths got lost in these tables. It is this error which shows that all Byzantine versions are derived from the same archetype ([1]). The Ethiopic tables, however, are based on the correct tradition. V_3 shows the following arrangement of the Ethiopic months in seven columns:

	V	VI	VII	VIII	IX	
IV						X
	III	II	I	XII	XI	

which is correct, since IV and X are the solstitial months (IV = Tâḫśâś, roughly julian December, X = Sanê, roughly June) which

---

([1]) Cf. Neugebauer [1962] p. 36 ff.

remain " single " (**ለበሕቲቱ**). The texts B_2 and V_4 have again only seven columns but the arrangement is unreasonable:

VIII	IX			I	II	III
		X	IV			
XII	XI			VII	VI	V

in B_2 and

I	II	III		VIII	IX	
			IV			X
VII	VI	V		XII	XI	

in V_4. All the other tables operate with 12 single columns. W and B_1 begin with month IV, Ab, V_1, and V_2 with month V.

In V_1 and V_2 the columns for the shadow lengths are preceded by a column which gives the consecutive hours from 1 to 12 (¹). This column is moved one line too high up, such that the first hour is actually written in the line of the column titles for the consecutive months. The numbers of the hours are followed by the letter *sa*, obviously an abbreviation of *sa'ât* " hour ". The numbers in the table itself are given without addition, but the last line, for the 12th hour, shows in all columns the letter *rô*. This, however, is no abbreviation but the numeral 70, as is explicitly stated in the explanatory text which says that the shadow for the 12th hour, at the time of sunset, is 70 feet long (²). This senseless statement can be explained if one retranslates it into Greek. Obviously the original Greek source had for the last hour the entry, ō, that is " zero " — with the meaning " nothing, no shadow " — but misinterpreted as omicron, meaning 70 (³).

(¹) In W and B the numbers of the hours, running from 1 to 11, are written in red in each line of each column.

(²) In V_3 the hours are named individually from morning (**ነግህ**) to noon (**ቀትር**), to vesper (**ዓሲረ ፡ ሞርኽ**). The list of these names fills the first of seven columns, therefore no column was available for month X and its content was written in three lines below the table.

(³) The majority of the extant Greek tables give no entry for the 12th hour, two — (*a*) and (*b*) in my list — show *a* which is a misreading of δ which, in turn, is an abbreviation of δύσις ἡλίου, correctly given in Par. gr. 22 fol. 1.

O. Neugebauer

V_1 and V_2 have a last column, headed ሐመዓለት "for the day", giving the (Arabic) names of the seven planets, beginning with the sun, in decreasing order of their distances, cyclically arranged. Below the table we have another line, written upside down. It is headed የሌሊት "nights", followed by the planetary names, beginning with Jupiter. The whole arrangement is shown in Table III. The

Table III

	V	VI	VII	VIII	IX	X	XI	XII	I	II	III	IV	days
1^h	28	25	25	24	23	22	22	22	23	24	25	27	☉
2^h	18	15	15	14	13	12	12	12	13	14	15	17	♀
3^h	14	11	11	10	9	8	8	8	9	10	11	13	☿
4^h	11	8	8	7	6	5	5	5	6	7	8	10	☾
5^h	9	6	6	5	4	3	3	3	4	5	6	8	♄
6^h	8	5	5	4	3	2	2	2	3	4	5	7	♃
7^h	9	6	6	5	4	3	3	3	4	5	6	8	♂
8^h	11	8	8	7	6	5	5	5	6	7	8	10	☉
9^h	14	11	11	10	9	8	8	8	9	10	11	13	♀
10^h	18	15	15	14	13	12	12	12	13	14	15	17	☿
11^h	28	25	25	24	23	22	22	22	23	24	25	27	☾
12^h	70	70	70	70	70	70	70	70	70	70	70	70	♄
	♄	☿	☾	♃	♂	☉	☉	♄	☿	☾	♃	♂	nights

lists of the planets have nothing to do with the shadow tables and probably concern the astrological rulership of the planets over the hours of day or night, but the details escape me.

The duplications, and consequently the omissions, of columns are probably due to a rearrangement into 12 columns of a faulty Greek table of 6 double columns, omitting the column with the noon shadow 6 [1]. The same table is also missing in the Ethiopic version, whereas the Greek duplication of 2 and 8 is here replaced by additional tables for 2 and 5.

The tables in W and B have a slightly different arrangement. The extremal noon shadows are not 2 and 8, but zero (አልቦ) and 9. The longest shadow is not associated with month V but with month

[1] Cf. Neugebauer [1962] p. 36(a), p. 37(d).

IV and there are no omissions or duplications of columns. On the other hand the difference between consecutive noon shadows is not constant 1 but alternatingly 1 and 2:

month:	IV	V	VI	VII	VIII	IX	X	XI	XII	I	II	III
noon shadow:	9	7	6	4	3	1	0	1	3	4	6	7

But this change of differences is only apparent. If one reads the explanatory text one finds (fol. 22^r I,12 to II,3) the following enumeration for the shadows of the first hour:

month:	IV	V	VI	VII	VIII	IX	X
shadow at 1^h:	29	27½	26	24½	23	21½	20
month:		III	II	I	XII	XI	

This shows that the shadows do not increase by the constant difference 1 but by 1 and ½ (ወመንፈቅ). Only in the numerical table are all fractions omitted, excepting B_2. Finally we have in Ab a table of 12 columns, from month V with 27½ feet as shadow for the first hour, decreasing by 1 ½ feet each month to 20 feet in month X and increasing again to 29 in month IV. In this table the words " and one-half feet " are written out in each line of each column where it should occur, i.e. in all odd months. For the 12th hour, ፸ እግር " 70 feet " is always given.

On the whole we have here a modification of the Greek original also insofar as the extremal noon shadows are no longer 2 and 8 but 0 and 9, which induces a corresponding change in the equinoctial noon shadow from 5 to 4. We will meet the same trend in the Coptic tables (¹).

Also in the textual sections the versions V_1 and V_2 are markedly different from W and B. In V_1 and V_2 the text is called " Book of the measure of the hours " and is rather short. It begins with a reference to the investigations of this topic by the " Greek kings ". Grébaut and Tisserant emend (²) " kings " to " masters " — perhaps incorrectly since also Greek versions refer to a royal patron in the introduction to the tables (³). Then follows the description of the

(¹) Below p. 67.
(²) Vat. Catal. p. 483.
(³) Cf. Neugebauer [1962] p. 35.

difference scheme: 10 feet increment from first to 2nd hour, 4 feet from 2nd to 3rd, etc., with equality of shadow lengths for the 5th and 7th hour, 4th and 8th, etc.; finally 70 feet for the 12th hour (¹). The concluding list was perhaps a list of the shadow lengths for each month (specifically mentioning also the epagomenal days which do not figure in the table) but I cannot make sense of these values which vary irregularly between 4 and 19.

The text of W and B is much more elaborate. It also begins with the enumeration of the increments of the lengths of shadow, cast by a person standing upright, from hour to hour. Then follows the list of the shadows for the first hour for months IV (winter solstice) to X (summer solstice), decreasing from 29 to 20 with the difference $1\frac{1}{2}$ feet (²). This section concludes with the list of symmetries for the remaining months.

Then follows a section concerning the hours of night which are subjected to the same difference scheme 10, 4, 3, 2, 1, between first and sixth hour, but I do not understand how this is done.

After this we return again to the shadow lengths of the hours of the day by listing the hours before and after noon which have equal length of shadow, concluding again with the statement that the shadow of the 12th hour is always 70 feet.

Then follows a variety of remarks concerning fasting (9th hour), shadows and hours, mentioning again the increment of $1\frac{1}{2}$ feet, equinoxes, solstices and the epagomenal days. Toward the end we have two long paragraphs which enumerate the equalities between the hours of night (e.g. of month IV) and daylight (e.g. of month X) and between hours of night (e.g. for the months IX and XI).

In W, but not in B, one finds a concluding list of dates:

II,20   IV,12   VI,4   VII,26   IX,28   XI,10   epag. 2 or 3

which represents intervals of exactly 52 days except for the cyclically closing interval from epagomenal day 2 to II,20 which amounts to 53 days and presumably in leap years another such interval of 53 days from XI,10 to epagomenal day 3. The purpose of this seven-division of the year is not clear to me.

I can only point to a list of numbers (in W fol. 32ᵛ I,5 to 10), denoted as kêkrôs and related to the zodiacal signs (given with their

(¹) Cf. above p. 63.
(²) Cf. above p. 65.

Arabic names). Properly combining months and zodiacal signs, the resulting variation of these numbers seems to reflect the solar velocity with respect to the 52-day intervals. Although the details are not clear the kêkrôs appear again to be related to Islamic influence.

5. Coptic Shadow Tables

Since the Ethiopic calendar is based on the Coptic-Alexandrian calendar one may expect that also the shadow tables should have Coptic predecessors. This is indeed the case. U. Bouriant has published three texts of this type ([1]) in transcription and translation ([2]), one (A) an inscription from the monastery of St. Simeon (Deir Anbâ Sim'ân) ([3]) on the west side of the Nile, opposite Aswân, one parchment book (G) from the French excavations in Asyût ([4]), and another parchment (MF) purchased by Bouriant.

The text A gives entries only for the hours 6, 9, and 10 but the differences are always correctly 6 (= 1 + 2 + 3) and 4 (except for some scribal errors). Similarly MF gives only the hours 1, 6 (difference correctly 20), 9 (difference 6), 10 (difference 4), and 11 (difference 10). Thus the underlying scheme for the variation of the shadow lengths is identical with the Greek and the Ethiopic ones. The noon shadows deviate a little from the Greek scheme. Following the order of the months from Thoth (I) to Mesore (XII) ([5]) we have ([6])

month:	I	II	III	IV	V	VI	VII	VIII	IX	X	XI	XII
A:	4	[5]	[6]	[7]	6	5	4	[3]	2	1	2	3
MF:	4	5	7	8	7	5	4	3	2	1	2	3

([1]) As the fourth he included the Greek inscription from the Nubian temple in Taphis, (c) in my list.

([2]) Bouriant [1898].

([3]) The monastery has been abandoned since the 13th century; cf. De Morgan, Catal. pp. 129 and 138. The inscription A is reproduced on p. 137.

([4]) Now Cairo Museum No. 8030; cf. Crum, Catal. p. 13.

([5]) No mention is made of the epagomenal days.

([6]) Bouriant's restoration are not always correct since he did not realize the strictly schematic character of the texts. There are also discrepancies between numbers from text A given in De Morgan, Catal. p. 137, and in the publication of 1898. Also the short section of G reproduced by Crum, Catal. p. 13 (concerning Phaophi) is not identical with Bouriant's version.

The arrangement in Text A is strictly linear, as in the Greek proto-type, but uses 7 and 1 as extrema, 4 as equinoctial shadow, compared with the Greek 8, 2, and 5 respectively. Text MF inserts after 5 the difference 2, thus reaching the Greek maximum 8, but agreeing with A in the minimum 1 and the equinoctial value 4.

Text G is complete for all months and all hours but marred by many scribal errors and omissions. Its main feature is the insertion of half foot lengths, thus being a parallel to the Ethiopic texts Ab and W (¹). As the text stands the noon shadows are:

G: 4   5½   7   9   7   5   4   2½   1   ½   1   2½

This sequence would be again linear with a constant difference $1\frac{1}{2}$ as in the Ethiopic texts Ab and W, if we would emend 9 to 8½ (an unpleasant emendation), 5 to 5½ (practically certain because it is also required by symmetry), and if we could interpret the minimum value as shadow in the opposite direction, as would be proper for a place in the Sudan:

$[G:]$ 4   5½   7   [8½]   7   5[½]   4   2½   1   [−]½   1   2½
W: 4½   6   7½   9   7½   6   4½   3   1½   0   1½   3.

These lists suffice to show that slight adaptations were made of the Greek scheme to fit more southerly climates, at least so far as the shortest noon shadow is concerned — a wise step to take, since other-wise the monks could have enjoyed noon twice during the days of the summer. For their assembly at the 10th hour (²) it was, of course, irrelevant whether it occurred always at the same fraction of the length of daylight or not (³).

Added in proofs

The manuscript Aethiop. 84 of the Deutsche Staatsbibliothek in Berlin is a maṣḥafa ḥasâb (cf. Dillmann, Verzeichniss, p. 73), written in Sanê A. M. 6916, year of Grace 353, i.e. in June A.D. 1424. This

(¹) Cf. above p. 65.
(²) Bouriant [1898] p. 592.
(³) By ignoring the schematic character of the texts and by consid-ering one number (4) of the most disorderly text as if it were the result of a precise astronomical observation, Ventre-Bey obtained for G the date A.D. 1372 by making use of the secular variation of the inclination of the ecliptic! For A and MF he suggested properly inclined planes. As usual, trigonometry is not a substitute for common sense.

text contains on fol. 15ᵛ a shadow table, similar in arrangement and contents to the above-mentioned table B_2 (extremal noon shadows 0 and 9, difference 1 ½). Then, on fol. 16ʳ/16ᵛ, a second table is given for all 12 months separately. The numbers are badly garbled but show the same sequence for the noon shadows which is found in V_1 and V_2 except for the fact that the Berlin version adds the fraction ½ to all numbers for the months VII to X and for XII and I. This shows that this group of tables originated from a contamination of tables with the extrema 2 and 8 and tables with 0 and 9 as extremal noon shadows. As if to put his ignorance in writing, the scribe concluded these tables with the remark " the 12th hour for each month and the epagomenal days (corresponds to a shadow of) 70 feet ". The short explanatory text (fol. 15ᵛ/16ʳ) is the same as in V_1 and V_2.

Also the list of Greek shadow tables can be increased by one item from Berlin, cod. 173 (= CCAG 7 cod. 26) fol. 176ʳ. A copy of this text is Paris. suppl. gr. 1148 (Bibl. Nat., Cat., manuscr. gr., Astruc-Concasty, p. 306, 85 and CCAG 8,3 p. 86) fol. 186ᵛ/187ʳ. After a short introduction, there follows the table for four pairs of months, two pairs (III, X and IV, IX) being omitted. The extremal noon shadows are 2 and 8 as usual, but there are many errors in the remaining entries.

I should also have mentioned that in 1823, A.-J. Letronne gave an almost correct restoration of the tables in the Nubian temple in Taphis (¹) (Nouv. Ann. des voyages 17 = Œuvres choisies 2. sér. t. I p. 93).

A late Islamic survival of the here discussed shadow tables is found in the calendar of Ibn al-Bannâ (Morocco, 14th cent.) who gives a list of shadows for the noon prayer for each month, following the classical scheme with the extrema 2 and 8 except for some errors in the last three months (²).

A Syriac version was brought to my attention by my colleague Prof. A. Sachs. It is found in the astrological sections appended to the " Syrian Anatomy " edited and translated by Budge (³). The arrangement of the months, beginning with Kanûn I, is correct and so are the shadow lengths for the first hour which vary between 28

(¹) Text (c) in my list [1962].

(²) H. P. J. Renaud, Le calendrier d'Ibn al-Bannâ de Marrakech (Paris 1948) p. 27.

(³) Vol. II, Oxford 1913, p. 607 f.

and 22 feet; the rest is badly garbled. Another Syriac version of the same scheme is preserved in Bar Hebraeus' " L'Ascension de l'esprit" II ch. 4,1 (¹), written A.D. 1279, without error except for one incorrect number. This version leads back to the original Greek sources by quoting as authority " Dionysius who was with king Philip " — exactly as in Vat. gr. 1056 (²). Perhaps Nau was right when he considered the identification with the known astronomer Dionysius who lived under Ptolemy II Philadelphus (³).

REFERENCES AND ABBREVIATIONS

Abbadie, Catal.: Antoine d'Abbadie, Catalogue raisonné des manuscrits éthiopiens... Paris 1859.
Bouriant [1898]: U. Bouriant-[A. F.] Ventre Bey, Sur trois tables horaires coptes. Mémoires présentés à l'Institut Égyptien, t. 3, Le Caire 1898, p. 575-604.
Br. Mus.: W. Wright, Catalogue of the Ethiopic Manuscripts in the British Museum, London 1877.
CCAG: Catalogus Codicum Astrologorum Graecorum.
Chaine, Catal.: M. Chaîne, Catalogue des manuscrits éthiopiens de la collection Antoine d'Abbadie, Paris 1912.
Charles, Enoch: R. H. Charles, The Book of Enoch, ... Translated from the editor's Ethiopic Text ..., Oxford 1912.
Conti Rossini [1913]: C. Conti Rossini, Notice sur les manuscrits éthiopiens de la collection d'Abbadie. Journ. Asiat. sér. 11 vol. 2 (1913) p. 5-64.
Crum. Catal.: E. W. Crum. Coptic Monuments. Catalogue général des antiquités égyptiennes du Musée du Caire, Nᵒˢ 8001-8741. Le Caire 1902.
De Morgan, Catal.: J. de Morgan, U. Bouriant, G. Legrain, C. Jéquier, A. Barsanti, Catalogue des monuments et inscriptions de l'Égypte antique. Prem. sér.; Haute Égypte. T. 1: De la frontière de Nubie a Kom Ombos. Vienne, Holzhausen, 1894.
Dillmann, Verzeichnis: Die Handschriften-Verzeichnisse der königlichen Bibliothek zu Berlin. Bd. 3. Verzeichniss der abessinischen Handschriften. Berlin 1878.
Grébaut [1918]: Sylvian Grébaut, Table des levers de la lune pour chaque mois de l'année. Revue de l'Orient Chrétien 21 (1918-1919) p. 422-428.
Neugebauer [1945]: O. Neugebauer, The History of Ancient Astronomy, Problems and Methods. Journal of Near Eastern Studies 4 (1945) p. 1-38.

(¹) Translation Nau, Paris 1899, p. 159.
(²) CCAG, 5,3 p. 76; p. 37 (d) in my article of 1962.
(³) Observations quoted in the Almagest between 272 and 241 B.C.

Neugebauer [1962]: O. Neugebauer, Über griechische Wetterzeichen und Schattentafeln. Österr. Akad. d. Wissensch., Philos.-hist. Kl., Sitzungsberichte 240, 2 (1962) p. 27-44.

Parker, Calendars: R. A. Parker, The Calendars of Ancient Egypt. The Oriental Institute of the University of Chicago, Studies in Ancient Oriental Civilizations 26. Chicago 1950.

Rhodokanakis [1906]: N. Rhodokanakis, Die äthiopischen Handschriften der k. k. Hofbibliothek zu Wien. Sitzungsberichte d. kaiserl. Akad. d. Wissensch. in Wien. Philos.-hist. Kl. 151, 4 (1906) p. 1-92.

Vaillant, Hénoch: A. Vaillant, Le livre des secrets d'Hénoch, texte slave et traduction française. Textes publiés par l'Institut d'Études slaves 4, Paris 1952.

Vat. Cat.: S. Grébaut-E. Tisserant, Codices Aethiopici Vaticani et Borgiani. I. Bybl. Vaticana 1935.

Reprinted from
Orientalia
33 (1), 49−71 (1964)

Reprinted from
"VISTAS IN ASTRONOMY"
(ED. ARTHUR BEER), Vol. 10, pages 89—103
PERGAMON PRESS · OXFORD & NEW YORK · 1968

On the Planetary Theory of Copernicus

O. Neugebauer

Brown University, Providence, R. I., and
Institute for Advanced Study, Princeton, N. J., U.S.A.

IN MEMORY OF ROBERT OPPENHEIMER

In 1958 A. Koyré spoke about "l'abandon de l'équant, ce haut titre de gloire de l'astrono-mie copernicienne", adding that in the lunar theory "Copernic réussit la simplification la plus grande en la débarassant de l'équant (ce qui nous donne la mesure de son génie mathé-matique)."[1]

About three and a half centuries earlier, Vieta held a different opinion when he said:[2] "Ptolemy and Copernicus, who is always paraphrasing him, did not show themselves as good geometers at the determination of the apsides, the eccentricities, and the epicycle radii from 3 mean and 3 true positions; they assumed the problem settled and therefore solved it in an unfortunate fashion.[3] And Copernicus not only admits his unprofessional way but shows it in Chapter IX of Book III of *De Revolutionibus*, where he tries to deter-mine the maximum equation of the equinoxes from observations of Timocharis, Ptolemy, and al-Battānī as well as the epochs of the anomaly from the limit of the slowing down.[4] More a master of the dice than of the (mathematical) profession he asks to rotate the circle until the error which admittedly comes from his ungeometrical procedure might, with good luck, be compensated."

Both pronouncements are quite characteristic of their times: on the one hand, the ever increasing modern tendency toward hero worship on the basis of "ideas" and disrespect for technicalities; on the other hand, the aggressiveness of Renaissance scholarship, which did not hesitate to point out weaknesses wherever they could be found. But the reader will notice that neither one of the above-mentioned statements is concerned with the alternative geocentric versus heliocentric universe but with the mathematical achievements and abilities of Copernicus. It is only this latter aspect which is the theme of the present paper.

I am aware of the fact that much of the following is not new, at least not to the small group of scholars who during the past decade have uncovered the Islamic antecedents of the Copernican methods[5] nor to those who are familiar with the technical procedures of

---

[1] In Taton, *Histoire générale des sciences*, vol. ii, p. 64.

[2] In his "Apollonius Gallus" (Paris 1600), *Opera math.*, p. 343; also in Kepler, *Werke* vol. iii, p. 464 (ad p. 156, 11).

[3] Cf. below, p. 102, referring to an iteration method of approximations.

[4] Cf. below, p. 96.

[5] Cf. V. Roberts, The solar and lunar theory of Ibn ash-Shāṭir, a pre-Copernican Copernican model, *Isis* vol. 48 (1957), pp. 428–32; E. S. Kennedy and V. Roberts, The planetary theory of Ibn al-Shāṭir, *Isis* vol. 50 (1959), pp. 227–35; Fuad Abbud, The planetary theory of Ibn al-Shāṭir: reduction of the geometric models to numerical tables, *Isis* vol. 53 (1962), pp. 492–9.

7   89

the *Almagest* which were so consistently paraphrased by Copernicus.[1] Nevertheless it seems to me useful to present some of the most central technical features of the Copernican theory of planetary motion and to look at their relation to the *Almagest* not from the viewpoint of philosophical principles but of elementary mathematics. The basic identity of the Copernican methods with the Islamic ones needs no special emphasis in each individual case. The mathematical logic of these methods is such that the purely historical problem of contact or transmission, as opposed to independent discovery, becomes a rather minor one. As I said before, all this could (or rather should) be well known. That in fact it is not needs no documentation.

1. Let me first make clear the technical basis on which we operate, a basis common to the planetary theory in the *Almagest* (about A.D. 150) and in the *De Revolutionibus*.

A practical simplification consists in the separation of the theory of longitudes from the consideration of latitudes by ignoring at first all orbital inclinations. Only after the longitudes are known were latitudes determined by tilting the respective planes into their proper positions. We shall discuss this latter part of the theory only in passing.[2] We will also not go into great detail about the theory of the Sun (including precession) and of the Moon.[3]

To simplify our presentation we speak about a "planet" when we mean the three outer planets and Venus. In this way we need not mention in every case the modifications which are required for the peculiar theory of Mercury; we shall deal with it more conveniently at the end (p. 98).

Finally I remind the reader that before Tycho Brahe a theory was considered adequate when its results agreed with observations within about 10 minutes of arc. In general we shall ignore here the problem of agreement of the ancient theories with the empirical facts as known to us and shall focus our attention almost exclusively on the relation between the mathematical methods of Ptolemy and Copernicus.

2. Let us for a moment assume that the orbit of a planet is strictly circular with respect to the Sun. Since it is our goal to predict the geocentric longitudes λ of a planet, it is convenient to transform the Earth to be at rest. Then it is trivial that the geocentric orbit of Venus is epicyclic; it requires only the construction of one parallelogram to see that the same holds also for the outer planets. The observable tracks of the planets confirm this general cinematic picture.

What is not obvious, however, and is a matter calling for much ingenuity and patient observation is the problem of determining the parameters of such a planetary model. The case of Venus should be the most simple: the radius r of the epicycle is directly obtainable from the maximum elongation θ of the planet from the (mean) Sun.[4] Assuming already the existence of a definite theory of solar motion which provides us for any given moment with the equation of center for the Sun, we know θ and hence r from $r = R \sin \theta$ where R is normed by Ptolemy as 60, by Copernicus as 10^4, representing the radius of any planetary

[1] Cf., e.g., D. Price, Contra-Copernicus: a critical re-estimation of the mathematical theory of Ptolemy, Copernicus, and Kepler. *Critical Problems in the History of Science*, ed. M. Clagett, Madison, Univ. of Wisconsin Press, 1959, pp. 197–218.

[2] Cf. below, p. 103.

[3] Cf. below, p. 96 and p. 100.

[4] The modern reader should be warned that the "mean Sun" of ancient terminology moves with mean velocity *in the ecliptic*.

deferent. In the case of Venus it turned out that θ is not constant but depends largely on the solar longitude. Since it is plausible to assume that r and R are constant—in consequence of the circularity of orbits that was adopted *a priori*—one must conclude that the observer O is located eccentrically with respect to the deferent of center M. Through a systematic sequence of observations of maximum elongations, Ptolemy and his immediate predecessor Theon determined the solar positions for which θ appears as a maximum or as a minimum, hence locating the apsidal line and finding the amount of the eccentricity $e = OM$ (in terms of $R = 60$).

In the solar and lunar theory a simple eccenter (or a cinematically equivalent epicycle) seemed to suffice for the explanation of the inequality in the length of the seasons and of the intervals between lunar eclipses. But already Hipparchus (about three centuries earlier) had realized that this simple lunar model was defective in the quadratures though he was not able to bring order into the seemingly inconsistent empirical data. Here Ptolemy succeeded and established the laws which govern the inequality which is now known as "evection". It is well known that he spoiled his discovery by a hopelessly inadequate cinematic explanation.[1] But the interest paid to conditions in quadratures led to another

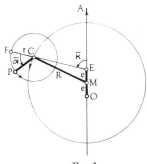

FIG. 1.

important discovery in planetary theory. Knowing for Venus the apsidal line, eccentricity and epicycle radius, one can easily predict the maximum elongation to be expected when the center C of the epicycle (or the mean Sun) is in quadrature to the apsidal line, assuming naively uniform rotation of C on the deferent, i.e. with respect to its center M. Ptolemy found, however, that the observed elongations require uniform rotation not about M but about a point E of the apsidal line located such that $EM = MO = e$. This point E is the famous *equant* (using a late mediaeval terminology), i.e. the point of the apsidal line from which the motion of C appears uniform (cf. Fig. 1).[2] For the modern reader it is not surprising that this concept played a crucial rôle in Kepler's attempts to account for the motion of Mars.

Ptolemy adopted the same cinematic principle also for the outer planets, of course combined with the trivial condition that the radius CP must always be parallel to the direction $\overline{OS}$ from the observer to the mean Sun. The eminently successful prediction of planetary positions computed on the basis of this model can rightly be considered as its justification, even if one had no longer such direct observations at one's disposal as in the case of Venus.

---

[1] For which cf. below, p. 100.

[2] None of our figures is drawn to scale; in particular eccentricities are greatly exaggerated.

7*

3. Without, for the moment, entering upon the question how to determine the parameters e and r for an outer planet (we will have to come back to this later (p. 102)), we may anticipate the results concerning the epicycle radii.[1]

As far as we know, Copernicus was the first clearly to understand that these radii are only different because the radii R of the deferents are all taken as unit. If one, however, uses the radius a of the Earth's (or Sun's) orbit as unit, then $a = R/r$ for an outer planet, and $a = r/R$ for an inner planet, provides us with the heliocentric distance of each planet. In this way we can compare both systems:

TABLE 1

r	Alm.	Cop.	mod.
Saturn	6;30	6;32	6;17
Jupiter	11;30	11;30	11;32
Mars	39;30	39;29	39;22
Venus	43;10	43;10	43;24
Mercury	22;30	22;35	23;14

TABLE 2

a	Alm.	Cop.	mod.
Saturn	9·231	9·175	9·539
Jupiter	5·217	5·219	5·203
Mars	1·519	1·520	1·524
Venus	0·719	0·719	0·723
Mercury	0·375	0·376	0·387

Table 2 represents the main contribution of Copernicus to astronomy: it opened the way to the determination of the absolute dimensions of our planetary system.

Surprisingly enough the problem of heliocentric distances is not at all emphasized by Copernicus. Of the numbers listed in Table 2 only the distance of Mars is explicitly mentioned (in V, 19); for Saturn and Jupiter one has to compute the mean distance from the extreme values. For Venus one finds only Ptolemy's value of r, and for Mercury one can compute a mean value $\bar{r}$ for the variable radius of the planet's orbit (cf. below, p. 99). A contemporary reader could scarcely get the impression that here lay the central core of the "Copernican System".

4. We now come to "l'abandon de l'équant". For the Moon one gets into trouble because no equant in the proper sense exists in Ptolemy's model. The mean motion takes place with respect to the Earth, the removal of which might be difficult, even for Copernicus.

As for the planets we shall now demonstrate that it was the goal of Copernicus' cinematic arrangements to *maintain* the equant, by no means to eliminate it.

For an outer planet Copernicus prescribes the following motion (cf. Fig. 2): the planet P moves on an epicycle of radius r' such that PC makes with CM the angle $\bar{\varkappa}$ when CM makes the same angle $\bar{\varkappa}$ with the apsidal line ($\bar{\varkappa}$ increases with the rate of the sidereal mean motion of the planet). The center M of the deferent has the eccentricity e_1 with respect to the mean Sun $\bar{S}$ about which the observer O rotates on a circle of radius r.

---

[1] For sexagesimal numbers I use a semicolon to separate integers from fractions, a comma for the separation of sexagesimal digits.

In order to relate this model to the Ptolemaic one, we transfer O to be at rest, using the familiar parallelogram construction, repeatedly applied by Copernicus and, of course, well known to all astronomers at least since Apollonius. Figure 3 shows in heavy solid lines the resulting structure, which differs from the Ptolemaic model only insofar as the planetary epicycle of radius r does not move with its center C_2 on the deferent and as the equant seems to be missing. In fact, however, the rule for the motion of C_2 is such that a point E on the apsidal line (cf. Fig. 4) at a distance r' from M' will always see C_2 at an angle $\bar{\varkappa}$ from the apsidal line. Hence E is the equant for C_2. Since C_2 is the center of the planet's epicycle the Copernican model would be *identical* with Ptolemy's if the path of C_2 were a circle.

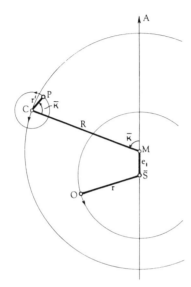

FIG. 2.

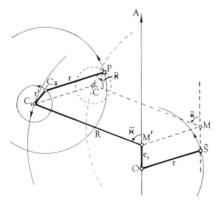

FIG. 3.

Copernicus proves that this is not the case,[1] as is easy to see if one considers, e.g., the situation at quadrature. But he does his very best (without saying so) to make the orbit of C_2 agree as closely as possible with the Ptolemaic deferent. To this end one must obviously require that

$$OE = e_1 + r' = 2e \qquad (1)$$

where e is the Ptolemaic eccentricity $OM = ME = e$. Furthermore one will have exact agreement with Ptolemy's deferent in the apsidal line if (cf. Fig. 5)

$$(R - r') + (e - r') = R$$

i.e. if

$$r' = \tfrac{1}{2} e \qquad (2)$$

and hence, because of (1),

$$e_1 = \tfrac{3}{2} e. \qquad (3)$$

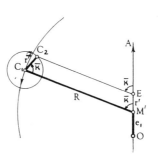

Fig. 4.

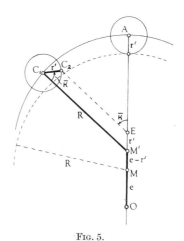

Fig. 5.

---

[1] Ostensibly in order to disprove an opinion of "the ancients". But in antiquity no such model was proposed and the only persons who would have been interested in this problem before Copernicus are the Muslim astronomers who invented this model. Indeed one finds in aṭ-Ṭūsī (about A.D. 1270) the same proof as with Copernicus (cf. Tannery (below, p. 99 note 4), p. 351).

The relations (2) and (3) are indeed the relations which Copernicus (as well as aṭ-Ṭūṣī and ash-Shāṭir) prescribes in relation to Ptolemy's eccentricities.

It is easy to see that this Copernican orbit of C_2 is in quadratures only about $e^2/2R$ wider than Ptolemy's deferent.[1] The angular displacement of C_2 as seen from O remains well below one minute for all planets. When Copernicus' recomputations of Ptolemaic longitudes occasionally result in differences of more than one minute, then the cause lies in the inaccuracy of the trigonometric procedures, not in the principle of the models.

In the case of Venus, Copernicus assumes that $\bar{S}$ is the mean Sun (cf. Fig. 6), C_1 at a distance e_1 from $\bar{S}$ is the center of a circle of radius r' which carries the center C_2 of the

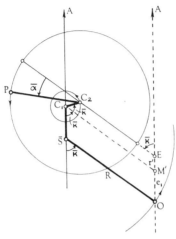

FIG. 6.

planet's orbit; all angles $\bar{\varkappa}$ increase proportional with time. Since $\bar{S}C_1C_2 = 2\bar{\varkappa}$ we see as before that E, at the observer's apsidal line OA, is the equant which controls the motion of the planet's orbital center C_2. This point, in turn, moves practically on the Ptolemaic deferent because

$$r' = 104 \quad e_1 = 312 \quad e \approx 208{\cdot}3 \ [2] \quad \text{for} \quad R = 10^4$$

satisfy the conditions (2) and (3).

Since we shall find (cf. p. 98) that Copernicus preserved the equant also for Mercury we can now say that his aim was by no means to abolish the concept of equant, but, exactly as his Islamic predecessors, to demonstrate that a secondary epicycle is capable of producing practically the same results (thanks to the smallness of the eccentricities) as Ptolemy's equant. Though the resultant deferent is unfortunately not a circle, each component motion is uniform and circular. Both ash-Shāṭir and Copernicus considered this as their main achievement, even if the model had become more complicated than Ptolemy's.

Kepler was less philosophically prejudiced, and he not only reintroduced the Ptolemaic equants in the planetary theory but took the heliocentric approach seriously and hence

---

[1] Kepler remarked (*Werke*, vol. iii, p. 75) that he would not mind this construction if it only would make the deferent narrower, not wider, than a circle.

[2] For $R = 60$ $e = 1{;}15$ (*Alm*, x, 3).

provided also the (circular) orbit of the Earth with an equant,[1] an improvement which increased the accuracy in his determinations of the positions of Mars. Ptolemy's discovery of the "equant" was not only never abandoned but proved of the greatest importance for the construction of the "oval" orbit of Mars and hence of the Kepler ellipse.

5. The identity of the Copernican cinematic model with the Ptolemaic is obscured by many secondary features. Instead of using tropical longitudes, Copernicus always operates sidereally, with γ Ari as zero point, simply because this star is the first zodiacal star listed in the star catalogue of the *Almagest* (with $\lambda = 6;40$ $\beta = +7;20$). Hence all Ptolemaic longitudes are reduced by 6;40 and augmented by precession, which, however, is vitiated by a trepidation term. Furthermore the "mean Sun" (our $\overline{S}$) in the center of the Earth's circular orbit is not quite the mean Sun but rotates slowly about another center which, finally, has a fixed distance from the real Sun.[2] The motion of $\overline{S}$, which modifies the solar eccentricity and apogee, is unfortunately regulated by the same parameter which controls the oscillations of the true vernal point and simultaneously the obliquity of the ecliptic. The two latter motions are supposedly the consequence of a motion of the celestial poles along a figure-eight-curve made of two small contacting circles, the point of contact being the mean pole. In fact, however, this supposed motion of the polar axis does not induce the simple harmonic motion of the vernal point which Copernicus finally assumed (and which made Vieta so angry). Hence it is not at all simple to find out what the Copernican equivalent of a Ptolemaic coordinate should be and such a transformation is made still more arduous by the countless small computing errors, inaccuracies, and inconsistencies which mar all discussions in *De Revolutionibus*. Frequent shifts from sexagesimal parameters to decimals and back again do not increase accuracy. Finally angles which have a simple geometric significance are moved to $\overline{S}$ or C (of course simply parallel) and given new names. Hence it is not surprising that it is not at all apparent that the Copernican planetary tables are the direct equivalent of the Ptolemaic ones. But it is geometrically clear that this must be the case since we know that the Copernican model preserves the equant of the Ptolemaic theory (cf. Fig. 7).

In order to find the longitude λ of a planet P as seen from O (with respect to γ Ari or to $\Upsilon\,0°$) one needs the angle η which appears at C as well as at $\overline{S}$. The position of the planet on its epicycle is defined by its "mean anomaly" $\overline{\alpha}$ or its "true anomaly" $\alpha = \overline{\alpha} + \eta$, which are called "parallactic anomalies" when counted at $\overline{S}$. The equant guarantees that $OCP\overline{S}$ always forms a parallelogram and hence we have the same angle θ at O as well as at P. Hence λ will be given in both versions by

$$\lambda = \lambda_A + \overline{\varkappa} + \eta + \theta,$$

of course with proper signs for η and θ.

The tables in the *Almagest* (xi, 11) contain 8 columns, the first two for the arguments, the remaining six for functions which we denote as c_3 to c_8. The corresponding tables of Copernicus (v, 33) give four functions C_3 to C_6. Between these functions c and C there exist simple correspondences. Already Ptolemy abolished in this *Handy Tables* the tabulation

---

[1] Incidentally, Kepler was not the first to design a solar theory with an equant. Ibn ash-Shāṭir (around 1350) assumed a secondary epicycle which carries the Sun at an angular distance $2\varkappa$ from the direction of the apsidal line (cf. Roberts, *Isis*, vol. 48, p. 429, Fig. 1). This is exactly the same device used in the theory of Mercury for the motion of the center C_2 of the planet's orbit (cf. below, p. 99, Fig. 9) and produces the same result, i.e. an equant located between the observer and the center of the deferent.

[2] It may be remarked that here Copernicus introduced into the solar theory exactly the same mechanism against which he polemicized on philosophical grounds in Ptolemy's lunar theory.

of two separate components c_3 and c_4 which together give the angle η as function of $\bar{\varkappa}$. Copernicus follows, of course, the same practice, thus

$$\eta(\bar{\varkappa}) = c_3(\bar{\varkappa}) + c_4(\bar{\varkappa}) \approx C_3(\varkappa).$$

With $\alpha = \bar{\alpha} + \eta$ as argument Ptolemy forms

$$\theta(\alpha, \bar{\varkappa}) = c_6(\alpha) + c_8(\bar{\varkappa}) \, c_5(\alpha) \quad \text{if} \quad c_8 \leqq 0$$

or

$$\theta(\alpha, \bar{\varkappa}) = c_6(\alpha) + c_8(\bar{\varkappa}) \, c_7(\alpha) \quad \text{if} \quad c_8 \geqq 0.$$

Here $c_8(\bar{\varkappa})$ is a coefficient of interpolation which increases in nearly sinusoidal fashion from -1 at $\bar{\varkappa} = 0$ to $+1$ at $\bar{\varkappa} = 180$; $c_6(\alpha)$ gives the angle θ when the epicycle is at mean distance, $c_6(\alpha) - c_5(\alpha)$ at maximum distance, $c_6(\alpha) + c_7(\alpha)$ at minimum distance. The above formulae indicate how θ is found for intermediary distances.

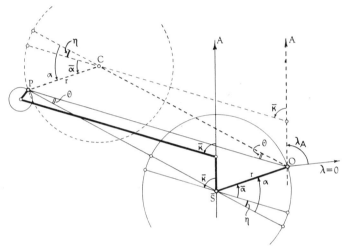

FIG. 7.

Copernicus modified this procedure by adopting a principle developed for Islamic tables, only to use positive corrections. Hence he finds θ from

$$\theta(\alpha, \bar{\varkappa}) = C_5(\alpha) + C_4(\bar{\varkappa}) \, C_6(\alpha)$$

where $C_4(\bar{\varkappa})$ increases from 0 at $\bar{\varkappa} = 0$ to $+1$ at $\bar{\varkappa} = 180$ because $C_5(\alpha) = \theta$ at maximum distance, $C_5(\alpha) + C_6(\alpha) = \theta$ at minimum distance. Hence

$$c_6(\alpha) - c_5(\alpha) \approx C_5(\alpha)$$
$$c_5(\alpha) + c_7(\alpha) \approx C_6(\alpha).$$

Table 3, excerpted from the tables for Saturn, shows how closely these relations are satisfied.

Obviously the Copernican tables will produce practically the same results as the Ptolemaic ones. Also the number of steps in computing a planetary longitude is the same in both systems.

TABLE 3

	Alm. xi, 11			Revol. v, 33		
	$c_3 + c_4$	$c_6 - c_5$	$c_5 + c_7$	C_3	C_5	C_6
30	3;6	2;42	0;19	3;6	2;42	0;19
60	5;29	4;49	0;35	5;29	4;49	0;35
90	6;31	5;53	0;41	6;31	5;52	0;42
120	5;49	5;21	0;42	5;49	5;22	0;42
150	3;24	3;13	0;26	3;24	3;13	0;26

6. Ptolemy's model for Mercury (cf. Fig. 8) assumes that the center N of the deferent rotates uniformly about a point M of the apsidal line whereas the center C of the epicycle is seen moving uniformly from an equant E which lies halfway between O and M. The resulting path of C brings the epicycle for $\bar{\varkappa} = \pm 120°30'$ (or a value very close to it[1]) nearer to O than at the "perigee" II at $\bar{\varkappa} = 180$. In order to preserve these features Copernicus first sets out to keep the equant in its proper position between O and M. For this reason he has now to let the center C_2 of the planet's orbit (cf. Fig. 9) rotate with a phase 180° different from the other cases. Applying the same type of argument which we used before (p. 94) it can be shown that again the conditions (2) and (3)

$$r' = \tfrac{1}{2} e \qquad e_1 = \tfrac{3}{2} e = 3r' \tag{4}$$

should be satisfied if not only E but also A and II should be kept in place. And indeed it is this relation (4) which Copernicus prescribes in *Revol.* v, 25, for his model of Mercury.

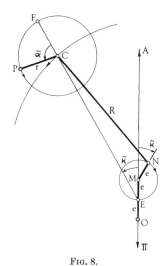

FIG. 8.

---

[1] W. Hartner has shown that this path is nearly elliptical; cf. *Vistas in Astronomy*, ed. A. Beer, vol. 1 (1955), p. 109. See also Hartner's article on "Mediaeval views on cosmic dimensions" in *Mélanges Alexandre Koyré*, vol. II, p. 268, footnote 25.

But if one looks at the subsequent calculations one finds that Copernicus uses not (4) but[1]

$$r' = 212 \quad e_1 = 736 \quad \text{Ptolemy: } e = 500.$$

It is possible to detect the reason for this change of parameters. In order to determine the remaining parameters of the model one could require the preservation of additional geocentric distances of the Ptolemaic model, e.g. at quadratures or at $\bar{\varkappa} = \pm 120$. It is not difficult to see, however, that this leads to unpleasant conditions for the motion of the planet with respect to the center of its orbit. Hence Copernicus abolished his original

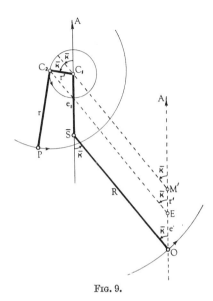

FIG. 9.

procedure (without saying so) and adopted an approach which is preferable also in principle. Ptolemy determined the parameters of his model from observations of maximum elongations, much in the same fashion as with Venus. What Copernicus now required for his model was that it should produce for $\bar{\varkappa} = 0$, 180, and ± 120 the same maximum elongations as Ptolemy's theory. Of course, Ptolemy's rotating deferent had to be replaced by a variation of the radius of the planetary orbit, or, in Copernicus' terminology, the planet had to perform a motion of libration along the radius of its orbit between fixed limits $\bar{r} + \tau$ and $\bar{r} - \tau$. For the mean radius $\bar{r}$ Copernicus took a close approximation of Ptolemy's epicycle radius r; then it is easy to show that the maximum elongations taken from the *Almagest* for the above-mentioned values of $\bar{\varkappa}$ determine the amplitude τ of libration. The motion of libration itself is again simple harmonic (as for the vernal point),[2] supposed to be generated by uniform rotations in the form devised by Proclus[3] or aṭ-Ṭūsī.[4]

---

[1] For unknown reasons sometimes also $211\frac{1}{2}$ and $736\frac{1}{2}$ respectively.

[2] Cf. above, p. 96.

[3] Commentary to *Euclid* I, Defin. IV. Cf. also transl. *Ver Eecke*, p. 96, n. 4.

[4] Cf. P. Tannery, *Recherches sur l'astronomie ancienne* (Paris 1893), p. 348. Also Kennedy-Roberts, *loc. cit.*, p. 231f.

Neither Ptolemy's nor Copernicus' machinery for the motion of Mercury could be considered plausible representations of physical facts; Copernicus himself had speculated in a loose fashion about an alternative mechanism.[1] It is difficult to see how devices of this type ever could have been taken for more than mathematical similes of no other significance than to guide the computations. I realize that one is supposed to be disgusted with Osiander's preface which he added to the *De Revolutionibus* (in keen anticipation of the struggle of the next generations) in which he, in the traditional fashion of the ancients, speaks about mere "hypotheses" represented by the cinematic models adopted in this work. It is hard for me to imagine how a careful reader could reach a different conclusion.

7. Ptolemy's model for Mercury is obviously inspired by the mechanism which he had invented in order to explain for the Moon in quadrature the increased equation of center (cf. above, p. 91). But the crank mechanism which properly enlarged the effect of the epicyclic anomaly also increased the lunar parallax and the Moon's apparent diameter to almost twice its actual value. Ptolemy kept silent about this obvious deficiency of his theory which accounted so nicely for the longitudes. But in late Islamic astronomy this defect was no longer accepted without question, and we now know that almost two centuries before Copernicus the device of a secondary epicycle was used by Ibn ash-Shāṭir.[2] The determination of the corresponding radius is trivial[3] since one has to do nothing more than to make the diameter of the second epicycle so large that the maximum equation increases from about 5° at syzygies to about 7;40 at quadratures, again simply accepting Ptolemy's data. Copernicus, operating on the same premises, reached of course the same result as the Muslim astronomers.[4]

The new lunar model had the great advantage over Ptolemy's that it kept the parallax under all conditions nearly within the limits prevailing at syzygies. Copernicus confirmed the new parallaxes by showing[5] that an occultation of Aldebaran by the Moon, observed in Bologna in 1497, was accounted for by his parallax. If one checks, however, Copernicus' computations, one finds errors in practically every step, even such obvious ones as a total of less than 180° for the angles of a spherical triangle. Fortunately the number of steps is large enough to make the total error insignificant. But the ancient lunar theory, assuming a fixed maximum latitude of 5°, also accepted by Copernicus, produced in the present case a latitude of only about −4;35, instead of −4;47. For the latitudinal parallax Copernicus found about −0;30 which moved the Moon down to Aldebaran, which he placed at a latitude of −5;10, a coordinate taken right out of Ptolemy's catalogue of stars. Had he checked it by observation he would have found the star at about −5;30 and his theory would have made the lower rim of the Moon pass almost 10 minutes above the star. Even worse, Ptolemy's latitudinal parallax would have moved the Moon from Copernicus' position right down to the star, hence supporting the ancient theory against the Copernican. In fact it was only a wrong Copernican lunar latitude in combination with a wrong Ptolemaic stellar position which "confirmed" the (essentially correct) Copernican parallax.

As is well known, Ptolemy's solar parallax was wrong by a factor of about 20. Since no direct measurement could possibly be made with the instruments of antiquity one followed a method, invented by Hipparchus, based on eclipses. Ptolemy had assumed that the Moon at maximum distance covers the Sun (at mean distance) exactly, both appearing under an

---

[1] *Revol.* v, 32.
[2] Cf. V. Roberts, quoted above, p. 89, note 4.
[3] Hence no *"mesure de... génie mathématique"*. Cf. also the very simple discussion in *Revol.* iv, 8.
[4] Roberts, *loc. cit.*, p. 431.
[5] *Revol.* iv, 27.

angle of $0;31,20°$. From a careful discussion of lunar eclipses he had derived for the diameter of the shadow a ratio 13/5 to the diameter of the Moon at maximum distance, the latter being at $64;10$ earth radii.

The parameters accepted by Copernicus required some changes insofar as a common tangent to Moon and Sun would occur at a distance of the Moon of only 62 earth radii instead of Ptolemy's $64;10$. For the shadow he considered a ratio 403/150 as more "convenient" than 13/5. Naturally one will ask: "convenient" for what? And why these special numbers? It is abundantly clear that Copernicus had no eclipse observation at his disposal giving him new information about the diameter of the shadow cone. Not only does he not adduce any such evidence but he gives in iv, 18 a very sketchy summary of Ptolemy's method how to deduce from eclipse magnitudes data for the shadow's diameter. The numerical data which Copernicus mentions are round numbers chosen *ad hoc* and so carelessly that they are excluded by Copernicus' own theory. Hence the ratio 403/150 must come from somewhere else. Indeed, since the distance of the Moon was changed from $64;10$ to 62 Copernicus simply multiplied 13/5 by $\dfrac{64;10}{62} \approx \dfrac{31}{30}$ to get a "convenient" shadow. And the convenience of this transformation lies in the fact that this change is required to obtain practically the same distance for the Sun as before with Ptolemy's parameters. No wonder that Copernicus' conveniently doctored data produced a solar distance of 1179 Earth radii as compared with Ptolemy's 1210. Had he used his parameters without corrections he could have easily ended up with a considerably greater distance for the Sun than Ptolemy had found and this would have been rather unpleasant for a heliocentric system which had to face the absence of any fixed-star parallax.

Having obtained the conventional order of magnitude for the solar distance Copernicus happily reverted to the classical ratio 13/5 for the shadow. One can hardly think of a greater contrast in methodology as between Copernicus and Tycho Brahe, only one generation later.

8. Vieta, in his criticism quoted at the beginning, referred to the determination of planetary parameters from 3 mean and 3 true positions. Indeed all the astronomical models between Apollonius and Kepler had to solve a problem of this type. Since the circularity of the basic motion was taken for granted three points had to be determined to characterize a circle. Three observations provided two time differences which furnish two angles ($\bar{\delta}_1$ and $\bar{\delta}_2$) of mean motions, whereas the observer recorded two angles of true motion (δ_1 and δ_2). Then one faces the following problem: find the position of an observer who sees three points on a circle under angular differences δ_1 and δ_2 while he knows that they would appear from the center of mean motion under the angles $\bar{\delta}_1$ and $\bar{\delta}_2$. If the latter center coincides with the center of the circular orbit the problem has a unique solution obtainable by straightforward trigonometric operations. This is the case for the Sun (where δ_1 and δ_2 are right angles if one observes the equinoxes and a solstice) and for the Moon at lunar eclipses where the mean anomalies provide $\bar{\delta}_1$ and $\bar{\delta}_2$ at the center of the epicycle.

This simple situation no longer holds for a model with an eccentric equant of unknown eccentricity. Then it is easy to see that the four above-mentioned angles alone do not determine the problem. For the inner planets one need not worry, because the size of their orbits is directly observable at maximum elongation. This does not hold, however, for the epicycles of the outer planets. Only their centers are in principle observable at the moments of opposition of the planet to the mean Sun (cf. Fig. 10a). Unfortunately the mean distance $\bar{\delta}$ between two centers is measured at the equant E and $OE = 2e$ is one of the unknown quantities one wishes to determine.

The Copernican arrangement (Fig. 10b) faces exactly the same difficulty. The observations give only the angle δ between the directions $\overline{S}O_1P_1$ and $\overline{S}O_2P_2$. But $\overline{\delta}$ is measured at M' and we do not know $e_1 = \overline{S}M'$ or $r' = \tfrac{1}{3}e_1 = CP$.[1]

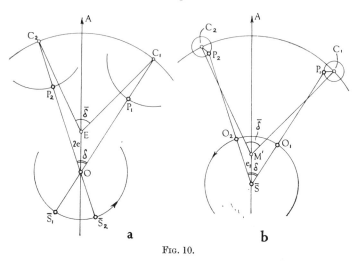

FIG. 10.

Ptolemy solves his problem by an iteration process. He assumes as first approximation that the equant coincides with the center M of the deferent and thus returns to the mathematical problem familiar from the lunar theory. This leads to an approximately correct position of the apsidal line and an approximate eccentricity. Now one can find by how much (ε) the observed angle δ had been falsified because of the identification of E and M. Hence a second approximation can be computed with $\overline{\delta}$ and $\delta + \varepsilon$ as given angles, and so forth until the results become stable. Ptolemy did not compute many steps: three for Mars, only two each for Saturn and Jupiter. He gives, of course, no proof of convergence and is satisfied to show that the last obtained parameters explain the observational data.

Copernicus admittedly did not bother to understand this iteration method which he simply characterizes as a "multitudo numerorum". All that he does is to repeat Ptolemy's final test with the parameters of his own model, obtaining exactly the same results (as was to be expected from the beginning).

A different situation arises when Copernicus sets out to repeat the determination of the parameters of his model on the basis of observations in his own time. As we have said before, he would have to face exactly the same mathematical difficulty as did Ptolemy. But such a systematic approach is foreign to him and hence he maintains as much as possible of the Ptolemaic parameters and modifies the position of the apsidal line by trial and error, knowing of course from contemporary astronomy how much displacement could be expected.

9. The rigid adherence to Ptolemaic methods deprived Copernicus of one advantage where the heliocentric approach is definitely superior to the geocentric one, i.e. in the

---

[1] It should be noted that the relative positions of the points $\overline{S}$, O, P are exactly the same in Figs. 10a and 10b.

theory of latitudes. Since the orbital planes go through the Sun, the assumption of deferent planes which go through the Earth produces inconvenient effects. Without eccentricities it would be correct to move the plane of the epicycle parallel to itself along the deferent at fixed inclination. In fact, however, a Ptolemaic eccentricity is the vector sum of the eccentricities of Earth and planet, and this vector lies neither in the ecliptic nor in the orbital plane. It is not difficult to see that this situation, unknown to Ptolemy, is the cause of the vibrations which he was forced to introduce into his theory of planetary latitudes in order to account for the observations. Since the Copernican theory is only a formal transformation of the Ptolemaic theory, Copernicus ends up with the same secondary vibrations of the orbital planes which he assumed to go through the mean Sun. As Kepler put it,[1] Copernicus did not know how rich he was.

It is surprising to see that it did not disturb the protagonist of a Universe in which the earth was only one of six planets that five of them entered in an "agreement with the center of the earth"[2] to nod with the frequency of the Earth's rotation. And because every celestial motion had to be mechanized by means of uniformly rotating circles, Copernicus attached to each orbital plane a perpendicular little circle inside of which rolled a second circle such that the orbits would move up and down in simple harmonic motion.

10. If one reads Copernicus only superficially and with the conviction that he had abolished, or at least greatly simplified, the Ptolemaic system, one will not be tempted to study the *Almagest* in any detail. Vieta, of course, still knew better. He must have been fully aware of the fact that there was not a single proof or mathematical procedure in the *De Revolutionibus* which did not have its exact replica in the *Almagest*. To Vieta as one of the leaders in the new trend of mathematics it must have appeared rather antiquated when Copernicus again and again demonstrated by numerical computation that his model agreed with Ptolemy's.

Modern historians, making ample use of the advantage of hindsight, stress the revolutionary significance of the heliocentric system and the simplifications it had introduced. In fact, the actual computation of planetary positions follows exactly the ancient pattern and the results are the same. The Copernican solar theory is definitely a step in the wrong direction for the actual computation as well as for the underlying cinematic concepts. The cinematically elegant idea of secondary epicycles for the lunar theory and as substitute for the equant—as we now know, methods familiar to a school of Islamic astronomers—does not contribute to make the planetary phenomena easier to visualize. Had it not been for Tycho Brahe and Kepler, the Copernican system would have contributed to the perpetuation of the Ptolemaic system in a slightly more complicated form but more pleasing to philosophical minds.

---

[1] *Werke*, vol. iii, p. 141, 3.

[2] Rheticus in the *Narratio Prima;* text, e.g., Kepler, *Werke*, vol.i, p. 125, 2f.; translation in Rosen, *Three Copernican Treatises* (1959), p. 183.

APPENDIX C

ASTRONOMICAL AND CALENDRICAL DATA
IN THE TRES RICHES HEURES
Notes by O. Neugebauer

It is not surprising that a work of the artistic excellence displayed in the *Très Riches Heures* also shows competence on the part of the makers of the calendar which precedes in traditional fashion the devotional sections. What "competent" means in this line of work in the early 15th century is perhaps not obvious to historians of art. We therefore hope to contribute a little to the appreciation of the *Très Riches Heures* when we draw attention to some aspects of medieval astronomy embedded in the tables and miniatures which open the book.

We begin with two strictly astronomical topics: the variation of the length of daylight during the year and the motion of the sun through the ecliptic. We then turn to the calendaric treatment of the lunar motion, both in its traditional ecclesiastic form and "nouvel." Finally we touch upon the "dies aegyptiaci" and the figure of the Zodiac Man, inserted after the calendar. That many questions remain unanswered may be taken as proof that it is worth while looking into these matters.

In the center of the following discussion always lie the *Très Riches Heures*. Occasionally, however, a comparison with the calendars in one of the other Books of Hours of the Duke of Berry is of interest. I refer to all these calendars by the conventional names:[1]

Petites Heures (*ca.* 1385)
Très Belles Heures de Notre Dame (*ca.* 1404/5)
Belles Heures (1406–1408)
Grandes Heures (1407–1409)
Très Riches Heures (1413–1416).

1. Length of Daylight

Following the list of saint- and feast-days the calendar pages of the *Très Riches Heures* show a column headed "Quantitas dierum" or "la quantite des jours" which gives day by day, in "hore" and "minuta," the length of daylight. The lowest value (m) is 8^h, listed for December 11 to 15, the highest (M) of 16^h is found for June 13 to 16.

Before going to the details of this table we can draw an immediate conclusion from the data mentioned. Since early Greek astronomy the ratio of longest to shortest daylight served to characterize a location on the

terrestrial sphere. With the advancement of spherical trigonometry the correct relation of this ratio to the terrestrial latitude ϕ was established and hence we find, e.g. in the Almagest, tables which relate M to ϕ. There we see[2] that M=16^h corresponds to ϕ=48;32°, a latitude which fits Paris (48;52°) better than Bourges (47;5°). Of course we should not assume that the Almagest was used by the calendar makers at the end of the 14th century; but the Alfonsine Tables (epoch 1252), among many other tables in circulation at the time, contain also a "Tabula Quantitatis Dierum" which gives the length of daylight for each degree of latitude between 36° and 55°. Here we find[3]

for $\phi = 47°$ $M = 15;42^h$
for $\phi = 49°$ $M = 16;0^h$.

Again Paris is clearly preferable to Bourges.

The bulk of the table for the "Quantitas Dierum" shows the effects of sloppiness common in mediaeval tables, in each copy increased by some new scribal errors.[4] At several occasions whole groups of numbers are out of place; for example the same value 9;45^h is given for January 31 and February 1 although the daily increment in that region is 3 minutes. Consequently all numbers until February 8 are 3 minutes too low. At that point the error was detected and the numbers change abruptly from 10;9^h to 10;15^h (which is correct for February 9). The whole correct sequence in inverse order is found between October 16 (10;15^h) and October 25 (9;45^h).

Errors of this type abound throughout the calendar tables. In order to reconstruct the correct table from which our calendar was ultimately derived one can make use of the symmetry which must exist in schemes of this type: the same sequence of numbers between winter solstice and summer solstice should appear in inverse order for the other half of the year. Inspection of the actual numbers shows furthermore for long stretches the use of constant increments, e.g., 4, 3, or 2 minutes for many days. Using this experience in combination with the principle of symmetry one can reconstruct the whole original table for the length of daylight. The result is shown in Fig. 1; the numbers written beside the graph show the constant increment (in minutes) valid in the corresponding section.

How a table of this kind had been computed is in principle well known. Again following ancient and medieval (Islamic and Byzantine) tradition astronomical tables usually contain a table of oblique ascensions for given geographical latitudes. The length of daylight for a given solar longitude is then the total of the oblique ascensions for a semicircle of the ecliptic from the position of the sun to its diametrically opposite point.[5] Since one can compute with the traditional methods the longitude of the sun for every day of the year one can also tabulate the length of daylight day by day.

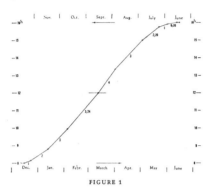

FIGURE 1

It is clear, however, that the table in our text was based only on a few accurately computed values whereas linear interpolation was used otherwise. Most of the constant differences had been chosen as convenient round numbers, a procedure which implies some doctoring of the few originally computed numbers. Where nevertheless fractions of minutes could not be avoided for the increments the results were given by truncation of the seconds.

It is of course impossible to say how far back this process of tabulation reaches. Most likely the table in the calendar of the *Très Riches Heures* was simply copied from some contemporary set of astronomical tables, adjusted for the latitude of Paris.[6]

In contrast to the astronomically meaningful scheme that describes in the *Très Riches Heures* the variation of the length of daylight one should remember that the *Petites Heures* as well as the *Belles Heures* display a much more primitive and utterly useless pattern. There we

find at the beginning of each calendar table a statement "la nuit a . . . heures le iour . . . " The numbers (always totalling 24) vary with the constant difference of 2 hours from month to month, the shortest daylight of 6 hours listed for December, the longest of 18 hours for June. Not only does the abrupt change at the solstices from increase to decrease and vice versa contradict all direct experiences but the ratio 3:1 for the extrema would only be correct at a geographical latitude of about 58° (almost the latitude of Stockholm). Linear schemes for the length of daylight go back to the early hellenistic period when Babylonian arithmetical methods made their influence felt on nascent Greek astronomy. At the fringes of the ancient world these primitive arithmetical methods survived for many centuries, often step by step modified into totally meaningless rules of procedure. Mediaeval astronomy, in India, in Ethiopia, and in western Europe provides us with many examples of this process.

2. Solar Motion

The table for the length of daylight provides us also with the dates assumed for the solstices and equinoxes. The entries for 12ʰ are found at March 12 and September 15. The shortest daylight is listed for the five days from December 11 to 15, the longest for June 13 to 16; hence we can accept December 13 and June 14/15 for the solstices.

The calendar tables, however, are not the only source for this information. The calendar miniatures which face the tables contain in the crowning semicircular field a host of astronomical information. The two outermost circles give degrees and zodiacal signs of the solar travel during the month in question. The innermost circle gives the consecutive days of the month; a radial line which separates the pictures of the zodiacal signs allows us to read off the date at which the sun transgresses from one sign into the next. In this way we find once more the dates for the cardinal points: entry into Aries and Libra on March 12 and September 15 respectively and December 12/13 and June 13/14 for the solstices in reasonably good agreement with the data derived from the calendar.[7]

Looking back to our graph for the length of daylight (Fig. 1, p. 422) we observe that the curve is not symmetric to the equinoxes. The reason for this phenomenon lies in the fact that our table uses days as independent variable and not solar longitudes. For equidistant solar longitudes the curve for the corresponding

MONTH	BOUNDARY		TEXT	DEGREE	
	BETWEEN	DAY		FIRST	LAST
I	♑ ♒	11 | 12	————	♑ 20	♒ 20
II	♒ ♓	10	Finis. graduum|aquarii|Initium|piscium|gradus.|19.	♒ 22	♓ 19
III	♓ ♈	12	finis.|. graduum.|piscium.|. Initium.|arietis.|. gradus.|.20.	♓ 20	♈ 19
IV	♈ ♉	12	————	♈ 20	♉ 18
V	♉ ♊	13		♉ 19	♊ 17
VI	♊ ♋	13 | 14	finis.|graduum.|.cancri.| Inicium.|.Leonis.|gradus.|.16.	♊ 19	♋ 16
VII	♋ ♌	15	Finis|graduum|Cancri.|Inicium|Leonis.|gradus. 16.	♋ 17	♌ 16
VIII	♌ ♍	15 | 16	————	♌ 17	♍ 16
IX	♍ ♎	15 | 16	Finis|graduum.|Libre|Inicium.|Scorpionis.|gradus. 15.	♍ 17	♎ 15
X	♎ ♏	15	Finis|graduum.|Libre|Initium.|Scorpionis.|gradus.|16	♎ 16	♏ 17
XI	♏ ♐	14	Finis|graduum|Scorpionis.|Inicium.|sagittarii.|gradus.|17.	♏ 17	♐ 17
XII	♐ ♑	12 | 13	Finis.|graduum.|sagittarii.|Initium.|capricorni.|gradus.|18.	♐ 18	♑ 19

TABLE I

NOTES TO TABLE I

I. ♑ 20: not a full space for 20; ♒ 20: should be 21; cf. next month

II. ♒ 22: preceded by a little blank space

III. gradus 20: should be gradus 19; ♓ 20: preceded by a little blank space; ♈ 19: followed by a little blank space

IV. ♉ 18: followed by a little blank space

V. ♊ 17: followed by about half a space, left blank; cf. next month

VI. ♊ 19: preceded by about half a space, left blank; one could assume that ♊ 18 was divided between the months V and VI. The text "finis graduum cancri, inicium leonis" belongs to the next month; the pictures give correctly Gemini and Cancer

VIII. ♍ 16: not quite a full space

IX. ♍ 17: preceded by one space left blank; nothing missing. ♎ 15: followed by a little space left blank

X. ♎ 16: space for 16 left blank. ♏ 17: 17 written into the last small space; actually 16 should be the final degree; cf. the text and the next month

XI. ♐ 17: followed by a little blank space

XII. ♐ 18: space for 18 left blank. The text "capricorni gradus 18" should be emended to gradus 19; cf. last degree in month XII and first degree in month I.

lengths of daylight would have been symmetric with respect to the equinoxes. The variation in the solar velocity, however, relates different intervals of solar longitudes to equidistant days and thus produces a deviation from the symmetric pattern.[8] Hence, as it had to be expected, the solar anomaly had been taken into account for the computation of the table of the length of daylight.

Again the calendar miniatures provide corroborating information about the anomalistic solar motion. We can determine the variable solar velocity in two ways: first, the dates of the entry into the consecutive signs tells us how many days are required to cover 30°; secondly we obtain from the outermost circle the number of degrees travelled by the sun in the number of days contained in each month.

Table I shows the data which we obtain from the miniatures. If two consecutive days are mentioned for the crossing of a boundary between signs the separating radius separates also the two day numbers of the innermost ring. Otherwise the radius leads (more or less accurately) into the middle of the space for one day. Similarly, in the outermost ring, we are shown the first and last degree for the sun in the given month; the last degree is also mentioned in the text of the second ring. Unfortunately the spaces are not drawn accurately enough that one could operate with fractions of degrees or days. This restriction to integers explains the irregularities in the solar motion deduced from these pictorial presentations.

Table II shows what we can conclude from Table I by using the time intervals Δt between entering of the sun into consecutive signs. Since each such interval corresponds to a travel of 30° in longitude the solar velocity during that sign is given by $30/\Delta t$.

Table III is based on the extremal points reached by the sun in each month. Hence we can determine the longitudinal travel $\Delta\lambda$ and find the velocity during this month by dividing Δλ by the number of days in the corresponding month.

The two results are schematically represented in Fig. 2. The solid line corresponds to Table III, the dotted line (for the sake of clarity set one half space to the right) gives the values from Table II.[9] It is clear that both tables are based on true solar longitudes which lead to a maximum velocity of about $1;2^{o/d}$, a minimum of about $0;57^{o/d}$.[10] No doubt the data in the *Très Riches Heures* were compiled by people who could competently handle the contemporary astronomical tables.

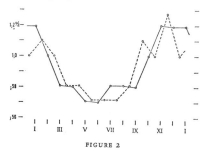

FIGURE 2

3. Golden Number

In the *Très Riches Heures* the "Golden Numbers" occur three times: in two different versions in the calendar tables and once more in the calendar miniatures. In the left column of the calendar tables we find Golden Numbers in the arrangement which is traditional in the late Middle Ages; the right column is headed "numerus aureus nouus" or "nombre dor nouuel." As will be shown presently the miniatures display also the "new" variety.

We begin our discussion with the traditional scheme because in it one can easiest follow the basic idea upon which is built the "perpetual" luni-solar calendar which regulated the dates for the movable feasts of the church.

THE 19-YEAR CYCLE

Let us assume that a new moon[11] falls on December 24, that is to say that we assume that December 25, the first day of the ecclesiastical year, coincides with the first day of a lunar month. Let us furthermore assume that this first lunar month is a full month, i.e., 30 days in length. Then the next conjunction will take place on January 23. A year of this type, i.e. a year for which the

ENTERING		Δt	30/Δt
SIGN	DATE		
≈	I 11/12		
		29;30^d	1;1,1^o/d
♓	II 10		
		30	1;0
♈	III 12		
		31	0;58,4
♉	IV 12		
		31	0;58,4
♊	V 13		
		31;30	0;57,8
♋	VI 13/14		
		31;30	0;57,8
♌	VII 15		
		31;30	0;57,8
♍	VIII 15/16		
		31	0;58,4
♎	IX 15/16		
		29;30	1;1,1
♏	X 15		
		30	1;0
♐	XI 14		
		28;30	1;3,9
♑	XII 12/13		
		30	1;0
≈	I 11/12		

TABLE II

MONTH	Δt	DEGREE		Δλ	Δλ/Δt
		FIRST	LAST		
I	31^d	♑ 20	≈ 21	32°	1;1,56^o/d
II	28	≈ 22	♓ 19	28	1;0
III	31	♓ 20	♈ 19	30	0;58,4
IV	30	♈ 20	♉ 18	29	0;58
V	31	♉ 19	♊ 18?	29;30?	0;57,6
VI	30	♊ 18?	♋ 16	28;30?	0;57
VII	31	♋ 17	♌ 16	30	0;58,4
VIII	31	♌ 17	♍ 16	30	0;58,4
IX	30	♍ 17	♎ 15	29	0;58
X	31	♎ 16	♏ 16	31	1;0
XI	30	♏ 17	♐ 17	31	1;2
XII	31	♐ 18	♑ 19	32	1;1,56

TABLE III

1

2 2/13	I 1 3	31 3	2	IV 1	V 1 11	31 11	30	30 19	29 19	28	28 8	27 8	27	26
3	2	II 1	3 11	2 11	2	VI 1	VII 1 19	31	30 8	29 8	29	28	28 16	27 17
4 10	3 11	2 11	4	3	3 19	2 19	2 8	VIII 1 8	31	30	30 16	29 16	29 5	28 6
5	4	3 19	5 19	4 19	4 8	3 8	3	2 16	IX 1 16	X 1 16	31 5	30 5	30	29
6 18	5 19	4 8	6 8	5 8	5	4 16	4 16	3 5	2 5	XI 1	XII 1	31 13		30 14
7 7	6 8	5	7	6 16	6 16	5 5	5 5	4	3	3 13	2 13	2 13/2	I 1 3	31 3
8	7	6 16	8 16	7 5	7 5	6	6	5 13	4 13	4 2	3 2	3	2	
9 15	8 16	7 5	9 5	8	8	7 13	7 13	6 2	5 2	5	4	4 10	3 11	
10 4	9 5	8	10	9 13	9 13	8 2	8 2	7	6	6 10	5 10	5	4	
11	10	9 13	11 13	10 2	10 2	9	9	8 10	7 10	7	6	6 18	5 19	
12 12	11 13	10 2	12 2	11	11	10 10	10 10	9	8	8 18	7 18	7 7	6 8	
13 1	12 2	11	13	12 10	12 10	11	11	10 18	9 18	9 7	8 7	8	7	
14	13	12 10	14 10	13	13	12 18	12 18	11 7	10 7	10	9	9 15	8 16	
15 9	14 10	13	15	14 18	14 18	13 7	13 7	12	11	11 15	10 15	10 4	9 5	
16	15	14 18	16 18	15 7	15 7	14	14	13 15	12 15	12 4	11 4	11	10	
17 17	16 18	15 7	17 7	16	16	15 15	15 15	14 4	13 4	13	12	12 12	11 13	
18 6	17 7	16	18	17 15	17 15	16 4	16 4	15	14	14 12	13 12	13 1	12 2	
19	18	17 15	19 15	18 4	18 4	17	17	16 12	15 12	15 1	14 1	14	13	
20 14	19 15	18 4	20 4	19	19	18 12	18 12	17 1	16 1	16	15	15 9	14 10	
21 3	20 4	19	21	20 12	20 12	19 1	19 1	18	17	17 9	16 9	16	15	
22	21	20 12	22 12	21 1	21 1	20	20	19 9	18 9	18	17	17 17	16 18	
23 11	22 12	21 1	23 1	22	22	21 9	21 9	20	19	19 17	18 17	18 6	17 7	
24 19	23 1	22	24	23 9	23 9	22	22	21 17	20 17	20 6	19 6	19	18	
25	24	23 9	25 9	24	24	23 17	23 17	22 6	21 6	21	20	20 14	19 15	
26 8	25 9	24	26	25 17	25 17	24 6	24 6	23	22	22 14	21 14	21 3	20 4	
27	26	25 17	27 17	26 6	26 6	25	25	24 14	23 14	23 3	22 3	22	21	
28 16	27 17	26 6	28 6	27	27	26 14	26 14	25 3	24 3	24	23	23 11	22 12	
29 5	28 6	27	29	28 14	28 14	27 3	27 3	26	25	25 11	24 11	24 19	23 1	
30	29	28 14	30 14	29 3	29 3	28	28	27 11	26 11	26	25	25	24	
31 13	30 14	III 1 3	31 3	30	30	29 11	29 11	28	27 19	27 19	26	26 8	25 9	

TABLE IV

"January Lunation" extends from the preceding December 24 to January 23, is given the Golden Number 1 thus being made the first year in a cycle of 19.

Continuing in this first year we make the second lunar month "hollow," i.e. 29 days long, and thus continue with alternating full and hollow months since we know that consecutive conjunctions are in the mean about $29\frac{1}{2}$ days apart. Our Table IV shows the new moon dates obtained in this way for a year with the golden number 1:

Jan. 23	May 21	Sept. 16
Febr. 21	June 19	Oct. 15
March 23	July 19	Nov. 14
Apr. 21	Aug. 17	Dec. 13

The arrangement of Table IV in columns of always 30 days length, regardless of the number of days in a julian month, makes it easy to distinguish between full and hollow months: if the golden number remains in the same line the lunation is full, if it moves one line up, it is hollow.

After the last lunation of the first year which is hollow and ends on December 13 we place again a full lunation which ends on January 12 of the second year. Thus we find the Golden Number 2 associated with January 12 and then, operating as before with alternating full and hollow months, with February 10, March 12, etc. This procedure cannot be continued indefinitely since the length of the mean lunation is a little more than $29\frac{1}{2}$ days. We therefore find in 19 years 7 cases in which a full month is inserted after a full month (cf. Fig. 3). This is done in all those years in which 13 conjunctions take place, i.e. in the years No. 3, 5, 8, 11, 13, 16, and 19 of the cycle. In the last year, however, we find not only a duplication of a full month but also twice a pair of hollow months.[12] This brings us back from No. 19 Dec. 24 to No. 1 Jan. 23 of the next 19-year cycle.

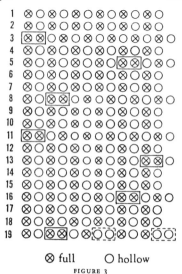

1	⊗○⊗○⊗○⊗○⊗○⊗○
2	⊗○⊗○⊗○⊗○⊗○⊗○
3	⊗⊗○⊗○⊗○⊗○⊗○⊗○
4	⊗○⊗○⊗○⊗○⊗○⊗○
5	⊗○⊗○⊗○⊗○⊗⊗○⊗○
6	⊗○⊗○⊗○⊗○⊗○⊗○
7	⊗○⊗○⊗○⊗○⊗○⊗○
8	⊗○⊗⊗○⊗○⊗○⊗○⊗○
9	⊗○⊗○⊗○⊗○⊗○⊗○
10	⊗○⊗○⊗○⊗○⊗○⊗○
11	⊗⊗○⊗○⊗○⊗○⊗○⊗○
12	⊗○⊗○⊗○⊗○⊗○⊗○
13	⊗○⊗○⊗○⊗○⊗⊗○
14	⊗○⊗○⊗○⊗○⊗○⊗○
15	⊗○⊗○⊗○⊗○⊗○⊗○
16	⊗○⊗○⊗○⊗○⊗⊗○⊗○
17	⊗○⊗○⊗○⊗○⊗○⊗○
18	⊗○⊗○⊗○⊗○⊗○⊗○
19	⊗○⊗⊗○⊗○○⊗○⊗○○

⊗ full   ○ hollow

FIGURE 3

Fig. 3 allows us to rapidly determine the numerical basis of this cycle. The number of lunations in the whole cycle is obviously

$$19 \cdot 12 + 7 = 235 \text{ months.}$$

The number of days can be found as follows: each pair of full plus hollow months represents 59 days and we have 6 of such pairs in each year, hence

$$19 \cdot 6 \cdot 59 = 6726 \text{ days.}$$

To this we must add seven full months, i.e., 210 days. In year 19, however, one full month was replaced by a hollow one (cf. Fig. 3); hence we have to subtract one day. This gives the total

$$6726 + 210 - 1 = 6935 \text{ days.}$$

Since $6935 = 19 \cdot 365$ we can say that the cycle as represented in Table IV would exactly fit the length of 19 Egyptian years. Our calendars assume, of course, Julian years which assign every fourth year 29 days to February. By simply deciding that the golden numbers remain the same also for Julian leap years we add implicitly 19/4 days to the above total for a cycle of 19 Julian years. One can express this rule also in the form that four cycles, i.e. 76 Julian years, contain

$$4 \cdot 6935 + 19 = 27759 \text{ days.}$$

We shall come back presently to this 76-year cycle (p. 429).

VARIANTS

There are more variants among the five calendars of the Duke than one should *a priori* expect. As far as simple scribal errors are concerned the two closely related calendars of the Petites and the Grandes Heures are by far the worst[13] whereas the Très Belles Heures de Notre Dame and the Belles Heures both show only few scribal errors[14] and agree very closely with the Très Riches Heures.

There exist, however, also variants which are not evidently copyist mistakes. The Belles Heures and the Très Riches Heures agree in assigning to December 2 two cycle numbers, 2 as well as 13. The Petites and the Grandes Heures change this to 13 for Dec. 1 and 2 for December 2; the Heures of Notre Dame assign 13 to December 2 and 2 to December 3. These changes influence the sequence of full and hollow months as shown in Fig. 3. The Heures of Notre Dame would make the last month in line 2 full and therefore the first in line 3 hollow. Similarly the last pair in line 13 interchanges full and hollow in the arrangement of the Petites and the Grandes Heures. I do not know what motivated these changes.

More variants exist for the cycle year 19. In this case we know that the mediaeval computists were puzzled by the above-mentioned necessity (p. 426) to subtract 1 day from the 235 lunar months which constitute the 19-year cycle. A simple solution consists, of course, in making one of the seven intercalary months hollow. Where such a reduction, the "saltus lunae," should be localized within the cycle is a question on which the mediaeval calendar makers did not reach a final agreement. Consequently we find even among our five calendars variants in the placing of the golden number 19, a fact which also increased the chance for errors, again exemplified in our material.

Obvious errors are found in the Belles Heures where 19 is associated with August 28—September 28—October 26 and in the Grandes Heures which give 19 to August 28—September 28—October 27. In the first instance a lunar month of 31 days would be followed by 28 days, in the second case at least the second interval of 29 days would be permissible. A simple emendation of the first case would be the replacement of September 28 by September 27, the date found in the four remaining calendars.[15] The second case can also be repaired by the same correction. Finally a plausible correction can be made in the Petites Heures where all entries from February 4 to 13 are one day too late in comparison with the remaining calendars. If we therefore move the golden number 19 from February 4 to 3 we obtain the customary beginning: full—hollow (instead of full—full).

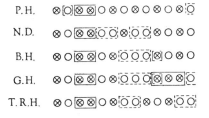

19

P.H.

N.D.

B.H.

G.H.

T.R.H.

G.

FIGURE 4

Fig. 4 illustrates the different ways in which our five calendars distribute the last golden number of the cycle. Not two of these patterns are identical. Only one of the two patterns (G) which Ginzel considers to be standard is represented here (Belles Heures).

APPLICATIONS

What we have obtained so far is a pattern for a sequence of 235 full and hollow months which repeats itself every 19 Julian years. Each year in the cycle is assigned a number between 1 and 19, the "Golden Number," which characterizes the position of the year in the cycle. Because of the strict periodicity of this pattern one speaks of an "eternal" Julian calendar which furnishes the dates of all new moons: if one knows, e.g., that the current year has the golden number 5 we see from Table IV that January 9 should be a new moon, followed by February 7, March 9, etc. In the next year January 28, February 26, etc. should be new moons.

To apply this scheme one has to know, however, the Golden Number (g) of the year (N) in question. One of the several equivalent rules for the determination of g is the following one: g is the residue of the division of N—531 by 19.[16]

According to this rule A.D. 532 has the Golden Number $g = 1$; consequently January 23 should be a new moon—as is indeed the case.[17] This year is the first year of the Easter Tables of Dionysius Exiguus; hence the counting of A.D. 532 as a year 1 of the cycle.

According to our rule also the year A.D. 1387 receives the Golden Number 1 because $1387 - 531 = 856 = 45 \cdot 19 + 1$. Checking again the astronomical facts one

finds, however, that not January 23 but January 20 was a new moon. This is not surprising because 19 Julian years are only approximately the equivalent of 19 mean synodic months while the cyclic determination of the Golden Numbers assumes exact equality. In spite of the slowly accumulating error the liturgical calendar remained based on the strict application of the cyclic computation long after it had become clear that the cyclically computed new moons would more and more deviate from the observable facts.

NOMBRE D'OR NOUVEL

The last column of the calendar pages of the *Très Riches Heures* gives a set of numbers under the heading "numerus nouus" or "nombre dor nouuel." The numbers, shown in our Table V, form again a pattern with a 19-year period.

The calendar miniatures give under the title "primaciones lune" in the innermost ring a sequence of letters which can be easily transposed into numbers by their alphabetical order:

a b c d e f g h i k l m n o p q r s t
1 2 3 4 5 6 7 8 9 10 11 12 13 14 15 16 17 18 19

Wherever the calendar miniatures are completed[18] the resulting numbers agree exactly with the nombres d'or nouvelles in the tables.

In the same way as before one can derive from Table

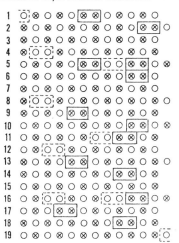

FIGURE 5

XII 1 10														
2	I 1 19	31 19	2 8	IV 1 8	V 1	31 16	30 16	30	29 5	28 5	28 5	27	27 13	26 14
3 18	2	II 1 8	3	2 16	2 16	VI 1	VII 1	31 5	30	29	29 13	28 13	28	27
4 7	3 8	2	4 16	3	3 5	2 5	2 5	VIII 1	31 13	30 13	30	29	29 2	28 3
5	4 16	3 16	5 5	4 5	4	3	3 13	2 13	IX 1	X 1 2	31 2	30 2	30	29 11
6 15	5 5	4 5	6	5 13	5 13	4 13	4	3 2	2 2	2	XI 1	XII 1 10	31 10	30
7 4	6	5 13	7 13	6	6	5 2	5 2	4	3 10	3 10	2 10	2	I 1 19	31 19
8 12	7 13	6	8	7 2	7 2	6	6 10	5 10	4	4	3 18	3 18	2	
9	8	7 2	9 2	8	8 10	7 10	7	6 18	5 18	5 18	4	4 7	3 8	
10 1	9 2	8	10 10	9 10	9	8 18	8 18	7	6	6 7	5 7	5	4 16	
11	10 10	9 10	11	10 18	10 18	9	9	8 7	7 7	7	6 15	6 15	5 5	
12 9	11	10 18	12 18	11	11	10 7	10 7	9	8 15	8 15	7 4	7 4	6	
13 17	12 18	11	13	12 7	12 7	11	11 15	10 15	9 4	9 4	8	8 12	7 13	
14	13	12 7	14 7	13	13 15	12 15	12 4	11 4	10	10 12	9 12	9	8	
15 6	14 7	13	15 15	14 15	14 4	13 4	13	12 12	11 12	11	10	10 1	9 2	
16	15 15	14 15	16 4	15 4	15	14 12	14 12	13	12	12 1	11 1	11	10 10	
17 14	16	15 4	17	16 12	16 12	15	15	14 1	13 1	13	12 9	12 9	11	
18 3	17 4	16	18 12	17	17 1	16 1	16 1	15	14 9	14 9	13	13 17	12 18	
19	18 12	17 12	19 1	18 1	18	17	17 9	16 9	15	15 17	14 17	14	13	
20 11	19 1	18 1	20	19 9	19 9	18 9	18	17 17	16 17	16	15	15 6	14 7	
21 19	20	19 9	21 9	20	20	19 17	19 17	18	17 6	17 6	16 6	16	15 15	
22 8	21 9	20	22 17	21 17	21 17	20	20 6	19 6	18	18	17 14	17 14	16	
23	22 17	21 17	23	22	22 6	21 6	21	20 14	19 14	19 14	18	18 3	17 4	
24 16	23	22	24 6	23 6	23	22 14	22 14	21	20	20 3	19 3	19	18 12	
25	24 6	23 6	25	24 14	24 14	23	23	22 3	21 3	21	20 11	20 11	19 1	
26 5	25	24 14	26 14	25	25	24 3	24 3	23 11	22 11	22 11	21	21 19	20	
27 13	26 14	25	27	26 3	26 3	25	25 11	24	23	23 19	22 19	22 8	21 9	
28	27	26 3	28 3	27	27 11	26 11	26	25 19	24 19	24 8	23 8	23	22 17	
29 2	28 3	27	29 11	28 11	28	27 19	27 19	26 8	25 8	25	24 16	24 16	23	
30	29 11	28 11	30	29 19	29 19	28 8	28 8	27 16	26 16	26 16	25	25	24 6	
31 10	30	III 1 19	31 19	30 8	30 8	29	29 16	28	27	27	26 5	26 5	25	

TABLE V

V the sequence adopted for full and hollow months (cf. Fig. 5). For the number of days in the cycle one finds exactly as before the total of 19 · 354 + 209 = 6935 = 19 · 365 days to which 19/4 days may be added by using the same golden numbers for all Julian years, ordinary and intercalary.

The reason for introducing "new" golden numbers becomes clear when one remembers that the ecclesiastic scheme assumes January 23 as the new moon's date for a cycle year 1 whereas the new moon in A.D. 1387 (which is a year 1 in the Dionysian cycle) occurred actually on Jan. 20.[19] Table V shows that the "new" golden number 1 was given to Jan. 19. Since the moment of the conjunction had to be found by computation an error of a few hours is not surprising. Hence it is clear that the new golden numbers were introduced in order to restore better agreement with observable facts.

Rather senseless, however, is the change in the distribution of hollow and full months (cf. Fig. 5). It lies in the very nature of any perpetual scheme for lunations that one must operate with mean synodic months, i.e. for the main part with simple alternations between full and hollow months as shown in Fig. 3 (p. 426). Hence one could have simply copied the pattern of Fig. 3 with the only difference that it should start with 1 at January 19 instead of at January 23.

Our reformer introduced, however, many pairs of full and of hollow months which in most cases cancel each other out (cf., e.g., in Fig. 5 the years 5, 11, 16). There exist, of course, in reality sequences of two hollow or two full months but no 19-year cycle can reproduce periodically such sequences of true lunations by the simple reason that the variation in the lunar velocity has not a period of 19 years. The complicated pattern of the nombres d'or nouvelles is therefore a valueless and arbitrary modification of the canonical scheme.

HISTORICAL REMARKS

The nombres d'or nouvelles of the *Très Riches Heures* do not represent the only available evidence for attempts to restore to the 19-year cycle agreement with the actual course of the moon, though none of the other Books of Hours of the Duke of Berry contains more than the traditional golden numbers in its calendar.

Best known is the reform suggested by Grosseteste, Bishop of Lincoln.[20] In order to distinguish his golden numbers from the traditional ones he introduced letters in alphabetical order in the same fashion as we have seen them used in the calendar miniatures of the Très Riches Heures. Nevertheless the nombres d'or nouvelles do not follow the system of Grosseteste which is based on a 76-year cycle, not on a simple 19-year cycle as in the Très Riches Heures. As we have seen (above p. 426), the 19-year cycle implicitly assumes that 235 lunar months contain $6935 + 19/4 = 6939\frac{3}{4}$ days. In order to avoid fractions of days Grosseteste operated with a cycle of $4 \cdot 19 = 76$ years $= 27759$ days and correspondingly 940 lunations[21]—repeating the shift from the Metonic to the Callippic cycle suggested in the fourth century B.C.

Grosseteste furthermore refined his system by determining the dates of the conjunctions not by simple alternation between full and hollow months (and occasional additions of one day) but by operating with hours. This increase in numerical accuracy requires a four column table of golden numbers, one column for each year in a julian quadruple of years. The adjustment to the real new moons seems to consist in assigning the number 1 to September 12 of 1208, indeed the date of a new moon. Our Table V shows that the nombres d'or nouvelles of the Très Riches Heures associate 1 with September 13, this time a little later than in reality.[22] But it is again clear that our text does not follow accurately the Lincolnian scheme. It is interesting to see that attempts to correct the position of the golden numbers within the traditional 19-year cycle begin so soon after the invention of this device around 1200.[23]

A few remarks may be added on the 19-year cycle itself on which the concept of the golden number ultimately rests. The fact that 235 lunar months correspond closely to 19 solar years was established by Babylonian astronomers in the fifth century B.C. Hence a definite system of intercalations was adopted in which 7 times in 19 years a 13th lunar month was added to the ordinary 12, thus bringing the beginning of the year (Nisan)

again near to the vernal equinox. This rule of intercalation prevailed in Mesopotamia not only during the late Persian period but also under Seleucid and Parthian rule. Naturally this calendaric system had also its influence on the date of the Jewish Passah Feast and then on its Christian version.

The Babylonian astronomers of the last four centuries B.C. were fully aware of the approximate character of the 19-year cycle; for the determination of the true lunations they used very refined, and of course non-cyclic, methods. Only for the civil calendar the 19-year cycle remained a very convenient arrangement. Greek mathematical astronomy from Hipparchus in the second century B.C. to Ptolemy in the second century A.D., and to Pappus and Theon in the fourth, completely ignored the problems of the lunar calendar. The Fathers of the Church, however, in their desire to maintain the lunar definition of Easter without copying exactly the Jewish rules argued violently, and without real understanding of the underlying astronomical facts, about the theological implications of various rules of thumb until one accepted eventually the cycles proposed by Dionysius exiguus and other Computists of the early Middle Ages. The fact that the spacing of the seven intercalary years in the scheme for the golden numbers in the early 15th century is still exactly the same as in the Mesopotamian calendar is one of several cases which testify to the influence Babylonian astronomy had on later ages.

The 19-year cycle and its Golden Numbers are not the only residue of reckoning by lunar months in our calendars. Each one of the five calendars begins each month with a sentence of the type "Feurier a 28 jours Et la lune 29." The first number is always the number of days corresponding to the Julian calendar. The second, however, obviously concerns lunar months, either full or hollow.

Astronomically it is utterly meaningless to associate a lunar month with each Julian month since any such direct correlation would rapidly become invalid since 12 lunar months add up only to about 354 days. If one should, however, give for some antiquarian reason a sequence of lunar months the only reasonable arrangement would be alternation between full and hollow months. Fig. 6 shows that traces of such alternating lengths are visible in all calendars but most of the time the scribes gave simply arbitrary numbers with the effect that no two calendars agree and that even the total of 354 days is nowhere obtained. Obviously there is no calendaric topic that could make use of these "lunar months."

P.H. N.D. B.H. G.H. T.R.H.

	P.H.	N.D.	B.H.	G.H.	T.R.H.
I	⊗	⊗	○	⊗	⊗
II	○	○	⊗	○	○
III	⊗	⊗	⊗	⊗	⊗
IV	○	⊗	○	○	○
V	⊗	⊗	⊗	⊗	⊗
VI	○	⊗	○	○	○
VII	○	⊗	○	⊗	○
VIII	○	⊗	⊗	⊗	⊗
IX	○	⊗	⊗	⊗	⊗
X	⊗	⊗	⊗	⊗	⊗
XI	○	⊗	⊗	⊗	⊗
XII	○	⊗	○	⊗	⊗
total	352	359	355	357	356 days

FIGURE 6

	N.D.	B.H.	G.H.	T.R.H.
I	1, 25	1	—	—
II	4 VIII hora, 26	5, 26	—	—
III	1, 28	1, 28	—	1, 28
IV	10 XX h., 20 XI h.	10, 20	—	10, 19
V	3	3, 25	—	—
VI	—	10, 16	—	—
VII	13, 22	13	—	13, 22
VIII	1, 30	1, 30	—	—
IX	3, 21	3, 21	3, 21	—
X	3	4, 22	—	—
XI	5, 28	5, 28	—	—
XII	7	7, 22	7	—

TABLE VI

4. Dies Aegyptiaci

With a few days we find in the calendar tables the letter D after the saint's name, indicating a "dies aegyptiacus,"[24] i.e., a day unfavorable for actions and, e.g., for bleeding.[25] Concept and terminology go back to late Roman antiquity[26] but the principle of selection is not clear. The fact that one usually has 25 Egyptian days in a year, two each month with one third one in January, looks like some lunar scheme of new and full moons (and epagomenal days?) but no clear pattern can be deduced even from the earliest extant sources.

The "dies aegyptiaci" were noted only for three months in the Très Riches Heures: March 1 and 28, April 10 and 19, July 13 and 22. These are precisely the months which use Latin terminology: "quantitas dierum," "numerus aureus," ". . habet dies . . . , luna habet dies . . ." Obviously the Latin copy had D-s, the French not.

By the end of the Middle Ages there existed already many variants for the dates of the Egyptian days. As far as one can tell from the published sources the Belles Heures agree with the Très Riches Heures, though all headings are in French.

This pattern, however, is not valid for the other calendars. The Belles Heures are all in French, the Heures of Notre Dame all in Latin and both calendars give a rather complete list of Dies Aegyptiaci. On the other hand the Petites Heures and the Grandes Heures are both in French, the calendar of the first has no D's, the second only for September and December. Table VI gives a concordance of the dates which show a D, obviously based on a common pattern if one ignores a few omissions and copyist errors.

An unusual feature is found in the Heures of Notre Dame. For February 4 and April 10 and 20 hours are added to the D-dates. These hours must be equinoctial hours since April 10 gives "XX hora." The epoch of these hours is probably midnight since VIII, XI, and XX after midnight seem to be more reasonable than hours after noon which would bring XI deep into the night.

5. Melothesia

The last calendar table, for December, occupies the first page of the third signature, fol. 13ʳ. After two blank pages follows on fol. 14ᵛ a beautiful miniature of a "Zodiac Man." The remaining pages of the quire, fol. 15ʳ to 16ᵛ, are again blank; with fol. 17ʳ begins the main part of the Book of Hours.[27]

The Zodiac Man in the center of the miniature on fol. 14ᵛ presents the well known pattern of medieval iatromathematical diagrams which illustrate the association between the signs of the zodiac and the parts of the body.[28] The second figure, however, seen from the back, raises many problems. Harry Bober in his excellent study of this miniature[29] thinks of another figure from the iatromathematical literature, the "Vein Man." At least to my knowledge this is the only interpretation proposed so far which connects also this part of the miniature with otherwise attested iconographical features. What speaks against it is the absence of a text which describes the location of the veins.

From the astronomical or astrological point of view the figure of the Zodiac Man poses no problems. Its origin lies in the theory of "Melothesia" of Hellenistic-

Roman astrology,[30] handed down, practically unaltered, through the Middle Ages. Another conventional astrological element are the short textual sections in the four corners of the present miniature, reading as follows:

Aries. leo. sagittarius. sunt calida et sicca collerica masculina. Orientalia.
Taurus. uirgo. capricornus. sunt frigida et sicca melancolica feminina. Occidentalia.
Gemini. aquarius. libra. sunt calida et humida masculina sanguinea. meridionalia.
Cancer. scorpius. pisces. sunt frigida et humida flemmatica feminina. Septentrionalia.

These are traditional descriptions of the qualities of the triplicities, again going back to classical astrology.[31] It is not clear, however, why these data were given here with the Zodiac Man. According to astrological lore the trigona are of vital importance in connection with planetary positions. This could indicate that also a planetary melothesia was contemplated to be represented in a second pair of figures but this time showing the planets on the body of the man facing the reader.[32]

The zodiac which in the shape of a mandorla surrounds the Zodiac Man has an outermost ring divided into 12 signs of 30 degrees each. The signs themselves are arranged in counterclockwise order, Aries to Virgo on the left side, Libra to Pisces on the right. An inner ring shows the days of the twelve months, each full space representing two days, with half spaces at the end of the months of 31 days. In spite of this apparent accuracy of the divisions the relation between ecliptic degrees and days is purely schematic: the boundaries between signs are always lined up with the midpoints of the months. This implies that equinoxes and solstices are simply taken to fall on the 15th of their respective months, in contrast to the astronomical accuracy underlying the calendar miniatures (cf. above p. 423).

The zodiacal figures here and in the calendar miniatures show some peculiarities worth notice, though they are not without parallels in other medieval sources.[33] For example, the Gemini are represented as man and woman, thus unrelated to the Dioscuri; Virgo is holding two ears of corn though only one would be needed for Spica. Capricorn has gone a long way from its Mesopotamian prototype, the "Goat-Fish." The twisted tail of late ancient and most medieval pictures has become a snail's house (twice in the Melothesia and in the December miniature) or a conch from which the goat emerges in the January miniature.[34] Aquarius is pouring water from a jar on his shoulder in the January miniature

while he holds the vessel at his hip in the Melothesia and in the February miniature. The lack of a systematic study of the iconography of the zodiacal constellations makes it impossible to say which branch of the medieval tradition inspired the artists who designed these subsidiary elements in the Books of Hours for the Duke of Berry.

6. Summary

Even the most superficial comparison of the five calendars which preface, as it were, the Books of Hours of the Duke of Berry must recognize that the *Très Riches Heures* strive for a definite advance beyond the traditional calendars. Not only is the primitive and totally valueless linear scheme for the length of daylight (in the *Petites* and *Belles Heures*) replaced by an astronomically meaningful table, adjusted for the latitude of Paris, but also to the traditional set of "golden numbers" is added another set of "nombres d'or nouvelles" and this not only in the calendar tables but also prominently displayed on the rings of the calendar miniatures. The tendency to be "modern" is unmistakable.

That modernization need not be a factual progress holds for the Middle Ages as well as for our times. The column for the length of daylight could be handled by any user of the calendar to tell him with reasonable accuracy the time of sunrise or sunset for the day in question. The nombres d'or nouvelles, however, are nothing but a display of empty learning. Even supposing that these new golden numbers are in better agreement with the actual lunations (which is true only in a very limited sense) they serve no purpose. The only place where the schematic lunar calendar, inherited from antiquity, is of any interest is the liturgical calendar; and here no attention was paid to the corrections suggested by the new golden numbers.

In fact the inclusion of traditional as well as of new golden numbers in a calendar for a layman's use could hardly ever have been of practical value. Even assuming the knowledge of the position of the current year within the 19-year cycle one has very little gain from then knowing in principle the dates of the new- or full-moons. One may rightly doubt that a non-professional user ever could determine from it the dates of Easter and other movable festivals. And of course the days of the lunar months (cf. Fig. 6, p. 430) are without relation to any real lunar phase.

The assembly of unrelated and often meaningless material is in no way an otherwise unknown feature of mediaeval astronomical tables. Tables which display trigonometric functions computed with high accuracy (of Islamic origin) and for small increments of the argument are perfectly able to give also one of the primitive Indian "Kardaga" tables which proceed in steps of 15°. Huge solar, lunar, and planetary tables can be assembled in the same work without realizing that contradictory elements had been used in different tables. It is not before the Renaissance (Peuerbach, Regiomontanus, Copernicus) that one felt the need to understand the astronomical reason and the empirical basis for technical tables of any kind. For many centuries it was enough in the West (and often also in Muslim circles or in Byzantium) to follow the computational rules which would, *deo volente*, produce a numerical answer in some relation to the question proposed.

How little our calendars were meant for serious use is drastically demonstrated by the "dies aegyptiaci." We have seen (p. 430) that the scribe who followed a Latin archetype included these days, omitted in the French version. Table VI makes it abundantly clear that the total disregard in the Petites Heures, the complete listing in the Heures of Notre Dame and the Belles Heures, and the arbitrary scattering of Egyptian Days over the calendars of the *Grandes* and *Très Riches Heures* has absolutely nothing to do with a changing evaluation of these critical days. We may be sure that the Duke was bled not on the consultation of his calendars but under the careful supervision of his physician who had certainly a rule book of his own.

As far as we know the assumption of inherent danger at specific days has no astrological basis but is simply one of the many manifestations of a belief in lucky and unlucky days, existing among all the ancient civilizations. But even if the Egyptian Days had any connections with astrology it would not have been considered improper to have the name of a Saint followed by a warning *D*. Miniatures of zodiacal signs are most common in calendars, associated naturally with the seasons and the labors of the months. To us the melothesia miniature at the end of the calendar in the *Très Riches Heures* seems purest astrological doctrine. To the Middle Ages these relations between parts of the body and solar or planetary positions were probably not much more than to us considerations of "environment" in the widest sense on human nature and health. In every aspect it is well-established tradition, sometimes rational and perhaps more often purely conventional, which dictated the choice of astronomical and calendaric matters which were combined with superb artistic skill in the illuminated calendars of the late Middle Ages.

NOTES TO APPENDIX C
by O. Neugebauer

1. For the dates and arrangement cf. the text of this book and of the *Late XIV Century*.

2. Almagest II 13 (ed. Heiberg, I p. 186 f.)—Notation: a semicolon separates degrees or hours from minutes, or in general, integers from sexagesimal fractions.

3. I quote from a printed edition of the Alfonsine Tables (Venice, 1518) in my possession.

4. E.g. 8;46 instead of 8;47 (Jan. 11) or 50 instead of 40 in the entries for April 27 to 29.

5. Cf., e.g., my Exact Sciences in Antiquity[2] p. 158 f.

6. A table of this type is, e.g., preserved in the set of the "Alfonsine Tables" Bib. Nat. Lat. 7295 A (fol. 135/6) of which my colleague G. J. Toomer has a copy. The numerical details differ, however, for the majority of entries, in spite of the identity of the main parameters.

7. These dates are essentially correct; cf., e.g., Ginzel, Handbuch d. math. u. techn. Chronologie III, p. 114, who gives for A.D. 1400 for the equinoxes March 11 and September 14, December 13 and June 13 for the solstices.

8. Since the apsides of the solar orbit nearly coincide with the solstices no appreciable asymmetry is caused by the solar anomaly for the length of daylight with respect to the solstices; hence the same curve in Fig. 1 represents both halves of the year.

9. It would be easy to correct the obviously wrong data in Table II for the months IX to XII. The error is caused by the difficulties in the relative spacing of the divisions in the outermost and innermost ring.

10. The maximum equation in the Alfonsine tables is 2;10° (p. 59 b of the printed edition, Venice, 1518) and explains the slightly exaggerated variation in the solar velocity.

11. "New Moon" means here and in the following always conjunction, not first visibility as in most of the ancient lunar calendars.

12. We shall presently return to the specific situation prevailing in the last year of the cycle.

13. Examples: In the Petites Heures the entries for February 4 to 13 and for April 7 to 15 are all one line too low. The same holds in the Grandes Heures for April 7 to 13. Both calendars write 14, 3, 11 instead of 15, 4, 12 in September 12 to 15 and in November 10 to 13.

14. Scribal errors in Notre Dame: May 6 and 12; in the Belles Heures: Sept. 24 and Oct. 4.

15. This also agrees with the alternative pattern given in Ginzel, Hdb. III, p. 136.

16. A list of the Golden Numbers for all years from A.D. 300 to 1794 is given in Ginzel, Handb. d. mathem. u. techn. Chronologie vol. III p. 393–405.

17. Cf. H. H. Goldstine, New and Full Moons, 1001 B.C. to A.D. 1651. Amer. Philos. Soc., Memoirs 94 (1973).

18. Left blank in the miniatures for January, April, May, and August. A scribal error occurred for September 8 with f = 6 instead p = 15.

19. Around noon time in western Europe.

20. Cf. Van Wijk, Le nombre d'or, La Haye, 1936, p. 39 ff.

21. This leads to a mean synodic month of about 29.53 days which is quite accurate (sexagesimally: 29;31, 51,3, . . .).

22. Cf. above p. 428: January 19 instead January 20.

23. Neither the date of the invention of the "golden numbers" nor of the terminology is certain; cf. Van Wijk, pp. 29–33.

24. The "Belles Heures" write once (Jan. 1) "D. eg.".

25. Cf., e.g., Thorndike, History of Magic . . . I, pp. 685–688; p. 695 f.

26. There exist many references to "Egyptian Days"; cf. for the standard literature e.g., Degrassi, Fasti et Elogia (Inscriptiones Italiae, vol. 13, 1963), p. 362 f. Furthermore W. E. Van Wijk, Le nombre d'or (1936) pp. 96–98; R. Dozy-Ch. Pellat, Le Calendrier de Cordoue (1961), p. 26 note 2 and p. 34 note 3.

27. Cf. the diagram given by H. Bober, J. of the Warburg and Courtauld Institutes 11 (1948), p. 29.

28. In the *Très Riches Heures*, 1969, this figure is assumed to be female. This not only contradicts the whole iconographic tradition of the Zodiac Man but also the representation of women here and in the calendar miniatures where women are always shown with long hair (cf., e.g., Gemini and Virgo).

29. Cf. note 27.

30. Cf., e.g., papyri from the second century A.D. (P. Mich. 149; PSI 1289) or the commentary of Porphyri (3rd cent.) to the Tetrabiblos (CCAG 5, 4 p. 216 f.). For the Latin literature cf. e.g., J. de Vreese, Petron 39 und die Astrologie (Paris, 1927) pp. 198–202.

31. Cf. Bouché-Leclerq, L'astrologie grecque (Paris,

1899) p. 154, p. 169; Boll-Bezold-Gundel, Stern-glaube und Sterndeutung[4] (Leipzig, 1931) p. 54; Ptolemy, Tetrabiblos I, 18 (Loeb Classical Library, 1940). I do not know, however, a treatise which enumerates the qualities of the triplicities exactly in the order adopted here.

32. Cf., e.g., the two figures shown in Bober's article Pl. 4, c and d.

33. Useful references not only to ancient but also to medieval representations are found in Boll-Gundel's article "Sternbilder . . . bei Griechen und Römern" in Roscher, Lexikon der griechischen und römischen Mythologie VI (1937), col. 867–1072.

34. Also in the Belles Heures for December.

Reprinted from
The Limbourgs and Their Contemporaries, Meiss, 421–481
The Pierpont Morgan Library, 1974

Ṭentyon

O. Neugebauer – Providence, R.I.

In the "Computus" of the Ethiopic church an important role for the determination of the dates of the movable festivals is assigned a number t, called ṭentyon or ṭentêwon. It presents the weekday of the first day of the year (Maskaram 1), using the norm Wednesday = 1, Thursday = 2, etc. until Tuesday = 7. Obviously to given t the weekday of any day in the year is easily found since the Ethiopic calendar is based on a "julian" pattern, inherited from the Alexandrian calendar, with 12 times 30 days plus 5, or 6, intercalary days at the end.

The question which concerns us here is the etymology of the term ṭentyon. Modern Orientalists agree that the ending -yon or -êwon indicates a Greek loan word [1]. Also the native authors of the "ḥasab", the "Computus", felt the need for an explanatory gloss [2], saying "ṭentyon means beginning", in particular beginning of the computus at the creation of sun and moon when reckoning of time was initiated. The basis for this explanation is obvious since ṭent (from waṭana) means "beginning".

To the pious calendar makers this etymology had an added appeal. In order to find t for a given year n in some era one must add to $n-1 + n/4$ the ṭentyon t_0 valid for the beginning of the era. Since the Ethiopic Era of Grace is actually the Era Diocletian in which the first year began (on Thoth 1) with a Friday one has $t_0 = 3$. The computists, however, found that the formula

$$t = n + n/4 + 2$$

can fittingly be interpreted by the situation at creation. Obviously this lent strong support to their etymology "beginning".

When Ethiopic studies were started by Scaliger (in *De emendatione temporum*, 1583) [3] he assumed, of course, a foreign origin of the term ṭentyon and looked for a Greek word of similar calendaric significance. Unfortunately he embarked on a real *tour de force*. Instead of ṭe (Ṭ) he read pe (Ṭ) which is mainly used in foreign words [4] and then emended pentyon to πλινθίον. He had, however, no evidence whatsoever for a calendaric use of πλινθίον but he found a passage in Cedrenus (around A.D. 1100) [5] who equated πλίνθος with *laterculum*; furthermore Isidore of Seville (around A.D. 600) [6] connected *laterculum* with

---

[1] Cf., e.g., Dillmann, *Lex.* 1392.
[2] E.g. in Vat. Aeth. 119 fol. 88ʳ I, 11/12; Berol. Aeth. 84 fol. 3ʳ I,11-14.
[3] I quote from the edition of 1629.
[4] Dillmann, *Gramm.*, 15.
[5] Hist. Comp. P 169/170 (Bekker I, 296, Migne PG 121 col. 336).
[6] Etymol. VI 17, 3f. (Lindsay).

calendaric cycles. "Itaque (says Scaliger) et Computatores Latini veteres cyclum hebdomadicum laterculum vocant". Ever since, ṭentyon has been considered to be derived from πλινθίον [7].

In fact this extremely implausible construction has nothing to support it. Nowhere in Greek astronomy is πλινθίον used as a calendaric term, neither for weekday nor for anything similar. Hence we have to read ṭe, just as the computists do, drop Scaliger's λ, and look for a Greek term for weekday. Ironically Scaliger mentioned the very passage which solves our problem. He says [8] that Paulus Alexandrinus in his "Isagoge", in an example that concerns the year Diocletian 94 (A.D. 378 Feb. 14) gives exactly the above mentioned formula for the computation of the weekday t. Scaliger adds " et vocat ἡμέρας Θεῶν". Actually the text says ταύτας εἶναι τῶν θεῶν λέγομεν [9]. Obviously ṭentêon or ṭentyon is simply a rendering of τῶν θεῶν. The first vowel of the Ethiopic term is presumably due to the folk etymology mentioned above.

In conclusion it may be mentioned that the ἡμέραι τῶν θεῶν are listed, long before Paulus, in a graffito in Pompeii (thus before A.D. 79) [10]. Hence a widespread use of this terminology cannot be doubted. Its connection with the planetary gods of Greek astrology was long forgotten by the scribes who introduced Alexandrian calendaric methods into Ethiopia, centuries later, just as the weekdays in western christianity preserved the pagan terminology.

---

[7] E.g. Dillmann, *Lex.* 1392, referring to Scaliger; Chaine, *Chronol.*, 107 and note (1) there.

[8] P. 688 III.

[9] Ed. Boer p. 40,8, adding a gratuitous (τὰς ἡμέρας). Similarly, p. 40,15, again adding ἡμέρας which is not in the text.

[10] Cf. Colson, *Week*; from *ZNW* 6 (1905) 27. Also in Athanasius' Easter Letters.

Reprinted from
Orientalia
44 (4), 487f (1975)

Ethiopic Easter Computus

by

O<small>TTO</small> N<small>EUGEBAUER</small>

To the Memory of Eduard Schwartz (1858-1940)

1. "Easter" being defined as the first Sunday after the first Full Moon after the vernal equinox would require, if taken astronomically seriously, not only the determination of the length of the tropical year but also a solution of the highly complex problem of predicting the moments of the full moons.

Historically this intricate definition originated from the connection of the Christian feast with the Jewish Passover date, the 14th of Nisan, i.e. with a lunar calendar, that is to say a "year" based on lunar months whose first days are the days of first visibility of the new crescent. As is well known these days were determined in Jerusalem, before the destruction of the Temple, by direct observation. The "full moon" was then schematically defined as the 14th day of the lunar month (in good Babylonian tradition) and the relation of this lunar calendar to the solar year was regulated (again following Babylonian example) by using a 19-year cycle in which twelve "years" were given 12 lunar months, while the remaining seven ("intercalary") years had 13 lunar months. The resulting pattern indeed keeps the beginning of a given month in a fixed neighborhood of the vernal equinox, i.e. it prevents the lunar year from "rotating" with respect to the "solar year". Hence the determination of Passover in Jerusalem had been a simple affair. For the Jews in the Diaspora, however, the situation was quite different. Direct observation of the new crescent from some other locality, e.g. Alexandria or Rome, need not result in identical dates with Jerusalem, nor could one ignore the existence of a civil calendar which regulated the lives of the majority of the population.

One way out of this dilemma would consist in applying the best available astronomical theory of the solar and lunar motion and of the lunar visibility for given geographical conditions to the determination of the evenings of first lunar visibility. One such highly sophisticated attempt is well known to us. In the late 12th century Maimonides discussed this problem on the basis of the Ptolemaic lunar theory (for the accurate determination of the conjunctions of sun and moon) combined with a theory of lunar visibility closely reminiscent of Babylonian methods known from the Seleucid-Parthian period (though not identical in details).

For the early Christian period we have no treatise that would inform us about the theoretical background of the Easter computus. We know much about the "Easter Controversy" between Roman and Alexandrian dates and we know, from the time of Constantine on, of the frequently expressed condition for the Christian feast to avoid any coincidence with the Jewish Passover. But we know pratically nothing about the festival calendar of the Alexandrian Jews during the early centuries of Christianity.

It is at this point that it seems reasonable to look at Ethiopic sources. We know of the existence of large tables displaying dates for Easter and Passover and of numerous related treatises, from a few paragraphs in length to many folios of diverse contents. Since the Ethiopic calendar is identical with the Alexandrian (as established by Augustus) one may well hope to find here also Hellenistic-Roman material preserved in the Ethiopic isolation, as is the case with various religious literature (e.g., the "Book of Enoch" and similar works). Many modern scholars have expressed their opinion that the Alexandrian Easter computus represents the last flowering of "Alexandrian Science", which indeed reached during the early Christian centuries its highest development as far as astronomical science and methodology is concerned. The second century saw the publication of Ptolemy's "Almagest", and around 400 the "Handy Tables", in Theon of Alexandria's version, appeared — two works which constituted the basis of all Arabic and Western astronomy of the Middle Ages. Indeed, it would be worthwhile to establish the connection between the origin of the Christian Easter Computus and the contemporary Alexandrian astronomical techniques, well known to us from excellently preserved sources.

Some fifteen years of increasingly detailed study of Ethiopic sources have led me to a very different evaluation of the situation. Not that it could be doubted that the Ethiopic tables and treatises reflect the Alexandrian computus of early Alexandrian Christianity. But there is no trace of Alexandrian "science" in this whole procedure, which turns out to be of the utmost simplicity. Its foundation, however, is the (equally simple) Jewish Passover computus which is nothing but a simple adaptation to the Alexandrian calendar of the 19-year cycle of a schematic lunar calendar. One single rule determines the Alexandrian Easter Computus: Easter is the first Sunday after Passover. Since Passover is by definition a "full-moon" date, Easter follows the full moon, and a proper definition of the date of the "Vernal Equinox" in the Jewish 19-year cycle introduces automatically the corresponding Christian rule. In return for discarding the myth of Alexandrian science in the Easter reckoning, we obtain a clear picture of the festival calendar of the Alexandrian Jews during the early Christian period. This calendar is also of the utmost simplicity, exclusively based on a 19-year cycle of the most elementary structure (and not to be taken as the equivalent of its Babylonian ancestor). The Ethiopian texts have indeed provided new insight into Alexandrian calendaric conditions but only at the price of disspelling the traditional picture of a connection between early Christianity and contemporary pagan science.

2. In order to see our discussion of the Easter computus in the proper perspective, it will be useful to sketch the background of Ethiopic "astronomy" in general. We know of only two groups of problems which have no direct relation to the calendrical tables: one concerns the reckoning of hours of daytime by means of the shadow cast by a man standing upright; the other tabulates the rising amplitudes of the moon during 12 consecutive months of a lunar year. Both of these problems are dealt with by the simplest arithmetical patterns. For example, the noon shadow in 7 consecutive months is supposed to be (measured in feet) 2 3 4 5 6 7 8 and then back again from 7 to 3 in the remaining 5 months. In each month the hours before and after noon are found from the noon shadow by adding 1, 2, 3, 4, and 10 feet to its predecessor (thus, e.g., for the first six hours 22, 12, 8, 5, 3, 2 before noon in the first case). Since in all cases the day is divided into twelve hours, these hours are "seasonal hours", a common type in Alexandrian Egypt. On the other hand the same treatises that contain shadow tables operate also with "equinoctial hours" by assigning variable lengths of daylight to each month, either in the ratio 15:9 or 2:1. These internal contradictions are of only minor significance, however, as compared to the fact that the shadow lengths of such ratios are totally excluded for Ethiopia with its almost equatorial latitude. It must be admitted, however, that similar tables were copied, century after century, from Byzantium to monasteries in northern France, without the slightest chance for practical usefulness and at a time when the correct determination of the variation of the length of daylight and of shadow lengths had long been found in Greek astronomy.

Also the "gates" of moonrise are determined by similar arithmetical patterns. From the "Book of Enoch" and from the Dead Sea Scrolls, it seems likely that these schemes originated in the Palestinian area some time in the last centuries before our era. In the present context it is of interest that the "months" during which the moon traverses once back and forth six sections of the horizon (the "gates") add up, alternatingly, to 30 and 29 days. Here we have a typical example of a schematic "lunar year" of 6 "full" and 6 "hollow" months, thus 354 days in length. Again, this is a widespread schematic description of the variability of the synodic months, convenient for use but far removed from the actual facts which were analyzed in great detail and with remarkable success in the cuneiform ephemerides computed in Mesopotamia during the centuries from about — 300 into the first century A.D. And again, the basically correct analysis of the variability of the lunar months and their application to the Babylonian lunar calendar had not the slightest effect either on the Greek lunar calendars or on the "Gates" and other concepts in the Book of Enoch and similar compositions of enormous popularity.

3. A short description of the Ethiopic sources of our study seems desirable in the present context. I know of only one consistent work on calendaric-chronological matters : the computus by Abū Shāker, a book of 59 chapters, written in the 13th century in Egypt and translated in the 16th century from the Arabic original (now lost) into Ethiopic. It has never been edited or translated — a fate shared with almost all of the other calendaric sources — in spite of the fact that it is considered to be the basis of all Ethiopic "computus". References to it are not rare in other short treatises but without much justification, as far as I can see. Another famous computus was supposedly written in A.D. 213/14 by Demetrius, the 12th Patriarch of Alexandria. Actually there is no treatise preserved which would safely be related to one author. Finally, some late additions to our material are taken (usually out of context) from Arabic astronomy, always written in Amharic, and without any influence on the traditional treatises.

Excepting Abū Shāker, the texts which concern the computus consist of a chaotic mixture of short sections that deal more or less directly with concepts needed for the construction of the tables whose final goal is the determination of the dates of all moveable feasts, Jewish as well as Christian. One could imagine that we have here countless scattered fragments and excerpts from some larger treatise explaining the structure and usage of the calendarical tables, somewhat similar in purpose to the introductions to the "Handy Tables" in Greek astronomy.

This material, as we have it, consists of many variants of texts, many times senselessly distorted by repeated copying, and usually not understood by the scribes. The general tendency is "didactic", i.e. the mechanical compilation of rules which ordinarily are simple consequences of another rule formulated in some other paragraph a little earlier or later. The chaos is increased by the desire to incorporate into sections based on the Alexandrian calendar and the Jewish Passover computus also the wisdom of the "Enoch" tradition, that means to consider "years" of 364 days, or "seasons" of 91 days each. Later scribes might then improve on such passages by adding a new layer of Julian data onto Enochian passages — the so-called "Slavonic Enoch" shows nice examples of this process which has also bewildered modern commentators.

In spite of this chaos of fragmentary treatises, it is quite possible to bring sense into the methods which were used for the computation of the tables and for the control of the numerical data. Having once understood the structure of the tables, the bulk of the texts makes sense, even if marred by a lack of distinction between a few basic rules and a host of rather obvious consequences which any user could have derived by himself*.

4. Before turning to a description of the Ethiopic tables it is convenient to mention a mathematical terminology (introduced in 1801 by Gauss, who, by the way, also wrote an article about the numbertheoretical structure of the modern Easter canon, based on the "reform" of 1582 under Gregory XIII).

We say that two numbers a and b are "congruent modulo c" (written $a \equiv b \mod c$) if the difference $a - b$ is divisible by c. For example, $39 \equiv 1 \mod 19$ because $39 - 1 = 2 \cdot 19$. In particular $a \equiv 0 \mod c$ means that a is a multiple of c; e.g., $76 \equiv 0 \mod 19$ but also $76 \equiv 0 \mod 4$ (because $76 = 19 \cdot 4$).

Almost all calendaric operations can be conveniently expressed as "congruences". For example, the "Enoch-year" is 364 days long and $364 \equiv 0 \mod 7$. This implies that the position of the weekdays remains always the same in this type of year. Obviously this was the very purpose of creating such a year and all our texts confirm that it was never modified. Modern scholars tried to discover some hidden intercalation system because they could not imagine that one could live with a "rotating" calendar. Apparently they do not know, e.g., about the rapidly rotating lunar calendars of the Assyrians or of the Islamic calendar.

The "Egyptian year" of 365 days is congruent 1 mod 364. Consequently the Ethiopic texts speak of an "extra" day when one goes from an Enoch year to a 365 day year. But $365 \equiv 5 \mod 30$, hence the 5 days in excess of 12 civil months of 30 days each are called epagomenal days. Finally, $365 \equiv 1 \mod 7$. Consequently if, e.g., Jan. 10, 1978 was a Tuesday then this same date in 1979 will fall on a Wednesday, in 1980 on a Thursday. But the year 1980 has 366 days, $\equiv 2 \mod 7$; thus the next year Jan. 10 will jump to Saturday.

Four Alexandrian (or four "julian") years total a number of days which are congruent $3 \cdot 1 + 2 = 5 \mod 7$ and since $5 \equiv -2 \mod 7$ we can also say that weekdays in the Alexandrian calendar recede 2 days in each quadruple of years.

These different forms of "years" are intended to agree more or less with the "solar year" i.e. with the climatic seasons. "Lunar years", however,

---

* For details see my monograph "Ethiopic Astronomy and Computus" (Oesterr. Akad. d. Wiss., Phil.-Hist. Kl., S.B. 347, 1979).

can produce such an association only by switching occasionally from a year
of 12 months to a year with 13 months. Consequently such years, operating
either with accurate lunar months (as the computed Babylonian months)
or with schematic months (of 29 or 30 days) cannot produce "years" of fixed
lengths.

The months themselves in the years of Enoch, in the Egyptian and in the
Alexandrian calender, are no longer related to the moon but are fixed
at 30 days of length. Only the theoretically determined lunar months of
the Babylonian ephemerides vary between full and hollow according to the
highly complicated factual variation of the dates of first visibility. Indeed
these data are by no means simple (or even regular) alternations between full
and hollow months. And since the character of each month depends on the
moon's visibility that defines the first day of a month the Babylonian
calendar days begin in the evening. This "evening epoch" of "lunar days"
was also taken over by the Jewish calendars. The Egyptian days, however,
and with it the days of the Alexandrian calendar, are counted in "morning
epoch". This then has also become the norm in the Ethiopic tables.

We have colophons in Ethiopic texts, or dates in documents or annals,
which give two days. A book may have been finished, according to a literal
translation of the colophon, "on the 6th at the beginning of night, at the
10th at the beginning of day". Such obvious nonsense has disturbed few
translators. In fact, "beginning of night" must mean "days which begin
in the evening", i.e. simply "lunar dates" in contrast to the Alexandrian
dates in morning epoch or simply "civil dates". We have extensive rules
in our texts on how to find the lunar dates from civil dates, and vice versa.
All Jewish feasts have not only civil dates but also lunar dates, in particular
Passover has the lunar date 14, i.e. (schematic) full moon. If p is the civil
date of Passover, f of Easter, then the lunar date of Easter is simply $14 + f - p$.
In mediaeval terminology this is the "luna", the "age of the moon", of
Easter which by definition must be more than 14 since f must be later
than p. It is one of the points of controversy in the contest between Alexandria
and Rome whether it is permissible that the luna of Easter is as low as 15
(Alexandrian norm), i.e., that Easter can be a Sunday following a Passover
that falls on a Saturday. The Ethiopic tables show that they followed the
Alexandrian norm in giving dates as low as $f = p + 1$.

5. Babylonian astronomy is built on the experience that astronomical
phenomena repeat themselves periodically. Lunar eclipses, for example,
return in the same magnitude in a cycle of 18 years. Saturn returns to
the same region among the fixed stars in about 30 years, Jupiter in 12.
Consequently these two planets will be in the same position relative to each
other in $2 \cdot 30 = 5 \cdot 12 = 60$ years. By combining characteristic periods

in this way it is easy to predict (or to exclude) situations of a more complex character. For a people who lived with a real lunar calendar it was only natural to observe also the return of the new moon, that is, the conjunction of sun and moon, to the same position in the sky. The result then is a number of "synodic" months that corresponds to a number of ("sidereal") years. It turns out that with a high degree of accuracy 235 is this number of months. Since $235 = 19 \cdot 12 + 7$ one can add 7 "intercalary" months to 19 ordinary "lunar years" and will obtain agreement with 19 "solar" years. This interval is called the "19-year cycle" or the "Metonic cycle" (because it was proposed, perhaps independently — and unsuccessfully — by Meton in Athens). We shall meet a simplified version of this cycle in the Ethiopic tables.

A cycle without any real astronomical background is the 7-day week. If we combine it with the 4-year cycle of the Alexandrian intercalation, we see that only after $7 \cdot 4 = 28$ years an Alexandrian year will begin with the same weekday. In our treatises this cycle is called the "solar cycle". If we wish to combine weekdays, Alexandrian calendar, and lunar phases, we must seek a common period of the solar cycle with the 19-year cycle. Since 19 and 28 have no common factor, the shortest period which comprises 7, 4, and 19 is the product 532 of these three periods. This number 532 is at the foundation of the whole Easter computus.

The 532-year cycle is well known to the mediaeval computists, from the Greek East to the Latin West. When the Monk Dionysius Exiguus in the middle of the 6th century introduced our present era he related the year 532 of his new "Christian era" with the then current era of Diocletian by equating A.D. 532 with Diocletian 248. The Ethiopic eras, based on Alexandrian prototypes are arranged slightly differently. The era W of the "World" (or "from Adam") is related to the era J of the "Incarnation" by the relation $J\ 0 = W\ 5500$. The era of the Incarnation is connected with the Era Diocletian (or the "Era of the Martyrs") by $D\ 0 = J\ 276$ (hence $D\ 248 = J\ 524$ and J 0 corresponds to A.D. 7/8). Finally, an era of "Grace" or "Mercy" is defined by $G\ 0 = D\ 76$. The reason for this norm of the most commonly used era is simply that $G\ 0 = W\ 5852 \equiv 0 \bmod 532$. Hence the beginning of the era G coincides with the beginning of a 532 year cycle (the 12th) of the era W. But all these eras are based on the era Diocletian and thus on the Alexandrian calendar as established by Augustus.

6. We now can turn to the Ethiopic calendaric tables. Their most important type, preserved in many copies (but unpublished) consists of 28 tables, each of which covers one 19-year cycle; we therefore call these tables the "532-year tables". They are usually based on the era G and therefore concern one of the three cycles that begin ($k = 1$) with the years W 5853, 6385, and 6917.

All of the existing manuscripts were written in the last, 14th, cycle (from A.D. 1424/5 into the 20th century). There exist several types of shorter tables, all of which have periods of 19 years, or of multiples of 19, and are therefore implicitly contained in the 532-year tables, though in different arrangement, e.g. by weekdays.

The main type of the 532-year tables contains about 20 columns but there exist larger tables with about 30 columns. Most of these tables contain a first column, headed "tārik", i.e. "history". The next two columns count the lines either from $c = 1$ to 19 in each individual table or from $k = 1$ to 532 in the whole set. The column "tārik" mentions events of Biblical history or of contemporary history, (e.g. the death of Patriarchs or Kings), without formal distinction of these dates with respect to the cycle to which they belong. For example, the entry "baptism" at $k = 211$ refers to the baptism of Christ in the year W $5531 = $ G 211 (in the 11th cycle).

c	e	m	yk	tb	p
1	0	30	9	14	10
2	11	19	28	3	29
3	22	8	17	22	18
4	3	27	6	11	7
5	14	16	25	30	26
6	25	5	14	19	15
7	6	24	3	8	4
8	17	13	22	27	23
9	28	2	11	16	12
10	9	21	30	5	1
11	20	10	19	24	20
12	1	29	8	13	9
13	12	18	27	2	28
14	23	7	16	21	17
15	4	26	5	10	6
16	15	15	24	29	25
17	26	4	13	18	14
18	7	23	2	7	3
19	18	12	21	26	22

"Table XIX"

The remaining columns refer to the dates of feast days, Jewish and Christian, culminating in the last column "fāsikā", i.e. "Easter". Several columns show for all years the same number and are headed "beginning of night", i.e. lunar date. They always belong to a neighboring column which gives the civil dates of a Jewish feast. Thus the lunar date of matqe'e ("trumpet", i.e. the Jewish New Years Day) is 1, of Yom Kippur 10, of Tabernacle 15, and of

Passover 14. Looking more closely at the civil dates of the Jewish feasts one will notice that they are the same in each 19-year table. Excerpting these data from the larger tables we obtain the above shown "Table XIX" which is repeated 28 times. Here e denotes the "epact" which is related to the date m of the New Year by $e + m = 30$. All numbers in the last four columns increase by 19 (mod 30) every year, hence e must decrease by 19, or modulo 30, increase by 11. A date of Passover printed in italics indicates that it belongs to month VII of the Alexandrian calendar (Phamenoth = Ethiopic Magābit). All other Passover dates belong to VIII (Pharmouthi = Miyāzyā). It follows from the arithmetical structure of Table XIX that the dates of all Jewish feasts are known as soon as the date of one of them is known. For example, $m \equiv p + 20$, $yk \equiv p — 1$, $tb \equiv p + 4$, always mod 30.

It should furthermore be noted that the periodicity of this table requires that the transition from the line $c = 19$ to $c = 1$ requires for e the addition of 12 days instead of the usual 11. Correspondingly all other dates increase by only 18 instead of the ordinary 19. This specific situation is described by the medieval computists as the "saltus lunae", the object of much empty speculation. In fact it represents a very simple matter. A schematic lunar year has a length of 354 days, hence receding 11 days each year with respect to 365 days. These 11 days are called the "epact" in Greek and Western medieval astronomy. Continued application would remove a lunar date, e.g. m of the Jewish New Year, from its general location in the solar year and thus a full month of 30 days will be added. This explains our sequence m in Table XIX in which we add the proper month numbers of the civil calendar :

$e =$	$m =$	$p =$	$c =$
0	I 30	VIII 10	1
11	I 19	VII 29	2
22	II 8	VIII 18	3
3	I 27	VIII 7	4
14	I 16	VII 26	5
25	II 5	VIII 15	6

This also shows that the rule $p \equiv m + 10$ mod 30 results from a fixed distance $p = m + 190$. The same holds for all moveable festivals. It is the same to say that the dates of m are restricted to the interval I 15 $\leq m \leq$ II 13 and, similarly, Passover to VII 25 $\leq p \leq$ VIII 23.

Incidentally it may be remarked here and for all that follows that it is of primary importance to express all arithmetical rules in the system in which they were developed, i.e., in the Alexandrian calendar with its 30-day months. Introducing our "julian" calendar with its perverse disorder of month-lengths completely obscures the arithmetical simplicity of all structures.

7. The above rules for the determination of the civil dates of the Jewish festivals give us a complete insight into the meaning of "19-year cycle" in this procedure. In Babylonian astronomy the 19-year cycle assumed that 19 sidereal years are equal to 235 mean synodic months. In our present tables, however, 19 Alexandrian years are equated with 235 schematic lunar months. Furthermore, in the Babylonian calendar the months followed closely the complicated pattern of the intervals between evenings of first visibility. The Ethiopic cycle knew nothing of such refinements. It simply assumed an epact of $11 = 365 — 354$ days which was a crude but convenient estimate for the slippage of the lunar phases, and adjusted this 11-day epact (with the help of the "saltus lunae") so that it returned to the same civil dates after 19 Alexandrian years without concern for the location of the Alexandrian intercalations. Since this scheme is extended over 532 years exact periodicity is granted also with respect to intercalation. In fact, this is already the case after $4 \cdot 19 = 76$ years, an interval which appears frequently in calendaric treatises. If the interest is centered on weekdays $7 \cdot 19 = 133$ years are significant. But 532 years remains as the shortest cycle for all parameters under consideration. If one wishes to convince oneself of the quality of this cycle, one can remark that $532 \cdot 365;15$ days are assumed to be equal to $28 \cdot 235$ synodic months, which gives for one month the length of about $29;31,51,4$ days, which is a very good approximation. This illustrates the fact that very good results can be reached (often accidentally) by extremely simple arithmetical procedures.

Modern scholars cherished the idea that occasional "observations" of full or new moons were applied to "correct" the results of cyclic computation. The high quality of the approximation of the cycle during five centuries shows that such empirical corrections were not at all necessary. On the contrary : the occasional comparison with some true conjunction or opposition would have only introduced errors to the full amount of the considerable difference that can occur between "mean" and "true" syzygies.

8. Having reached complete insight into the pattern for the dates of the Jewish festivals, we can obtain the same for the Christian feasts without further difficulties. Exactly as in the preceding case all Christian dates are known from any one of them. For example, "Beginning of Fast" $\equiv$ "Nineveh" $(n) + 14$; "Mount Olive" $\equiv n + 11$; "Palm Sunday" $\equiv n + 2$, and Easter $(f) \equiv n + 9$, always modulo 30. The proper months are determined from respective limits, for example VII $26 \leq f \leq$ VIII 30.

The really crucial rule concerns Easter. It is simple enough : Easter is the Sunday following Passover. Since p was limited by VII 25 as the earliest date and since VII 25 is considered to be the date of the vernal equinox (how accurate astronomically is of no concern), we have now established Easter in the canonical fashion : after equinox, after full moon (i.e. Passover) and a Sunday.

We thus see that the whole Christian calendar was made dependent on the Passover date, which in turn is a simple application of the epact computus of the schematic "19-year cycle". It is indeed as a text expressed it : "matqe'e and epact are the foundation of the whole computus".

There remains only one little step to be clarified : obviously we need now to know the weekday of Passover. But since Passover = matqe'e + 190 and since $190 \equiv 1 \bmod 7$, it suffices to determine the weekday of m. But the weekday of any Alexandrian year, or of a year of the era Diocletian, has been known since Antiquity. Hence our table simply lists, in a column headed "ţentyon", the weekdays of the first of Thoth. For example, we have for the first day of each 532 year cycle of the era G the weekday "Tuesday". Since column m gives us the civil date of the matqe'e we can immediately determine the weekday of m and thus of Passover and finally the date of the next Sunday. This solves our problem.

9. Example : find the date of Easter for the year $k = 118$ in the era G. Since $118 \equiv 4 \bmod 19$, we have $c = 4$; hence (from Table XIX) $m = $ (I) 27 and $p = $ (VIII) 7. The 532-year tables give for $k = 118$ Monday as the weekday of I 1. Now I 27 = I 1 + 26 and $26 \equiv 5 \bmod 7$; thus weekday of $m = $ Monday + 5 = Saturday. Hence the weekday of $p = $ Saturday + 1 = Sunday. And hence Easter is 7 days later, i.e., $f = p + 7 = $ VIII 7 + 7 = VIII 14; the "luna" of Easter Sunday is 14 + 7 = 21.

Check with modern tables : $k = 118$ corresponds to W 6916 + 118 = W 7034 = J 1534 = A.D. 1542. For this year Alexandrian VIII 14 = April 9, which is indeed the Easter Sunday for 1542. The preceding astronomical full moon, however, was on March 31, i.e. 2 days before the date of Passover.

In the above computation we used only the basic elements of the 532-year tables. Many of these tables give, however, columns both for the weekday of m and of p. Hence we would have seen from the table that $p = $ VIII 7 fell on a Sunday, and hence $f = $ VIII 14. Only after the Gregorian reform in 1582 would the Ethiopic tables no longer be useable for the determination of the Catholic Easter dates.

10. In principle we have now reached our goal to explain the method by which the Ethiopic 532-year tables furnished the dates of Easter year after year. It is the purpose of the subsequent sections to discuss the historical background of these tables and related treatises.

First of all we should elaborate somewhat on our main result — the location of Easter Sunday in the week immediately following Passover. The correctness of this relationship can be demonstrated in three ways: first, by passages in our calendaric treatises stating explicitly this rule; secondly, by purely arithmetical proof on the basis of the structure of the relevant columns; thirdly, by simply exerpting from the tables the date f of Easter as a function of the date p of Passover. The result of this last, most direct proof is shown in the subsequent table. Column c gives the cycle number in each of the 28 19-year cycles; column p is the same as in our previous "Table XIX" (p. 94); column f shows all attested Easter dates correlated with the same pair of number c and p. Obviously f ranges from $p + 1$ to $p + 7$, i.e., the space of one week after $p + 1$ as required by the fundamental rule, which is thus fully demonstrated.

c	p	f						
1	10	11	12	13	14	15	16	17
2	29	30	1	2	3	4	5	6
3	18	19	20	21	22	23	24	25
4	7	8	9	10	11	12	13	14
5	26	27	28	29	30	1	2	3
6	15	16	17	18	19	20	21	22
7	4	5	6	7	8	9	10	11
8	23	24	25	26	27	28	29	30
9	12	13	14	15	16	17	18	19
10	1	2	3	4	5	6	7	8
11	20	21	22	23	24	25	26	27
12	9	10	11	12	13	14	15	16
13	28	29	30	1	2	3	4	5
14	17	18	19	20	21	22	23	24
15	6	7	8	9	10	11	12	13
16	25	26	27	28	29	30	1	2
17	14	15	16	17	18	19	20	21
18	3	4	5	6	7	8	9	10
19	22	23	24	25	26	27	28	29

It should be noted that our table gives only $7 \cdot 19 = 133$ values for f. Hence each Easter date must occur four times in 532 years. The explanation of this multiplicity lies, of course, in the fact that our rule does not contain any statement about the Alexandrian intercalation which produces four different possibilities for each combination p, f.

If one investigates the occurrence of these four cases within the 532-year tables one finds (either by arithmetical theory or by inspection) an important phenomenon: these cases are always $95 = 5 \cdot 19$ years apart. For example, the combination $p = 10, f = 11$ occurs in the years $k = 115, 210, 305$, and 400. Similarly, the earliest possible Easter date $f = $ (VII) 26 (at $c = 16, p = $ (VII) 25) occurs in the years $k = 54, 149, 244$, and 491 which is $\equiv 54 - 95$ mod 532.

The author (or authors) of the 532-year tables was fully aware of this law of distribution for equivalent values within a 532-year cycle and statements to this effect are also found in our treatises and will not surprise anyone who actually computes a complete 532-year table. Medieval Latin computists were also aware of this "periodicity" which explains why, for example, the tables of Dionysius cover the 5 19-year cycles from A.D. 532 to 626. Modern writers on medieval computus missed this point, stating correctly that 95 is not a period in the 532-year cycle but ignoring the fact of the unavoidable multiplicity of data in groups of 95 years.

Recognition of the 95-year intervals is not the only procedural element that spilled over from the Alexandrian computus to the Latin one. As we have seen, the determination of Easter Sunday requires knowledge of the weekday of the first of Thoth (śarqatito in Ethiopic). As noted before, this day is given in our tables in the column headed "tentyon" (t) (which is a distortion of the term [ἡμέραι] τῶν θεῶν, used for "weekday", e.g., by Athanasius). This number counts the weekdays (modulo 7) so that 1 = Wednesday. Latin scribes, however, used a norm for the "feria" in which 1 = Sunday. Now it so happens that $t = 1$ (Thoth = Maskaram 1 Wednesday) always corresponds to March 24 = Sunday = feria 1. Thus all rules that involve t are numerically identical with rules which use the feria of March 24, a number which the medieval computists honored with the special name concurrentes.

11. Our calendaric treatises are full of invective against the "impious Jews", stressing over and over again the purpose of the rules concerning Easter to avoid contamination by Passover. Nevertheless they allowed, as we have seen, an approach to the very next day. The "Romans", eager to follow rules of their own, and opposed to Alexandrian superiority, insisted on a two-day minimum, i.e., on a lowest "luna" 16 as against the Alexandrian 15. Of course neither one of these norms has any astronomical basis whatsoever and is simply a matter of arbitrary choice for the boundaries of a parameter.

This is not the only object in the bitter Easter controversy between Alexandria and Rome. Since the Alexandrians had the good luck to adopt the Jewish 19-year cycle, the Romans insisted on some other cycles based mainly on the "octaeteris" which relates 8 years to $8 \cdot 12 + 3 = 99$ months, corresponding to a mean synodic month of about $29;30,54^d$, which by any standard of ancient astronomy is of clearly inferior accuracy. Hence the necessity of repeated corrections of the cycle — not, of course, by "observation" but by adjustment to the Alexandrian norm. It was only in the sixth century that Dionysius broke the impasse by accepting the Alexandrian pattern on the authority of a (spurious) decree of the Council of Nicaea, and by replacing the simple data of the Alexandrian calendar by the Roman calendar with all its pagan relics of calends, ides, and nones. Furthermore, by transforming the years of the Diocletian era to the years "A.D." he became the father of our present calendaric system.

12. There remains one more point to clarify. We have repeatedly referred to Alexandrian procedures from evidence in the Ethiopic tables and treatises. To what extent are we justified in doing so, even if it would appear *a priori* unlikely to assume Ethiopic innovations in these texts which abound in Greek terminology and concepts? One must nevertheless admit that our data are chronologically fixed for only mod 532 years and it could be possible that any one of these cycles represents historically the first one.

Here a lucky accident comes to our rescue. In 1976 Ephraim Isaac (of Harvard University) published a catalogue of Ethiopic manuscripts in the library of the Armenian Patriarchate in Jerusalem. This catalogue mentioned among others two manuscripts of evident interest to our discussion. One was obviously a shadow table of a well-known type; the other manuscript suggested a 532-year table. By courtesy of His Eminence, the Patriarch, I received photographs of these manuscripts which confirmed my initial conjecture concerning their contents. But the 532-year table contained an unexpected variant beyond some slight changes in arrangement: it contained a column giving the "indictio" of the year.

As is well known, this parameter refers to a 15-year cycle introduced by Diocletian for administrative purposes. But somehow this number acquired the role of a short-term era, frequently used in all kinds of documents from Byzantine domains to the medieval West. The application of this count also in the 532-year tables is of primary interest to our problem. Since 15 has no common divisor with 4, 7, or 19, it repeats itself only with the same line of data in $15 \cdot 532 = 7980$ years. In other words: the indictio listed in one of our tables fixes its date uniquely.

The time scale of the indictions is of course well known from ancient and medieval documents. For us it is enough to mention that Athanasius regularly gives the indictio of the year in his Easter messages. The subsequent table gives in its upper section a transcription of the Ethiopic table.

	D	i	ep	t	j	e	m	$\bar{m}$	yk	$\bar{yk}$	tb	$\bar{tb}$	bf	$\bar{bf}$	p	$\bar{p}$	pw	f	$\bar{f}$
Arm. 3483 194ᵛ/195ʳ	44	1	6	1	4	25	5	1	14	10	19	15	24	22	15	14	4	19	18
	45	2	5	2	5	6	24*	1	3	10	8	15	16	25	4	14	1	11	21
	46	3	5	3	6	17	13	1	22	10	27	15	29	19	23	14	7	24	15
	47	4	5	4	7	28	2	1	11	10	16	15	21	22	12	14	4	16	18
	48	5	6	6	2	9	21*	1	30*	10	5	15	12	24	1	14	2	7	20
	49	6	5	7	3	20	10	1	19	10	24	15	2*	25	20	14	1	27	21
	50	7	5	1	4	1	29*	1	8	10	13	15	17	21	9	14	5	12	17
Athanasius	44	1		1		25												19	18
	45	2		2		6												11	21
	46	3		3		17												24	15
	47	4		4		28												16	18
	48	5		6		9												7	20
	49	6		7		20												20	15
	50	7		1		1												12	17

We need not describe the single columns since all we need are the parameters already defined in the preceding pages. Only the last column $\bar{f}$
should be mentioned since it contains the values of the "age" of the moon,
i.e. the "luna" $14 + f - p$. The era in the first column is the era D of Diocletian.
Changing it to the era $G = D - 76$ one obtains exact agreement of all columns
with the ordinary 532-year tables. But when this agreement normally would
give only the date of these tables modulo 532, we can now say — because of
the presence of a column with the indictio (i) — that D 44 can only mean
the year Diocletian 44. But exactly for these years we have also the elements
quoted by Athanasius: the years of Diocletian, the indictio, the "ṭentyon",
i.e. the weekday of I 1, the epact e (thus also $m = 30 - e$), the dates of Easter
(f) and the age of the moon ($\bar{f}$), thus also the date of Passover $p = 14 + f - \bar{f}$.

Hence Athanasius' dates give all elements underlying both the Jewish and
the Christian calendars that are necessary for the determination of Easter.
And it is now rigorously proved that the Ethiopic tables are identical in
substance with the Alexandrian Easter computus of the time of Athanasius.

13. One may rightly say that this result is not surprising, though one may
also remark that there is always a certain difference between historical
plausibility and a mathematical proof that does not imply anything but
numerical data, comparable to the data of a sharply defined solar eclipse.

But we also have gained independent historical information. Knowing
now in all details not only the Ethiopic computus but also the methods of

the Christian Easter computus of the 4th century, we can say that these methods contain absolutely nothing of contemporary Alexandrian astronomy which at that time had just reached its final development, of fundamental importance for the next thousand years of mathematical astronomy. The architects of the Alexandrian Easter tables did not use a single concept of pagan astronomy and borrowed all their rules from the simple Jewish procedure to relate the remnants of the Babylonian 19-year cycle by means of "epact" and "saltus lunae" to the Alexandrian civil calendar.

And we now also see how the Jews in the Diaspora in Alexandria regulated their "lunar" calendar during the first centuries of our era. The fierce antagonism against Judaism which is evident in so many ways in our texts guarantees that the data of the Jewish feasts, in particular Passover, were the actual data of contemporary Jewish customs — otherwise the whole construction of the Christian rules would be pointless. This situation changed only centuries later when the Latin West adopted the Alexandrian rules while rabbinical scholarship (in the early 6th century) developed a lunar calendar of much higher astronomical and legal sophistication; in other words, in principle a return to the mentality of the Babylonian astronomers (though not to their level of insight). Since that time Christian and Jewish calendars no longer have had causal connections and the fear of contamination has subsided.

In the introduction to his "Histoire du peuple d'Israël", Ernest Renan wrote : "Pour un esprit philosophique … il n'y a vraiment dans le passé de l'humanité que trois histoires de premier intérêt : l'histoire grecque, l'histoire d'Israël, l'histoire romain". Having not a philosophically inclined mind, I may perhaps differ from this restriction of interests. For the history of the Easter computus, however, Renan's formulation is unusually well fitted. The mutual antagonism and distrust between the three cultural spheres of Judaism, Alexandrian and Roman episcopats shaped the arguments which are responsible for the form in which the Easter computus still exists today (March 26, 1978).

Reprinted from
Oriens Christianus
63 (4), 87–102 (1979)

Also by O. Neugebauer

A History of Ancient Mathematical Astronomy
(In Three Volumes)

"(This book) is a landmark, not only for the history of science, but for the history of scholarship. *HAMA* places the history of ancient astronomy on an entirely new foundation. We shall not soon see its equal."
— *Historia Mathematica*

"One can only hope that a future historian will be able to accomplish as much when the astronomy of the twentieth century has itself been reduced to a few odd books and some handfuls of fragments."
— *Bulletin of the American Mathematical Society*

1975/1456 pp./619 illus./
ISBN 0-387-06995-X
(Studies in the History of Mathematics and Physical Sciences, Vol. 1)